# EVOLUTION OF THE GEOLOGIC TIMESCALE

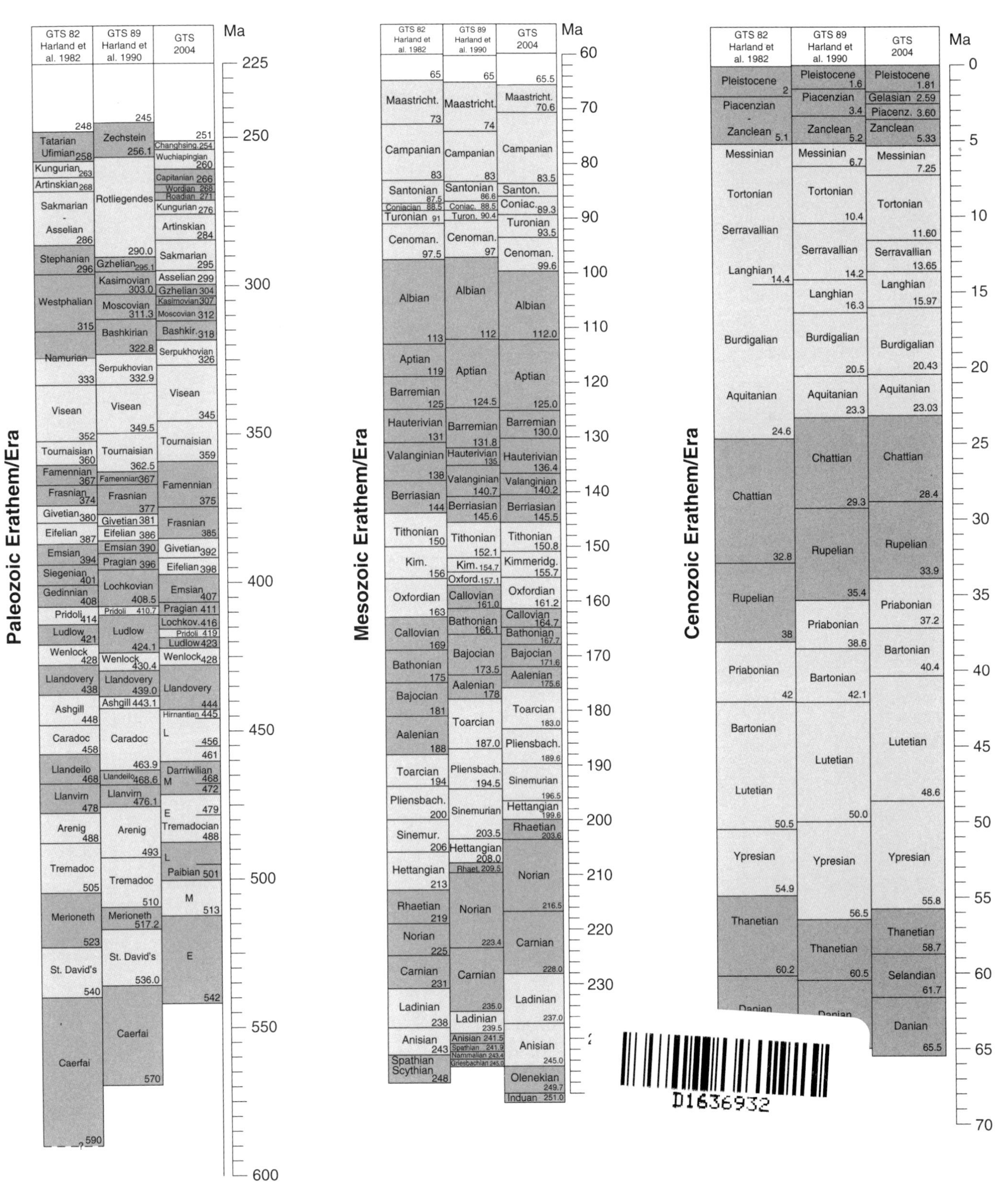


These columns illustrate the evolution of the geologic timescale (GTS), reflecting improvements in age-dating and global biostratigraphic correlations. Depicted here are changes in the durations and names of stages throughout the Phanerozoic Eon, based on three widely accepted versions of the timescale published in consecutive decades. Many of the older stage names, such as those for the Ordovician, remain in common usage, particularly when the changes to the timescale are fairly recent. Students are encouraged to use this figure as a reference to these names throughout the book. In addition, several of the figures in the book depict absolute ages as they were known when the sources of the figures were first published. (*Redrawn from figures originally drafted for the International Commission on Stratigraphy, available at http.//www.stratigraphy.org/down.htm*).

*Principles of Paleontology*

*Principles of*

# PALEONTOLOGY

THIRD EDITION

MICHAEL FOOTE

*University of Chicago*

ARNOLD I. MILLER

*University of Cincinnati*

W. H. FREEMAN AND COMPANY

New York

Publisher: Susan Finnemore Brennan
Acquisitions Editor: Valerie Raymond
Development Editor: Sharon Merritt
Media Editor: Amy Thorne
Marketing Manager: Scott Guile
Project Editor: Vivien Weiss
Photo Editor: Ted Szczepanski
Photo Researcher: Elyse Rieder
Design Manager: Diana Blume
Text Designer: Patrice Sheridan
Illustrations: Fine Line
Illustration Coordinator: Bill Page
Production Coordinator: Susan Wein
Composition: Prepare Inc.
Printing and Binding: RR Donnelly

Library of Congress Preassigned Control Number: 2006926757
ISBN-13: 978-0-7167-0613-7
ISBN-10: 0-7167-0613-X



Printed in the United States of America

First printing

W. H. Freeman and Company
41 Madison Avenue
New York, NY 10010
Houndmills, Basingstoke RG21 6XS, England
www.whfreeman.com

To Hope and Mary Jo

# CONTENTS

# FOREWORD

## *by David M. Raup*

Textbooks come in two flavors: *followers* and *leaders*. Books that follow a field or discipline are an effective means of digesting well-established material and presenting it in a form accessible to student and instructor alike. Books that lead their field do so by concentrating more on front-line research and currently emerging methods. While favoring the latter strategy, Foote and Miller have produced a good blend of the two approaches and I predict *POP3*, as it is affectionately called, will have a long and productive life.

In the 35 years since Steve Stanley and I published *POP1*, paleontology has changed dramatically. In 1971, we (and many others) were struggling to drag our discipline into the twentieth century. Rigorous quantitative methods were being introduced in morphologic analysis as well as biostratigraphy, computers were beginning to be used routinely to work with large databases, and many ideas from population biology were being applied to fossils for the first time. Perhaps more important, paleontology was beginning to interact effectively with evolutionary biology and with areas of geology beyond biostratigraphy, especially in sedimentology, biogeochemistry, and tectonics.

Now the challenge for Foote and Miller, and all users of *POP3*, is to push the field forward into the twenty-first century. As pointed out early in Chapter 1, paleontology is in the midst of a renaissance, with the introduction of all sorts of approaches and analytical techniques unknown 35 years ago. Because of this, and to its credit, *POP3* overlaps very little with the earlier *POPs*. Nevertheless, Foote and Miller have maintained the basic imperative that requires that all good paleontology be built on a foundation of careful description and classification of fossils and of the geologic settings in which they are found.

David M. Raup
Washington Island, Wisconsin
Spring 2006

# PREFACE

David M. Raup and Steven M. Stanley revolutionized the teaching of paleontology during the 1970s with the publication of the first two editions of *Principles of Paleontology*. Students were challenged and encouraged by Raup and Stanley—and a generation of instructors who used the textbook—to think creatively about important ways in which paleontological data could be put to effective use. Many of these students have now matured to make their own contributions to the science, inspired directly by the examples that Raup and Stanley presented.

Paleontology has thrived and broadened as a discipline in the years since the publication of the second edition of *Principles of Paleontology* in 1978. Although many aspects of paleontology are still appropriately represented in the second edition, it has been clear for some years that an update is past due.

## CONTINUED EMPHASIS ON GENERAL PRINCIPLES

Given the dizzying progress of the past three decades, it seemed to us that a substantial rewrite, rather than a more limited revision, was in order. Our goal, however, has been to maintain the forward-looking focus on paleontological concepts and analytical procedures pioneered so effectively by Raup and Stanley in the previous two editions. Therefore, like the earlier editions, the purpose of this book is to provide a thorough *conceptual* coverage of paleontology appropriate for an undergraduate course. This book is not a survey of the facts documented by paleontology but rather of the questions addressed by the field and the ways in which paleontologists go about their business.

## BALANCE OF APPROACHES AND LEVELS

The data and methods used to address paleontological questions vary along many spectra: from biological to geological; from individual organisms to larger biologic groups; from empirical to theoretical; from local to global; from qualitative to quantitative. Instead of focusing on a single set of approaches or a "school" of paleontology, we have attempted to strike a balance among the diverse ways that paleontology is carried out in practice, and we have kept our treatment at a level appropriate for undergraduates. We also hope that, outside of the undergraduate classroom, other readers will find our coverage of topics to provide a useful introduction to paleontological research.

We believe that it is not practical for one book of reasonable length to cover both the principles of paleontology and the systematics of all major groups of organisms, and still do justice to either subject. We therefore anticipate that this book will be supplemented by additional readings in systematic paleontology and anatomy, particularly if a paleontology course is accompanied by a laboratory devoted to hands-on work with fossils.

## NEW ORGANIZATION

In preparing this edition, we have deviated from the organization of the previous editions in three ways.

- First, we no longer divide the book into two parts; the previous editions highlighted separately the *Description and Classification of Fossils* and *The Uses of Paleontological Data*. As paleontology has evolved, methods and questions no longer appear quite so distinct. That is to say, investigations of compelling questions in paleontology now typically involve the development of new approaches to data collection and analysis in their own rights. In our treatment of paleontological questions throughout the book, we have therefore interleaved discussions of how paleontologists acquire and analyze data that are appropriate to the tasks at hand.
- Second, we have been struck by the number of instances over the past thirty years in which approaches highlighted throughout the book have been integrated in the investigation of a broad range of questions in geology, biology, and other fields. With this in mind, we have added a final chapter, *Multidisciplinary Case Studies in Paleontology*, in which we present several current avenues of investigation, each of which draws on diverse lines of inquiry covered in the earlier chapters.
- Finally, while the previous editions of *Principles of Paleontology* contained a number of *boxed features* highlighting important procedures in systematics and data analysis, this edition makes much more extensive use of boxes. Many of these provide extended treatments of material discussed in the main text and figures, including detailed descriptions of how various analytical procedures actually work. Our goal in providing these boxes is to help demystify many of the methods that have become part of the standard analytical toolbox for paleontology. We believe that any student who takes the time to work through the boxes will come away with an understanding of how paleontological analyses are done. In a sense, therefore, we view the boxes as providing a kind of "book within a book," and we recognize that some readers may choose only to skim the material contained therein.

Paleontology is by its nature an interdisciplinary science—an intersection of geology, biology, chemistry, and physics. In representing such a broad endeavor, it is inevitable that some arbitrary decisions be made as to which topics to exclude or to include only briefly. We have sought to emphasize areas that are essential to the practicing paleontologist and those in which paleontological data have been central in our understanding of a subject. We have treated some topics in less detail because they are rich and fully mature disciplines in their own right (for example, taphonomy); because they are largely in the realm of post-graduate research (for example, multivariate analysis); or because their importance for paleontology is still in an early stage of development.

To give just one example of this last point, we are excited about the remarkable progress in modern developmental genetics, which is clearly crucial to our understanding of evolution and the history of life. At the same time, the direct melding of paleontological data with information from developmental genetics is still in its infancy. For this reason, we have treated the subject to a limited extent, although we anticipate that the next edition of *Principles of Paleontology* may require an entire chapter devoted to this topic!

Michael Foote
Arnold I. Miller
May 2006

# ACKNOWLEDGMENTS

We owe a great debt to David M. Raup and Steven M. Stanley, without whom we would not have contemplated writing this book. Their vision of paleontology and of the paleontology textbook has influenced us as students, scientists, teachers, and authors.

A number of colleagues formally reviewed parts of this book and provided much needed direction. They are:

Ann F. Budd, *University of Iowa*
Sandra J. Carlson, *University of California, Davis*
Karl W. Flessa, *University of Arizona*
Thor Hansen, *Western Washington University*
Peter J. Harries, *University of South Florida*
Nigel C. Hughes, *University of California, Riverside*
Roger L. Kaesler, *University of Kansas*
Lance L. Lambert, *University of Texas at San Antonio*
Rowan Lockwood, *College of William and Mary*
Rosalie F. Maddocks, *University of Houston*
Jörg Maletz, *University of Buffalo*
Thomas D. Olszewski, *Texas A & M University*
Dena M. Smith, *University of Colorado*
James T. Sprinkle, *University of Texas at Austin*
Sally E. Walker, *University of Georgia*

Many other colleagues generously shared their ideas and acted as sounding boards in our writing of this book. In addition to the reviewers already mentioned and to the people, too numerous to list, who provided photographs and artwork, we would like to thank the following for their contributions: Jonathan Adrain, John Alroy, Richard Aronson, C. Kevin Boyce, Carlton E. Brett, Devin Buick, Katherine Bulinski, Matthew T. Carrano, Rex E. Crick, Douglas H. Erwin, Jack Farmer, Chad Ferguson, Daniel C. Fisher, Richard A. Fortey, Juan M. Garcia-Ruiz, Philip D. Gingerich, Shannon Hackett, Austin Hendy, Steven M. Holland, Gene Hunt, David Jablonski, Christian F. Kammerer, Susan M. Kidwell, Michael LaBarbera, Riccardo Levi-Setti, David R. Lindberg, Peter J.

Makovicky, Charles Marshall, Pamela Martin, Frank K. McKinney, David L. Meyer, Karl J. Niklas, Carl Simpson, Andrew B. Smith, Mark Webster, and Scott L. Wing. We are especially grateful to the late Jack Sepkoski, whose keen insights have helped to guide us throughout our preparation of the book.

Finally, we thank the editorial and production staff at W. H. Freeman and Company, in particular Valerie Raymond and Vivien Weiss, for guiding us and for efficiently moving this work along.

We ask readers to bring errors to our attention so that these can be rectified in future printings and editions.

# *Chapter 1*

# THE NATURE OF THE FOSSIL RECORD

## 1.1 NATURE AND SCOPE OF PALEONTOLOGY

- *Fossils are mainly found in sedimentary rocks, and so where we find fossils is largely a matter of where we find such rocks. Over 30 years ago, paleontologists documented a simple correspondence between the amount of preserved sediment and the number of fossil species known from a given period of geologic time (Figure 1.1). This raises the obvious (and still unanswered) question: How well do changes in biological diversity that we see in the fossil record reflect the true course of diversity in the history of life?*
- *The chances that an organism will be fossilized after it dies are quite small. This would lead us to expect that the full roster of species living in a community today will not ultimately be represented in the fossil record. Nonetheless, if we want to build a complete list of the species living in a given area, it is often possible to do so more effectively by sampling dead skeletal remains from the unconsolidated sediment than by sampling live individuals* [SEE SECTION 1.2]. *Why should this be?*
- *By studying individual species in great detail, paleontologists have found that many species show scarcely any evolutionary change over millions of years. If this is so, how can there be major evolutionary trends in the history of life—for example, toward greater body size, higher complexity, more efficient feeding, or increased intelligence* [SEE SECTION 7.4]*?*

These are but a few of the questions addressed by **paleontology,** which is the study of ancient life in its broadest sense. While this definition is clear enough, it does not convey the excitement that currently envelops the field. In fact, it can be said fairly that paleontology is in the midst of a renaissance, spurred on by a new generation of analytical techniques and approaches that unlock the fossil record for application to geological, biological, and even astronomical questions—which may surprise some students—related to the history, current state, and future of life on the planet.

The data of paleontology come from the form, chemistry, and spatial and temporal distribution of fossils. The term **fossil** (from the Latin *fodere*, to dig) was once used to refer to nearly any object dug up from the ground, but now refers more specifically to remains of past life. Fossils are mechanically or chemically extracted from rocks or unconsolidated sediments that crop out naturally at the earth's surface or are exposed by activities such as road building and mining. Historically, the fossil record has been most thoroughly studied near major population centers in the developed world. Following initial reconnaissance surveys, however, increasing efforts are being made to sample from remote regions such as Antarctica.

The relationship between observed diversity and preserved sediment just mentioned reflects the more general issue of how the imperfections of the fossil record affect our ability to study the life of the past. A central theme of this book is that the nature of the fossil record must be taken into consideration at all times but that its deficiencies can be rigorously addressed. One of our main goals is to help students interpret the data of the fossil record, often with the help of models and observations on how it forms, while avoiding two pitfalls. The first is the assumption, often tacit, that the fossil record is so complete that it can always be taken at face value. The

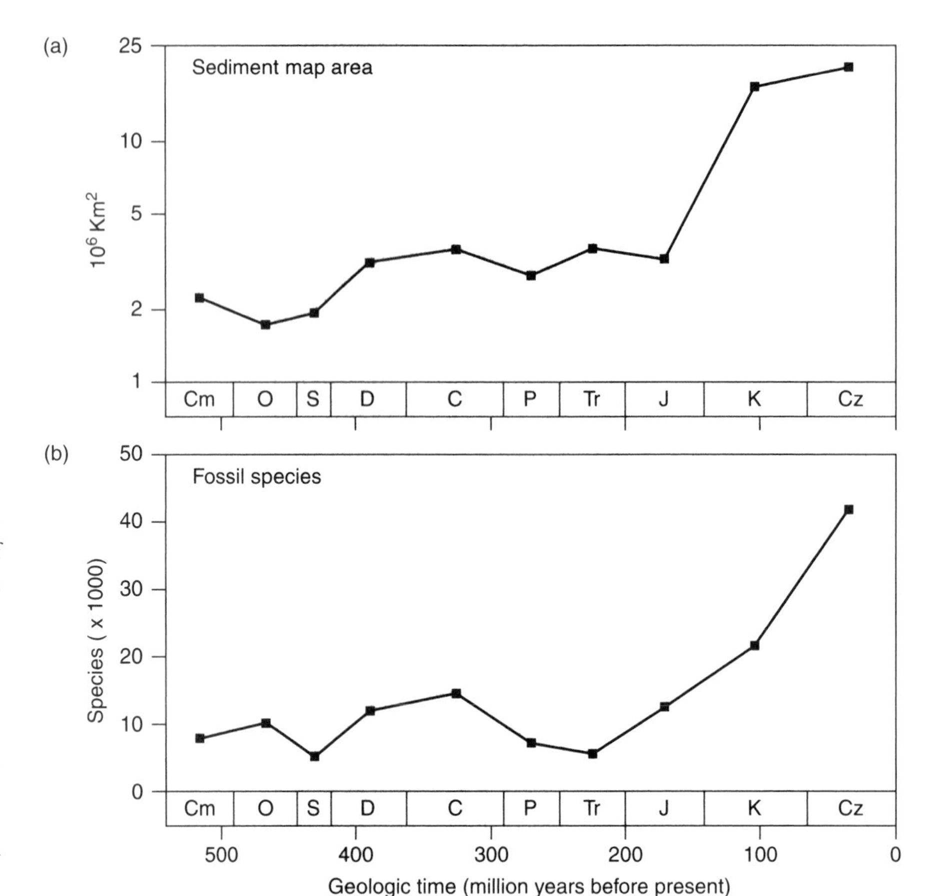


**FIGURE 1.1 Sedimentary rocks and fossil diversity through geologic time.*** (a) The outcrop area of sedimentary rock as measured from geological maps for periods of time in the Phanerozoic. (These data are plotted on a logarithmic scale, which is used to emphasize proportional variation among numbers. A unit difference on such a scale represents a constant ratio between numbers; for example, the difference between 1 and 2 is the same as that between 5 and 10.) (b) An estimate of the numbers of invertebrate species that were discovered and named between 1900 and 1975. Pleistocene and Holocene data are ignored in both figures. All data were compiled from global sampling. Periods of time with more sedimentary rock tend to have more fossil species as well. *(Data are from Raup, 1976a, b)*

* Please see the geologic timescale on the inside front cover of this book.

second pitfall, in effect the opposite of the first, is the assumption that the fossil record is so biased or incomplete that it is of little scientific use.

In fact, incompleteness often leads to very specific predictions about the distortion of patterns in the fossil record, and these predictions can be used to advantage. To take a simple example, calcite is a form of calcium carbonate that is more stable than an alternative form, aragonite [SEE SECTION 1.2]. Most **BRACHIOPODS**[1] have calcitic shells, whereas both aragonite and calcite are common in **BIVALVE MOLLUSCS.** We might therefore expect bivalves to have a lower preservation potential and to be more severely underrepresented in the fossil record. During the Mesozoic and Cenozoic Eras, however, the number of fossil bivalve species increased greatly relative to that of brachiopods. This observation goes against the prediction of the postulated bias and is therefore not a result of it. We can trust that bivalves really have diversified more than brachiopods over the past 250 million years.

[1] Please see the inside back cover for a summary of the paleontologically important groups of organisms.

Considering the incompleteness of the fossil record naturally leads to the question of how to interpret the absence of a species or a larger biologic group in the record of a particular time and place. Given the rarity of preservation, absence from the record does not necessarily imply that the organisms in question did not live there. One way to determine whether an absence is true or preservational is to use the notion of **taphonomic control.** (**Taphonomy,** as we will discuss below, is the study of fossilization processes.) If one species is not found but a preservationally similar species—the taphonomic control—is found, then we know the necessary conditions for preservation were present. In such a case, the absence of the first species is more likely to reflect a true absence than if the taphonomic control had not been found.

## 1.2 FOSSIL PRESERVATION

One of the remarkable aspects of the fossil record is the variety of ways in which a once-living organism can be preserved as a fossil. Intuition alone suggests that the possession of skeletal material (**hard parts**) should enhance the likelihood of preservation, and this is generally the case. However, the preservation of hard parts is not straightforward. Skeletal material typically undergoes mechanical and chemical alteration following the death of the organism and the microbial decay of soft tissues (**soft parts**). Most mineralized skeletons have an associated **organic matrix** that is subject to rapid degradation; this may compromise the post-mortem durability of the skeleton. Moreover, soft parts can be preserved in the fossil record under unusual circumstances. Combining experiment with observation, paleontologists have come to understand many of the intricacies of fossil preservation and how they relate to the questions we treat in this book.

## General Considerations

***Post-Mortem Degradation*** Environments in which life can exist are generally teeming with organisms. Death is quickly followed by scavenging, decay of organic tissue, and use of hard parts as substrates by other organisms. Thus, a number of biological processes reduce the chances of fossilization, and removal to an area of lower biological activity, by transport or burial, can enhance preservation. Nonetheless, biological activity can actually improve the prospects of preservation. Organisms that encrust shelly material may shield the shells from dissolution. Experimental work also shows that colonization by bacteria shortly after death plays an important role in the preservation of soft parts by changing local water chemistry in a way that favors mineral precipitation. Finally, organisms, such as certain shrimp, that continuously move sediment through their burrows in order to process it for food may significantly accelerate burial of skeletal material (see Figure 9.13).

Physical factors of degradation include wind and freeze–thaw cycles in subaerial environments, and current and wave activity in subaqueous environments. The erosive force of wind comes primarily from the sediment suspended in the air, whereas in the water, the energy of the moving water itself is important in addition to the effect of erosive sediment. As we will see below, field observations and experiments intended to simulate physical transport show that all but the most robust skeletons can be quickly degraded when they are moved by water currents.

Although many materials are known to be biologically produced (Table 1.1), the most important constituents are organic compounds, carbonates, phosphates,

TABLE 1.1

Paleontologically Important Groups of Organisms and the Principal Inorganic and Organic Components They Produce (● indicates a major component and ○ a minor component)

| | Inorganic | | | | Organic | | | | |
|---|---|---|---|---|---|---|---|---|---|
| **Group** | **Carbonates** | **Phosphates** | **Silica** | **Iron Oxides** | **Chitin** | **Cellulose** | **Lignin** | **Collagen** | **Keratin** |
| Prokaryotes | ● | ○ | | ○ | | ○ | | | |
| Algae | ● | | ○ | | ○ | ● | | | |
| Plants | ○ | | ○ | ○ | | ● | ● | | |
| Unicellular eukaryotes | ● | | ● | ● | ○ | ○ | | | |
| Fungi | ○ | ○ | | ○ | ● | ● | | | |
| Porifera | ● | | ● | ○ | | | | ● | |
| Cnidaria | ● | | | | ○ | | | ○ | |
| Bryozoa | ● | ○ | | | ● | | | ○ | |
| Brachiopoda | ● | ● | | | ● | | | ○ | |
| Mollusca | ● | ○ | ○ | ○ | ○ | | | ○ | |
| Annelida | ● | ● | | ○ | ○ | | | ● | |
| Arthropoda | ● | ● | ○ | ○ | ● | | | ○ | |
| Echinodermata | ● | ○ | ○ | | | | | ● | |
| Chordata | ○ | ● | | ○ | | ○ | | ● | ● |

*SOURCE:* Towe (1987)

and silica. Organic materials are composed mainly of carbon, hydrogen, nitrogen, and oxygen. They include **chitin** (a major component of **ARTHROPOD** cuticles and **FUNGI**), **cellulose** and other **polysaccharides** (major components of cell walls in **ALGAE** and plants), **lignin** (a constituent of conductive tissue in **VASCULAR PLANTS**), **collagen** (forming much of the connective tissue in animals), and **keratin** (a protein constituent of horns, claws, bills, and feathers). **Carbonates** include calcium carbonate, $CaCO_3$, which is secreted by a diverse array of organisms. **Calcite** is thermodynamically the more stable form of $CaCO_3$; **aragonite** is less stable and tends to dissolve or to convert to calcite over time. **Phosphates** include calcium phosphate, one form of which is **apatite,** $Ca_5(PO_4, CO_3)_3(F, OH, Cl)$. This class of minerals is important in **VERTEBRATE** teeth and bones, in the shells of some brachiopods, and in the jaw elements of **ANNELIDS.** Hydrous **silica,** or **opal,** $SiO_2 \cdot H_2O$, is less common than carbonates and phosphates, but it is important in such groups as the **SPONGES** and single-celled **DIATOMS** and **RADIOLARIA.**

Basic chemistry dictates the conditions under which biological materials tend to be stable. For example, organic compounds are generally unstable under oxidizing conditions. This fact reflects chemical reaction with oxidants as well as scavenging and other activities of organisms that live in oxygen-rich environments. Thus, the presence of well-preserved organics tells us that conditions were not strongly oxidizing. Carbonates tend not to persist below a **pH** of about 7.8, whereas phosphates and silicates are stable under slightly more acidic conditions.

Consideration of organic preservation naturally leads to the subject of genetic material. **DNA** is not an extremely stable molecule, so most confirmed cases of preserved DNA involve desiccated or frozen organisms less than one million years old. Contamination by other sources of DNA, including microbes and human lab workers, is a major problem, and rigorous tests are required to establish the authenticity of ancient DNA. To take just one example of what can be done with ancient genetic material, DNA from extinct mammoths (*Mammuthus*) has been compared with that of Asian and African elephants (*Elephas* and *Loxodon*). Recent analyses of DNA sequences suggest that the DNA of *Elephas* is more similar to that of *Mammuthus* than it is to *Loxodon*, and that therefore mammoths and Asian elephants are more closely related in an evolutionary sense [SEE SECTION 4.2] than are the living elephants with each other. There is some uncertainty here, and earlier analyses had suggested a closer affinity between *Mammuthus* and *Loxodon.* Regardless of how this particular case is resolved, the number of reliable instances of ancient DNA is increasing rapidly, and analysis of this material will continue to play an important role in a variety of evolutionary studies.

***Biological Traits that Enhance Preservability*** Considering the major biological, physical, and chemical factors of degradation, we can predict the kinds of organisms and parts of organisms that should stand the best chance of becoming fossilized. Because the organic matrix of skeletal materials often degrades quickly after death, a higher ratio of mineral to organic material tends to enhance preservability. The **TRILOBITE** cuticle has a much higher proportion of calcium carbonate than does the cuticle of **MALACOSTRACAN CRUSTACEANS.** Accordingly, trilobites stand a greater chance of fossilization. Similarly, the dense teeth of vertebrates are generally more durable than their bones.

The number of skeletal elements and how they are joined also influence preservability. Sponges often contain a large number of isolated spicules in an organic matrix, while many **CORALS** consist of a single, robust skeletal element. Corals therefore tend to preserve more readily. The mineral composition is also significant. As mentioned earlier, calcite is a more stable form of calcium carbonate than is aragonite. Thus, aragonitic species are sometimes found with their shells dissolved in the same deposits in which the shells of calcitic species are preserved intact. Organic molecules also differ in their stability. Lignin is less likely to decompose than cellulose, for example. As a result, lignin-rich vascular plants generally are more likely to be preserved than nonvascular plants. The waxy cuticle that covers the surfaces of vascular plants is also resistant to decay.

In addition to structural features of organisms, aspects of their ecology also influence fossilization potential. Perhaps most important is habitat. The land is an area of net sediment erosion while lakes, seas, and parts of river systems are areas of net sediment deposition. For this reason, aquatic organisms stand a better chance of being buried shortly after death and removed from biological activity; terrestrial organisms are most likely to fossilize if they are transported to subaqueous environments. Overall, then, the marine realm has a richer and more complete fossil record than the terrestrial realm. All else being equal, we might also expect species with greater numerical abundance to have better chances of fossilization. This is not a straightforward expectation, however, because species with more individuals tend to have smaller body size, while larger body size in many cases enhances the chances of at least partial preservation.

***Time Averaging*** Fossil assemblages are generally **time averaged;** that is, they represent an assemblage of skeletal material that has accumulated over some span of time, typically tens to thousands of years but in some cases extending to millions of years. Instantaneous deposits are in fact quite rare. Time averaging results from a number of factors but depends mainly on the rate of production of preservable skeletal material and on the rate of sediment accumulation. For a given rate of skeletal production, a lower rate of sedimentation will allow more generations to be represented in a given thickness of rock, which will therefore be more time averaged. This is complicated by the fact that very low sedimentation rate can lead to long-term exposure and therefore destruction of skeletal material. Sedimentary beds can also become time averaged through **bioturbation.** This is the normal churning and reworking of unconsolidated sediments that occurs as a by-product of such organismic activities as burrowing and ingesting sediment to extract food.

Many living species exhibit a patchy spatial distribution [SEE SECTION 9.3]. Therefore, an instantaneous collection of living individuals from a single locality would not contain all the species living in the larger area which that locality represents. Over time, however, the spatial distributions of species fluctuate as a result of both chance variation in colonization and temporal variation in the distribution of particular habitats. Consequently, a time-averaged sample of fossil organisms from a particular locality may provide a far more complete representation of species that lived in an area than could have been obtained by sampling the living biota over the same lateral extent. This benefit comes at the cost of decreased temporal resolution, however, and in some cases a loss of information about fine-scale spatial patchiness.

The importance of time averaging depends on the timescale of the process we are studying. If we are interested, for example, in reconstructing local communities that lived in the past, a significantly time-averaged assemblage may include species that never lived together. In contrast, evolutionary changes in biologic form often occur over hundreds of thousands to millions of years—spans of time substantially longer than the typical scales of time averaging. The analysis of such changes, therefore, is barely affected by this process. Much has yet to be learned about the scale of time averaging in different geologic situations, and paleontologists are now attempting to measure time averaging by radiometric dating and other methods that reveal the ages of dead shells [SEE SECTION 10.6].

## Modes of Fossilization

Paleontologists have recognized different modes by which individual organisms can become fossilized. These form a spectrum of preservation that ranges from most complete (including preservation of soft parts or easily degraded hard parts such as chitin) to least complete (preservation of only indirect traces of the organism). Preservational modes near the top of the following list are far less common than those near the bottom.

1. *Freezing* (Figure 1.2a). In rare circumstances, ancient organisms, such as woolly mammoths, have been found virtually intact, frozen in permafrost regions of Siberia and elsewhere. These specimens are only a few thousand years old and, thus, may be on the fringes of what we would define as fossils. Their preservation is truly remarkable nonetheless, and they have provided unique opportunities to study the species in question. For example, DNA and gut contents have been extracted from frozen animals.
2. *Preservation in amber* (Figure 1.2b). This is one of the primary means through which **INSECTS** and **SPIDERS** are preserved as fossils. Relatively small organisms sometimes become trapped in highly viscous resin secreted by various trees. When the resin hardens, the trapped organisms are preserved relatively intact in a transparent medium. Incidentally, ancient air bubbles have also been trapped in amber. Geochemists have studied these for clues about the composition of the earth's atmosphere in the past.
3. *Carbonization* (Figure 1.2c). Soft parts of organisms may be preserved as carbon films through **distillation** under heat and pressure, which preferentially removes hydrogen and oxygen. Therefore, even if we are fortunate enough to recover organic material from the fossil record, its original chemical form is often substantially altered. Nevertheless, carbonization can preserve exquisite details of soft anatomy [SEE SECTION 10.2]. Leaves in coal and fine-grained sediments provide a good example.

(a)

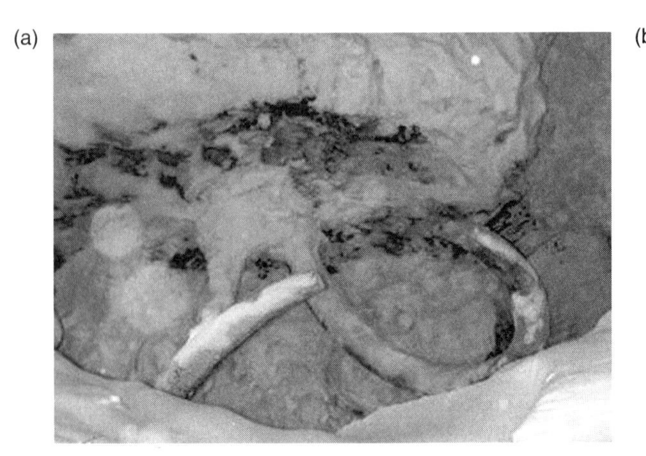

(b)

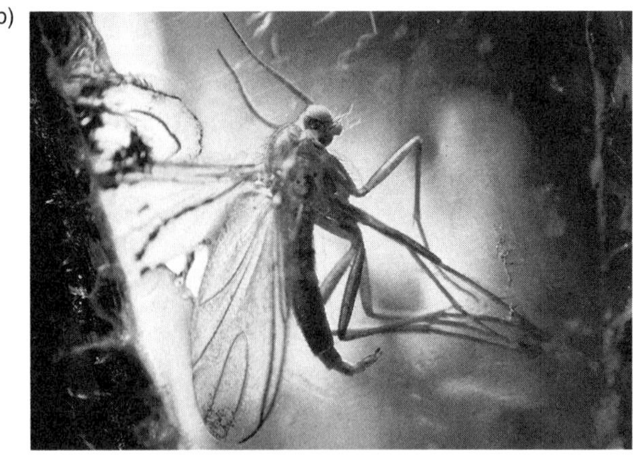

**FIGURE 1.2 Common modes of fossilization.** (a) Freezing: Wooly mammoth extracted from Siberian permafrost in 1999. (b) Preservation in amber: Eocene insect (midge) from the Baltic region of Europe (magnification ×9). (c) Carbonization: Triassic **FERN** from the Carnic Alps. (d) Permineralization and petrifaction: Triassic petrified log from the Painted Desert of Arizona. (e) Replacement: Pyritized Ordovician trilobite (*Triarthrus eatoni*) from New York. (f) Recrystallization: Paleozoic **BRYOZOAN** from Tennessee. (g) Internal mold: **GASTROPOD** steinkern. (h) Cast: Upper Carboniferous **LYCOPSID** rooting structure (*Stigmaria*) from West Virginia. *(a: Discovery Channel/Handout/X00561/Reuters/Corbis; b: Alfred Pasieka/Photo Researchers, Inc.; c: John Cancalosi/Peter Arnold, Inc.; d: Eric and David Hosking/Corbis; e: Thomas Whitely; f: Unrug et. al., 2000; g: R. A. Paselk, HSU Natural History Museum; h: West Virginia Geological and Economic Survey)*

4. *Permineralization* (Figure 1.2d). As suggested earlier, the buried hard parts of organisms are not impervious to alteration. **Pore water** that percolates through a fossil-bearing unit can dissolve skeletal material, in some cases many years after it was buried by sediment in the first place. However, the pore water may be laden with dissolved materials that precipitate from solution in the spaces within the skeletal material. Through this process, substances such as silica, phosphate, and pyrite permeate the skeletal material, thereby hardening it, while preserving such fine structural details as growth bands, skeletal pores, and shell layers. A closely related process is **petrifaction,** which is the conversion of organic material to mineral material. Permineralization and petrifaction are both important in the preservation of plant tissue.
5. *Replacement* (Figure 1.2e). This process is similar to permineralization, except that the original skeletal material is itself replaced by the permeating materials, sometimes molecule-for-molecule, again preserving fine-scale structure. The exact nature of replacement depends on the details of pore-water chemistry. Examples include replacement by pyrite, $FeS_2$ (pyritization), silica (silicification), and phosphate minerals (phosphatization).
6. *Recrystallization* (Figure 1.2f). This is a very common process in which skeletal material that is subjected to elevated temperature and pressure converts spontaneously to a thermodynamically more stable form (e.g., aragonite to calcite and amorphous silica to quartz). At a macroscopic scale, a recrystallized skeletal element may be difficult to distinguish from the original, but fine-scale structures may be virtually eliminated, as the element takes on the crystal structure of the new mineral.
7. *Molds* (Figure 1.2g) and *Casts* (Figure 1.2h). Molds are negative impressions of hard parts. Even when all of the original skeletal material has been dissolved away by pore water, an excellent replica of the hard part may still be preserved as a mold in the sediment that encases it, provided that the sediment is sufficiently fine grained. Some paleontologists have injected epoxy resin into carbonate rocks that contain molds and have then dissolved away the rock with acid to yield positive casts of the hard part. Casts also occur in nature when the original material first dissolves away, leaving a void that is filled subsequently with a secondary mineral substance or sediment. Of course, only surface features are preserved in molds and casts, but the level of detail preserved can be quite striking. Where sediment has filled in the empty skeleton of an organism, an internal mold or **steinkern** results. Here the internal features, such as scars showing muscle attachment, may even be preserved.

(c)

(d)

(e)

(f)

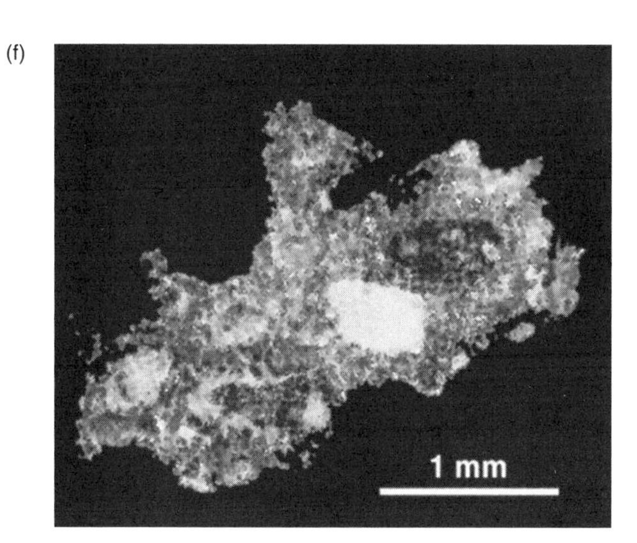


(g)

(h)

(a)

(b)

(c)

**FIGURE 1.3 Examples of trace fossils.** (a) Probable worm burrows from the Jurassic of England. (b) Resting trace (*Rusophycus*) of a trilobite from the Ordovician of Australia. (c) Dinosaur footprints from the Painted Desert of Arizona. (*a: Mike Horne, FGS, Hull, UK; b: From the collection of the Australian Geological Survey Organization, Canberra; c: Tom Bean/Corbis*)

8. *Trace fossils* (Figure 1.3a, b, c). The modes of fossilization discussed up to now refer to **body fossils,** the remains of actual parts of organisms. Some organisms that are not preserved directly nevertheless leave behind traces of their activity. The most common examples of **trace fossils** include burrows and footprints. In most instances, it is difficult to know for certain the organism that made the trace, but the producers of some trace fossils can be identified. For example, Figure 1.3b illustrates a trilobite resting trace. Trace fossils may provide information concerning the behavior of organisms that is not available from body fossils alone. For example, large numbers of parallel trackways have suggested that certain dinosaur species traveled in herds. Evidence of activity by organisms can also be found directly on body fossils. Such traces include bite marks and boreholes.

## Pseudofossils and Artifacts

The geologic record presents us with many inorganic structures resembling biological remains, such as mineral growths that branch like ferns, sediment degassing and dewatering features that look like animal trails and jellyfish, and sedimentary rip-up clasts that can be mistaken for arthropod fragments (Figure 1.4). It is essential to distinguish such **pseudofossils** from true remains of organisms. Morphological complexity, symmetry, and close resemblance to undoubtedly biologic remains are generally reliable, if not foolproof, criteria for recognizing true fossils.

The problem of pseudofossils has been especially prominent in the search for microbial life in the geologic record of the Archean and Proterozoic Eons. This is because inorganic, microscopic structures may falsely resemble microbes [SEE SECTION 10.7]. Paleontologists who

(a)

(b)

**FIGURE 1.4 Examples of pseudofossils.** (a) Pyrite rosette in a sandstone, resembling a fossil cnidarian (Silurian of Pennsylvania). (b) Various mineral structures in thin sections of quartzite, resembling microfossils (Archean of Greenland). *(a: From Cloud, 1973; b: From Schopf & Walter, 1983)*

study microbial life therefore look for a narrow size range within single species, for cells preserved in the act of dividing, and for other aspects of cellular structure (Figure 1.5). Although all organisms are subject to postmortem degradation, particular problems arise with microbes. Experiments with living forms have shown that false "cells" and features can result as artifacts of preservation. Of particular interest is the fact that the preservation of **PROKARYOTES** (organisms composed of small, simple cells lacking a nucleus and other organelles) can produce artifacts that resemble the organelles of the cells in more complex **EUKARYOTES.**

## Taphonomy

An understanding of fossilization processes is a great aid to paleontologists in interpreting the biological meaning of collected materials. Put another way: We know that a fossil assemblage will inevitably yield data that differ in quantity and quality from those that would have been available from the living assemblage from which it was drawn. But if we understand these differences, we can adjust our interpretations of the record accordingly. These considerations fall within the field of taphonomy. Two broad aspects of fossilization are included under this umbrella: **biostratinomy,** in which the focus is on processes that affect a dead organism prior to burial; and fossil **diagenesis,** the processes that affect it after burial. Of course, an organism may be buried and exhumed several times after its death, so it may be subject to several phases of biostratinomic and diagenetic processes.

***Experimental Approaches*** The direct study of fossilization processes is a central part of the field of **actuopaleontology,** a term derived from a German word meaning *paleontology of the present day.* To better understand the route to preservation, paleontologists have monitored the death, disintegration, and burial of

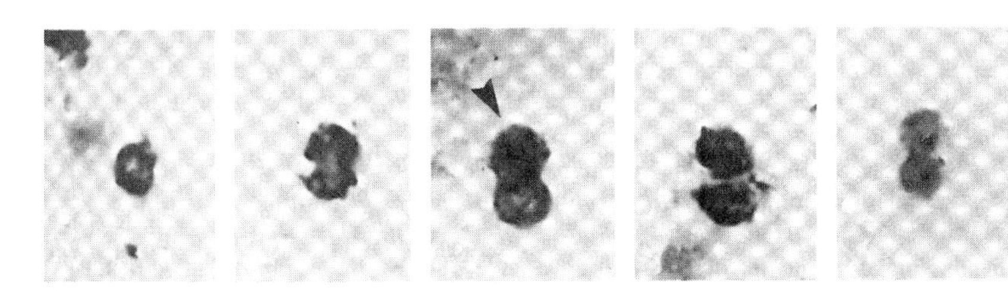

**FIGURE 1.5 Photomicrographs of fossil cells in various stages of cell division.** Material is from the Archean of South Africa. Magnification ×1600. *(From Knoll & Barghoorn, 1977)*

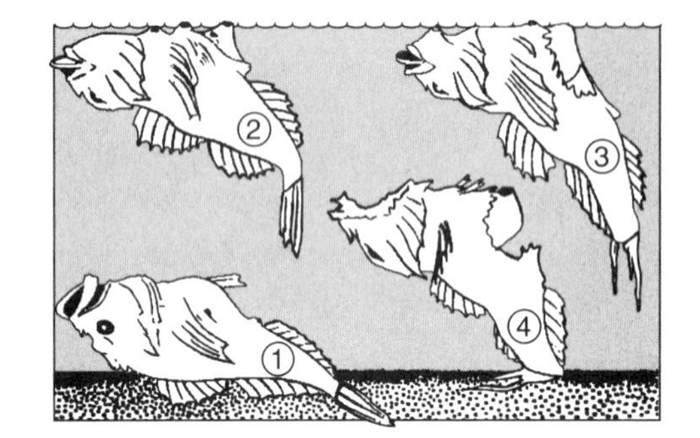


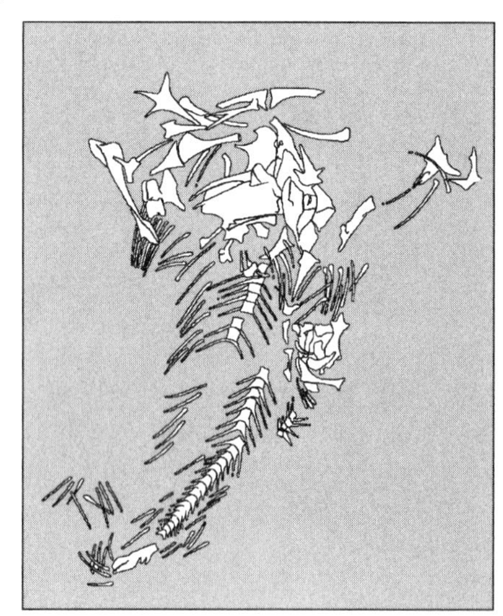


**FIGURE 1.6 An example of death and disintegration.** (a) Early stages (first four days) of decay in a carcass of the sea scorpion *Myoxocephalus scorpius*. An inflated stomach and gas bubbles from decomposition initially cause the carcass to float. Subsequently, gas escapes through tears that develop, and the carcass sinks. (b) Skeletal remains of *M. scorpius* on the sea floor after three months of exposure in agitated waters. *(From Schäfer, 1972)*

many kinds of organisms in the field (Figure 1.6). In the laboratory, they have simulated processes of destruction in ways that permit the control of experimental conditions. A common laboratory approach is to place skeletal material in tumblers with abrasives, such as pebbles, and to turn the tumblers to simulate transport and other agents of mechanical destruction.

For example, Susan Kidwell and Tomasz Baumiller (1990) ran a series of tumbling experiments on two species of the **ECHINOID** genus *Strongylocentrotus* after first allowing carcasses to decay for varying lengths of time in a number of temperature and oxygenation conditions. They found that variations in the degree of oxygenation during decay had little effect on the tendencies of these echinoids to disintegrate during tumbling. However, variations in temperature dramatically affected tumbling results (Figure 1.7). The number of hours of tumbling required to cause near total disintegration of *S. purpuratus* carcasses was significantly greater among specimens allowed to decay in cold water (11°C) than it was among species allowed to decay in warmer water (23°C or 30°C). Colder conditions evidently retarded the rate of

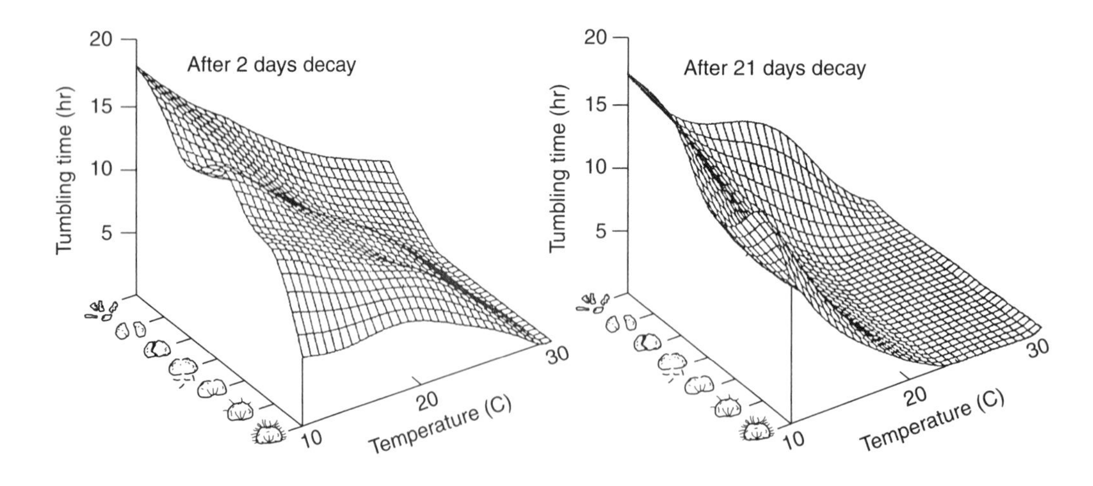


**FIGURE 1.7 Experimental results from tumbling experiments on the echinoid *Strongylocentrotus purpuratus*.** Surfaces show combinations of temperature, state of disintegration (indicated by the icons), and tumbling time for experiments in which echinoids were allowed to decay for 2 days and for 21 days prior to tumbling. Note that the pattern for the experiment involving 21 days of decay is not substantially different from that involving 2 days of decay. In both cases, the number of hours of tumbling required for major disintegration increases as the temperature decreases. *(From Kidwell & Baumiller, 1990)*

decay, thereby promoting the retention of soft tissues that help prevent disintegration. This was true whether the period of decay prior to tumbling was two days or 21 days, suggesting that the slowdown in the rate of decay at 11°C was so appreciable that a significant amount of soft tissue remained even after several weeks. These experimental results predict that there may be latitudinal and **bathymetric** trends in the preservation of echinoids, with better preservation, for the most part, in higher latitudes at moderate depth, where the water is cool and burial by storms is most common.

In another tumbling experiment, Benjamin Greenstein (1991) compared the durability of four echinoid genera. For each tumbled specimen, Greenstein calculated a *coefficient of breakage* (*CB*) based on assessment of the fragments greater than 2 millimeters (mm) in size that remained. The 2-mm cutoff was used as a practical limit below which fragments would probably not be recognized in the fossil record. The coefficient is calculated as:

$$CB = \frac{\text{Number of fragments} > 2\text{ mm}}{\text{Weight of fragments} > 2\text{ mm}} \times \frac{1}{\text{Weight percent of fragments} > 2\text{ mm}}$$

Because a highly fragmented skeleton consists of a large number of pieces that collectively account for relatively little weight, such a skeleton will have comparatively high values for both the first term and the second term. Greater values of the *CB* therefore imply a greater degree of breakage of the skeleton. The second term helps to compensate for cases in which the skeleton is comparatively small to begin with or becomes highly pulverized. A pulverized skeleton, with perhaps only a single remaining fragment larger than 2 mm, would yield a low *CB* if only the first term were used. However, because the weight percent of this one remaining fragment would likely be very small, the inclusion of this second term increases the *CB*.

Greenstein's experimental results are depicted in Figure 1.8. The four genera exhibited strikingly different degrees of breakage, with *Diadema* showing the most significant disintegration and *Echinometra* remaining intact. In light of his experimental results, Greenstein then considered the fossil records of the four families to which these genera belong. Paradoxically, a much greater percentage of species belonging to the family Diadematidae are preserved with tests intact than might be expected based on the tendency of *Diadema* to disintegrate rapidly. Greenstein showed, however, that fossil preservation in this family was strikingly bimodal: In most cases, specimens are preserved either in a highly fragmented state or nearly intact, with intermediate states of preservation barely represented. This is testimony to the fragility of the diadematid skeleton. A diadematid must be buried rapidly to avoid rapid disintegration; in cases when it is buried rapidly, it is preserved in a relatively complete state.

Taphonomic experiments have also assessed the chemical transitions associated with the decay and preservation of soft tissue [SEE SECTION 10.2]. For example, Stephen Grimes et al. (2001) subjected twigs of the plane tree to a variety of chemical environments meant to promote the precipitation of pyrite in association with decaying plant matter. Most experiments failed

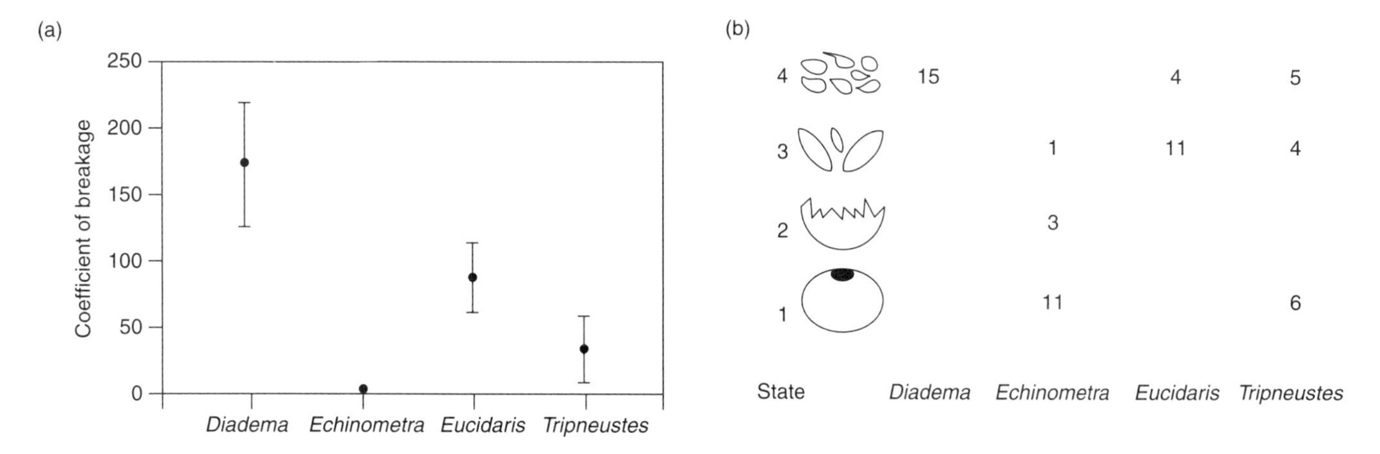


**FIGURE 1.8 Experimental results from echinoid tumbling experiments.** (a) Mean coefficient of breakage of four different echinoid taxa, each represented by 15 tests. Error bars show two standard errors of the mean (see Box 3.1). Schematic illustrations of the extent of breakage exhibited by the tumbled tests and the corresponding number of tests. *(From Greenstein, 1991)*

to yield pyrite. Rather, very particular combinations of sulfur, iron, oxygen, and organic concentrations were required in the lab. Presumably the requirements in nature would also be very specific. Significantly, one of the experiments that yielded no pyrite was one in which bacteria were not introduced. This is one of many studies to show that the chemical changes brought on by microbial decay—including depletion of oxygen—facilitate mineral precipitation within and on soft tissues.

***Assessment of Recent Subfossil Assemblages*** Paleontologists have also directly evaluated the extent to which **subfossil** assemblages that are accumulating today faithfully represent the living assemblages from which they

*Box 1.1*

## LIVE–DEAD COMPARISONS

In a broad survey of previous studies of living communities and their subfossil counterparts, Kidwell (2001) assessed whether the same roster of species tended to be present in life and death assemblages drawn from the same location, and whether they exhibited similar abundances. Figure 1.9a depicts one of these comparisons, from a tidal creek in California. Eleven species were found; their abundances in the live sample are plotted on the *x* axis, and their abundances in the sample of dead shells within the sediment are plotted on the *y* axis. Three of the species, plotted as zeroes on the *x* axis, were found as dead shells but were not present in the live sample. It is clear from this plot that the more abundant a species is in the live sample, the more abundant it tends to be in the dead sample; that is, the live and dead abundances are positively correlated. It is also evident that the dead abundances are generally higher than the live abundances.

Various correlation coefficients [SEE SECTION 3.2] are used to measure the strength of association between two variables. In this case, the correlation between live and dead abundances was measured with a rank-order coefficient that considers only the relative order of the variable; abundances of 0, 5, 6, and 100, for example, would be represented by the ranks 4, 3, 2, and 1. Correlations [and many other statistics; SEE SECTION 3.2] are commonly expressed by their corresponding *p*-values. The *p*-value estimates the probability that a correlation could be as high as observed, due to sampling error, if there were in fact no association between the two variables. The lower the *p*-value, the more reliable the inference that there is a true correlation in the data. It is conventional to regard *p*-values of .05 or less as statistically significant, in other words, indicating a true correlation.

The comparison of Figure 1.9a has 3500 live individuals and yields a correlation of +0.64 (on a scale of −1 to +1), with a *p*-value of about .04. This comparison is shown in Figure 1.9b, where the number of live individuals is on the *x* axis and the *p*-value on the *y* axis. This figure also includes 42 other comparisons from other localities. The horizontal dashed line marks the *p*-value of .05; results below this are considered statistically significant. Many of the comparisons do not yield significant correlations between live and dead abundances.

In the comparisons of Figure 1.9b, however, shells were collected with nets and sieves that have a mesh size of 1 mm or less; thus, shells at the smaller end of the size spectrum were included in these samples. Smaller shells are more susceptible to postmortem destruction and transport. Moreover, live samples that include small shells are more likely to be sensitive to whether or not there has been a

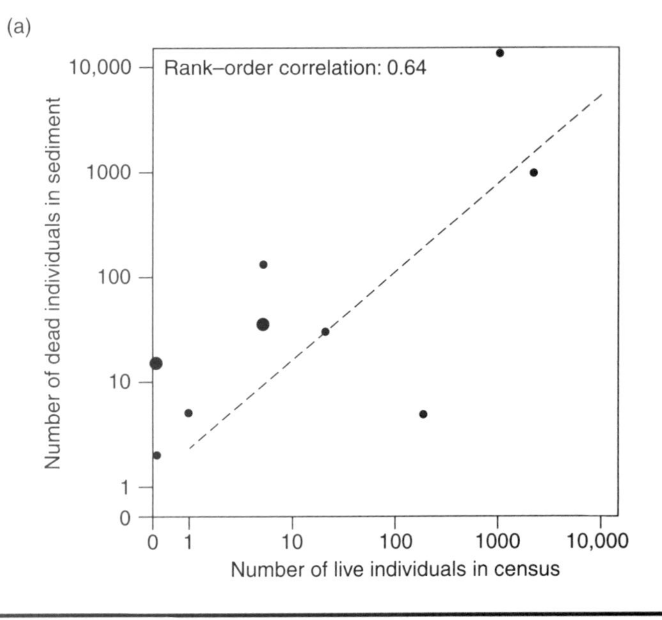

were drawn. Collectively, these so-called live–dead investigations have sought to establish the degree of spatial and temporal resolution likely to be preserved in fossil assemblages. Live–dead comparisons thus provide one of the principal ways to assess the quality of the fossil record.

In cases where data are available to compare the species composition of skeletal remains with live samples drawn from the same settings, the relative abundances of species in living communities (**life assemblages**) tend to be maintained rather well in associated subfossil accumulations (**death assemblages**). This was demonstrated by Susan Kidwell (2001) in a comprehensive comparison of live and dead mollusc shells in a variety of marine settings (Box 1.1).

recent influx of larvae into the population. Figure 1.9c thus depicts a second set of comparisons, those that excluded the smallest shells by sampling with mesh sizes greater than 1 mm. These comparisons on the whole yield lower $p$-values and fewer $p$-values above .05. In other words, there is a closer correspondence between live and dead abundances. Most of the comparisons that fail to yield significant correlations involve small data sets, with live abundances less than 100.

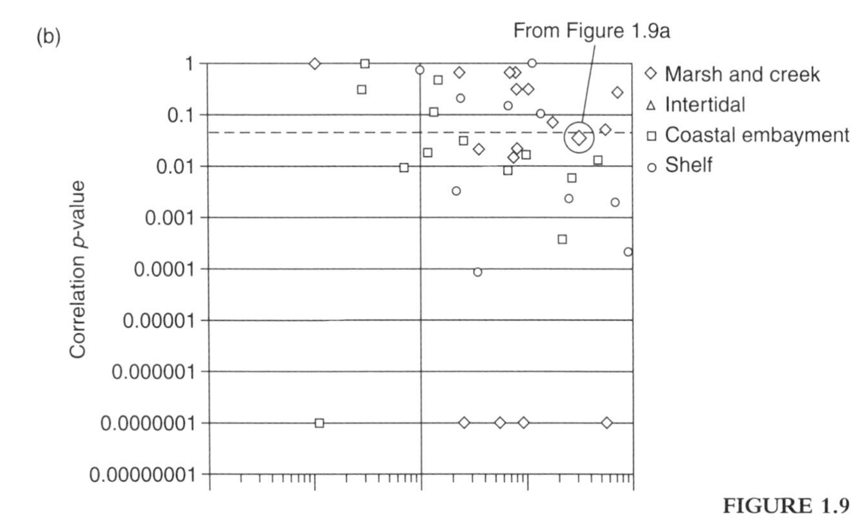


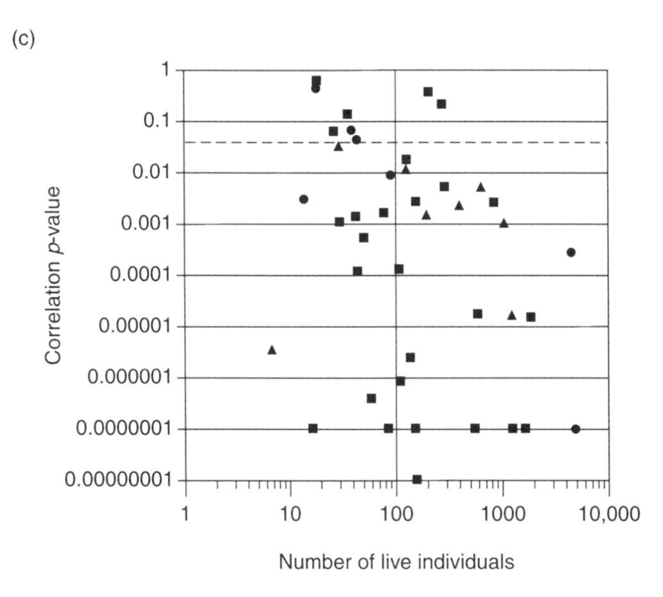


**FIGURE 1.9 Kidwell's comparison of life assemblages versus death assemblages for 85 collections representing a variety of marine settings.** (a) Comparison between live abundances and dead abundances of 11 species at a tidal creek locality in California. Small points represent one species each; larger points each represent two species with the same live and dead abundances. The dashed line shows the trend in the relationship between live and dead abundances for species found in both the live and dead samples. Species that are more abundant in the live sample tend to be more abundant in the dead sample as well. As explained in the text, the correlation between live and dead abundances is statistically significant with a $p$-value of .04. Parts (b) and (c) illustrate live abundances and $p$-values for 85 comparisons like that of part (a). (b) Forty-three collections made with sieve sizes of 1 mm or less. The comparison in part (a) is indicated by the large circled diamond. (c) Forty-two collections made with sieve sizes greater than 1 mm. For additional discussion, see text. *(a: Data from MacDonald, 1969; b and c: Data from Kidwell, 2001)*

Earlier we cited time averaging as one reason that a death assemblage often contains more species than the corresponding life assemblage. The analysis of Box 1.1 is fairly typical in illustrating a related reason: The death assemblage often contains more individual specimens than the life assemblage and is therefore more likely to capture the rarer species.

Comparisons between life assemblages and subfossil accumulations also indicate that death assemblages tend to retain excellent environmental fidelity, reflecting variation in species composition even at spatial scales as fine as tens of meters. However, the fidelity of death assemblages holds only for the readily preservable elements of the assemblage, and sometimes more strictly only for a single biologic group such as molluscs. It is unlikely that a fossil assemblage will provide a faithful rendition of the entire living assemblage from which it is derived. As discussed earlier, the loss of soft-bodied organisms, as well as organisms with fragile skeletons, is usually inevitable.

***Assessment of Ancient Assemblages: Taphofacies*** The term **facies** in general refers to the characteristics of sedimentary rocks. **Taphofacies** are suites of fossils characterized by particular combinations of preservational features. In their pioneering discussions of taphofacies, Carlton Brett and Gordon Baird (1986) brought together factors such as post-mortem transport, degree of exposure, water oxygenation, sedimentary chemistry, skeletal robustness, and the number of articulated elements that compose the skeleton. The taphofacies approach provides an opportunity to assess the extent to which a life assemblage is altered during the formation of a fossil assemblage, but it also provides a diagnostic tool for paleoenvironmental analysis.

The application of the taphofacies concept is illustrated here for trilobites of the Middle Devonian Hamilton Group in New York. Stephen Speyer and Brett assessed several taphonomic attributes of sampled trilobite assemblages that included the trilobites *Phacops rana* (Figure 1.10a) and *Greenops boothi*. Characteristics that were evaluated included the proportion of skeletal parts oriented in a convex-up direction, the degree of skeletal articulation, the proportion of enrolled individuals, and the proportion of skeletal remains that were associated with molted skeletons. On this basis, a suite of trilobite taphofacies can be recognized, which are related to a depth gradient and the degree of **terrigenous** sediment influx (Figure 1.10b). In general, shallower assemblages, which are characterized by coarser sediment, contained more highly fragmented material that tended to be oriented by current-related processes (e.g., taphofacies 1A in Figure 1.10b), except where sedimentation rates were comparatively high and skeletal material was buried rapidly or episodically. Articulation was more frequent in deeper water, with greater concentrations of skeletal material, including molts, in settings where there was not an overwhelming supply of terrigenous sediment (e.g., taphofacies 4A in Figure 1.10b).

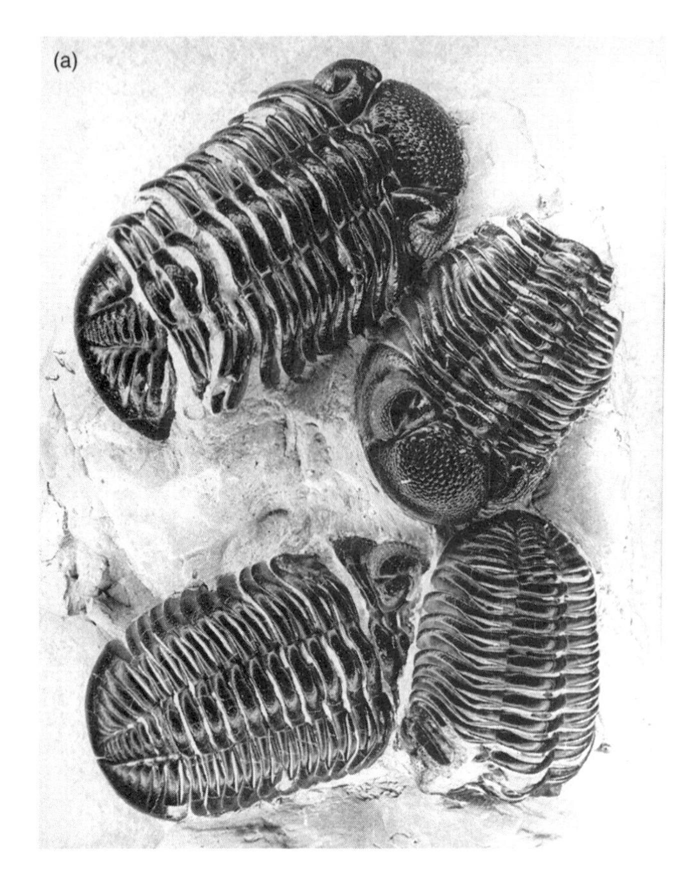


**FIGURE 1.10 Speyer and Brett's (1986) assessment of trilobite taphofacies in the Middle Devonian Hamilton Group of New York.** (a) Several well-preserved specimens of the trilobite *Phacops rana*. (*Reprinted from Levi-Setti, 1975*)

## Exceptional Preservation

Although the quality of preservation varies along a continuum, a handful of deposits have such exquisite preservation of organic and skeletal material that they are often discussed separately as fossil **Lagerstätten,** a

(b)

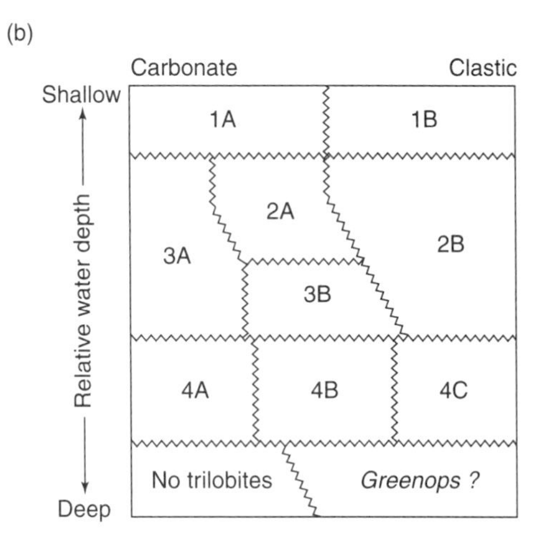


| Taphofacies | Biostratinomic indices | | | | |
|---|---|---|---|---|---|
| | FRG | CNV% | ART% | ENR% | MLT% |
| 1A & 1B | aa | 60–80 | <1 | p | np |
| 2A & 2B | a–p | ~50 | 10–25 | 40–80 | 30–50 |
| 3A | p–c* | 20–30 | 20–30 | 60–90 | 30–70 |
| 3B | c–p* | 60–70 | <15 | p | 85–95 |
| 4A | p | 25–40 | 60–75 | >80 | >95 |
| 4B | np | 25–40 | 40–60 | 70–100 | 20–40 |
| 4C | np | p | >95 | >70 | 20–30 |

*Fragmentation of *Greenops* cephala common

**FIGURE 1.10 (*cont.*)** (b) Delineation of a suite of Hamilton taphofacies based on biostratinomic indicators, and their paleoenvironmental locations with respect to water depth and the nature of sedimentation. In general, specimens tend to be less fragmented and more complete (higher-numbered taphofacies) in deeper water. (*Key:* aa = very abundant; a = abundant; c = common; p = present but rare; np = not present. FRG = degree of fragmentation; CNV% = percent of specimens oriented convex-upright; ART% = percent of articulated individuals; ENR% = percent of specimens that were enrolled; MLT% = percent of articulated specimens that were preserved molts.) *(From Speyer & Brett, 1986)*

German mining term. These deposits have been very important in revealing aspects of biology that are ordinarily not preserved, such as the nature of arthropod limbs and other soft parts (Figures 1.11, 1.12), the chemical composition of vascular and other tissues in land plants (Figure 1.13), the feathers of early **BIRDS** (Figure 1.14), and, in extremely rare cases, embryos (Figure 1.15). They have also opened windows onto whole communities during critical periods of time, such as the Middle Cambrian, an early phase in animal diversification that has been revealed by the Burgess Shale of British Columbia and other deposits elsewhere [SEE SECTION 10.2].

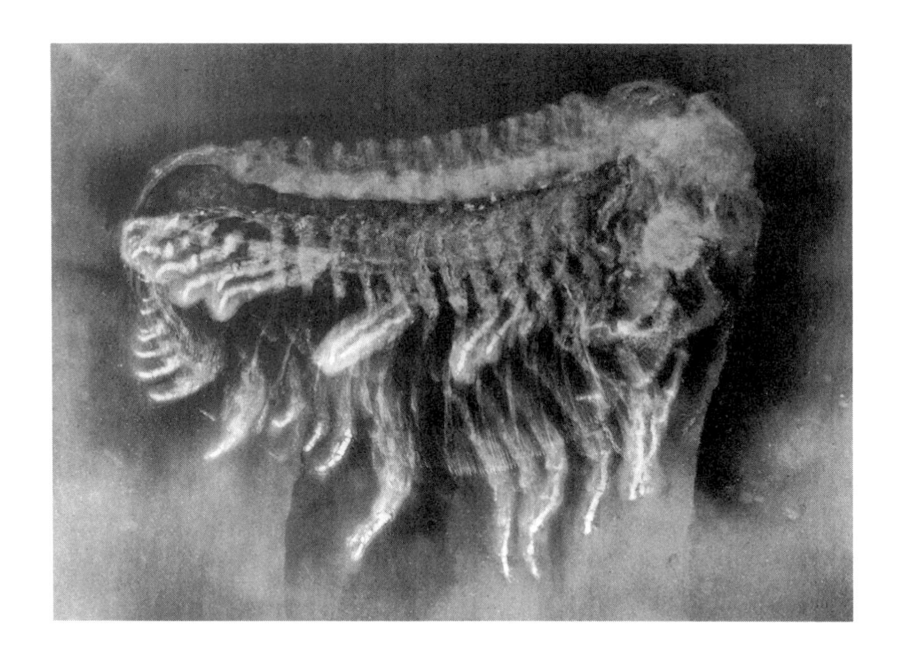

**FIGURE 1.11 Radiograph of a trilobite from the Hunsrück Slate (Devonian of Germany).** The specimen is unusual in showing an almost complete array of appendages. *(X-ray taken by W. Stürmer, Erlangen, Germany)*

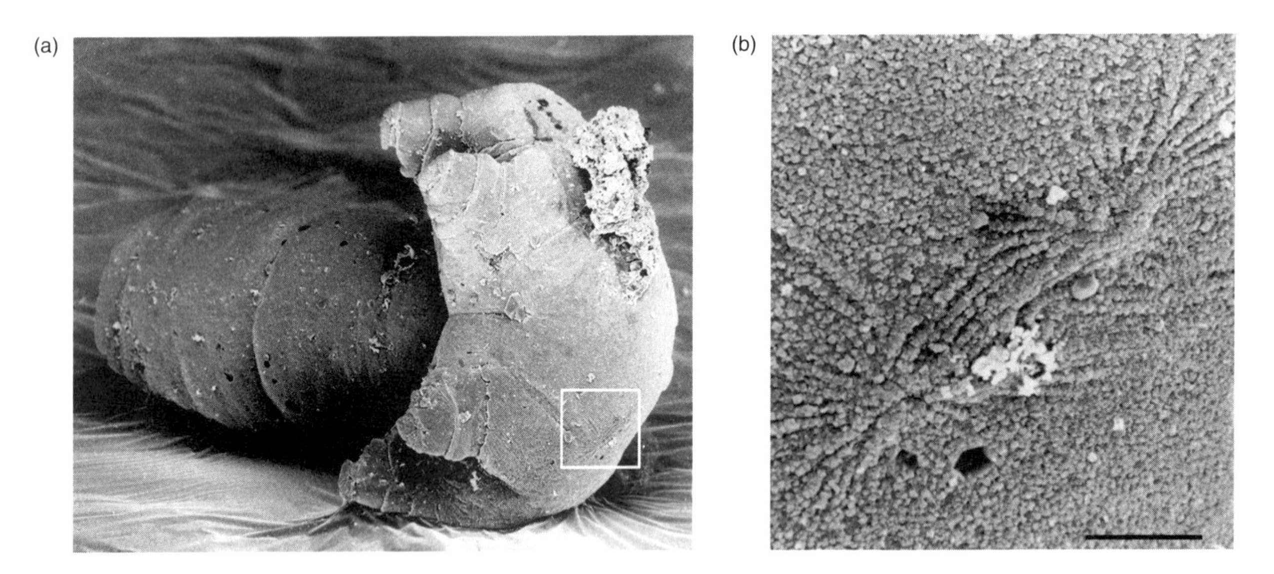


**FIGURE 1.12 Specimen of the pentastomid *Heymonsicambria kinnekullensis*, from the Upper Cambrian of Sweden.** The material is phosphatized and shows exceptionally fine morphological details. (a) Anterior view of head. (b) Magnification of boxed area on Figure 1.12a, showing detail of papillae (sensory organs). Scale bar is 10 $\mu$m. *(From Walossek & Müller, 1994)*

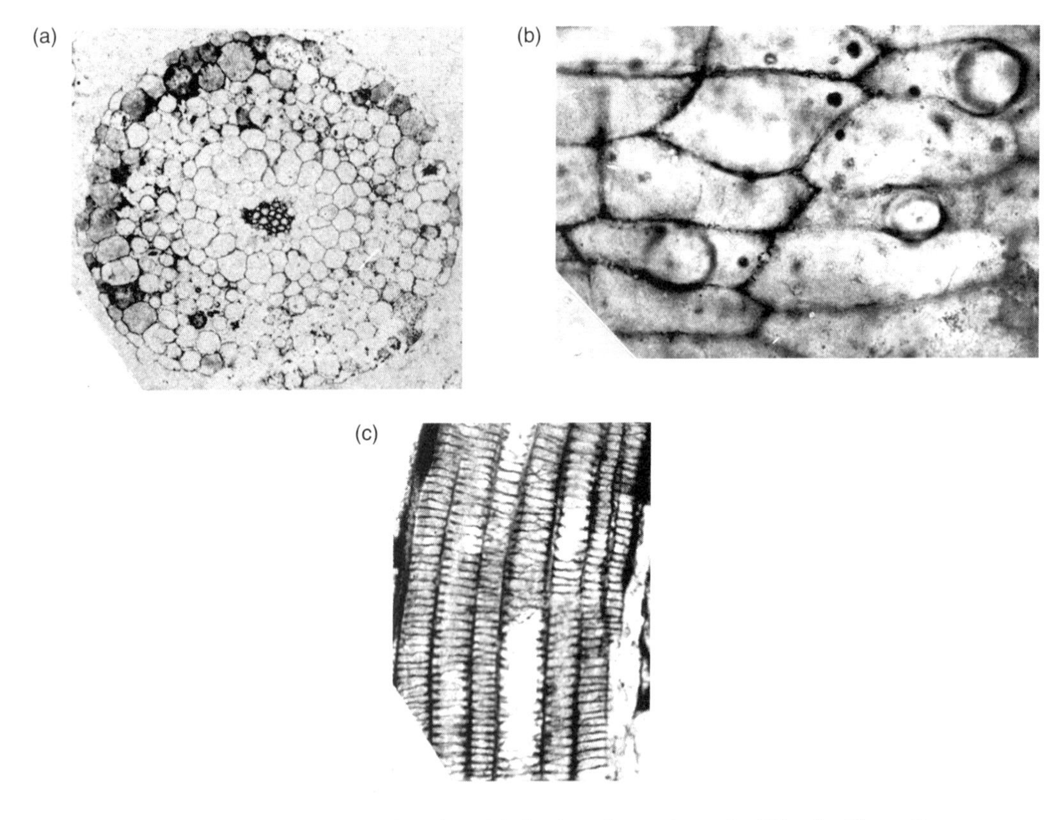


**FIGURE 1.13 Specimens of the vascular plant *Trichopherophyton* from the Rhynie Chert (Lower Devonian of Scotland).** (a) Cross section showing individual cells (magnification ×41). The innermost dark cells are the conductive tissue xylem. The conductive phloem consists of the cells immediately surrounding the xylem. Most of the remainder of the specimen consists of cortex cells. (b) Magnification (×187) of cells. (c) A longitudinal section of xylem (magnification ×216). Preservation in the Rhynie Chert is by petrifaction and permineralization. Original carbon is preserved, and this has been chemically analyzed to understand the evolution of vascular tissue by determining which tissues contained lignin (Boyce et al., 2003). *(From Lyon & Edwards, 1991)*

**FIGURE 1.14** ***Archaeopteryx*, the oldest known bird, from the Solnhofen Limestone (Jurassic of Bavaria).** This fossil shows impressions of feathers. (*Louie Psihoyos/Corbis*)

## 1.3 SAMPLING OF THE FOSSIL RECORD

Rather than asking whether the record of a group of organisms is complete—for it never is—it is useful to consider whether the record is *adequate* for a particular purpose (Paul, 1982). Two issues must be addressed for any sample, each of which makes sense only in the context of specific paleontological questions. First, is the sample random (unbiased)? Second, are the quantities we measure sensitive to the size of the sample itself, even if it is unbiased?

Sampling inevitably involves error, but the science of statistics tells us how to deal with this. Suppose we collect a sample of 100 specimens from the Middle Cambrian Wheeler Formation in Utah and we find that 40 of them belong to the trilobite species *Elrathia kingi*. Provided that we do not preferentially collect the larger, more complete, or more attractive specimens—that is, provided that we sample randomly—our best guess is that the true proportion of *E. kingi* among the fossils in this formation is 40 percent. There would nevertheless be a well-defined margin of error associated with this estimate, and we would not be surprised if, upon collecting a second sample of 100 specimens, we found that as few as 35 or as many as 45 of them belonged to *E. kingi*. The larger the initial sample, the smaller the margin of error.

We can take pains to ensure that our sampling of the record is unbiased, but the record itself is strongly biased in favor of organisms that preserve more readily

(a)

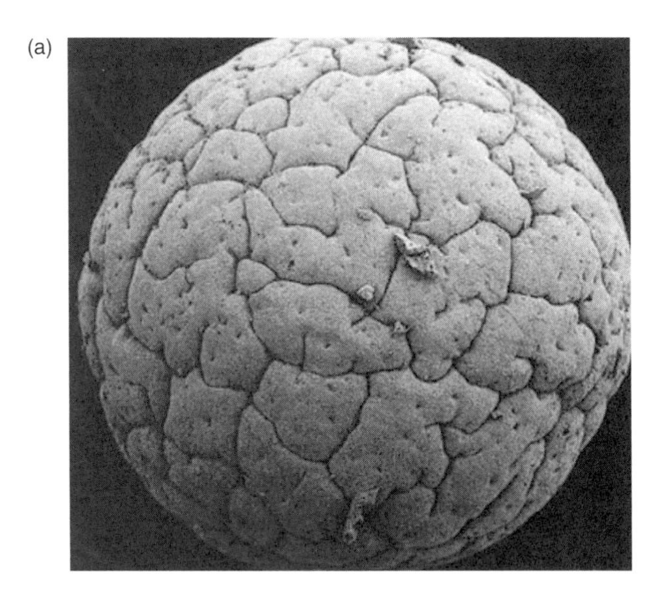

(b)

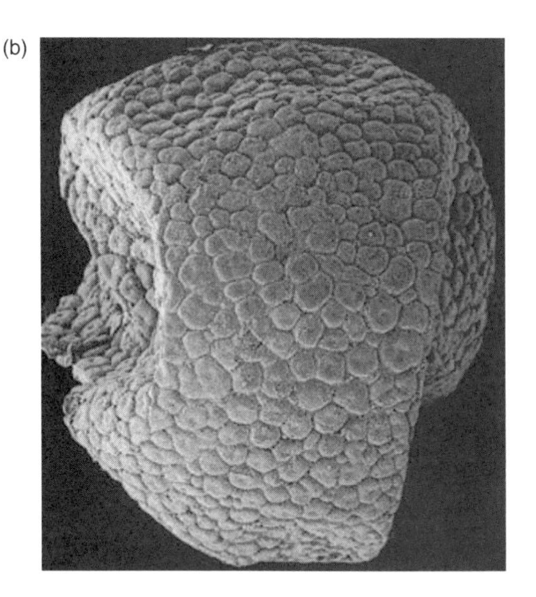

**FIGURE 1.15 Phosphatized animal embryos from the Neoproterozoic of Guizhou, China.** The width of the field of view is about 650 $\mu$m in part (a) and about 450 $\mu$m in part (b). *(From Xiao & Knoll, 2000)*

and that live where fossilization has occurred. Coming back to *E. kingi*, it would not be sound to conclude that 40 percent of all individuals that lived during the Middle Cambrian in the area of present-day Utah belonged to this species. Trilobites have mineralized exoskeletons, yet there would have been numerous soft-bodied species that left few if any fossil remains. The importance of such biases depends on the question we hope to address.

For estimating the relative proportions of individuals in an ancient community, differential preservation can represent a severe bias. (Another Middle Cambrian formation, the Burgess Shale, represents an unusual instance of soft-bodied preservation in the fossil record [SEE SECTION 10.2]. It has been estimated that, of well over 100 species known from this formation, less than 15 percent have hard parts that would be preserved under typical circumstances of fossilization.) For many questions, however, among-group variation in preservation potential is not so important. For example, if what interests us is evolutionary changes in body size within *E. kingi*, the preferential preservation of trilobites compared with that of other groups is irrelevant.

What about biases related to the size of the sample? Whether this matters depends on what we hope to measure from the sample. The average proportion of individuals in a sample that belong to a certain species generally does not depend on the number of individuals sampled. By contrast, the number of species recovered from a locality depends on the number of individuals sampled, as does the maximum body size recorded for a given species.

Ideally, the effects of sample size are reduced by standardizing the nature and extent of sampling as part of a study design—for example, collecting the same number of specimens from different formations if the goal is to compare the number of species among those formations. Often this is not feasible, as when we are analyzing data that have already been collected for other purposes; the standardization of the sample must therefore be performed statistically. A simple procedure for this after-the-fact standardization is **rarefaction** (Box 1.2), which estimates the number of species or other taxa that would have been found if a smaller number of individuals had been sampled.

All organisms, fossil and living, are classified not only into species but also into higher groupings or categories [SEE SECTION 4.3]. At the higher end of the classificatory scale are taxonomic groups such as phylum and class. At the lower end of the scale are groups such as family and genus. We can also apply rarefaction to estimate how many

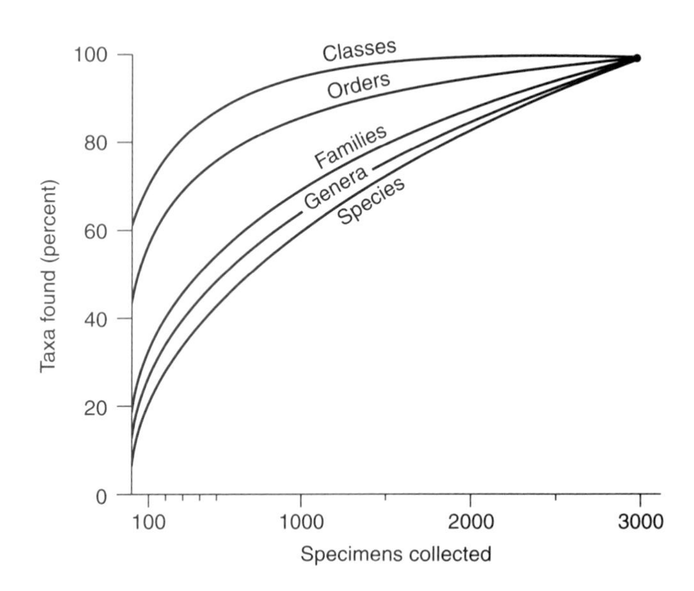


**FIGURE 1.16 Rarefaction curves for molluscan fossils found in a well sample of Miocene age in Denmark (based on data from Sorgenfrei, 1958).** The point at the upper right represents the actual sample. The curves estimate how many taxa would have been found had the sample been smaller.

genera, families, or other taxa would have been found if a smaller sample of individuals had been collected.

Figure 1.16 illustrates the rarefaction method with an example from the Miocene of Denmark. Table 1.2 shows the number of groups at each taxonomic level that were recovered from a sample of nearly 3000 individuals. Because the author of this study was interested only in molluscs, only one phylum was recorded. Three molluscan classes were present (Bivalvia, **GASTROPODA,** and **SCAPHOPODA**). At the other end of the taxonomic scale, there were 86 species. As is almost always the case, each lower taxonomic level yielded a

**TABLE 1.2**

Numbers of Taxa Found in a Molluscan Sample from Arnum Formation (Miocene of Denmark)

| 2954 Specimens | |
|---|---|
| Phyla | 1 |
| Classes | 3 |
| Orders | 12 |
| Families | 44 |
| Genera | 64 |
| Species | 86 |

*SOURCE:* Sorgenfrei (1958)

*Box 1.2*

## RAREFACTION METHOD

To compute a rarefaction curve such as the species curve in Figure 1.16, it is necessary to know the number of individuals in the sample ($N$), the number of species ($S$), and the number of individuals found for each species ($N_i$, where $i = 1, 2, \ldots, S$). There are several of ways to compute the expected number of species that would be found in a smaller sample of $n$ individuals, denoted $E(S_n)$. A simple approach is to program a computer to grab $n$ individuals at random from the entire collection. For example, suppose there are $N = 25$ individuals distributed among three species with $N_1 = 15$, $N_2 = 8$, and $N_3 = 2$. The list of individuals is shuffled like a deck of cards, and the first $n$ on the list are chosen. If we identify each individual with its species number, the initial list looks like this: 1 1 1 1 1 1 1 1 1 1 1 1 1 1 1 2 2 2 2 2 2 2 2 3 3. We randomly shuffle the list and end up with: 2 1 1 2 2 2 1 1 1 1 1 2 2 1 1 1 1 1 3 1 1 3 1 2 2. If we want to take a sample of $n = 10$ individuals, we simply read the first 10 off this list, and we end up with six individuals of species 1, four individuals of species 2, and zero individuals of species 3. Thus, two species were found in this sample of $n = 10$ individuals.

There is clearly random variation associated with this procedure. For example, another shuffling of the list yields: 3 2 2 1 2 1 1 1 2 2 1 1 3 1 1 1 2 1 1 1 2 2 1 1 1, with the result that all three species are found in the first 10 individuals on the list. For this reason, the randomization is carried out hundreds or thousands of times, and the results are averaged together. There is also an exact equation that produces the same results directly (see Raup, 1975).

In practice, the calculations are carried out for an arbitrary series of $n$ values (all less than $N$) and each $E(S_n)$ thus produced yields one point on the rarefaction curve. Or if several samples are to be compared at a standard $n$, that $n$ can be used for a single computation for each sample. It is also possible to compute the uncertainty attached to the estimated species numbers, that is, the variance of $E(S_n)$ (see Raup, 1975; see also Box 3.1).

In the case of the species rarefaction in Figure 1.16, $N$ was equal to 2954 and $S$ was 86. The most common species was represented by 818 individuals; 40 of the species had only one specimen each. A few of the computed values of $E(S_n)$ and its variance are given below:

| Number of Specimens ($n$) | Expected Number of Species $E(S_n)$ | Variance of $E(S_n)$ |
|---|---|---|
| 2500 | 79.66 | 5.42 |
| 2000 | 71.93 | 9.89 |
| 1500 | 63.14 | 12.59 |
| 1000 | 52.52 | 13.59 |
| 500 | 38.56 | 12.23 |
| 100 | 19.05 | 6.64 |
| 50 | 14.05 | 5.05 |
| 10 | 6.24 | 2.88 |

If rarefaction is to be done at higher taxonomic levels (as in Figure 1.16), data for genera, families, and higher categories are simply substituted for the species data and the same procedure is used.

larger number of taxa. If fewer than 2954 specimens had been collected, the number of taxa recovered would have been smaller, as shown by the rarefaction curves. This effect would have been more pronounced at the lower taxonomic levels. For example, the rarefaction equation predicts that, if 1000 individuals had been sampled, only about 60 percent of the 86 species would have been recovered, but over 80 percent of the orders would have been found.

This last result reflects an important and general aspect of sampling of the fossil record: *Sampling is more complete at higher taxonomic levels.* This is a necessary consequence of the nesting of taxonomic groups within one another. Each genus contains one or more species, each family contains one or more genera, and so on. Therefore, there will tend to be more individuals in any genus than in one of its component species, and more individuals in any family than in one of its component genera.

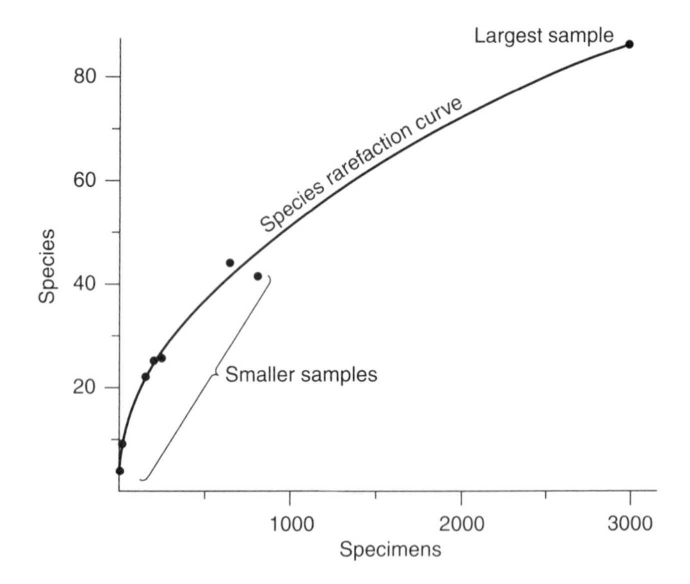


**FIGURE 1.17 Effect of sample size on the number of species found in a fossil assemblage.** The solid line is the species rarefaction curve from Figure 1.16, and the points represent other samples collected from above and below the main sample. *(Based on data from Sorgenfrei, 1958)*

In Figure 1.17, the effect of sample size on species number is shown in another way for the Danish data. The sample of 2954 individuals was the largest of eight samples taken from one formation. The other seven samples came from different stratigraphic intervals above and below the large sample. The numbers of specimens and species for each are plotted in Figure 1.17, along with the rarefaction curve for species in the largest sample, taken from Figure 1.16. The points for the small samples fall very close to the theoretical rarefaction curve, suggesting that the differences in numbers of species are just the result of sample size differences.

Figures 1.16 and 1.17 illustrate the general point that increased sampling tends to yield diminishing returns. Doubling the number of individuals sampled from a formation generally results in substantially less than a doubling of the total number of species recovered. The main reason for this is that a small number of species usually account for the vast majority of individuals. Therefore, repeated sampling tends to produce the more common species again and again; additional, rarer species are much less likely to be recovered.

It is important to keep in mind two points that are often misunderstood about rarefaction. First, it cannot be used for extrapolation—to estimate how many species would have been found had a larger sample been taken. Second, although rarefaction curves often give the appearance of leveling off, the apparent flatness of a curve does not necessarily imply that the true quantity of interest, such as the number of species that actually lived at some time in the past, has nearly been reached. The most we can conclude from a nearly level rarefaction curve is that we may be unlikely to obtain much more information with modest amounts of additional sampling; enormous efforts could be required to find the rarest preserved species.

## Measuring Completeness of the Fossil Record

When we discussed the observed diversity of brachiopods and bivalves earlier, we noted that knowing something of the relative completeness of different groups of organisms can help determine whether evolutionary differences among them are likely to be artifacts of differences in preservation potential. It is therefore important to have some means of estimating **paleontological completeness.** Completeness can be expressed as the probability of sampling a given taxon within a specified interval of time, or as the probability of sampling the taxon at least once within its entire duration.

The simplest way to measure the probability of sampling per time interval is to compare the number of time intervals in which a taxon is actually sampled with the number of intervals during which we know it existed and therefore had the opportunity to be sampled (Box 1.3). An interval of time during which a taxon existed but from which it is not sampled is a **gap** in the stratigraphic range of the taxon [SEE SECTION 6.4].

The tabulation of gaps in sampling requires a substantial amount of information, namely the presence or absence of each taxon in each interval of time. If these data are not available, there are a number of indirect ways to estimate sampling probability. One is based on the expectation that a lower sampling probability will lead to a greater proportion of taxa being sampled from only one interval of time, regardless of how long-lived they may in fact have been (Box 1.3). Table 1.3 presents some sampling estimates based on this approach. For skeletonized animals, sampling probabilities range from about 10 percent to over 90 percent per genus per 5-million-year time interval.

To estimate completeness summed over the durations of taxa, we can simply compare the number of living taxa within some group with the number that

**TABLE 1.3**

Estimated Completeness of Genera Within Some Paleontologically Important Groups

| Group | Probability of Preservation per Genus per Time Interval |
|---|---|
| Sponges | 0.4–0.45 |
| Corals | 0.4–0.5 |
| Polychaetes | 0.05 |
| Malacostracan crustaceans | 0.2–0.35 |
| Ostracodes | 0.5 |
| Trilobites | 0.7–0.9 |
| Bryozoans | 0.7–0.75 |
| Brachiopods | 0.9 |
| Crinoids | 0.4 |
| Asterozoans | 0.25 |
| Echinoids | 0.55–0.65 |
| Bivalves | 0.45–0.5 |
| Gastropods | 0.4–0.55 |
| Cephalopods | 0.8–0.9 |
| Graptolites | 0.65–0.9 |
| Conodonts | 0.7–0.9 |
| Cartilaginous fishes | 0.1–0.15 |
| Bony fishes | 0.15–0.3 |

*SOURCE:* Foote & Sepkoski (1999)

*NOTE:* Time intervals are roughly 5 million years long on average. Estimates are based on the principle that the probability of preservation is likely to be lower in groups where a higher proportion of genera are confined to a single time interval (Box 1.3). Details of the calculation are found in Foote and Sepkoski (1999).

**TABLE 1.4**

Proportion of Living Taxa with a Fossil Record

| Group | Taxonomic Level | Percent |
|---|---|---|
| Sponges | Family | 48 |
| Corals | Family | 32 |
| Polychaetes | Family | 35 |
| Malacostracan crustaceans | Family | 19 |
| Ostracodes | Family | 82 |
| | Genus | 42 |
| Bryozoans | Family | 74 |
| Brachiopods | Family | 100 |
| | Genus | 77 |
| Crinoids | Family | 50 |
| Asterozoans | Family | 57 |
| | Genus | 5 |
| Echinoids | Family | 89 |
| | Genus | 41 |
| Bivalves | Family | 95 |
| | Genus | 76 |
| Gastropods | Family | 59 |
| Cephalopods | Family | 20 |
| Cartilaginous fishes | Family | 95 |
| Bony fishes | Family | 62 |
| Arachnids | Genus | 2 |
| | Species | < 1 |

*SOURCES:* Raup (1979); Foote & Sepkoski (1999); Valentine et al. (2006). Data are global.

are known as fossils. Tables 1.4 and 1.5 present tabulations of this kind, which of course cannot be compiled for completely extinct groups. These tabulations show the expected effects of taxonomic level. A greater proportion of genera have a fossil record compared with the species of the same group, and likewise for families versus genera.

A second way to estimate the overall proportion of taxa sampled was suggested by paleontologist James Valentine (1970). This is to compare the total number of taxa known as fossils with an estimate of the number that have ever lived. For example, some 300,000 animal species have been described from the fossil record. How does this compare with the number of animal species that are likely to have lived over the Phanerozoic? This second number

**TABLE 1.5**

Proportion of Living Molluscan Taxa in the Californian Province with a Pleistocene Fossil Record

| Group | Taxonomic Level | Percent |
|---|---|---|
| Bivalves | Family | 91 |
| | Genus | 84 |
| | Species | 80 |
| Gastropods | Family | 88 |
| | Genus | 82 |
| | Species | 76 |

*SOURCE:* Valentine (1989)

*Box 1.3*

## SELECTED MEASURES OF PALEONTOLOGICAL COMPLETENESS

We first consider the probability of sampling a taxon per unit time interval (Paul, 1982). In the hypothetical data of Figure 1.18, gaps in sampling are seen as time intervals during which a species existed but left no known fossil record. The fewer such gaps, the higher the estimated probability of sampling. For example, species 1 persisted through four time intervals between its first appearance in interval 1 and its last appearance in interval 6. Thus, it had four opportunities to be sampled. Of these four intervening intervals, it is known only from interval 5; intervals 2, 3, and 4 mark gaps in the record of species 1. Its estimated sampling probability is therefore 1/4, or 25 percent per interval. Because there are few observations for each individual species, the margin of error of the estimated sampling probability tends to be quite high. By combining data for many species, a more reliable estimate of average sampling probability can be obtained. Gaps can also be tabulated for individual time intervals to determine how sampling probability varies over time.

Compilations of stratigraphic information often report only the times of first and last occurrence of fossil taxa, not the intervening occurrences shown in Figure 1.18. Such data can be used to estimate sampling probability with a method that relies on mathematical formalization of a simple, intuitive principle: Because incomplete sampling tends to shorten ob-

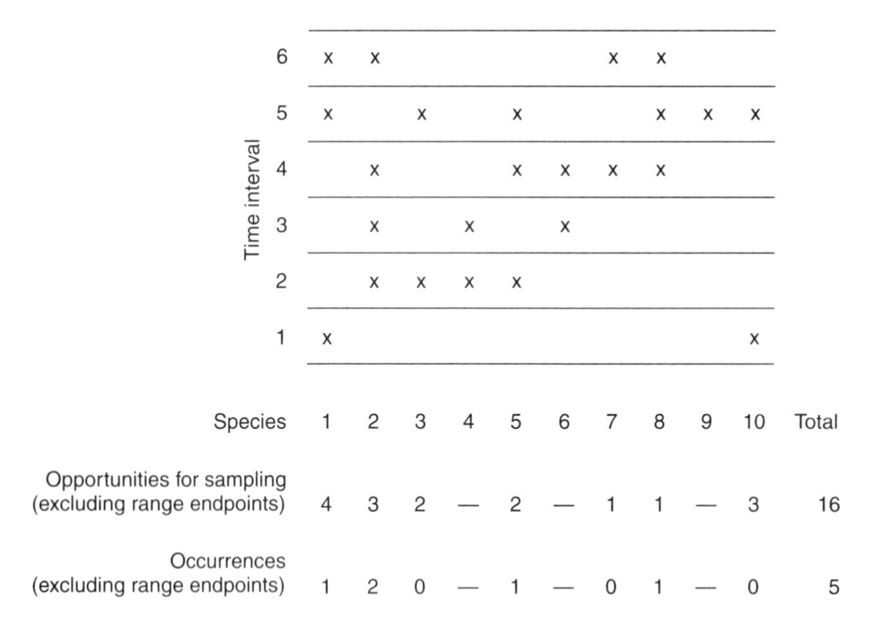


**FIGURE 1.18 Schematic illustration of gap analysis used to estimate average sampling probability of a group of species.** Each X marks a time interval in which the corresponding species is sampled; intervening blank spaces are gaps in the record of the species. The number of occurrences for each species is tabulated as the number of intervals in which the species is found, excluding the intervals of first and last appearance. This is compared with the number of opportunities for sampling—that is, the number of time intervals during which the species had a chance either to be sampled or not. This is simply the sum of time intervals between the first and last appearance. Because a species necessarily must be sampled in its intervals of first and last occurrence, these intervals are not included in the tabulations; to include them would overestimate the sampling probability. In this case, there are 16 opportunities for sampling and 5 occurrences; both numbers exclude first and last appearances. Thus, the estimated sampling probability of these species is 5/16, or 31 percent per time interval.

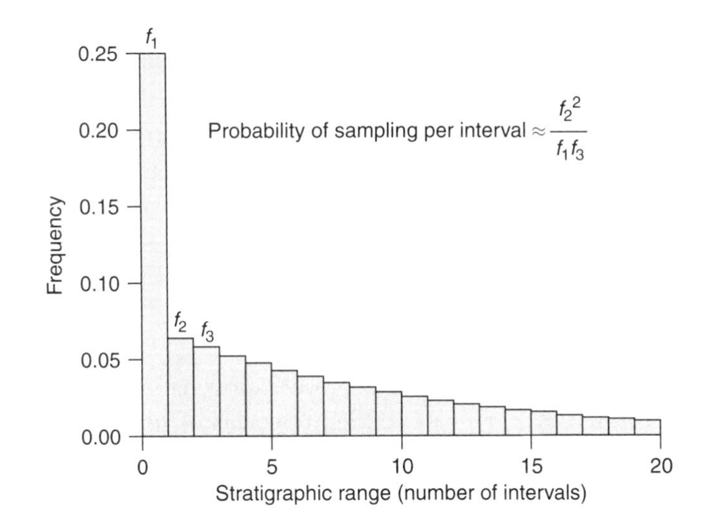


**FIGURE 1.19 Frequency distribution of stratigraphic ranges of taxa, assuming incomplete but uniform sampling.** Sampling probability per unit time interval can be estimated from the frequencies of taxa with ranges of one, two, and three intervals. *(After Foote & Raup, 1996)*

served stratigraphic ranges, the proportion of taxa that are known from only a single stratigraphic interval should be inversely proportional to the quality of sampling. Assuming that sampling probability is constant over time, the probability of sampling per unit time can be estimated as $(f_2)^2 \div (f_1 f_3)$, where $f_1$, $f_2$, and $f_3$ are the numbers or relative proportions of species with preserved stratigraphic ranges of one, two, and three intervals (Figure 1.19; Table 1.3). Further details can be found in Foote and Raup (1996).

To compare the number of known fossil species with the estimated number that actually existed, we restrict our consideration to the paleontologically important groups; strictly soft-bodied organisms are ignored. Estimating the total number of species over the course of the Phanerozoic Eon requires that we know the average longevity of species—which tells us how often existing species became extinct and were replaced by new species—as well as the level of diversity during the Phanerozoic. Using methods similar to those we will treat in detail in Chapter 7, we can estimate the typical longevity of marine invertebrate species as roughly 4 million years. In other words, about 25 percent of existing species became extinct every million years.

The number of living species that have been described is about 1.5 million, although there is great uncertainty in this number. If we focus on the paleontologically important groups, present-day diversity is about 180,000 species. The path from an initial diversity near zero to the current diversity of 180,000 is hard to know, because observed diversity severely underestimates the true number of species that were alive at any time in the past. We can, however, take an approach that almost surely overestimates past diversity, and therefore leads to a minimum estimate of the proportion of species sampled. Suppose we assume that the present-day level of diversity was attained immediately at the beginning of the Cambrian Period and has been maintained since then. Then 25 percent of 180,000 species, or 45,000 species, became extinct and were replaced by new species every million years. In rough terms, the Phanerozoic is 550 million years long. This leads to an estimate that there have been 180,000 + (45,000 × 550), or about 25 million species. Comparing this with the 300,000 described fossil species implies that between 1 percent and 2 percent of species are known as fossils.

depends on how the number of species has varied over geologic time [SEE SECTION 8.3], and the rate at which species have become extinct and have been replaced by new species [SEE SECTION 7.2]. Carrying out the relevant calculations (Box 1.3) yields estimates that around 20 million species in the paleontologically important groups have existed over the past half billion years. In other words, somewhat over 1 percent of all animal species in the readily preserved groups are known from the fossil record. At first glance, this may seem like a small percentage, but in fact, reliable statistical inferences in many fields are routinely drawn with much smaller samples.

To take just one example, the typical national presidential poll in the United States uses a sample of about 1500 voters, out of a voting-age population of over 200 million. Thus, the sample represents less than 0.001 percent of the eligible voters, yet such polls tend to be fairly accurate as predictors of election results.

Completeness, of course, varies from group to group. Groups such as trilobites, brachiopods, molluscs, and some classes of **ECHINODERMS** are much better represented in the fossil record than the 1 percent figure would suggest.

How can we reconcile the estimate of 1 percent completeness with the tabulations like that of Table 1.5—showing that nearly 80 percent of the marine molluscan species living in California today are known from the Pleistocene fossil record? One reason for the discrepancy is, of course, that the bulk estimate of animal completeness covers not just molluscs, which are relatively well preserved, but also groups such as **STARFISH,** crustaceans, and sponges, which are not so well preserved. A more important reason reflects the distinction between **local completeness** and **global completeness.** Where there is some preserved fossil record, we often find that its completeness is rather high. Fossiliferous rocks have a patchy geographic distribution, however. For any given interval of geologic time, most localities have left no sediments that are exposed today.

The data on modern and Pleistocene molluscs from California illustrate the point that entering the fossil record is only part of the picture. The record itself must escape subsequent erosion and metamorphism, and must remain unobscured by younger, overlying sediments for the species that are initially fossilized to contribute to our knowledge of the history of life. The fact that local completeness is often high is important for the study of evolution, for it implies that evolutionary patterns within locally preserved species may often be represented with considerable fidelity in the fossil record.

## 1.4 TEMPORAL CHANGES IN THE NATURE OF THE FOSSIL RECORD

Because many of the geologic and evolutionary processes studied with fossil data act over long timescales (tens to hundreds of millions of years), it is important to understand how the nature of the fossil record has changed over such timescales and how this may influence our interpretation of paleontological data. We already touched on this issue in discussing the relationship between the number of fossil species and the exposed area of sedimentary rock (Figure 1.1).

### Bioturbation

The intensity of marine bioturbation has evidently increased over the past 500 million years of animal life. Figure 1.20 shows one way that this has been documented. This figure illustrates the ichnofabric index, which provides a rough measure of bioturbation preserved in sedimentary rocks. (The Greek prefix *ichno-* refers to "trace.") Analysis of a sequence of Cambrian and Ordovician sedimentary rocks shows an increase in the average ichnofabric index, portrayed in Figure 1.21. This corresponds to the evolution and diversification of animal groups that exploited the sediment as habitat and as a food source.

The first step marks the appearance of trilobites within the Lower Cambrian. The reasons for the second step, between the Middle and Upper Ordovician, are not so clear. It does not coincide with the appearance of a major new group of hard-bodied animals, although it does come shortly after a pronounced diversification of marine taxa that would come to dominate Paleozoic marine settings [SEE SECTION 8.4]. It is therefore possible that this step marks a behavioral innovation in some skeletal group or the evolution of a new group of soft-bodied animals. Whatever their causes, it is likely that these increases in bioturbation resulted in an increase in time averaging. They may also have led to greater physical disturbance and disaggregation of carcasses and to a reduced probability of soft-part preservation.

### Skeletal Mineralogy

The relative abundance of various shell-forming minerals has changed over the history of life. One major change that is potentially important for the quality of

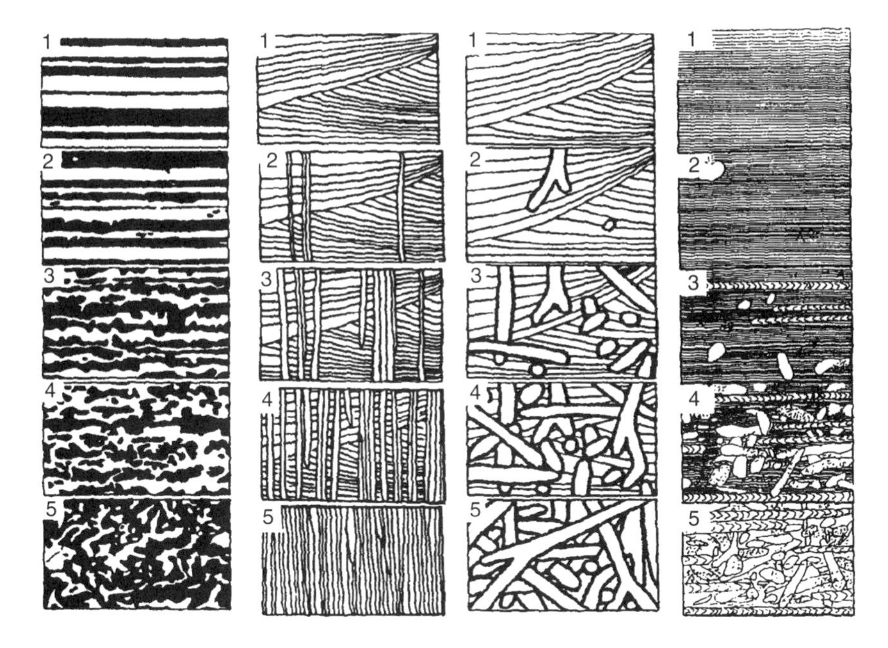


**FIGURE 1.20 Ichnofabric index standards for four different environmental settings.** Left to right: shelves; nearshore environments characterized by the vertical trace fossil *Skolithos*; nearshore environments characterized by the trace fossil *Ophiomorpha*; and deep-sea environments. In each setting, higher ichnofabric indices, toward the bottom, indicate a greater degree of bioturbation. *(From Droser & Bottjer, 1993)*

the fossil record lies in the abundance and diversity of marine organisms with calcitic versus aragonitic skeletons. Because aragonite tends to recrystallize to calcite over time, it is important to note that study of shell microstructure is often able to determine whether what is preserved as calcite was originally calcite or aragonite [SEE SECTION 1.2]. Although the importance of calcite versus aragonite has fluctuated over time, in very rough terms calcitic skeletons are more common in Paleozoic animals, whereas aragonitic skeletons are more common after the Paleozoic Era.

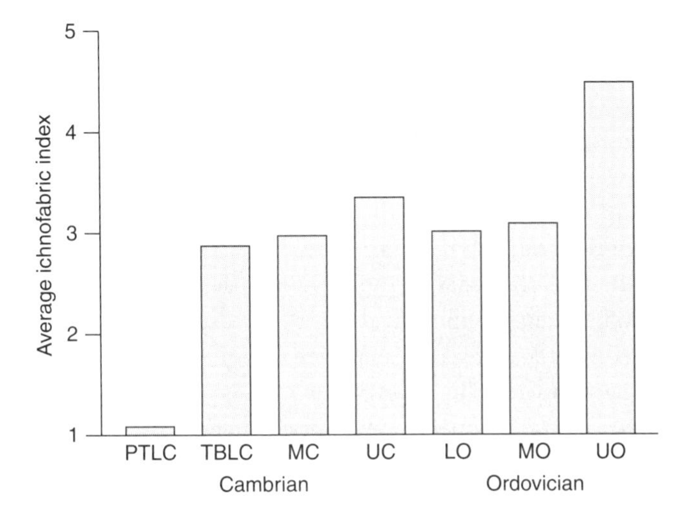


**FIGURE 1.21 Average ichnofabric index through the Cambrian and Ordovician, based on a field study of the Great Basin of the southwestern United States.** "PTLC" and "TBLC" refer to the pre-trilobitic Lower Cambrian and the trilobite-bearing Lower Cambrian. *(From Droser & Bottjer, 1993)*

While the reasons for temporal changes in skeletal composition are still under investigation, the potential importance of aragonite loss can easily be seen by comparing fossil deposits that represent paleoenvironments of the same time and place but that have undergone different styles of diagenesis. In some instances, the more poorly preserved deposits contain aragonitic species as molds only or as recrystallized calcite that has lost most of the original microstructural detail of the shell; the better preserved equivalents contain abundant aragonitic shells. To the extent that calcitic taxa are better represented than aragonitic taxa, the fossil record may prove to be more complete for those intervals of time when calcitic skeletons were truly more common.

## Geographic and Environmental Distribution of Fossiliferous Rocks

Most of the fossil record is marine in origin. It is therefore quite natural that the majority of paleontological research focuses on the marine realm, even if the terrestrial realm is more familiar to many students. Moreover, most of what we know of ancient marine life comes from relatively shallow deposits that were formed when oceans flooded parts of the continents and continental shelves. These deposits have been revealed either by a drop in absolute sea level or by tectonic uplift. Sedimentary rocks deposited in the deep oceans are sometimes scraped up onto continents at collision zones, and such rocks have also been sampled directly by drilling into the sea floor at great depths [SEE SECTIONS 9.1, 9.5, AND 10.4]. Most oceanic sedimentary rock older than about 180 million years has been subducted, however. The relatively sparse Paleozoic record of the deep seas, compared with that of the Mesozoic and Cenozoic, is the result of subduction of the oceanic crust.

For terrestrial organisms, the most commonly represented habitats are in the coastal lowlands. Habitats at higher elevations are generally in areas of net erosion that are unlikely to be preserved in the sedimentary record for long periods of time. Thus, we have a better record of uplands as we approach the present day.

The global sedimentary record suggests that widespread **epicontinental seas** were more common during much of the Paleozoic Era than they were afterwards. In addition, there has been a general movement of the northern continents (making up present-day North America and Eurasia) from predominantly tropical and subtropical during the Paleozoic to predominantly subtropical and temperate afterwards. As a result of these factors, the relative extent of shallow-marine tropical sediments has generally declined over the course of the Phanerozoic Eon. In contrast to the increasing quantity of the deep-sea record as we approach the present day, however, the general decline in epicontinental seas and the latitudinal shift in continental position are quite real. Geologic evidence shows that tropical, epicontinental seas really were more common in the Paleozoic; it is not the case that we merely have a more complete record of them for that interval of time.

## 1.5 GROWTH OF OUR KNOWLEDGE OF THE FOSSIL RECORD

As suggested in our earlier treatment of rarefaction, one way to assess the adequacy of the fossil record is to determine at what point the results we have documented no longer change appreciably as we sample more. If a result is stable in the face of improved sampling, we can have some confidence that we are seeing what the fossil record has to offer. Of course, this does not guarantee that we are seeing a faithful reflection of the biological or geological processes we hope to reveal. This is because the fossil record itself may be biased with respect to a particular question; merely increasing the size of the sample is not necessarily going to undo this bias. Moreover, the principle of diminishing returns means that it may take an enormous amount of additional sampling to add significantly new information to a study—for example, to sample the rarest species.

Sir Alwyn Williams, a specialist on brachiopods, was aware as early as the 1950s that tabulations of the number of taxa known from different periods of geologic time can be idiosyncratic, depending, for example, on the stratigraphic intervals in which particular paleontologists are interested and on the taxonomic concepts that they employ. He therefore conducted a thought experiment: Which stratigraphic intervals would show peaks in the number of brachiopod genera if, at various times in the history of our science, paleontologists had analyzed the data then available? Williams tabulated the number of genera by geologic time period for the data compilation published in 1894 by James Hall and J. M. Clarke, for that published in 1929 by Charles Schuchert and C. M. Le Vene, and for the data available as of 1956, largely the result of efforts by G. A. Cooper and Williams himself.

Successive data compilations are not independent. Instead, they are cumulative, building on previous knowledge and interpretations of fossil material. For example, consider the brachiopod genus *Finkelnburgia*. This genus was erected by Charles Walcott in 1905 on the basis of two species he collected and described from Upper Cambrian sandstones in Wisconsin and Minnesota. Hence, this genus is absent from the 1894 data compilation, but it is listed in Schuchert and Le Vene's 1929 compilation as occurring in the Cambrian. In 1932, Schuchert and Cooper assigned a few specimens from Lower Ordovician deposits to the genus *Finkelnburgia*, and in 1936 E. O. Ulrich and Cooper described several Lower Ordovician species of *Finkelnburgia* from North America. As a result, in the 1956 compilation this genus is present in both the Cambrian and Ordovician.

In fact, the story is more complicated than this. In 1865, Elkanah Billings had described a species from the Lower Ordovician of Canada and assigned it with some uncertainty to the genus *Orthis*. This species, *Orthis? armanda*, was assigned in 1932 by Schuchert and Cooper to the genus *Finkelnburgia*. Thus, material representing Ordovician *Finkelnburgia* was known to paleontologists 40 years before this genus was described and nearly 70 years before it was actually credited to the Ordovician. As this example shows, both the collection of new material and the taxonomic treatment of existing material can affect the inferences drawn from paleontological data.

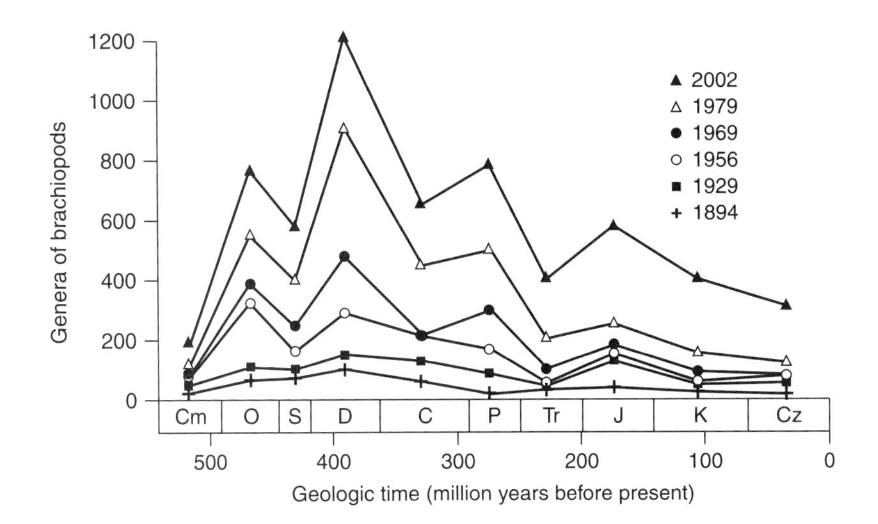


**FIGURE 1.22 Genus diversity in brachiopods over geologic time according to six different data compilations, each indicated by its date.** The total number of genera known from each geologic period is shown. This figure shows how certain features of evolutionary history start to stabilize as more data are collected. *(Data from Williams, 1957; Grant, 1980; & Sepkoski, 2002)*

Figure 1.22 shows Williams's tabulations of the total number of genera known from successive intervals of geologic time. The peak in brachiopod diversity for the 1894 data set falls in the Devonian, that for the 1929 data set also falls in the Devonian, but there is a post-Paleozoic peak in the Jurassic; and that for the 1956 data set falls in the Ordovician. Thus, the picture of taxonomic diversity seemed to Williams to be rather unstable, changing in striking ways with additional study.

We can make sense of the particular shifts by taking note of some of the predominant students of fossil brachiopods. James Hall, who was active for much of the second half of the nineteenth century, worked extensively on the Devonian and other Paleozoic rocks of New York State and other areas. Thomas Davidson, a contemporary of Hall, studied Jurassic brachiopods of Europe, but he had what is generally regarded as a conservative, or "lumping," taxonomic approach: He tended to describe forms as new genera only on the basis of rather substantial differences. In the early twentieth century, S. S. Buckman was one of the foremost students of European Jurassic brachiopods. He was more of a "splitter"—he tended to describe new genera on the basis of relatively subtle anatomical distinctions. This helps account for the large number of Jurassic genera in the 1929 data compilation. Finally, by the 1950s, a number of brachiopod specialists had begun to study lower Paleozoic formations in more detail. This helps account for the appearance of the Ordovician peak in the 1956 data set.

A similar study was carried out in 1980 by Richard Grant, who compared brachiopod diversity based on Cooper's 1969 compilation with his own 1979 data set. His results are also graphed in Figure 1.22. The diversity picture shows some interesting differences compared with the 1956 compilation. For example, the Devonian now appears again as the principal peak, just as it was in the 1894 and 1929 compilations. Moreover, the 1969 and 1979 data sets show a new, subsidiary peak in the Permian. This reflects, among other things, Cooper and Grant's extensive work on the silicified Glass Mountain faunas of west Texas. An important feature of Grant's study is that the 1979 data set, despite being much more extensive than that of 1969, shows essentially the same pattern of diversity over time.

What do the data on brachiopod occurrences tell us today, some 50 years after Williams's study? A 2002 compilation by J. J. Sepkoski [SEE SECTION 8.2], based on that of Grant as well as numerous additional sources, is also depicted in Figure 1.22. This shows that the picture today is actually much as Grant saw it nearly 30 years ago. Overall, there are many more known genera of fossil brachiopods, but the relative distribution of these throughout geologic time appears to have stabilized. Small differences, such as the relative diversity of brachiopods in the Ordovician versus the Permian, are likely to fluctuate slightly as more material is studied. However, major patterns, such as the overall peak in the Devonian and the generally higher level of diversity in the Paleozoic relative to the Mesozoic and Cenozoic, seem to be robust.

Similar studies have been done for diversity in other groups and for many other issues—for example, evolutionary relationships among taxa and the anatomical differences among them. The general outcome of such studies agrees with what has been found for brachiopod diversity: Finer-scale features in the fossil record are more likely to be overturned as additional data are collected, while large-scale features tend to be stable. However, there is no way of knowing in advance precisely which features will be reliable in any given case. It is therefore important to continue to consider to what extent our interpretations of the fossil record are sensitive to new discoveries.

As illustrated with the example of *Finkelnburgia*, comparisons between older and newer data compilations combine two sources of modification: increase in the sheer amount of data and change in the taxonomic opinions and practices of specialists on particular biologic groups. We will discuss taxonomic practice in more detail in Chapter 4. For now, we note that it is possible to isolate this second factor by starting with an existing data compilation and revising it according to a consistent set of taxonomic protocols—for example, how wide a range of species forms to include within a single genus. The procedure of adopting a consistent approach and scrutinizing existing data to ensure that they are in agreement with the adopted standards is referred to as **taxonomic standardization.**

Figure 1.23 shows some results of taxonomic standardization, in this case, focusing on genus diversity of Ordovician and Silurian trilobites. Part (a) plots the percentage of genera present in the unstandardized data for the given time interval that are not considered valid in the standardized data. This is a measure of the extent of disagreement between unstandardized and standardized data. Parts (b) and (c) depict, for both data sets, total diversity and the percent change in diversity from one interval to the next. It is clear that, despite pervasive disagreements between the two data sets, diversity and short-term change in diversity are actually in substantial agreement. In this example, the discrepancies between the data sets are, in effect, random noise that is averaged out.

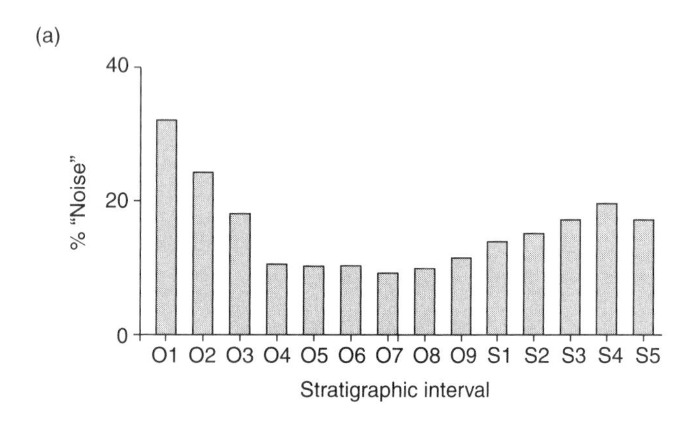


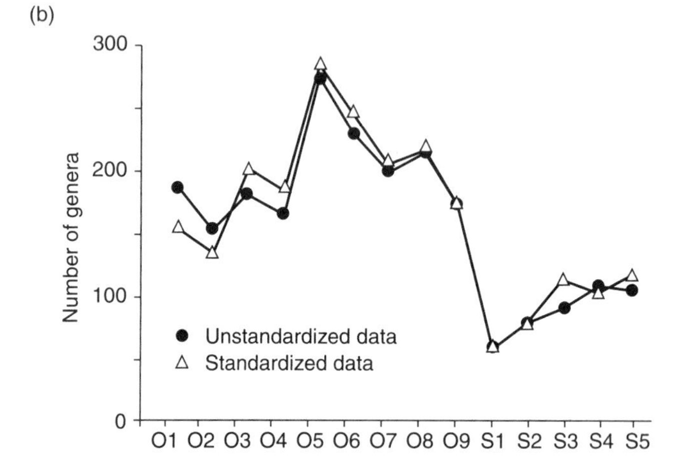


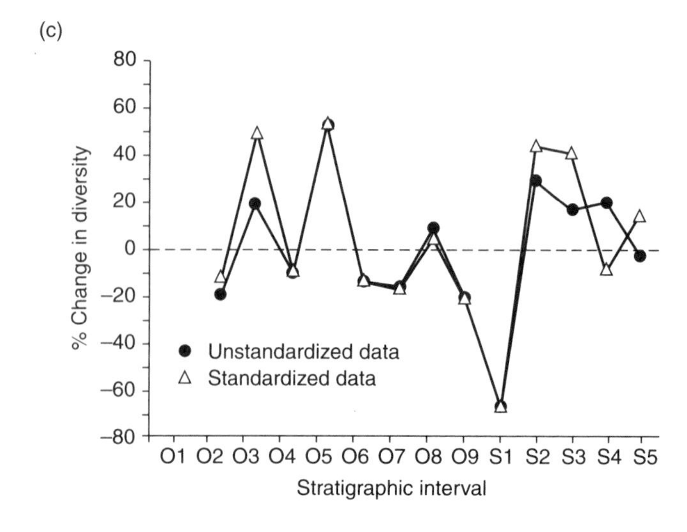


**FIGURE 1.23 Effect of taxonomic standardization on perceived diversity of Ordovician and Silurian trilobites.** Data compiled by Sepkoski (2002), who was not a trilobite specialist, are referred to as *unstandardized*. These data were then scrutinized by two trilobite specialists who produced the *standardized* data. (a) The percentage of genera in the unstandardized data, by interval, that are not valid in the standardized data. This percentage is referred to here as "noise." (b) Total diversity in the two data sets. (c) Percent change in diversity from one time interval to the next in the two data sets. Although part (a) shows that there are many discrepancies between the two data sets, the diversity patterns derived from them are nearly the same. *(From Adrain & Westrop, 2000)*

## 1.6 BIBLIOGRAPHIC SOURCES FOR PUBLISHED PALEONTOLOGICAL DATA

The previous discussion of data compilations and how they have grown over time naturally leads to a consideration of how one can keep track of such information. Although we will discuss the mechanics of data compilation throughout this book, a few preliminary comments are in order on the subject of bibliographic sources.

Paleontologic information (particularly taxonomic information) has been published in a vast literature extending back well into the eighteenth century. It is published in all major languages and a wide variety of publication media. Paleontologists are thus highly dependent on bibliographic aids.

Compilation and standardization of data are greatly aided by definitive monographs on either a specific taxonomic group or fossils found in a particular part of the geologic column. If the monograph has been well prepared, it includes reference to all important literature. The reader need then consult other bibliographic sources only for articles that have been published since the publication of the monograph. In using a given monograph, the reader must understand to what taxonomic categories the writer's definitive summary reaches. For example, the authors of the *Treatise on Invertebrate Paleontology*—a summary of the geologic and geographic occurrences, morphology, and classification of fossil invertebrates—attempt to be comprehensive in listing genera but generally do not cover species.

When no up-to-date summary treatment is available, paleontologists must turn to published bibliographies, such as the *Zoological Record*. Each volume is a reasonably comprehensive survey of the zoological and paleozoological literature published during the preceding year. The index includes a comprehensive list of all taxonomic names used in the papers cited. The *Zoological Record* is a valuable aid for a number of areas of work because it permits tracing the bibliographic citations to a genus or species year by year.

The *Zoological Record* is not complete, however. No such bibliography could be and still be issued within a reasonable time after the publication of the literature on which it is based. Therefore, the *Zoological Record* must usually be supplemented by other bibliographies such as *Biological Abstracts*, *GeoRef*, and more specialized sources for particular taxonomic groups.

Electronic bibliographic databases have grown enormously in recent years. These can be rapidly and automatically searched for specified taxonomic, stratigraphic, and geographic terms. Older literature is also incorporated into many such databases. For example, the electronic version of the *Zoological Record* now extends back to 1978, and *GeoRef* goes back to the eighteenth century. Electronic bibliographies have become an indispensable aid to the practicing paleontologist. Despite their great utility, however, all bibliographies are incomplete and imperfectly indexed, and the student should be aware that methods such as cross-referencing and browsing in the library are generally still necessary.

## 1.7 CONCLUDING REMARKS

The fossil record provides a very small sample of past life. Because so many species have lived in the past and because the amount of time covered is so vast, however, paleontologists have an enormous quantity of data with which to study the history of life. Almost since the beginning of paleontology as a science, a major concern has been how to study biological and geological processes in the face of paleontological incompleteness. In this chapter, we have emphasized two major points:

1. It is possible to design paleontological studies so that the imperfections of the fossil record do not dominate. Major approaches include: (a) focusing on the well-skeletonized fraction of species; (b) comparing fossil assemblages that come from similar environments with similar preservational conditions; (c) using taphonomic controls in order to judge whether the absence of a species from a particular time and environment is likely to be real or preservational; and (d) studying evolution and ecology at a local level, at which the fossil record is relatively complete.
2. Biases in the fossil record can be used to advantage by the paleontologist. An observed pattern that is the opposite of what is predicted by a bias such as differential preservation is likely to be more reliable than one that is predicted by the bias.

To be sure, data from the fossil record cannot always be taken at face value, but available data are often quite adequate for specific purposes. As we will see throughout this book, we can interpret the data of the fossil record in ways that take preservational bias and incompleteness into account.

# SUPPLEMENTARY READING


Allison, P. A., and Briggs, D. E. G. (eds.) (1991) *Taphonomy: Releasing the Data Locked in the Fossil Record*. New York, Plenum Press, 560 pp. [A comprehensive collection of papers on taphonomic processes and the quality of the fossil record.]

Benton, M. J. (ed.) (1993) *The Fossil Record 2*. London, Chapman and Hall, 845 pp. [A fairly complete survey of the families known from the fossil record.]

Bottjer, D. J., Etter, W., Hagadorn, J. W., and Tang, C. M. (2002) *Exceptional Fossil Preservation*. New York, Columbia University Press, 403 pp. [Overviews of fossils and geologic features of many of the most important fossil Lagerstätten.]

Donovan, S. K., and Paul, C. R. C. (eds.) (1998) *The Adequacy of the Fossil Record*. Chichester, Wiley, 312 pp. [A collection of essays on methods and case studies that concern completeness of the fossil record.]

Pääbo, S., Poinar, H., Serre, D., Jaenicke-Després, V., Hebler, J., Rohland, N., Kuch, M., Krause, J., Vigilant, L., and Hofreiter, M. (2004) Genetic analyses from ancient DNA. *Annual Review of Genetics* **38**:645–679. [Overview of progress in the extraction and analysis of ancient DNA.]

Raup, D. M. (1979) Biases in the fossil record of species and genera. *Bulletin of the Carnegie Museum of Natural History* **13**:85–91. [Discussion of some of the more important biases, with particular reference to the tabulation of diversity.]

Schäfer, W. (1972) *Ecology and Paleoecology of Marine Environments* (English translation). Chicago, University of Chicago Press, 568 pp. [Pioneering treatment of actualistic paleontology.]

# *Chapter 2*

# GROWTH AND FORM

The raw materials of paleontology are individual specimens that we collect and prepare for study. Virtually all questions addressed with fossil data, whether geological, ecological, or evolutionary, rest ultimately upon detailed observation of individual specimens. To cite just a few examples:

1. We may wish to test the hypothesis that body size tends to increase over evolutionary time [SEE SECTION 7.4]. If so, we need to know what we mean by size and how to measure it.
2. Studies in systematics [SEE SECTIONS 3.3 AND 4.1], biostratigraphy [SEE SECTION 6.1], and biodiversity [SEE SECTION 8.1] require that paleontologists be able to assign specimens correctly to species, to know whether two specimens belong to the same or to different species, and to determine how many species are represented in a collection of specimens. Such studies depend on detailed description and measurement of these specimens.
3. If we want to interpret the mechanics, physiology, and ecology of extinct organisms [SEE SECTIONS 5.2 AND 9.4], we must understand the principles by which function depends on size and shape change through growth.

Although chemical, mineralogical, and other kinds of data are often collected from fossil specimens, the principal information typically concerns their form. We therefore begin this chapter with an elementary treatment of **morphology,** the study of biological form and structure.

## 2.1 ASPECTS OF FORM

Whether we are studying a whole organism or just one of its parts, there are three components of form to consider: size, shape, and the relationship between size and shape. **Size** is perhaps the most obvious biological trait. Not only is size obvious, but it is also of fundamental importance to the organism. Many aspects of function—such as metabolism, reproduction, and locomotion—vary regularly with size. Biologists can therefore predict much about an organism just by knowing how big it is. Given that an elephant is a mobile terrestrial animal with a large body mass, a biologist need never have seen one to be able to state with some confidence that it must have stout limbs vertically oriented below its body.

The second major component of organic form is **shape.** Intuitively, shape reflects the relative proportions of different parts of an organism. The length and breadth of a bone are measures of its size. By contrast, the ratio of the length to breadth—relatively slender versus relatively stout—is a measure of shape. Strictly speaking, a shape measure is derived as a ratio among other measures that have specific dimensions—usually mass ($M$), length ($L$), area ($L^2$), and volume ($L^3$)—in such a way that the dimensions cancel out to yield a dimensionless number. (The dimensions are distinct from the specific units, such as grams for mass and centimeters for length.) Both bone length and breadth have dimensions of length, $L$. Therefore, the ratio between them has dimensions $L \div L$, which equals $L^0$ and so is dimensionless. Not all ratios among size measures yield measures of shape. For example, the ratio between mass and volume has the dimensions of density, namely $M \div L^3$ or $ML^{-3}$. Likewise, the ratio of bone cross-sectional area to bone length has dimensions $L^2 \div L$, which is equal to $L$.

An obvious aspect of shape is symmetry. Many familiar animals are bilaterally symmetrical (essentially identical from left to right) or radially symmetrical (similar in form in any direction perpendicular to the anterior–posterior axis). Other forms of symmetry exist, such as the fivefold symmetry of echinoderms. Basic symmetry is generally invariant within species, genera, and even higher taxa.

Shape typically changes as an organism grows, which brings us to the third component of form: the relationship between size and shape. The ways in which shape changes with size often reflect how organisms function, as we will see when we discuss the nature of growth and development.

## 2.2 DESCRIBING AND MEASURING SPECIMENS

What we measure and with what precision depend ultimately on the question being addressed. Prior experience studying a particular group of organisms is often essential in guiding the choice of measurements. It is therefore difficult to provide a single formula that will serve well for describing and measuring any specimen or species. Sometimes a simple measure of body size can discriminate two closely related species, but in other cases the species may be nearly identical in size, so that more detailed measurements will be necessary to distinguish them.

Which measures work best for which purposes will be developed by example during the course of this book and through the reading of case studies and examination of material in the laboratory. Two guidelines usually apply: (1) Within-species studies require higher precision than comparisons between different species, because the morphological differences are more subtle, and (2) the greater the level of detail in a system of description or measurement, the smaller the range of species to which that system can be applied. The simple height:width ratio can be compared among all shelled gastropods, for example, but measurements of particular aspects of ornamentation would apply only to those species that possess the ornament in question. Here we emphasize description and measurement of specimens that have already been collected and prepared for study. References at the end of the chapter and in the bibliography should be consulted for techniques of paleontological collection and preparation.

### Pictorial Description

Except for the smallest of fossilized microorganisms, nearly all initial observations on specimens are made with the naked eye. Rich detail can often be documented with nothing more sophisticated than paper, pencil, and patience. Nonetheless, the structures we wish to study may be so small that they require a magnifying lens, optical microscope, or electron microscope. Or they may not be visible on the exterior of the specimen. Sectioning allows observations on interior structures, but this is destructive. Radiography, computerized axial tomography (CAT) scanning, and related methods allow one to gain a visual image of the interior of a specimen without damaging it. Such methods typically detect contrasts in density or composition, such as that between actual skeletal material and the sedimentary matrix. Figures 1.11 and 2.1 show examples of nondestructive imaging of fossils.

Once we can see a specimen, there are many tools that help us keep a permanent record of that specimen for future study or to aid measurement. In addition to the photograph, one of the most useful tools is the **camera lucida.** This is an optical device, sometimes attached to a microscope, that facilitates drawing by allowing one to see the image of the specimen superimposed on the image of one's hand. The sketch or line drawing is still an extremely important tool in the description of specimens. In cases in which material has suffered substantial post-mortem disarticulation and distortion, for example, it is absolutely essential to draw the specimen in such a way that the relevant biological detail is emphasized to the exclusion of taphonomic features.

The importance of the line drawing is illustrated in Figure 2.2. Figure 2.2a shows a specimen of the arthropod *Marrella splendens*, the most abundant species known from the Middle Cambrian Burgess Shale of British Columbia [SEE SECTION 10.2]. The specimen is distorted and partially obscured by sediment. Moreover, the color contrast between the specimen and the sedimentary matrix is somewhat limited, making it hard to distinguish the specimen from the sediment. Figure 2.2b is a line drawing of the specimen, made with the aid of a microscope and camera lucida. This drawing has elements of both faithful representation

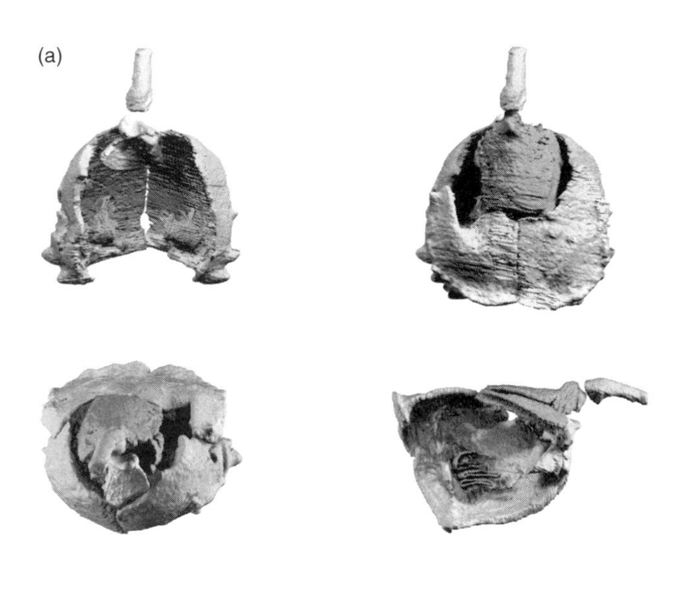

**FIGURE 2.1 Nondestructive imaging of fossils.** (a) A mitrate echinoderm from the Upper Carboniferous of Oklahoma, shown in four different orientations. Numerous, fine-scale X-ray images were taken to simulate serial sectioning without damaging the specimen. These serial images were then reconstructed in three dimensions via computer. (b) Magnetic resonance image (MRI) of the complex, three-dimensional trace fossil *Macaronichnus* from the Cretaceous of Alberta. *(a: From Dominguez et al., 2002; b: From Gingras et al., 2002, Permission by Society for Sedimentary Geology)*

and interpretation. The artist has tried to portray only the anatomical details that are actually evident. At the same time, given such problems as specimen distortion and obstruction by matrix, a certain degree of interpretation is inevitable. By studying many specimens preserved in different orientations, and making a line drawing of each, the reconstruction of Figure 2.2c was possible. Such a reconstruction is far more interpretative than any single line drawing; it portrays an idealized individual that is far more complete and symmetrical than any single specimen.

There are numerous systems that allow one to **digitize** specimens, that is, to store electronic images of

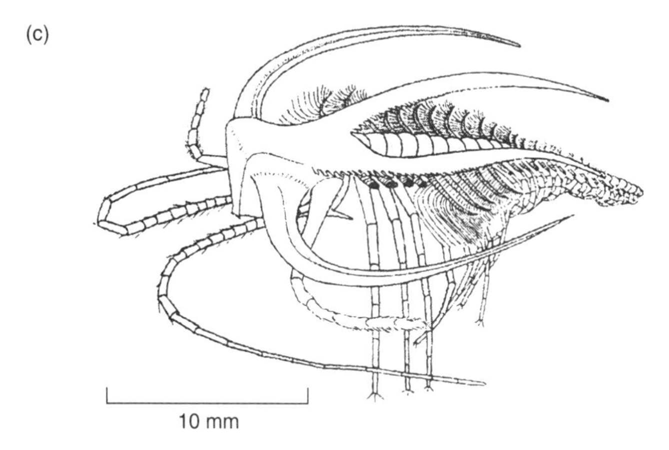

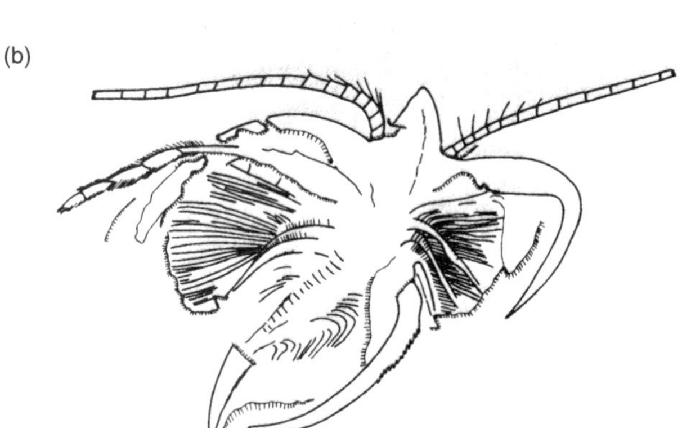

**FIGURE 2.2 Images of the arthropod *Marrella splendens* from the Middle Cambrian of British Columbia.** (a) Photograph of a specimen. (b) Line drawing of this same specimen, made with the aid of a camera lucida. (c) Reconstruction based on camera-lucida drawings of several specimens. *(From Whittington, 1971)*

(a)

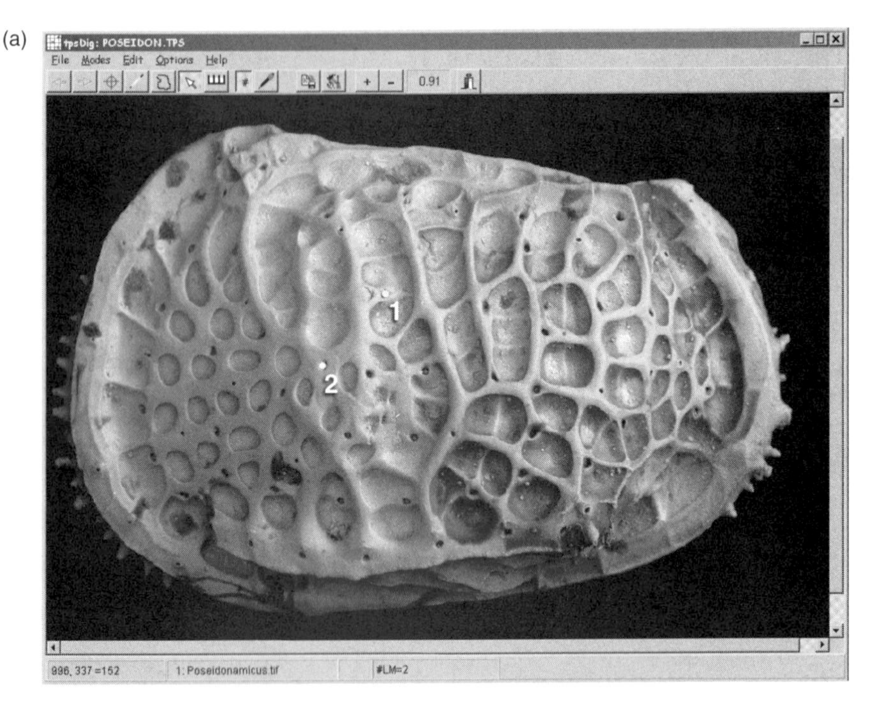


(b)

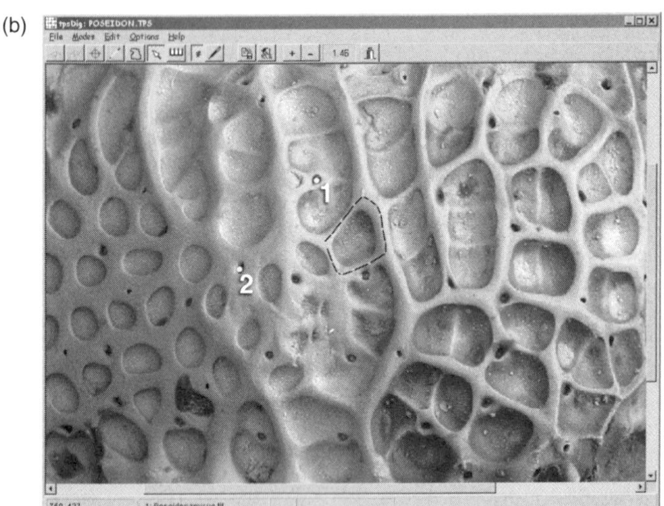


**FIGURE 2.3 Digitized image of the left valve of the OSTRACODE *Poseidonamicus*, displayed with the computer software tpsDig.** This image was produced with a scanning electron microscope. (a) Overview of a specimen showing two points, marked 1 and 2, that have been selected to have their $x$- and $y$-coordinates automatically recorded. Note the relative magnification of 0.91 indicated near the top of the computer screen. (b) Part of this same specimen at a relative magnification of 1.46. In addition to the points 1 and 2, a polygon has been traced around one of the fossae (the shallow depressions) on the valve. The area and geometric centroid of this polygon can be computed by the software.
*(Courtesy of Gene Hunt)*

them. Digitization can be manual, in which case a specimen is drawn with a special stylus whose motion is mechanically or electronically recorded in two or three dimensions. More commonly, an entire photographic image is stored electronically. Automated storage of three-dimensional images can be accomplished by digitizing many closely spaced, thin sections, as in Figure 2.1a. It is also possible to digitize pairs of photographs taken at slightly different angles—which simulates binocular vision—and to use computer algorithms to reconstruct the form trigonometrically. Manual digitization is often more laborious than automated methods, but, like the line drawing relative to the photograph, it has the advantage that only the desired detail is recorded.

Modern digitizing equipment has high optical resolution and can reduce the distortion that inevitably results from projecting a three-dimensional specimen into two dimensions. Moreover, there are many sophisticated and versatile computer programs to work with digitized images. Photographic images can be processed to remove unwanted detail and therefore to make measurement easier. For example, the contrast between a light-colored specimen and a dark background can be exploited to approximate the outline of the specimen so that this need not be traced by hand. Many standard measurements can be recorded with little effort using computer programs. In an important sense, therefore, having a properly digitized image is like having the actual specimen at hand (see Figure 2.3).

Digitization and computer software facilitate specimen description and measurement. They do not, however, solve the important problem of deciding *what* to describe and measure. We now turn to this question.

## Descriptive Terminology

By far the most common medium for describing a structure is the word. Well-designed terminology can powerfully and economically describe form, reducing a great deal of information to a single word or relatively few words. If carefully chosen, the words are self-explanatory and relatively easy to learn.

**FIGURE 2.4 Cretaceous gastropod *Calliomphalus conanti*.** See text for the formal description of this species. *(From Sohl, 1960)*

As an example of the effective use of descriptive terminology, the formal description of a gastropod species follows (from Sohl, 1960). Photographs of a specimen of the species are shown in Figure 2.4.

> Shell small, trochiform, phaneromphalous with nacreous inner shell layer; holotype with about $7^1/_4$ rapidly expanding whorls. Protoconch smooth on early whorl with coarse axial costae appearing at slightly more than one whorl, followed almost immediately by fine spiral lirae; suture impressed. Whorl sides slope less steeply than general slope of spire, giving an outline interrupted by overhang of periphery of preceding whorls; periphery subround to subangular; whorl side slopes steeply below periphery to broadly rounded base. Sculpture of axial and spiral elements same size; 8 spiral lirae on upper slope possess subdued tubercles where overridden by somewhat coarser and closer spaced axial cords; base covered by about 10 unequally spaced spirals with poorly defined tubercles and numerous axial lirae. Umbilicus narrow, bordered by a margin bearing low nodes. Aperture incompletely known, subcircular, slightly wider than high and reflexed slightly at junction of inner lip and umbilical margin.

To a person unfamiliar with gastropod morphology and its descriptive terminology, this description may be nearly unintelligible. For the person acquainted with the subject, however, the description should provide a convincing sketch of a group of specimens.

Any system of descriptive terminology can lead to difficulty. For example, note the word *small* in the first line of the description. This implies a size comparison with other organisms—but what other organisms? By convention, terms such as *small*, *large*, *wide*, and *narrow* imply comparison only with closely related organisms.

Descriptive terminology often includes a number of categories to which the principal forms in a biologic group can be assigned. Figure 2.5 shows some examples for gastropod shells. A common problem is deciding

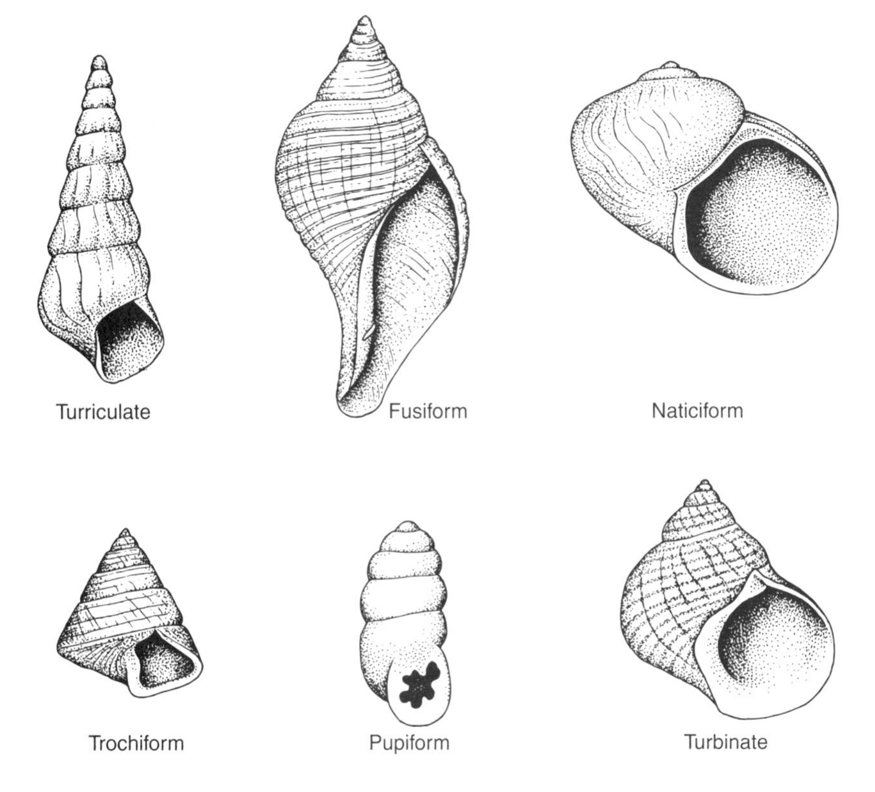


**FIGURE 2.5 Some of the terms for shape used in morphologic descriptions of gastropods.** The origin of such terms, some of which are several centuries old, is quite varied. Some terms are derived from descriptive words: "Turbinate" comes from the Latin *turbinatus*, meaning "top-shaped." Others not shown here are from geometry: conical, biconical, obconical, and so on. Still others are derived from the names of particular taxa—"naticiform" from the genus *Natica*, for example.

where one category leaves off and another begins. What do we do, for example, if we are faced with a shell that is somewhere between naticiform and turbinate? One reason that intermediate forms do not present even more problems than they do is that biologic form is not randomly distributed [SEE SECTION 5.3]. Certain shapes are much more common than others. If a system of descriptive terminology accurately reflects the natural clusters of predominant forms, then relatively few specimens will be found that are between two categories. Thus, the establishment of a system of descriptive terminology may be an important scientific contribution in itself, representing a fundamental interpretation of the natural world.

## Description by Measurement

The counting of **meristic characters** is a simple form of quantitative description. Examples include the number of ribs in a fish, the number of petals in a flower, and the number of segments in an arthropod. Such a tabulation may seem crude, but it can be quite informative. For example, it is possible to distinguish among many groups of trilobites based on the number of segments, and species within certain groups of trilobites are characterized by a specific number of pits on the fringe of the head.

Most quantitative description of fossils involves measurement in the strict sense rather than counting. As a simple example, consider the leaves portrayed in Figure 2.6, along with one shape measure, the length:width ratio. The leaves are nearly identical in this measure but obviously differ substantially in other aspects of shape. To describe form with nothing more than the length:width ratio implicitly assumes that a rectangle is a good approximation to the form—in other words, that it is reasonable to use the rectangle as an idealized model of form. In fact, any system of measurement assumes a model of some sort.

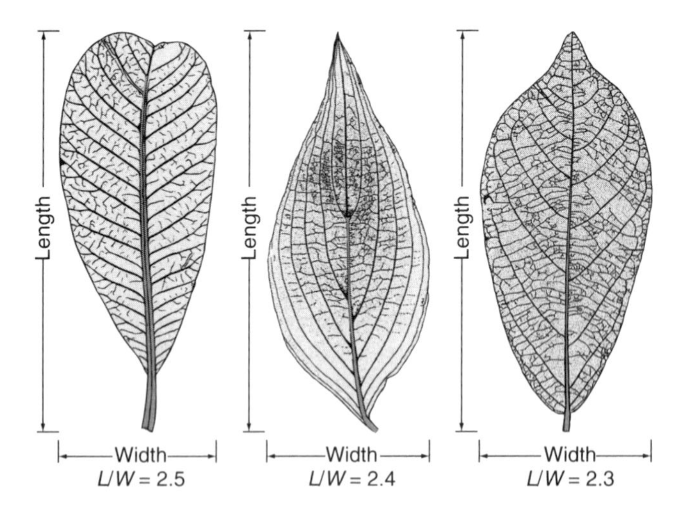


**FIGURE 2.6 Leaves of three different tree species, with length and width indicated.** These leaves have approximately the same length:width ratio, but they differ in other aspects of shape. *(From Leaf Architecture Working Group, 1999)*

Four hypothetical coiled **CEPHALOPODS** are shown in Figure 2.7. For each, the largest diameter, *d*, is indicated, which distinguishes the large forms from the small. The diameter, however, tells us little about shape. The model tacitly assumed is a circle. Any number of quite different spiral cephalopods can be inscribed in the same circle, and thus the diameter gives us little indication of anything except size. For each of the shapes in Figure 2.7, a pair of unequal radii ($r_1$ and $r_2$), each originating in the morphologic center of the spiral, gives us more information. For these hypothetical forms, the ratio of any two radii measured at equal angular intervals, in this case 180 degrees, is a constant value. This means that, with respect to these radii, the shape of the shell is not changing with growth. Many real shells conform approximately to an ideal model with constant ratios of radii. Based on the ratio of radii, we can distinguish three cephalopod forms. The ratio of the two radii in Figure 2.7a is larger than their ratio in Figure 2.7d, and this reflects some of the differences between the two shapes. The differences between Figures 2.7b and 2.7c are evident in the drawings, but these two forms are indistinguishable from each other by the ratios of radii.

The foregoing examples illustrate the simplest and still most commonly used approach to quantitative description: the measurement of a series of linear size dimensions, and the definition of shape variables based on ratios among these. This approach has a number of advantages. It yields measures that are intuitively satisfying and correspond to the aspects of form that commonly strike the eye of both the expert and the beginner. It can be carried out quickly and easily with only minimal equipment, such as a ruler or calipers. And it allows a wide range of forms to be compared, provided that the measures are carefully chosen. Perhaps the main shortcoming of this approach is that complex analyses, some of which we discuss in the next chapter, may be required to interpret the data if one has taken many measurements on each specimen in an effort to describe the form in great detail. Biologically realistic models, such as that for the coiled shell, can therefore be of great value in allowing an economical set of measurements [SEE SECTION 5.3].

Related to the measurement of linear dimensions is the recording of *x*-, *y*-, and *z*-coordinates of reference

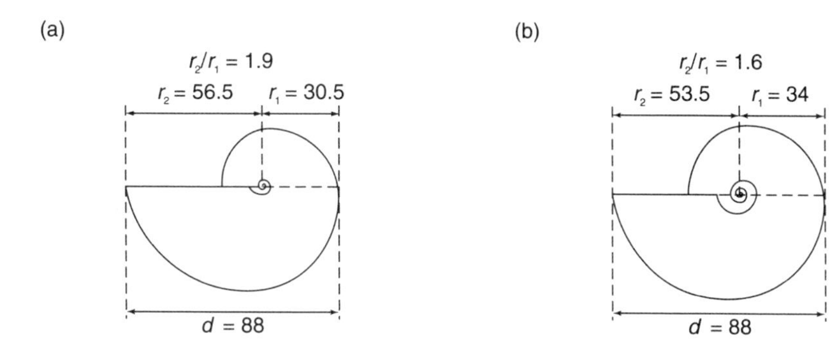


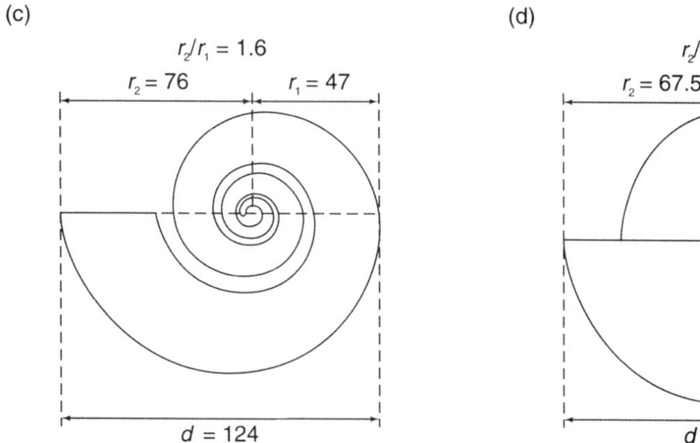


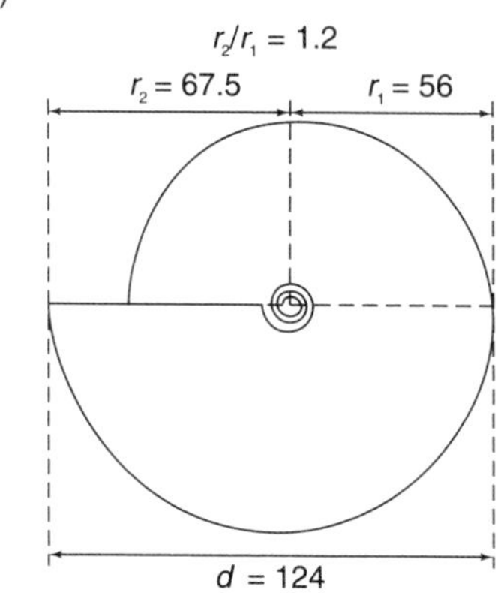


**FIGURE 2.7 Four generalized shell forms common in coiled cephalopods.** Parts (a) and (b) have the same diameter ($d$), as do parts (c) and (d). These diameters depict size but not shape. The ratio of radii, $r_1/r_2$, conveys considerable shape information. Parts (b) and (c) have the same ratio, however, even though they differ in other respects.

points or **landmarks** on the specimen. These landmarks may be, for example, the triple-point junctions of echinoderm plates or arthropod sclerites, the points of articulation of skeletal elements, or extreme points such as the tip of a limb. Figure 2.8 shows some examples of landmarks. In the case of the OSTEICHTHYAN fish (Figure 2.8a), the form is approximated by a two-dimensional projection. For the BLASTOID echinoderm (Figure 2.8b), a special microscope was used to record the coordinates of landmarks in three dimensions.

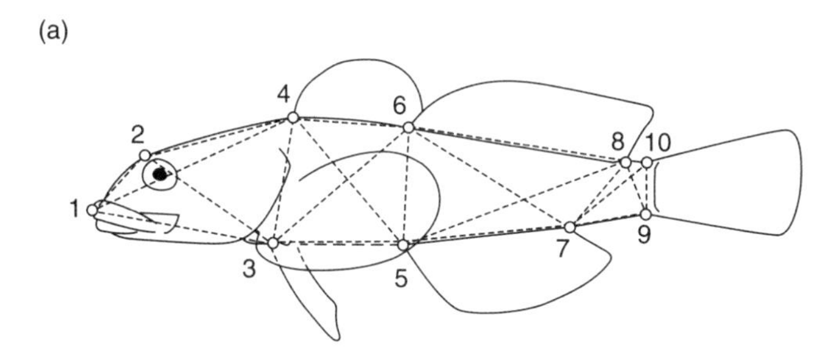


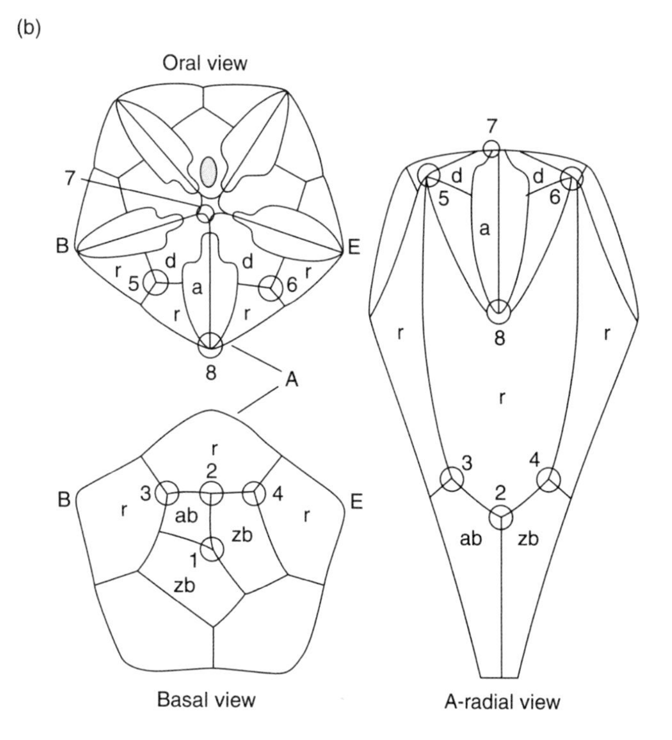


**FIGURE 2.8 Examples of landmarks.** (a) Tissue junctions (for example, 4, 6, and 7) and extreme points (for example, 1) around the perimeter of a fish. Dashed lines indicate some of the linear distances that can be measured between landmarks. (b) Plate junctions in a generalized blastoid echinoderm shown in three different views. Landmarks are numbered, and different kinds of plates are indicated by letters: d, deltoid; a, ambulacral; r, radial (with A, B, and E denoting different radials); ab, azygous (unfused) basal; zb, zygous (fused) basal. *(a: From Bookstein et al., 1985; b: From Foote, 1991)*

Box 2.1

## CENTROID SIZE AND SHAPE COORDINATES

Figure 2.9a depicts the cranidium of the Cambrian trilobite *Crassifimbra walcotti* with a number of landmarks indicated. These landmarks were chosen to give relatively uniform coverage to the cranidium. In this case, only landmarks on the midline and the right side are recorded. It is implicitly assumed that the form is bilaterally symmetrical—in other words, that deviations from perfect symmetry can be ignored because they are small. This example is typical in projecting the specimen into two dimensions and disregarding the third dimension for simplicity.

There are $n$ landmarks with coordinates $(x_1, y_1), \ldots (x_n, y_n)$. The mean $x$- and $y$-coordinates are calculated as $\bar{x} = \sum_{i=1}^{n} x_i/n$ and $\bar{y} = \sum_{i=1}^{n} y_i/n$, and the centroid is defined as the point having coordinates $(\bar{x}, \bar{y})$ (see Box 3.1). The centroid size is defined as the square root of the sum of squared differences between the observed points and the centroid:

$$\text{CS} = \sqrt{\sum_{i=1}^{n} (x_i - \bar{x})^2 + (y_i - \bar{y})^2}$$

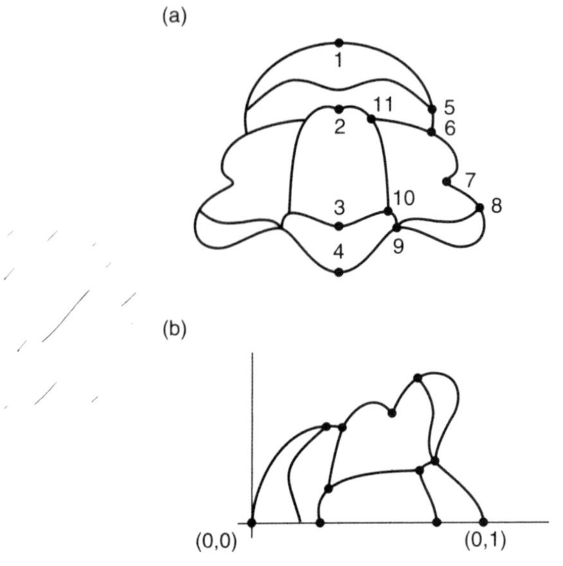


**FIGURE 2.9 Orientation of a specimen for the calculation of shape coordinates.** (a) Cranidium of the Cambrian trilobite *Crassifimbra walcotti*, showing a number of landmarks on the midline and right side. (b) Translation and scaling of this specimen so that points 1 and 4 define a baseline from (0, 0) to (0, 1). *(From Smith, 1998)*

Shape coordinates can be calculated from the landmarks of Figure 2.9a by a suitable choice of scaling, shown in Figures 2.9b and 2.10. A pair of points, typically approximating the long axis of the specimen, is used to define a **baseline.** This means that one point of the baseline will become the origin in a new coordinate system, and the other baseline point will be placed on the $x$ axis. Each additional point is treated as the vertex (point C) of a triangle, the other two vertices (points A and B) of which are the baseline points. The coordinates of these points are denoted $(x_A, y_A)$, $(x_B, y_B)$, and $(x_C, y_C)$. The stored image of the specimen is then expanded or contracted so

The dashed lines in Figure 2.8a show distances between landmarks that can be used as measures of size. The coordinates of landmarks are commonly combined in a single size measure called **centroid size,** based on the distances of each point from the centroid or arithmetic mean of all points (Bookstein, 1991) (Box 2.1). The landmarks may also be converted to shape measures called **shape coordinates** (see Figures 2.9 and 2.10 in Box 2.1). If the landmarks are well chosen, they can provide a representation of the form that includes many aspects of shape.

Many organisms have prominent shape features that are characterized neither by a geometric model such as the logarithmic spiral nor by a set of landmarks. Consider the living mussel *Mytilus edulis*, shown in Figure 2.11. The shape of its outline would not be approximated well by a model such as a rectangle. Moreover, although we could select a number of points on the outline, their biological equivalence among specimens would not be evident; they would not be landmarks such as those in the examples of Figures 2.8 and 2.9.

In cases like that of the mussel outline, we can take advantage of methods of curve fitting. One of the most common is harmonic analysis (Box 2.2). Its utility stems from the fact that any simple curve can be described mathematically as the sum of a series of sine and cosine curves, each one multiplied by a different number called the *harmonic coefficient.* The particular values of the harmonic coefficients depend on the shape of the curve.

that the origin point of the baseline has coordinates (0, 0) and the other baseline point has coordinates (1, 0). Thus, the baseline is given a length of 1 unit. This scaling is done in such a way that the *relative* positions of the points are unaffected, that is, so that the shape of the specimen is not changed. The three points of the triangle are now denoted A′, B′, and C′. The $x$- and $y$-coordinates of point C′ summarize all the shape information that is contained in the triangle. These coordinates are obtained by the following formulas that come from trigonometry:

$$x_{C'} = \frac{(x_B - x_A)(x_C - x_A) + (y_B - y_A)(y_C - y_A)}{(x_B - x_A)^2 + (y_B - y_A)^2}$$

and

$$y_{C'} = \frac{(x_B - x_A)(y_C - y_A) - (y_B - y_A)(x_C - x_A)}{(x_B - x_A)^2 + (y_B - y_A)^2}$$

The shape coordinates are calculated in the same way for each point, and the entire set of shape coordinates gives the shape information that is present in the chosen landmarks.

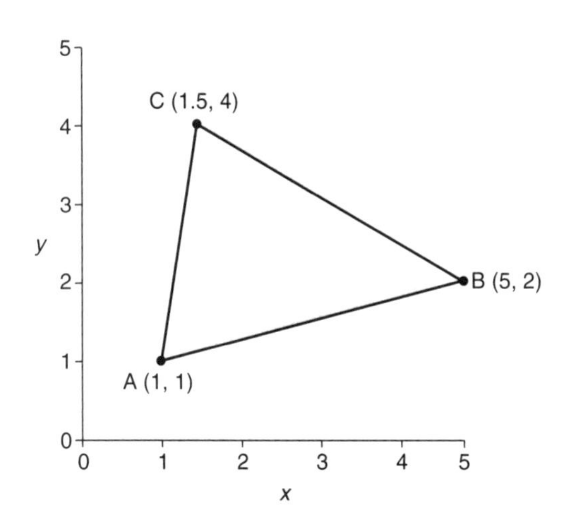


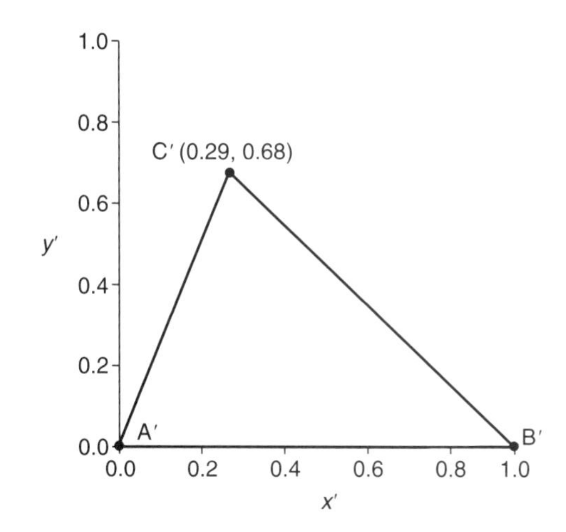


**FIGURE 2.10 Transformation of hypothetical, original coordinates (left) into shape coordinates (right).** Points labeled A and B are the baseline points, and point C is any other landmark. Numbers in parentheses are the $x$- and $y$-coordinates before and after transformation.

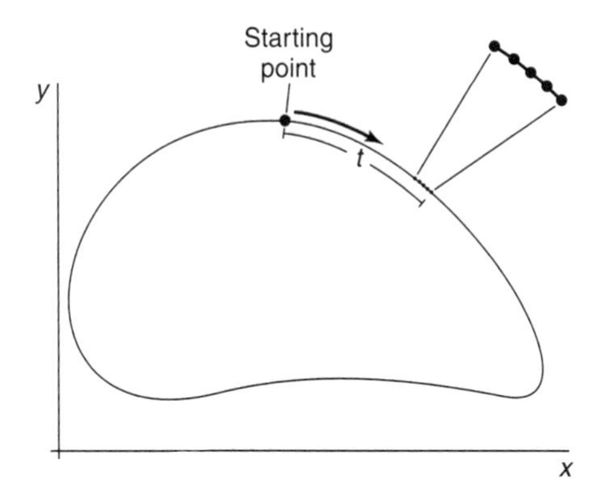


**FIGURE 2.11 Outline of a specimen of the mussel *Mytilus*.** Inset shows magnification of part of the outline with sampled points. The length along the curve from the starting point is given by $t$; the arrow shows the direction in which the curve is traced. In harmonic analysis, the $x$- and $y$-coordinates are analyzed as functions of $t$. *(After Ferson et al., 1985)*

In Figure 2.11, the outline of the mussel is digitized from a starting point and is represented by hundreds of stored points. Thus, it is in fact represented by a polygon—but one with so many vertices that, practically speaking, it is indistinguishable from a smooth curve. For each point, the $x$- and $y$-coordinates are recorded. The variation in $x$-coordinates and the variation in $y$-coordinates are treated as two distinct functions that depend on the length traversed along the curve from the starting point. Each function is described by a distinct set of harmonic coefficients.

Once the harmonic coefficients are calculated, the curve can be approximated by a subset of them, as shown in Figure 2.12. Clearly the outline corresponding

Box 2.2

## HARMONIC ANALYSIS

Description of the outline in Figure 2.11 uses harmonic analysis, a standard approach for studying many kinds of curves. Here we describe the steps involved in one method of harmonic analysis, and we apply the approach to a simple example.

Consider the data of Figure 2.13a, which depict hourly temperature readings for five years at a weather station in Farmingdale, New York. Annual variation in temperature is clearly visible as the low-frequency (long-wavelength) oscillation. Superimposed on this annual cycle is considerable variation whose source is not immediately evident.

In this analysis, time is treated as an independent variable and temperature as a dependent variable. It is essential that the dependent variable be a single-valued function of the independent variable. In other words, for every time there is exactly one temperature. Note that the converse is not true; for any given temperature, there may be numerous times that correspond to that reading. In the data of Figure 2.13, temperature is recorded at evenly spaced (hourly) intervals. The approach to harmonic analysis presented here assumes evenly spaced values of the independent variable, in this case, time.

The general harmonic equation has the form:

$$T = \sum_{k=0}^{\infty} A_k \cos(k\Theta) + B_k \sin(k\Theta)$$

This says that the function $T$ (here, temperature) can be expressed as the sum of an infinite number of periodic cosine and sine curves. The numbers $A_k$ and $B_k$ are the harmonic coefficients, the numbers by which the ideal cosine and sine curves are multiplied. The harmonic number $k$ refers to the period of the cosine and sine curves. For example, $k = 1$ implies a curve that repeats once within the total span of time; $k = 2$ implies a curve that repeats twice; and so on. Thus, higher harmonic numbers correspond to finer-scale variations—those with higher frequency or shorter wavelength. With finite data, it is not possible to calculate an infinite series of harmonics. If there are $n$ observations, the maximum harmonic number is no greater than $n/2$.

We place the $n$ observed values of temperature in sequence and denote them by $T_i$, where $i = 1, \ldots, n$. The harmonic coefficients $A_k$ and $B_k$ are calculated for each value of $k$ as:

$$A_k = \frac{2}{n} \sum_{i=1}^{n} T_i \cos(2\pi ik/n)$$

and

$$B_k = \frac{2}{n} \sum_{i=1}^{n} T_i \sin(2\pi ik/n)$$

These equations implicitly scale the total span of time to vary from 0 to $2\pi$, the fundamental repeat length for a periodic function with harmonic number $k = 1$. For each harmonic, the height of the wave form—its amplitude—is given by $\sqrt{(A_k^2 + B_k^2)}$, and the power is given by $(A_k^2 + B_k^2)/2$. The higher the power, the more information on variation in the data accounted for by the corresponding harmonic.

In the temperature data of Figure 2.13a, there are 1826 days (four years of 365 days plus one leap year of 366 days), and thus $n = 43{,}824$ hourly readings. The first three readings are $-4.4$, $-5.0$, and $-5.0$ degrees, and the last three are 1.1, 1.1, and 0.6 degrees. Applying these equations to the first harmonic we have:

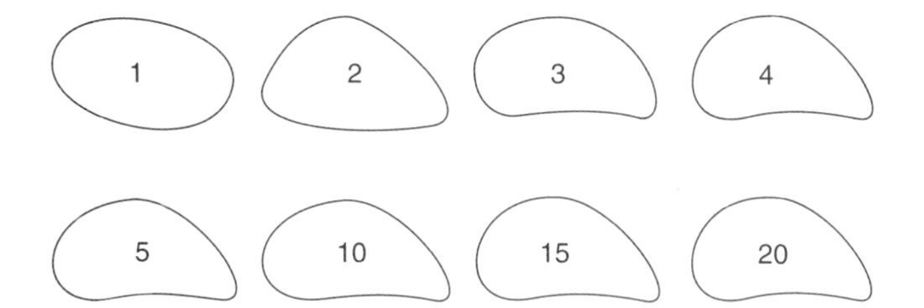


**FIGURE 2.12 Approximations to the outline of *Mytilus* using different numbers of harmonics.** The approximation to the outline of Figure 2.11 improves as more harmonics are included. *(From Ferson et al., 1985)*

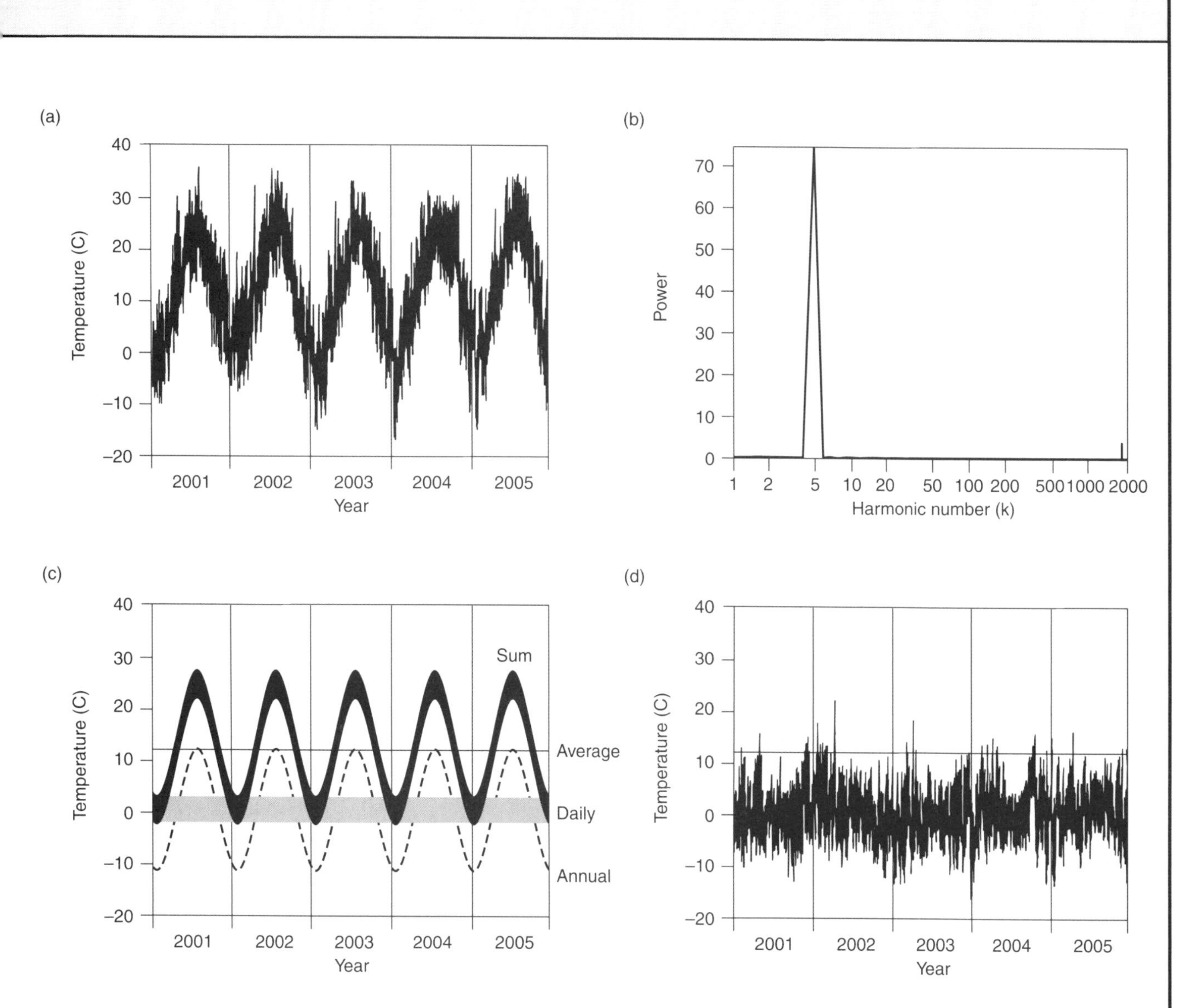


**FIGURE 2.13 Harmonic analysis of temperature data.** (a) Temperature in degrees centigrade, taken at hourly intervals for five years at Republic Airport, Farmingdale, New York. Missing readings are replaced by data from nearby weather stations. (b) Power spectrum, showing peaks at harmonics 5 and 1826, corresponding to annual and daily cycles. (c) Reconstructed temperature based on annual harmonic (dashed line), daily harmonic (gray band), overall average (horizontal line), and the sum of these three components (dark band). (d) Residual temperature variation (variation not accounted for by average, annual, and daily terms), calculated as the difference between the data in part (a) and the dark band in part (c). *(Data from National Oceanic and Atmospheric Administration [http://cdo.ncdc.noaa.gov/ulcd/ULCD, accessed 30 January 2006])*

*continued on next page*

to the first harmonic is a poor approximation to the original outline. Successively higher harmonics generally represent finer-scale aspects of the form. As more of the higher harmonics are added to the reconstruction, this reconstruction approximates the original form progressively better. Although it may not be immediately obvious that anything has been gained by the harmonic analysis, it is usually the case that a relatively small number of harmonics can provide a good approximation to the original form. Thus, the form is represented far more efficiently than would have been possible by simply recording the $x$- and $y$-coordinates of the entire outline. In the case of the mussel, hundreds of points were recorded, whereas only 20 harmonics are needed to provide the excellent approximation shown in Figure 2.12.

*Box 2.2 (continued)*

$$\begin{aligned} A_1 = {} & (2/43{,}824)\{(-4.4)\cos[2\pi(1)(\mathbf{1})/43{,}824] \\ & + (-5.0)\cos[2\pi(2)(\mathbf{1})/43{,}824] \\ & + (-5.0)\cos[2\pi(3)(\mathbf{1})/43{,}824] + \\ & \ldots \\ & + (1.1)\cos[2\pi(43{,}822)(\mathbf{1})/43{,}824] \\ & + (1.1)\cos[2\pi(43{,}823)(\mathbf{1})/43{,}824] \\ & + (0.6)\cos[2\pi(43{,}824)(\mathbf{1})/43{,}824]\} \end{aligned}$$

and

$$\begin{aligned} B_1 = {} & (2/43{,}824)\{(-4.4)\sin[2\pi(1)(\mathbf{1})/43{,}824] \\ & + (-5.0)\sin[2\pi(2)(\mathbf{1})/43{,}824] \\ & + (-5.0)\sin[2\pi(3)(\mathbf{1})/43{,}824] + \\ & \ldots \\ & + (1.1)\sin[2\pi(43{,}822)(\mathbf{1})/43{,}824] \\ & + (1.1)\sin[2\pi(43{,}823)(\mathbf{1})/43{,}824] \\ & + (0.6)\sin[2\pi(43{,}824)(\mathbf{1})/43{,}824]\} \end{aligned}$$

where the harmonic number is indicated in bold and the ellipsis (. . .) stands for the remaining readings between the first three and the last three. Similarly, for the second harmonic:

$$\begin{aligned} A_2 = {} & (2/43{,}824)\{(-4.4)\cos[2\pi(1)(\mathbf{2})/43{,}824] \\ & + (-5.0)\cos[2\pi(2)(\mathbf{2})/43{,}824] \\ & + (-5.0)\cos[2\pi(3)(\mathbf{2})/43{,}824] + \\ & \ldots \\ & + (1.1)\cos[2\pi(43{,}822)(\mathbf{2})/43{,}824] \\ & + (1.1)\cos[2\pi(43{,}823)(\mathbf{2})/43{,}824] \\ & + (0.6)\cos[2\pi(43{,}824)(\mathbf{2})/43{,}824]\} \end{aligned}$$

and

$$\begin{aligned} B_2 = {} & (2/43{,}824)\{(-4.4)\sin[2\pi(1)(\mathbf{2})/43{,}824] \\ & + (-5.0)\sin[2\pi(2)(\mathbf{2})/43{,}824] \\ & + (-5.0)\sin[2\pi(3)(\mathbf{2})/43{,}824] + \\ & \ldots \\ & + (1.1)\sin[2\pi(43{,}822)(\mathbf{2})/43{,}824] \\ & + (1.1)\sin[2\pi(43{,}823)(\mathbf{2})/43{,}824] \\ & + (0.6)\sin[2\pi(43{,}824)(\mathbf{2})/43{,}824]\} \end{aligned}$$

and so on for the higher harmonics.

Figure 2.13b depicts what is known as the power spectrum, a plot of power versus harmonic number $k$. There are two peaks in this spectrum, a large one at $k = 5$ and a smaller one at $k = 1826$. The coefficients for these two harmonics are ($A_5 = -10.8$, $B_5 = -5.5$), and ($A_{1826} = -2.0$, $B_{1826} = -1.6$).

The larger peak corresponds to a periodic function that repeats five times in five years; in other words, it is the annual cycle. The smaller peak is the daily cycle, which repeats 1826 times. If we compare the power of these two harmonics with the sum over all harmonics, we find that the annual and daily cycles account for 76.9 percent and 3.5 percent of the temperature variation for these five years. Thus, harmonic analysis allows a very efficient summary of information. Most of the information can be accounted for by just two harmonics, while less than 20 percent of the variation is spread throughout the remaining harmonics (of which there are nearly 22,000).

To reconstruct the predicted curve corresponding to a given number of harmonics, the following equation is applied:

$$\hat{T}_i = \sum_k A_k \cos(2\pi ik/n) + B_k \sin(2\pi ik/n)$$

We have summarized but a few of the available methods for morphological description. Many more are used routinely, and new ones are being developed all the time, often in fields far removed from paleontology. Historically, some of the greatest advances in paleontology have been made by those who have successfully adapted techniques from other disciplines. For example, crystallographers describe minerals with a system that includes their axes of symmetry (Figure 2.14a). Crystalline calcite in the rhombohedral habit has one principal axis of threefold rotational symmetry (termed the *c* axis) and three symmetry axes perpendicular to this (termed the *a* axes). Despite their intricate microstructure, echinoderm plates are, crystallographically speaking, single crystals of calcite. The spatial orientations of their crystal axes can be measured with equipment such as an optical goniometer,

where $\hat{T}_i$ is the predicted value corresponding to the observed value $T_i$, and the sum is taken only over selected values of $k$. For example, the predicted value of the first reading, based only on the overall average of 11.9 degrees and the annual cycle with $A_5 = -10.8$ and $B_5 = -5.5$, is equal to:

$$\hat{T}_1 = 11.9 + (-10.8)\cos[2\pi(1)(\mathbf{5})/43{,}824] + (-5.5)\sin[2\pi(1)(\mathbf{5})/43{,}824]$$

which is equal to 1.1. Thus, the observed temperature reading of −4.4 is a few degrees lower than would be predicted simply on the basis of the annual cycle. If we also added the daily cycle with $A_{1826} = -2.0$ and $B_{1826} = -1.6$, then we would have:

$$\begin{aligned}\hat{T}_1 = 11.9 &+ (-10.8)\cos[2\pi(1)(\mathbf{5})/43{,}824] \\ &+ (-5.5)\sin[2\pi(1)(\mathbf{5})/43{,}824] \\ &+ (-2.0)\cos[2\pi(1)(\mathbf{1826})/43{,}824] \\ &+ (-1.6)\sin[2\pi(1)(\mathbf{1826})/43{,}824]\end{aligned}$$

which is equal to −1.2. It makes sense that including the daily cycle leads to a lower predicted temperature, as the time in question is shortly after midnight. Even with the daily cycle taken into account, however, the reading of −4.4 is lower than expected.

Figure 2.13c shows curves corresponding to the annual and daily cycles. The daily variation is of such high frequency that it appears as a blur. The amplitude of the annual harmonic is 12.1 degrees, implying a temperature range of 24.2 degrees. This is about 5 times that of the daily harmonic, with an amplitude of 2.6 degrees. (The difference in power between annual and daily cycles is so much greater than the amplitude difference because power is proportional to the square of the amplitude.)

Superimposed on Figure 2.13c is the sum of the annual cycle, the daily cycle, and the average temperature. This combination of three terms tracks the original data reasonably well. Finally, Figure 2.13d shows the actual data with the average and the annual and daily cycles subtracted out. This is the residual variation in temperature for the recorded interval of time that is not accounted for by the average, annual, and daily terms. Weather services attempt, with occasional success, to predict the short-term pattern of residual variation in temperature that is not accounted for by simple daily and annual cycles.

For the outline of Figure 2.11, the length along the curve is treated as the independent variable, and the *x*- and *y*-coordinates are treated as two dependent variables. Two separate harmonic analyses are carried out, one for *x* and one for *y*, which result in four sets of harmonic coefficients. It may not seem obvious that this outline has a regular periodicity like the temperature data of Figure 2.13a. It must be periodic, however, because the starting and ending points are the same. In biological shapes such as that of Figure 2.11, there is often a clear relationship between the harmonic coefficients and major features of shape. An outline with strong four- or fivefold symmetry, for example, will tend to have high power in the fourth or fifth harmonics.

Harmonic analysis has many important applications in the earth sciences. For example, analysis of late Cenozoic climatic records has provided crucial empirical support for theoretical arguments that global temperature and other aspects of climate vary predictably, on timescales of tens of thousands of years, in response to variations in the earth's orbit and rotational axis. Harmonic analysis has also been used to detect cycles in biological diversity and extinction over geologic time [SEE SECTION 8.6].

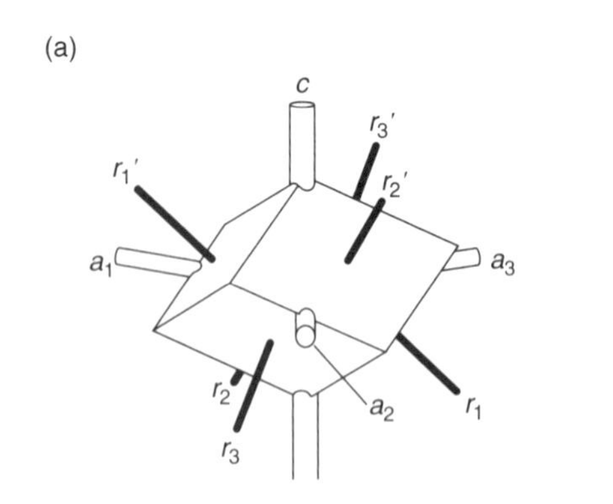


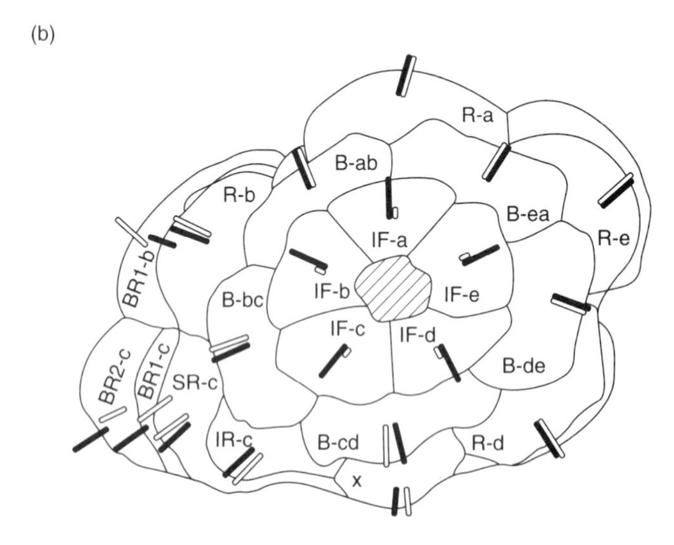


**FIGURE 2.14 Crystallographic axes of a calcite crystal and orientations of axes on a crinoid.** In part (a), the labels *c* and *a* indicate the *c* and *a* axes, and *r* indicates the direction that is normal to the crystal face. In part (b), the crinoid is in basal view, the lightly shaded rods indicate the direction of *c* axes, and the darker rods indicate the direction that is normal to the plate surface. Uppercase letters indicate different kinds of plates, and lowercase letters denote different positions of these plates (IF, infrabasal; B, basal; R, radial; IR, infraradial; SR, superradial; X, anal; BR, brachial). *(a: From Bodenbender, 1996; b: From Bodenbender & Ausich, 2000)*

and these axis orientations can be formally treated as shape variables.

Figure 2.14b shows typical crystallographic axis orientations in a **CRINOID** calyx. The light lines show the directions of *c*-axis orientation on a number of plates. For comparison, the dark lines indicate the directions of the vectors perpendicular to each plate surface. For many plates the two lines do not coincide; thus, the crystallographic axes provide additional information that is not present in the shape of the plate surface.

## 2.3 THE NATURE OF GROWTH AND DEVELOPMENT

Earlier we stated that one of the three major aspects of organic form is the relationship between size and shape. This relationship is expressed during the growth and development, or **ontogeny,** of individual organisms. Knowing how form changes with growth is often central to understanding biologic function. It is also necessary for a complete understanding of variation—for example, whether two dissimilar forms represent different species or just growth stages of the same species. Studying ontogeny is also essential to the study of evolution, because evolutionary change in adult form is effected through the evolution of ontogeny.

A special case of ontogeny is **astogeny,** the growth and development of colonial organisms. Animals such as **GRAPTOLITES, BRYOZOANS,** and corals typically consist of tens to thousands of genetically identical, **clonal,** units. They carry out life functions and in many ways resemble distinct organisms. In many cases, they can even be highly specialized to perform different functions—some dedicated to feeding, others to defense, and still others to reproduction. The soft "animal" of each unit is known as a *zooid* in graptolites and bryozoans, and as a *polyp* in corals. The corresponding skeletal unit is referred to as a *theca* in graptolites, a *zooecium* in bryozoans, and a *corallite* in corals.

The astogeny of a bryozoan colony is illustrated in Figure 2.15. The colony begins as a single zooid attached to a substrate such as a bit of shell or a rock. In this case, the zooids first proliferate horizontally to provide a base for the colony. Vertical growth then ensues, after which some zooids become thickly calcified and specialized for support, while others are dedicated to feeding and reproduction. Of course, each unit undergoes its own ontogeny in addition to being part of the astogeny of the larger colony. Because of this, and because both the single units and the colony must maintain function as they grow, colonial organisms are somewhat more complicated to study than are solitary organisms. Nevertheless, the tools used to study astogeny are the same as those used to study ontogeny.

### Types of Growth

Organic growth is extremely complicated; it usually involves several types of change, among which are changes in cell size, number of cells, number of cell types, and, especially in animals, relative positions of cells. During life,

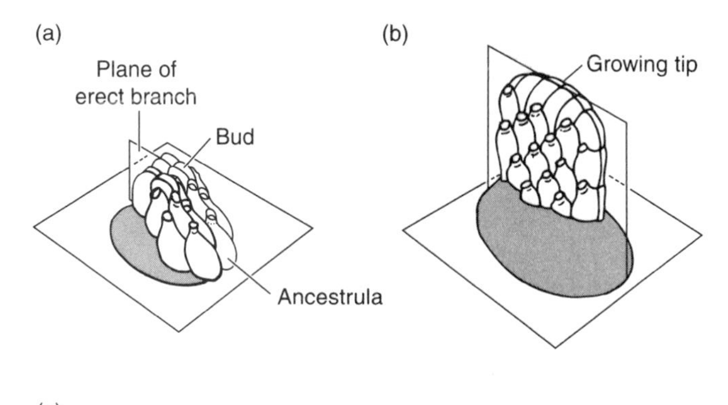


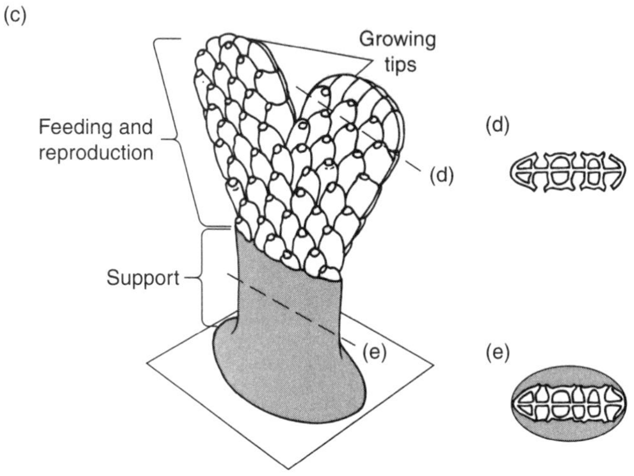


**FIGURE 2.15 Idealized astogeny in a living bryozoan colony.** Parts (a) through (c) show three developmental stages. The ancestrula is the founding member of the colony. A bud is a newly generated zooid. Parts (d) and (e) show cross sections through two locations at stage (c). *(From Cheetham, 1986a)*

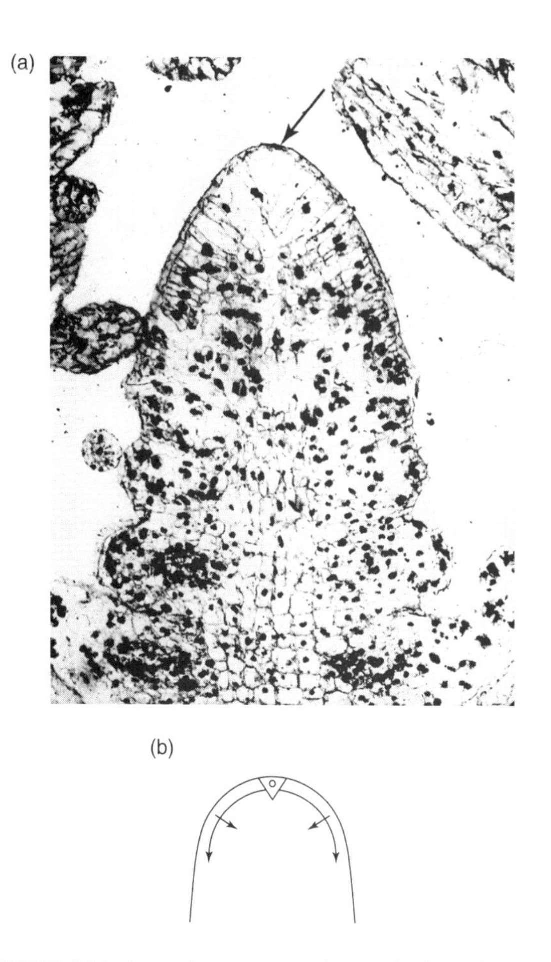


**FIGURE 2.16 Accretionary growth at apical meristems.** (a) Thin section of apical meristem of a Carboniferous sphenopsid, showing a single apical cell (arrow) from which the surrounding cells divided. Magnification is ×150. (b) Schematic diagram of accretionary growth at an apical meristem. Arrows show the directions of growth. This example has a single apical cell, but many other cellular configurations are known. *(a: From Good & Taylor, 1993; b: From Fahn, 1982)*

an organism may change in form abruptly or it may change gradually. Growth, especially of hard skeletons, takes place in a few major ways and in combinations of these.

***Accretion of Existing Parts*** Organisms in many biologic groups grow by adding new material at growth zones throughout ontogeny. For example, proliferation of new cells in vascular plants takes place at special regions called **meristems.** Apical meristems are at the growing tips and are responsible for increase in length. Figure 2.16a shows a magnified image of an apical meristem in a Carboniferous **SPHENOPSID.** A single apical cell is present, indicated by the arrow. As cells divide from the apical cell, they are pushed away from the tip, with the result that older cells are farther away. This process is illustrated schematically in Figure 2.16b, which shows the apical cell at the top; directions of accretionary growth are indicated by the arrows. In general, cells outside the meristems do not divide once they are produced, although they typically change in form to differentiate for specialized functions.

Figure 2.17 shows a cross section through the stem of a seed plant, with some tissue types indicated. Five major tissues are produced at the apical meristems: pith, xylem, phloem, cortex, and epidermis. These are referred to as primary tissues. Secondary tissues have evolved numerous times in plants. For example, in seed plants there is a layer of vascular cambium, a lateral meristem that produces secondary xylem and secondary phloem. Accretionary growth at the vascular cambium takes place in two opposite directions, resulting in a thickening of the stem. In many different kinds of plants, the epidermis is eventually shed, and the cortex and primary phloem are crushed as they are pushed

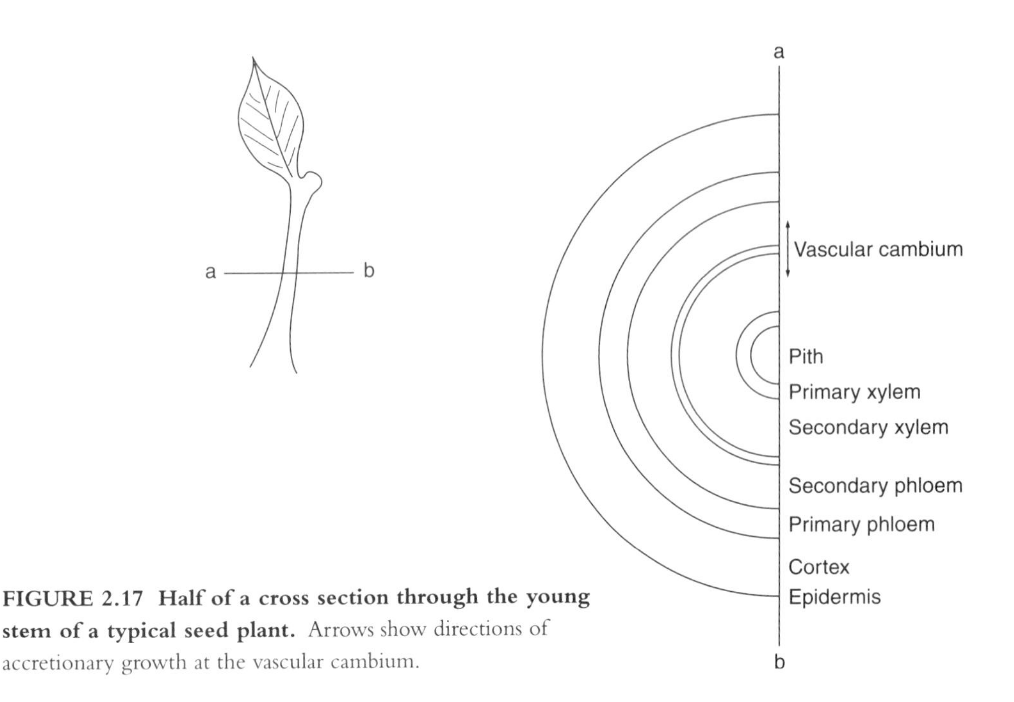


**FIGURE 2.17 Half of a cross section through the young stem of a typical seed plant.** Arrows show directions of accretionary growth at the vascular cambium.

outward, forming the dead outer bark. In these plants there is a secondary cortex as well.

Growth by accretion is especially common among animals whose shell is external and serves primarily for protection and muscle attachment—for example, brachiopods, gastropods, cephalopods, and bivalved molluscs. Figure 2.18 illustrates growth in a bivalve. The cross section of the shell is marked by growth lines that can be observed in the shell as well as on the outer surface. New skeletal material is added not only at the leading edge of

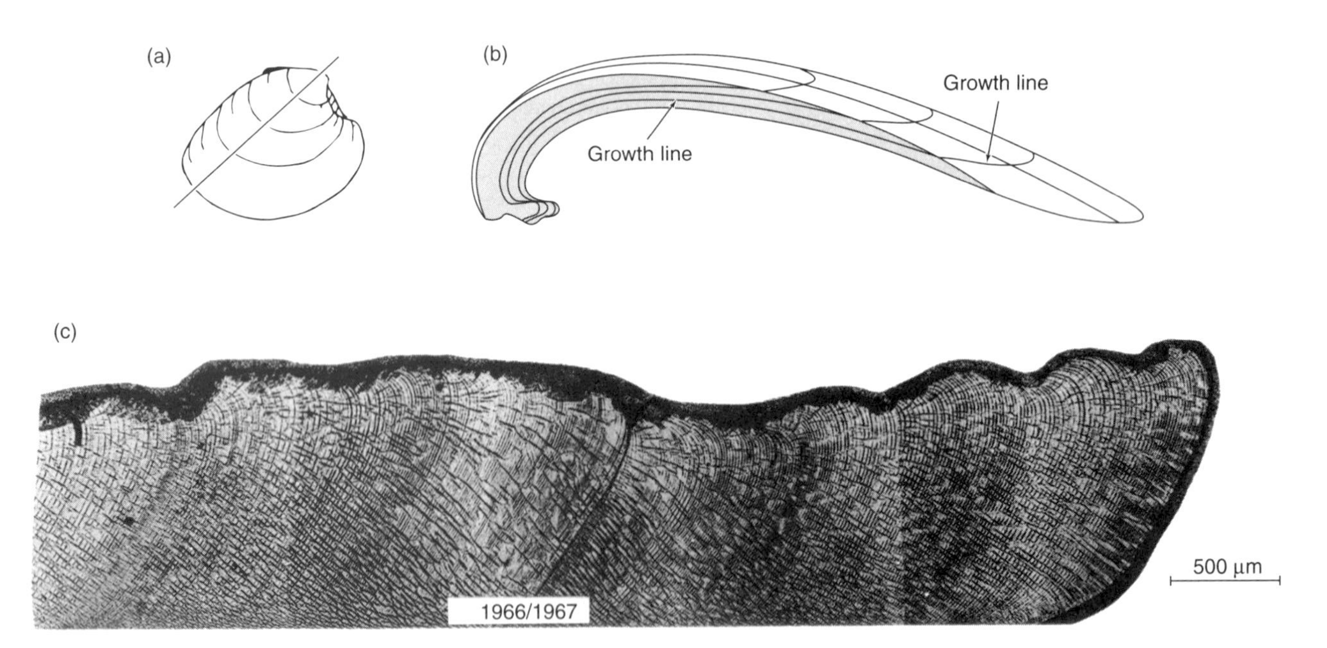


**FIGURE 2.18 Accretionary growth in bivalved molluscs.** (a) The exterior of one shell of the species *Mercenaria mercenaria*, with three prominent, concentric growth lines. The straight line gives the position of a section through the shell, shown in part (b). The photograph in part (c) shows approximately one year's growth in a specimen of this species. The leading edge of the shell is to the right. The dark band near the middle is an annual growth line, and the finely spaced bands that run roughly parallel to this are daily growth increments. *(From Pannella & MacClintock, 1968)*

the shell, but also as a covering over most of the interior surface of the shell. Figure 2.18 includes an enlarged photograph of a series of growth lines. In this instance, the shell was grown under controlled conditions, and it is possible to establish that the growth lines are in fact daily. The dark annual growth line, corresponding to the slowing of growth in the winter, can also be clearly seen. The ability to correlate growth lines with astronomical cycles is useful in many areas of research (see Figure 2.34 and Sections 9.5 and 10.5).

***Addition of New Parts*** A common method of skeletal growth for those organisms whose skeletons consist of many parts, either tightly articulated or fitting loosely in the soft tissue, is addition of new skeletal parts.

Figure 2.19 schematically shows some aspects of the growth of crinoid stems by addition of new columnals. In this type of growth, large columnals are added periodically at the base of the calyx (labeled A and B in Figure 2.19). They may or may not grow appreciably by accretion after their formation. As each new columnal is added at the calyx, the previous ones are displaced down the stem. At some distance away from the calyx, smaller columnals appear between the preexisting ones. They grow by accretion and, as they do, still more small columnals are intercalated. The result of this process (labeled E in Figure 2.19) is an orderly set of generations of intercalated columnals—with the generations being distinguishable by size.

Addition of new parts is integral to the growth of many other organisms. Figure 2.20 shows several stages in the ontogeny of a trilobite. One of the more obvious ontogenetic changes is the gradual addition of segments.

***Molting*** Like other arthropods, the trilobite shown in Figure 2.20 used another basic mechanism of growth, the periodic shedding or molting of the entire skeleton and formation of a new one to accommodate the increase in size of the soft parts. Figure 2.21 shows a plot of head width against length in an assemblage of the Middle Ordovician trilobite *Trinodus elspethi*. The measurements fall into clusters, each representing a molt stage, between which size increases in steps. Differences between points in a cluster represent minor differences in size and shape among the individuals of that stage.

***Modification*** In many groups, skeletal material is continually resorbed and redeposited as the skeleton grows.

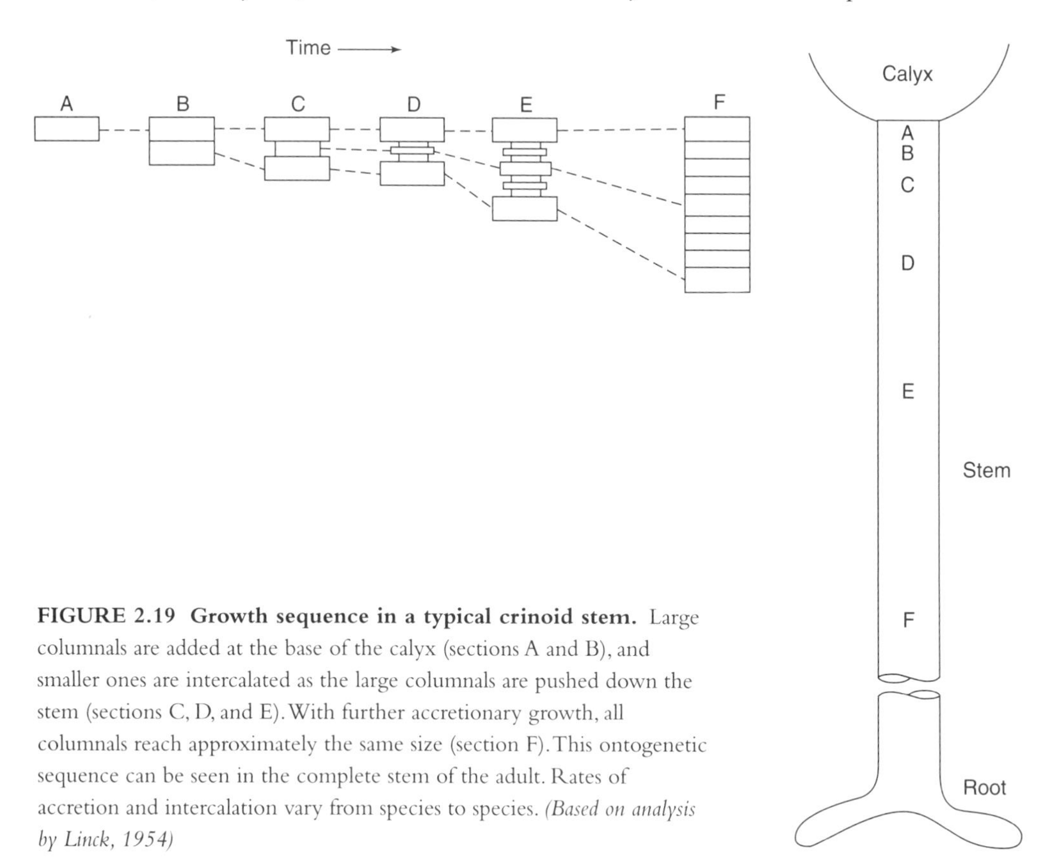


**FIGURE 2.19 Growth sequence in a typical crinoid stem.** Large columnals are added at the base of the calyx (sections A and B), and smaller ones are intercalated as the large columnals are pushed down the stem (sections C, D, and E). With further accretionary growth, all columnals reach approximately the same size (section F). This ontogenetic sequence can be seen in the complete stem of the adult. Rates of accretion and intercalation vary from species to species. *(Based on analysis by Linck, 1954)*

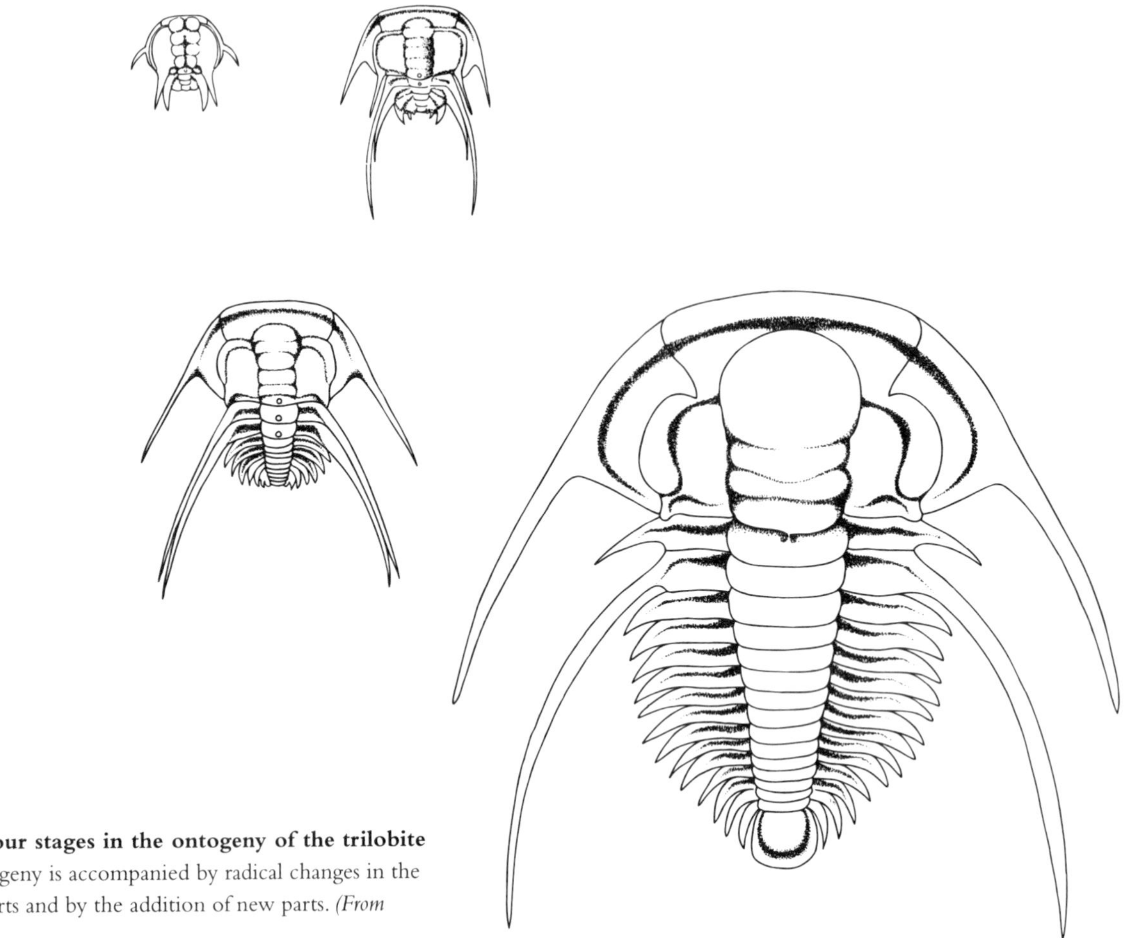

FIGURE 2.20 **Four stages in the ontogeny of the trilobite *Paradoxides*.** Ontogeny is accompanied by radical changes in the form of existing parts and by the addition of new parts. *(From Whittington, 1957)*

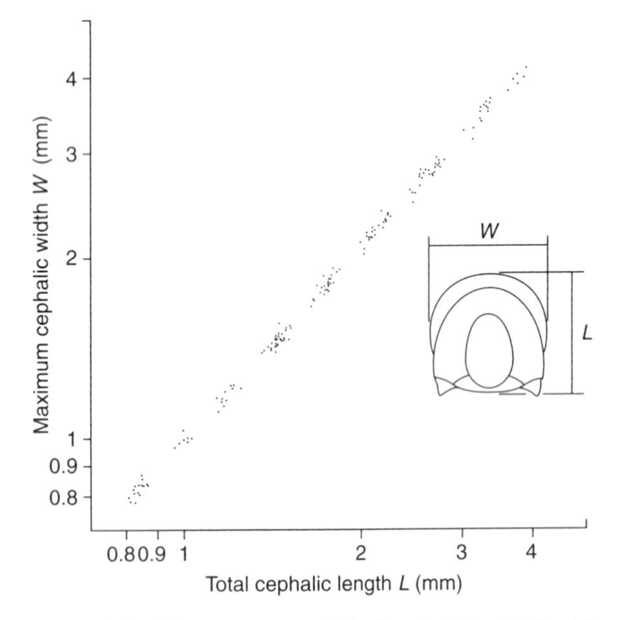


**FIGURE 2.21 The ontogeny of the head of the Ordovician trilobite *Trinodus* as measured by the development of length and width dimensions.** Clusters reflect molt stages, with the major increase in size taking place from one stage to the next. *(From Hunt, 1967)*

The result is that the entire form of the skeleton changes, not just a well-defined growing margin. Growth by modification is common in the bones of vertebrates (Figure 2.22), but it also occurs in organisms otherwise typified by accretionary growth, such as certain molluscs and echinoderms.

The modes of growth just outlined are idealized end-members. Many organisms in fact grow by a combination of these modes. The crinoid stem (Figure 2.19) combines addition of new parts and accretion of existing parts, and the trilobite (Figure 2.20) combines molting and addition of parts.

## Describing Ontogenetic Change

In living organisms we can observe growth as it occurs. Therefore, ontogenetic changes can be studied in detail, and the timing of changes can be determined. With fossil material, as with specimens of living forms in which growth has already taken place, ontogenetic change and its

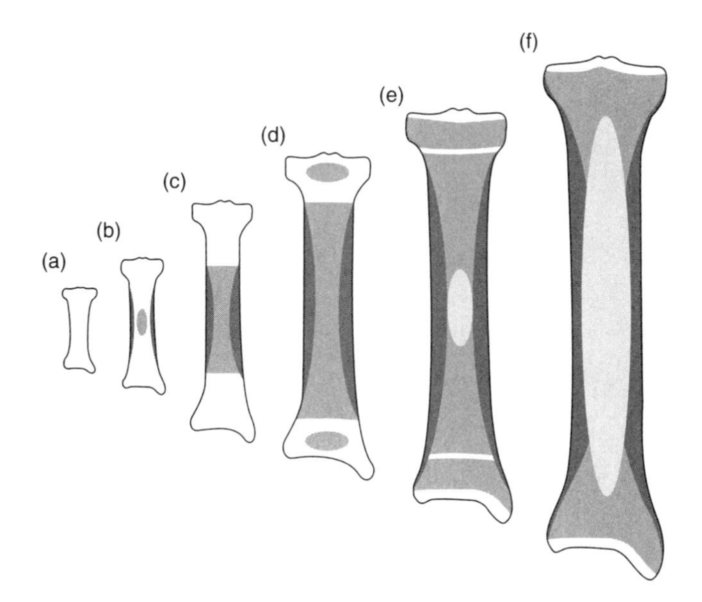


**FIGURE 2.22 Growth of a mammalian limb bone by modification.** (a) Initial deposition of nonmineralized cartilage (white). (b) Early replacement of cartilage by spongy bone (medium gray). (c, d) Further deposition of spongy bone and replacement of some spongy bone by compact bone (dark gray). (e) Formation of marrow (light gray) and continued modification of bone. (f) Expansion of marrow and continued bone modification. At this point, only the ends of the bone are cartilage. *(From Romer, 1970)*

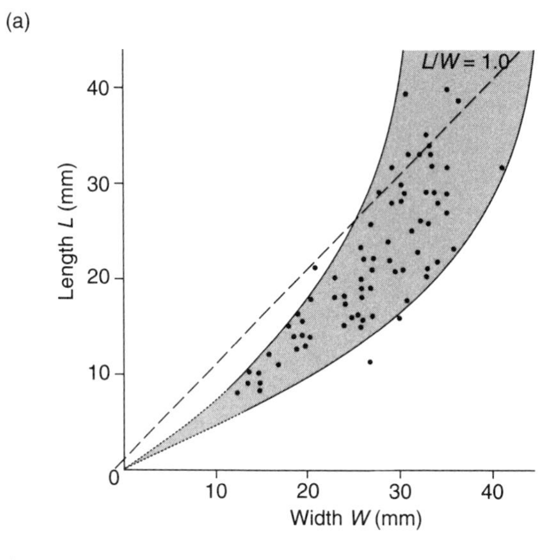


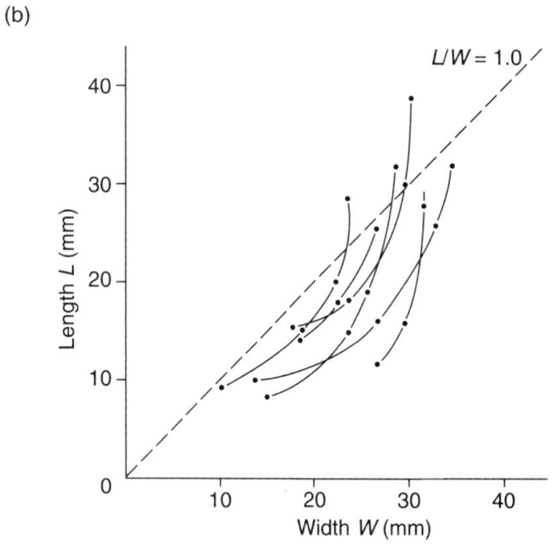


**FIGURE 2.23 Ontogenetic change in shape in brachiopods.** Plots are of measurements made on an assemblage of the Carboniferous brachiopod *Ectochoristites*. Length refers to the anteroposterior (front-to-back) length of the pedicle valve; width is measured along the hinge of the pedicle valve. Points connected by curved lines in the lower graph are based on growth-line measurements of single specimens. *(From Campbell, 1957)*

timing cannot be directly observed. There are two general approaches to studying ontogeny in fossils, **cross-sectional analysis** and **longitudinal analysis.**

In cross-sectional studies, a series of specimens at different sizes or growth stages is compared to study change in shape with increasing size. Because the growth of individual organisms is not followed, this is an approximate procedure. It is generally the only one available with fossils, however.

In longitudinal analysis, the ontogeny of a single organism is followed. With fossils, of course, this is possible only for organisms that keep a record of their growth. For example, the bivalve of Figure 2.18 is an adult, but its growth lines show the form of the margin at various times throughout its life. Its ontogeny is recorded in its adult shell. In general, longitudinal analysis of fossils can be applied only to organisms that grow by simple accretion at a margin and that do not radically modify the skeleton once it is laid down. For reconstructing ontogeny, longitudinal analysis is more direct than cross-sectional analysis and is therefore generally preferable if it can be carried out.

It is worth asking how well a cross-sectional analysis can approximate ontogenetic change. Figure 2.23 depicts the length and width of a number of shells of a brachiopod species. Figure 2.23a shows a cross-sectional analysis; each point is a distinct individual of a different size collected from the same assemblage. The scatter of points suggests a curved growth trajectory. Smaller shells are below the 45° line that marks equal length and width; they are wider than long. Larger shells are above this line; they are longer than wide. But the curved scatter of points does not necessarily imply that the growth trajectories of individual shells were also curved. In a cross-sectional study such as this, we assume that individual growth roughly follows that of the average curve. In this case, we can test the assumption directly, because the brachiopods grew by accretion. Each growth curve shown in Figure 2.23b was reconstructed by measuring a single adult shell at a series of points corresponding to growth lines. In this case, the longitudinal analysis confirms the suggestion of the cross-sectional study—that the individual growth pattern in this species was curved.

A different approach to describing ontogenies of entire organisms or parts of them is to use **coordinate transformation,** a method that also has been adapted to describe differences between individuals in a population and even between related species. This approach was first suggested by D'Arcy Wentworth Thompson in his classic work *On Growth and Form*. Since that time, several mathematical descriptions of coordinate transformation have been developed.

Perhaps the most promising of these is the method of the **thin-plate spline.** (A spline is simply a mathematical function that interpolates smoothly between points; it approximates the position of intermediate values between observed points. Physical splines, made of flexible wood, are used in graphic design to draw curves.) The two forms to be compared must have a set of corresponding landmarks whose $x$- and $y$-coordinates are recorded, as described earlier in this chapter. The deformation of one form to the other—the changes in the $x$- and $y$-directions that must be added to each of the landmarks of the first form in order to produce the second form—can be described as a combination of a few well-characterized mathematical functions. As with the sine and cosine functions used in harmonic analysis, the functions of the thin-plate spline are quite general; they take on different coefficients depending on the two forms being compared. The deformations are defined not only at the landmarks; as the name of the spline method suggests, they are also interpolated between points.

Figure 2.24 illustrates the thin-plate spline approach applied to growth in the Lower Cambrian trilobite *Olenellus fowleri*. Figure 2.24a is a photograph of a juvenile specimen with the positions of a number of landmarks indicated. The landmarks on this same specimen are shown on a grid in Figure 2.24b, and the positions of the corresponding points on an adult are shown in Figure 2.24c. Comparing the juvenile and adult landmarks, there are two prominent features of shape change through ontogeny. First is an inflation of the glabella, the central region of the head marked by points 3–16. Second is a lateral movement of the genal spines, marked by points 17, 19, and 21 on the left and by 18, 20, and 22 on the right. Figure 2.24d depicts the grid deformation involved in the shift of landmarks between juvenile and adult. The dilation of the grid near the center corresponds with the glabellar inflation, and the stretching of the posterior margin of the grid corresponds with the movement of the genal spines.

## Growth Rates

Some plants and animals grow very rapidly; others grow very slowly. About 20 years are generally required, for example, to complete human growth. Other organisms, such as insects, may go through a complete ontogeny in a matter of days or weeks. In nearly all organisms, the rate of growth varies with time; that is, it changes during ontogeny. One of the principal differences between organisms lies in whether growth eventually ceases. In organisms with **determinate growth,** such as most **MAMMALS,** a mature stage is reached in which structural growth stops even though the organism continues to live. (Humans show determinate growth, even though it is common to put on weight later in life.) In many plants and animals, however, growth is **indeterminate,** continuing throughout the life of the organism but at a greatly reduced rate, making it difficult to define a true adult stage. At best we can define a stage at which most of the growth will have taken place, but growth will not have ceased.

Different parts of an organism typically grow at different rates. We have already seen an example of this in brachiopods (Figure 2.23), in which length and width grew at different rates.

Nearly all paleontologic studies of ontogeny involve the measurement of change in one morphologic attribute in relation to change in another. We can define two basically different types of growth: **isometric** growth and **anisometric** growth. If the ratio between the sizes of two parts of an organism does not change during ontogeny, we have isometric

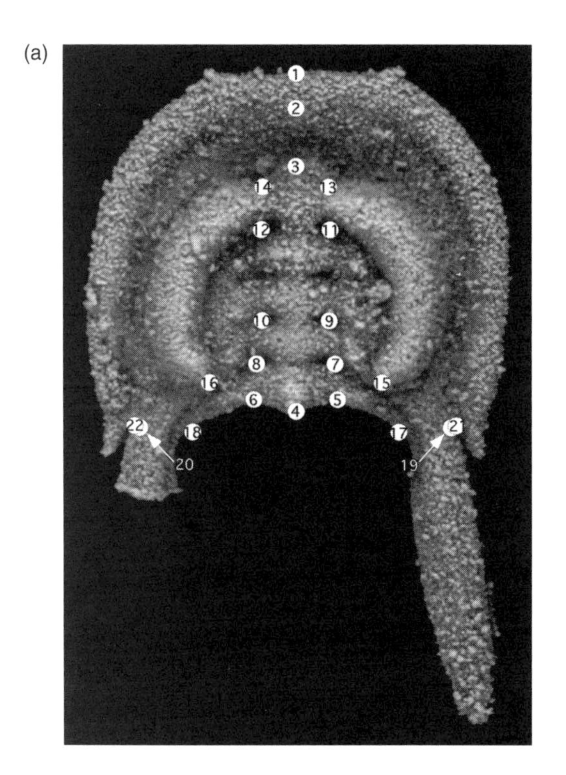


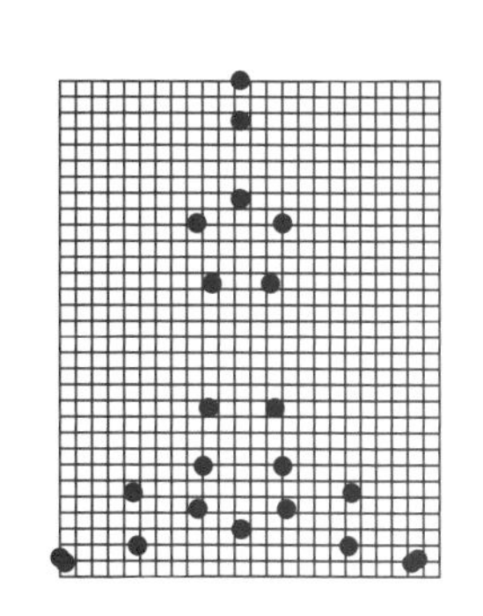


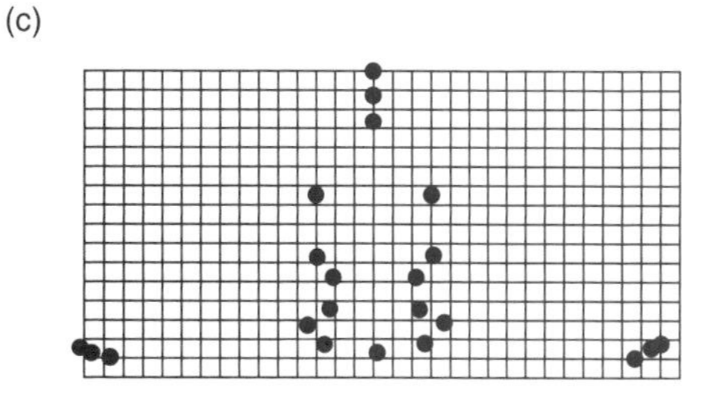


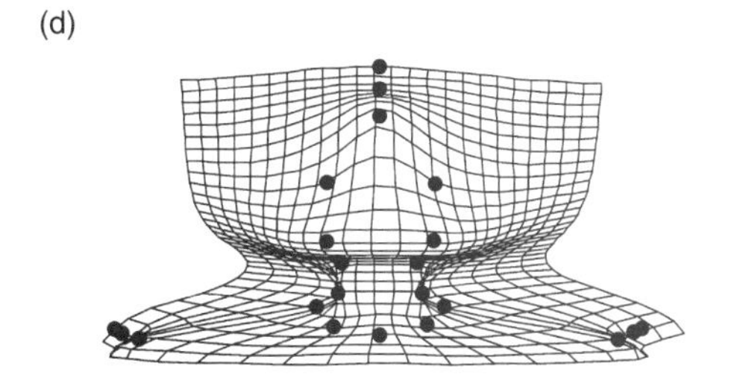


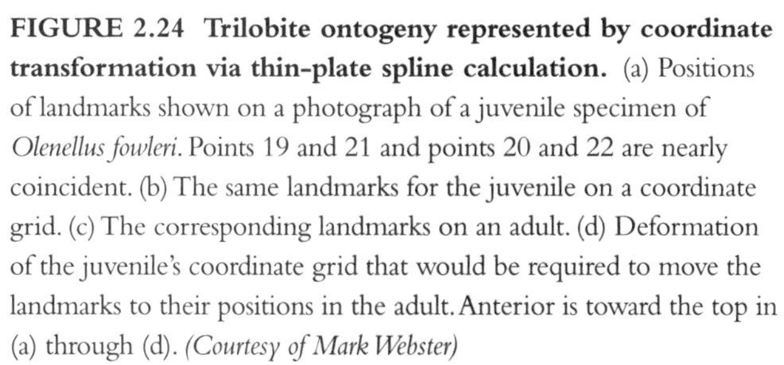

**FIGURE 2.24 Trilobite ontogeny represented by coordinate transformation via thin-plate spline calculation.** (a) Positions of landmarks shown on a photograph of a juvenile specimen of *Olenellus fowleri*. Points 19 and 21 and points 20 and 22 are nearly coincident. (b) The same landmarks for the juvenile on a coordinate grid. (c) The corresponding landmarks on an adult. (d) Deformation of the juvenile's coordinate grid that would be required to move the landmarks to their positions in the adult. Anterior is toward the top in (a) through (d). *(Courtesy of Mark Webster)*

growth; shape does not change as size increases (Figures 2.25a and 2.25b). If the ratio does change, we have anisometric growth; shape changes as size increases (Figures 2.25c and 2.25d). Anisometric growth is far more common than isometric growth. In other words, shape change during ontogeny is the rule rather than the exception.

Two kinds of isometric growth are shown in Figure 2.25. In the pattern of isometric growth plotted on graph (a), both parts grow at precisely the same rate. This yields a straight line at 45° to either axis. In the pattern of isometric growth plotted on graph (b), one part grows more rapidly than the other, but the ratio between them is constant. This results in a straight line that is separated from one axis by a smaller angle than from the other. In the pattern of anisometric growth plotted on graph (c), the *X* part grows more rapidly than the *Y* part at first, but this relationship is subsequently reversed. The plot of growth is a curve. The pattern of growth plotted on graph (d) is different from the other three patterns in that the plotted line would not pass through the origin if extended. In pattern (d) as in pattern (c), shape changes with growth. To summarize, growth is isometric if the plotted line is straight and passes through the origin (or would do so if extended). All other conditions produce anisometric growth.

When growth rates are expressed in absolute terms such as centimeters per year, larger dimensions generally increase more rapidly than do smaller dimensions. For example, the length of a femur may increase by 1 cm during the same span of time in which the width increases by only a few millimeters. It is therefore useful to consider **size-specific growth rates** (also known as

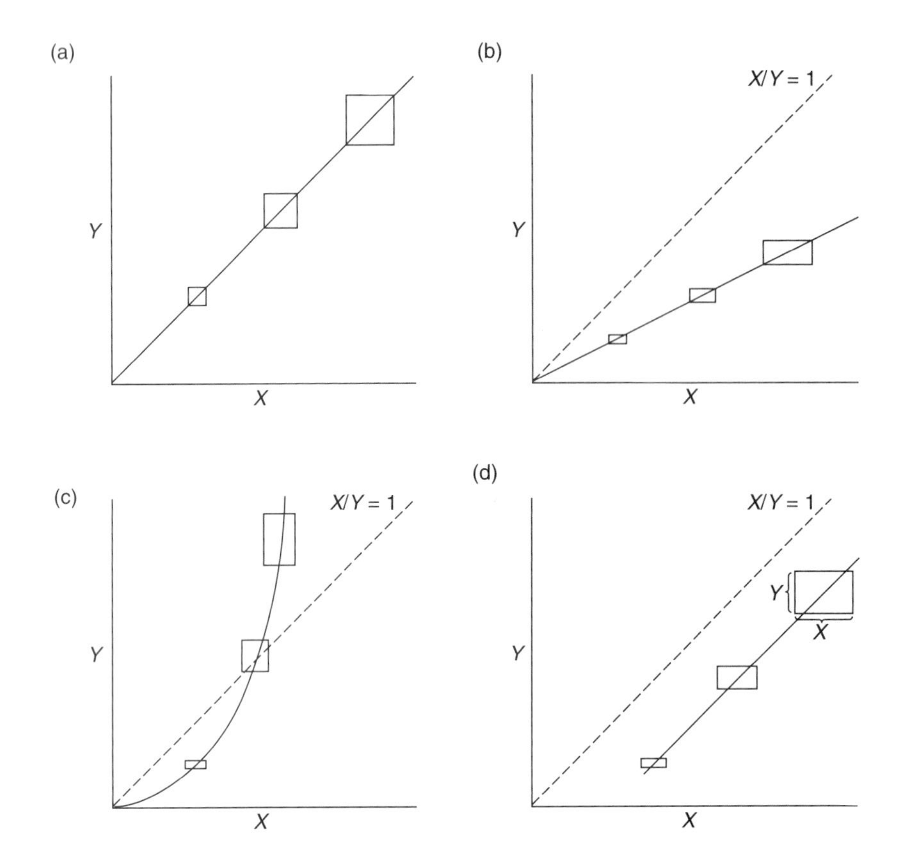


**FIGURE 2.25 Typical patterns of growth observed when two morphologic dimensions are plotted against each other.** In each graph, the shape of a hypothetical organism is shown by a square or rectangle at three ontogenetic stages. In parts (a) and (b), growth is isometric; there is no change in shape. In parts (c) and (d), growth is anisometric; shape is continually changing.

**relative growth rates**). A femur that grows in length from 10 to 11 cm in a year would have a size-specific growth rate of 1 cm per 10 cm per year, or 10 percent per year, in its length.

It is generally difficult to obtain measures of absolute age for growth stages of fossils. If we simply want to know whether two parts grew at the same size-specific rate, however, the time component cancels out when we take the ratio of relative rates. If femur length increases from 10 to 11 cm while femur width increases from 1 to 1.2 cm, the ratio of the two size-specific growth rates is 10 to 20 percent, or 1:2, even if we do not know how much time has elapsed between the smaller and larger size. Thus, it is possible to determine whether growth is isometric or anisometric, even if we cannot measure growth rates with respect to actual time. This, of course, is exactly what we saw in Figure 2.25.

## Reasons for Anisometric Growth

The two principal reasons why it may be advantageous for organisms to change shape as they grow are exactly opposite each other: either to change function or to maintain function.

Consider the metamorphosis of a frog, which develops into an adult through a succession of fairly gradual changes. The tadpole lives in and depends upon an aquatic environment: It extracts oxygen directly from the water. The adult frog, although partially dependent upon the proximity of water, is essentially a terrestrial organism. Some ontogenetic changes in frog anatomy are produced by the addition of new structures and the deletion of old; others are brought about by changes in relative rates of growth. The striking differences between a tadpole and an adult frog largely reflect differences in function.

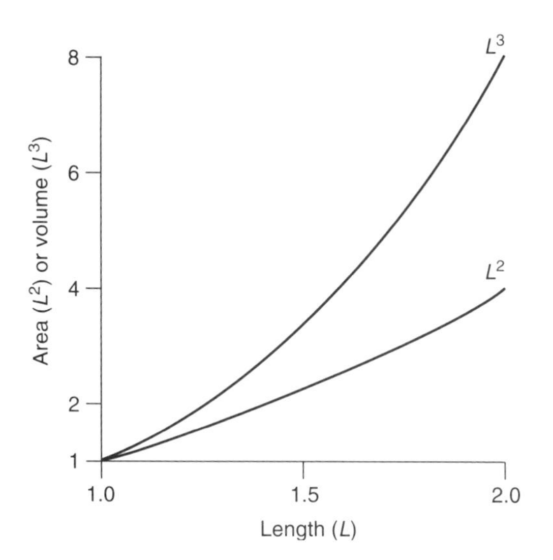


**FIGURE 2.26 Effect of isometric growth in linear dimensions ($L$) on area ($L^2$) and volume ($L^3$).** If growth is isometric, volume increases much more rapidly than does area.

Many organisms do not change function as radically as the frog does, yet they still change in shape substantially as they grow. The major reason is that, as size increases, it would not be possible to maintain function without a change in shape.

Consider a leg bone of a terrestrial vertebrate. The bone serves several functions, but one of the most important is to support the body. The strength of a bone (or any other supporting structure, for that matter) is approximately proportional to its cross-sectional area. Thus, a stout bone will be stronger than a slender bone regardless of its length. Now imagine that the entire organism grew isometrically (Figure 2.26), so that every linear dimension increased by the same proportional amount. For example, suppose every linear dimension doubled. In this case, the volume or mass ($L^3$) would increase eightfold ($2 \times 2 \times 2 = 8$), but the cross-sectional area ($L^2$) would increase only fourfold ($2 \times 2 = 4$). Thus, if growth were isometric, body mass would increase more rapidly than the cross-sectional area of the bone, and the bone would not be able to support the animal without breaking.

D'Arcy Thompson named this rule of scaling inequality the **principle of similitude.** To overcome such scaling problems, it is necessary for the shape of the bone to change with size. That is, the bone must become relatively stouter to carry the increased weight of the body. This is in fact what occurs in the ontogenetic development of many terrestrial vertebrates.

The principles of scaling that dictate anisometric change through growth in a single species also apply to size-related differences among species. Figure 2.27 shows the femurs of three species of pelycosaurian **REPTILES.** These are drawn at different scales; the one on the left is the longest and the one on the right is the shortest. Clearly, the longer the femur, the stouter it is. This is exactly what we would expect, given scaling of body mass and cross-sectional area.

Similar scaling principles have been deduced for many other structures as well. For example, the amount of gas that can be exchanged by a respiratory structure depends on the structure's surface area. The amount of gas exchange required by the organism depends roughly on the organism's mass or volume. Isometric growth would lead to an ever-decreasing surface:volume ratio, which would be functionally inadequate. Thus, respiratory structures in many organisms become more convoluted during growth so that respiratory surface area can keep up with body volume.

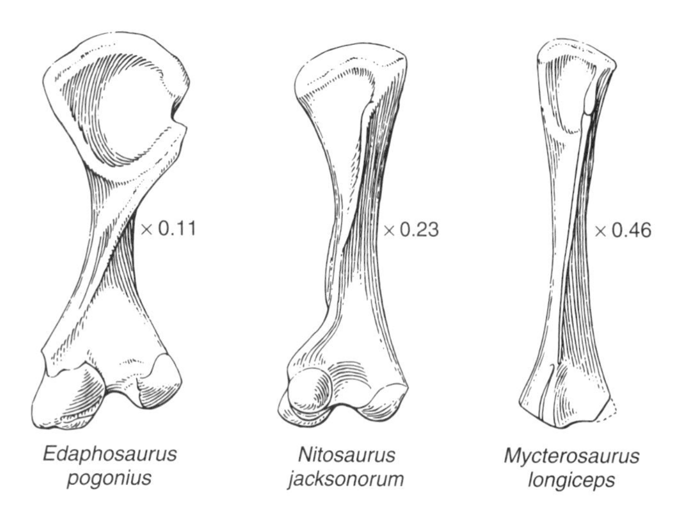


**FIGURE 2.27 Femurs of three pelycosaurian reptiles, illustrating variation in proportions with increasing size.** Actual size increases from right to left. Thus, longer bones are also stouter. *(From Gould, 1967)*

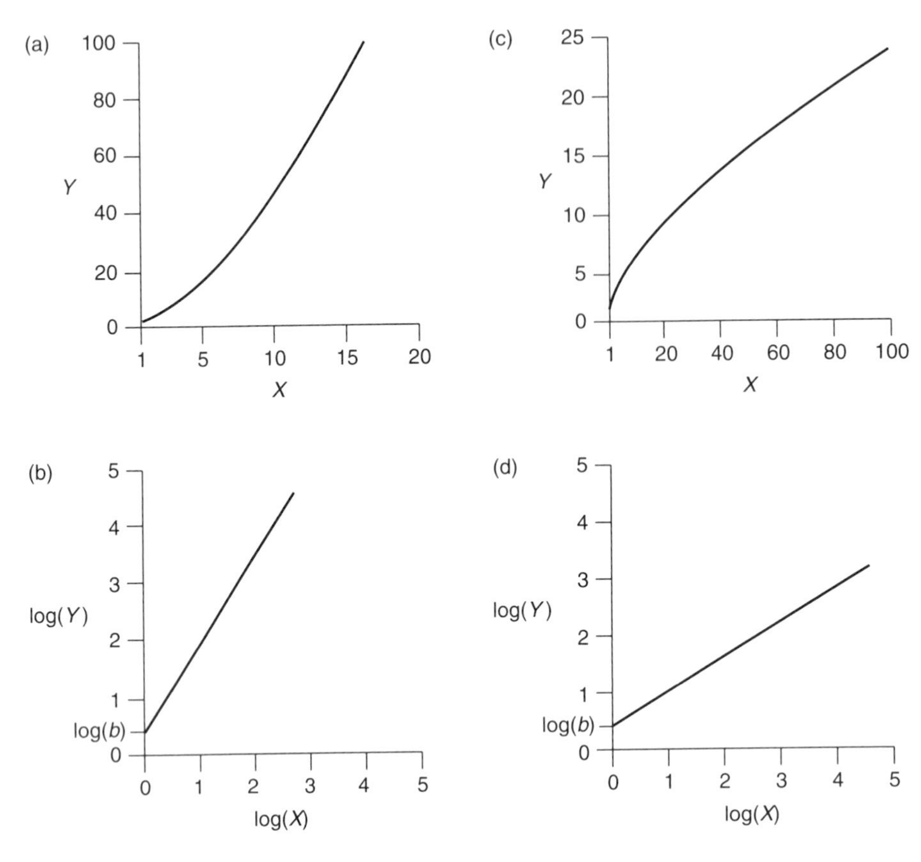


**FIGURE 2.28 Patterns of allometric growth in two morphologic dimensions.** Left-hand panels show positive allometry, in which $Y$ grows at a higher relative rate than $X$. Right-hand panels show negative allometry.

## Allometric Growth

The principle of similitude leads us to expect the prevalence of a special case of anisometric growth known as **allometry.** If $X$ and $Y$ are two size measures, then the general allometric equation is given by:

$$Y = bX^a$$

where $a$ and $b$ are constants. This equation is closely related to the size-specific rates we discussed earlier. If the growth of $X$ and $Y$ follows a simple form of scaling that we would expect under the principle of similitude, such as an area-to-volume relationship, then the ratio of size-specific growth rates should be constant during ontogeny; and growth will follow this equation. In fact, growth of this form is quite common. The constants $a$ and $b$ need to be estimated with measurement data, as shown in Figure 2.29. This is easily done with line-fitting methods [SEE SECTION 3.2]; thus, it is convenient to use an equivalent, linearized form of the allometric equation by taking the logarithm of both sides:

$$\log(Y) = a\log(X) + \log(b)$$

Although some workers use the terms *anisometry* and *allometry* interchangeably, for many purposes it is useful to recognize a distinct concept of allometry as defined here because of its simple, special, and broadly applicable formulation. Note that Figure 2.25d illustrated an example of anisometry that is not strictly allometric.

Figure 2.28 illustrates two cases of allometric growth of a trait $Y$ relative to a trait $X$. In part (a), the traits are plotted on an arithmetic scale. Clearly, $Y$ increases at a higher relative growth rate than $X$; $Y$ is said to exhibit **positive allometry** relative to $X$. The growth relationship is curved, with the curve becoming ever steeper as size increases. Part (b) shows the same growth relationship, this time with both variables plotted on a logarithmic scale. The relationship is now linear, and the slope of growth curve $a$ is greater than 1.

Parts (c) and (d) show an example of **negative allometry,** in which $Y$ increases at a lower relative growth rate than $X$. Plotted arithmetically, the growth curve becomes ever shallower as size increases. Plotted logarithmically, the growth curve is linear with a slope less than 1. Note that in both logarithmic plots, $\log(Y)$ takes on the value of $\log(b)$ when $\log(X) = 0$, that is, when $X = 1$.

We have focused so far on describing patterns of growth. An understanding of the principles underlying allometric growth can be used to test specific hypotheses about scaling as it relates to organismal function. Box 2.3 illustrates this approach with a case study.

## Other Allometric Relationships

The principles of allometry that we have applied to ontogeny often help us to interpret size–shape relationships at other scales—for example, the comparison among numerous species.

*Box 2.3*

### TESTING AN ALLOMETRIC HYPOTHESIS WITH AN ORDOVICIAN ECHINODERM

In the example that follows, measurements were taken to determine whether growth agrees with the expectations of a simple area:volume relationship. An unexpected result in this case has contributed to a better understanding of function.

Figure 2.29 shows a specimen of the **RHOMBIFERAN** echinoderm *Pleurocystites* from the Ordovician of the north-central United States. Several specialized structures known as *pectinirhombs* are circled in the photograph. These structures, which are common in rhombiferans, have a folded surface that suggests a respiratory function, and this interpretation seems quite reasonable. The respiratory function makes a clear prediction about allometric scaling. Assume that the amount of oxygen supplied to the body, or theca, is proportional to the total area of the pectinirhombs on an individual, while the amount of oxygen needed is proportional to the volume of the theca. Then we would expect the size-specific growth rate of rhomb length ($L_R$) to be greater than that of thecal length ($L_T$). Specifically, rhomb area increases only as $(L_R)^2$, while thecal volume increases as $(L_T)^3$. For rhomb area to increase in growth as rapidly as thecal volume does, $L_R$ would have to increase in proportion to $L_T$ raised to exponent 3/2, or 1.5. Because $[(L_R)^{3/2}]^2 = (L_R)^3$, such a scaling would allow thecal area to keep up with thecal volume. Thus, we would expect a logarithmic plot of rhomb length against thecal length to have a slope of 1.5. This is the same as saying a logarithmic plot of rhomb area against thecal volume should have a slope of 1.0.

The comparison between respiratory area and thecal volume is depicted in Figure 2.29b, showing a cross-sectional analysis in which each point is a measured individual of a different size. Rhomb area increases even more rapidly than is required by the simple scaling relationship; the double log plot of rhomb area against thecal volume has a slope of 1.3 rather than the expected 1.0. This suggests that some unknown factor limited the efficiency of respiratory pectinirhombs in larger individuals and that rhombs increased in size to compensate. A likely explanation is that oxygen was depleted from the sea water as it moved through the pectinirhombs, shown schematically in Figure 2.29c. By the time a parcel of water, initially rich in oxygen, passed through the entire length of a rhomb, much of its oxygen would have been removed. Thus, the ability of the downstream end of the rhomb to extract oxygen would have been substantially less than that of the upstream end. This effect would have been less important in smaller individuals with smaller rhombs.

*continued on next page*

Box 2.3 (continued)

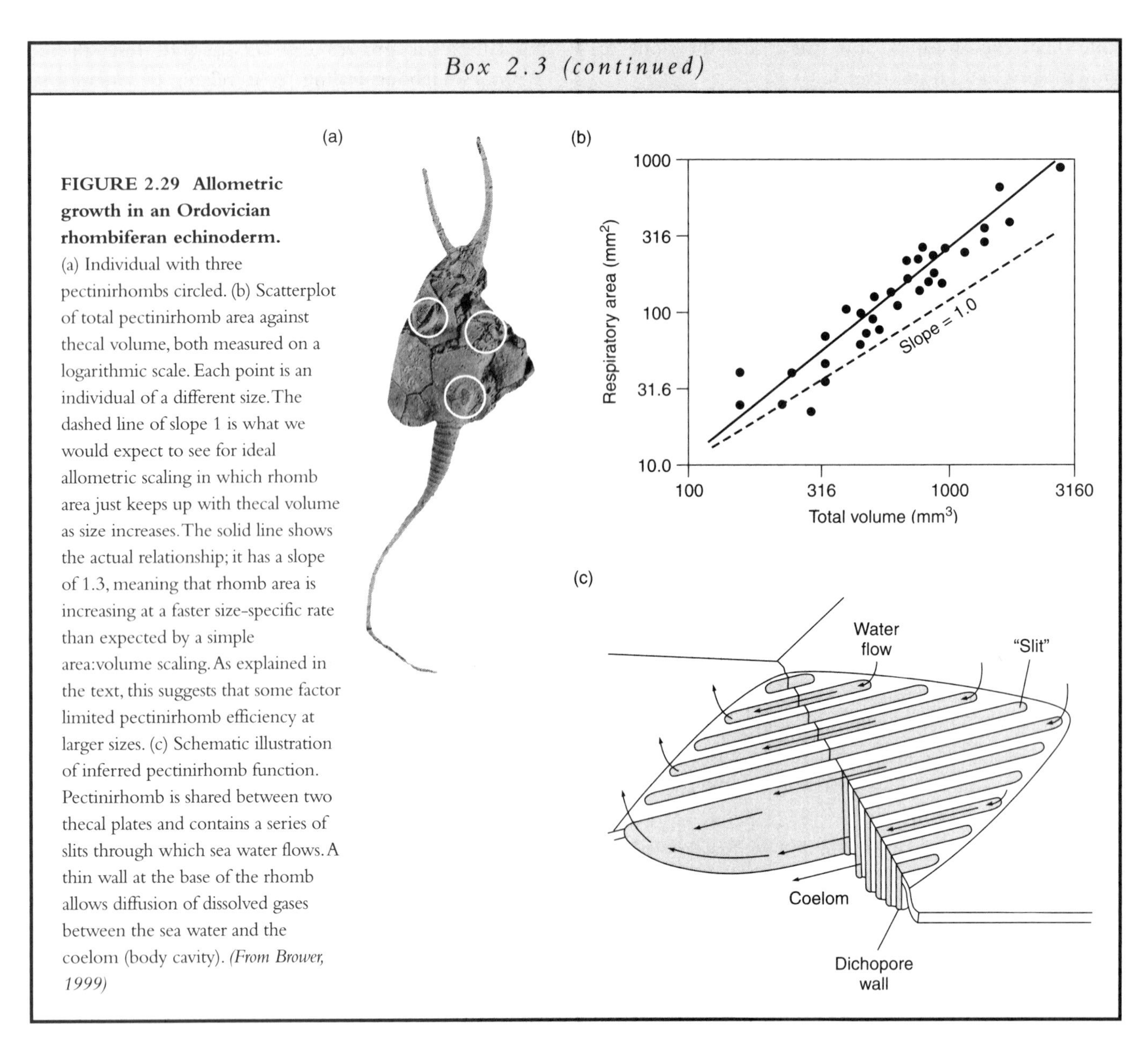


**FIGURE 2.29 Allometric growth in an Ordovician rhombiferan echinoderm.** (a) Individual with three pectinirhombs circled. (b) Scatterplot of total pectinirhomb area against thecal volume, both measured on a logarithmic scale. Each point is an individual of a different size. The dashed line of slope 1 is what we would expect to see for ideal allometric scaling in which rhomb area just keeps up with thecal volume as size increases. The solid line shows the actual relationship; it has a slope of 1.3, meaning that rhomb area is increasing at a faster size-specific rate than expected by a simple area:volume scaling. As explained in the text, this suggests that some factor limited pectinirhomb efficiency at larger sizes. (c) Schematic illustration of inferred pectinirhomb function. Pectinirhomb is shared between two thecal plates and contains a series of slits through which sea water flows. A thin wall at the base of the rhomb allows diffusion of dissolved gases between the sea water and the coelom (body cavity). *(From Brower, 1999)*

One of the best-known examples of an allometric relationship is that between brain mass and body mass in vertebrates. Figure 2.30 depicts this relationship schematically through human growth. There are two conspicuous features here. First, around the time of birth, the slope of the allometric curve decreases, which reflects a major decrease in the rate of cell division of neurons. Second, the slope after birth is substantially less than 1; with little or no neuronal cell division, growth of the brain is greatly outpaced by growth of the body. The relationship shows negative allometry: Brain mass increases more slowly than does body mass, and the ratio of brain mass to body mass therefore decreases steadily through ontogeny. (Of course, this will come as no surprise to anybody who has noted the proportionally large size of a human baby's head compared with that of an adult.)

We also see a negative allometry of brain mass relative to body mass if we study **interspecific allometry**—the changing relationship between two body parts considered across many species. Figure 2.31 depicts brain and body mass for a number of living vertebrate species, as well as lines showing the relationship between brain and body mass within several kinds of vertebrates. These lines have a slope of about 0.67. Although the reason for this is unclear, the fact that the slope is a simple ratio (2:3) suggests that it may be determined in part by an area-to-volume or other basic scaling relationship. The groups of points are also offset from one another. In terms of the allometric equation, they have the same value of the slope *a*, namely 0.67, but they have different values of *b*. What this means is that mammals and birds generally have larger brains for a given body size than do other vertebrates.

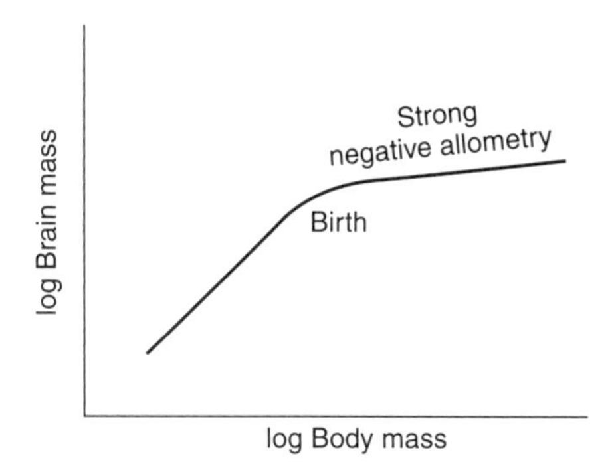


**FIGURE 2.30 Idealized relationship between log brain mass and log body mass during human ontogeny.** Initially, both masses increase at comparable size-specific rates, but the relative growth rate of brain mass is greatly reduced around the time of birth. *(After Lande, 1979)*

Species that fall above or below the line of allometry for their group have larger or smaller brains relative to what would be expected from their body mass alone; such species are referred to as more or less *encephalized.* Primates, for example, are highly encephalized among mammals, and humans are highly encephalized among primates.

Considering interspecific allometry can help us interpret differences in form between species when these differences are highly correlated with size. For example, large dinosaurs have a relatively small ratio of brain mass to body mass. This fact by itself, however, does not imply that dinosaurs were especially poorly encephalized. To determine the degree of encephalization, it is necessary to compare dinosaurs with the general trend for reptiles (Figure 2.32).

In living reptiles, brain mass and body mass can be measured directly; somewhat less direct means must be used for the extinct dinosaurs. Body mass can be estimated from skeletal measures because body mass and bone dimensions are correlated [SEE SECTION 3.2]. Similarly, brain mass can be estimated from cranial volume.

When dinosaurs are plotted along with other reptiles, it is clear that they represent a continuation of the allometric trend. The low ratio of brain mass to body mass for large dinosaurs is not a sign of unusually low encephalization but is instead exactly what one would expect for large reptiles in light of the negative allometry of brain size relative to body size.

## Heterochrony

The study of anisometric growth has been closely tied to the concept of **heterochrony,** or evolutionary changes in the timing of development. Important aspects of developmental timing include growth rate and the onset of sexual maturity relative to growth of the body.

The principal reason heterochrony is biologically significant is that it may be a way to achieve a large

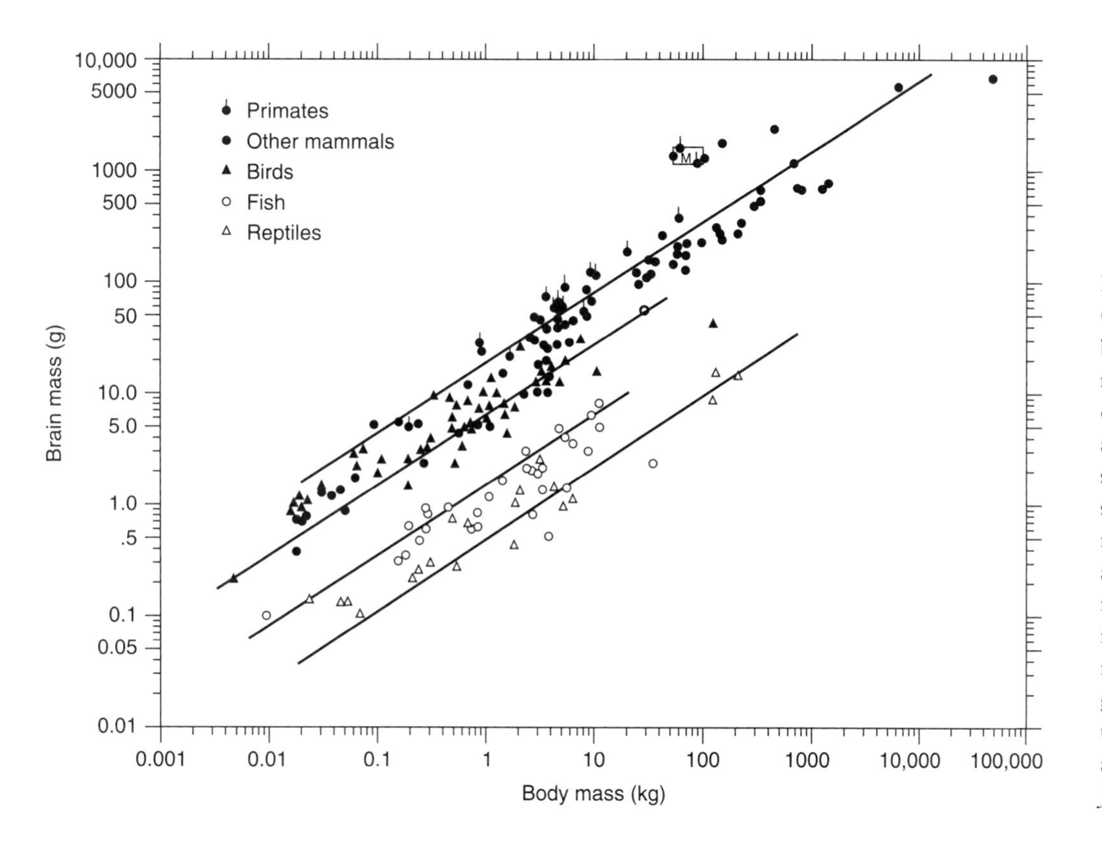


**FIGURE 2.31 Comparison between brain mass and body mass for some groups of vertebrates.** Both axes are logarithmic. Each species is represented by a single point, except for modern humans, which are represented by the range of variation indicated by the rectangle marked M. Within each group, the data show a trend with a slope approximating 2:3. *(From Jerison, 1969)*

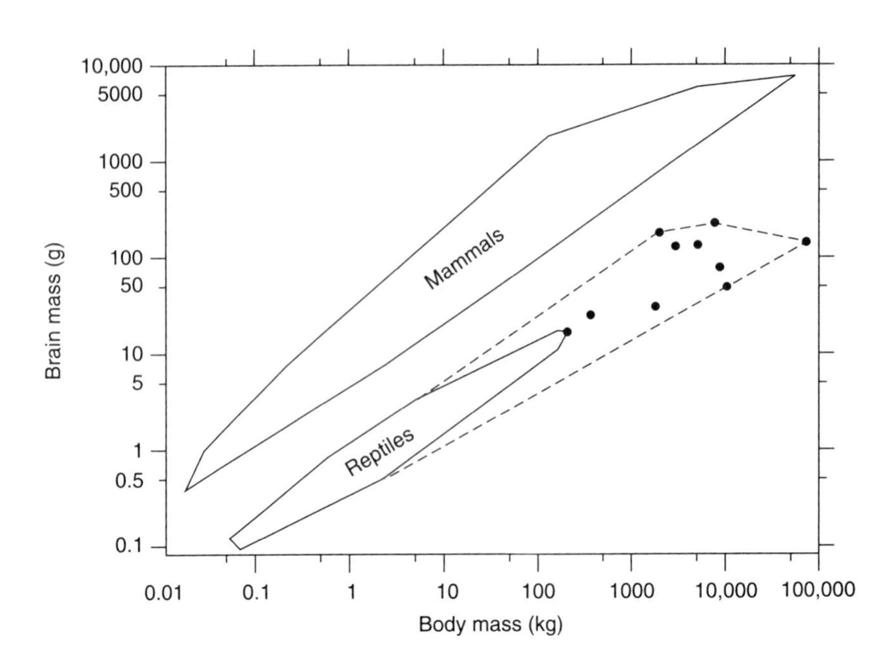


**FIGURE 2.32 Comparison between brain mass and body mass for living reptiles and for dinosaurs, with mammals included for comparison.** Both axes are logarithmic. The solid polygons enclose the range of variation for living species within mammals and living reptiles, as shown in Figure 2.31. The solid points are dinosaur species. The dashed polygon enclosing them can be interpreted as an extension of the field for living reptiles. *(From Jerison, 1969)*

evolutionary change in organismal morphology with a relatively small change in the underlying developmental pathway from juvenile to adult. If many traits are under common genetic control, natural selection [SEE SECTION 3.1] for size, growth rate, or some other underlying factor that affects many traits may bring with it numerous correlated changes in form.

Despite the attention paid to heterochrony as a potential mechanism for evolutionary change, it has been difficult to demonstrate it rigorously and to deduce the specific mechanisms that cause it. There are two main reasons for this. First, to be certain that a difference between two species reflects evolutionary change from one to the other, it is necessary to document ancestral–descendant relationships reliably (see Chapter 4). Second, a complete understanding of the mechanism of heterochrony often requires that we measure absolute growth rates with respect to age. This stands in contrast to our initial consideration of allometry, in which we compared the growth of two features to each other rather than to absolute age.

To see why this is so, consider the Lower Jurassic bivalve **lineage** *Gryphaea*, which shows the style of heterochrony known as **paedomorphosis.** In paedomorphosis, development evolves so that descendant adults resemble ancestral juveniles. (**Peramorphosis,** by constrast, is a kind of heterochrony in which the development of the descendant proceeds further than that of the ancestor, with the result that juvenile stages of the descendant resemble

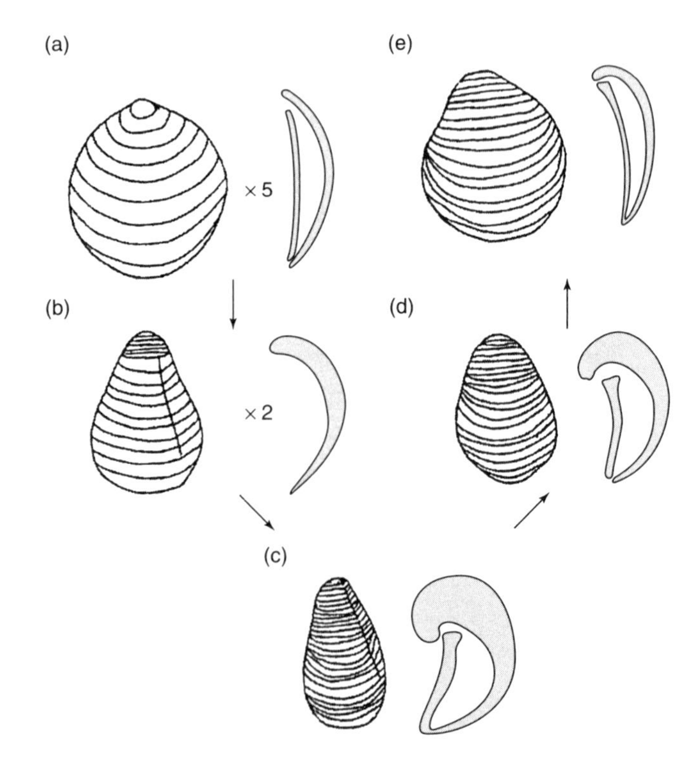


**FIGURE 2.33 Heterochrony in a lineage of Lower Jurassic *Gryphaea*.** The sequence labeled (a) through (c) shows ontogenetic development of the species *G. arcuata* from early juvenile through adult. The sequence labeled (c) through (e) shows evolutionary change in adult form, from (c) *G. arcuata* to (d) *G. mccullochi* to (e) *G. gigantea*. Each figure shows the exterior of the larger, left valve and a cross section through both valves. Note that the adults of *G. gigantea* resemble juveniles of *G. arcuata*. *(From Jones & Gould, 1999)*

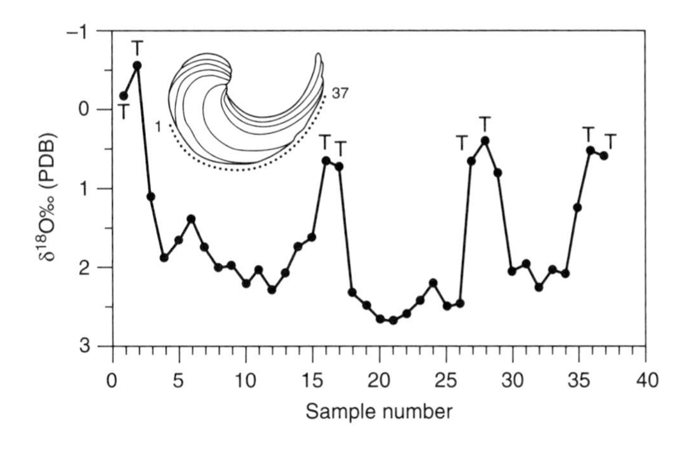


**FIGURE 2.34 Variation in the oxygen isotopic composition of shell carbonate throughout the growth of an individual of *Gryphaea arcuata*.** The inset shows a sketch of a shell, in which the direction of growth was to the right. The dots along the lower edge of this shell show 37 points at which shell samples were taken for chemical analysis. The plot compares measured oxygen isotopes of shell carbonate to the sample number. The notation "$\delta^{18}O‰$(PDB)" expresses the ratio of $^{18}O$ to $^{16}O$ with reference to a standard of known composition [SEE SECTION 9.5]. Higher values (toward the bottom of the graph) indicate a larger ratio of $^{18}O$ to $^{16}O$, which implies shell secretion at cooler temperatures. The sample points marked T are translucent shell increments that correspond to slow growth in the summer months. Three temperature cycles, indicating three annual cycles of shell growth, are shown here. *(From Jones & Gould, 1999)*

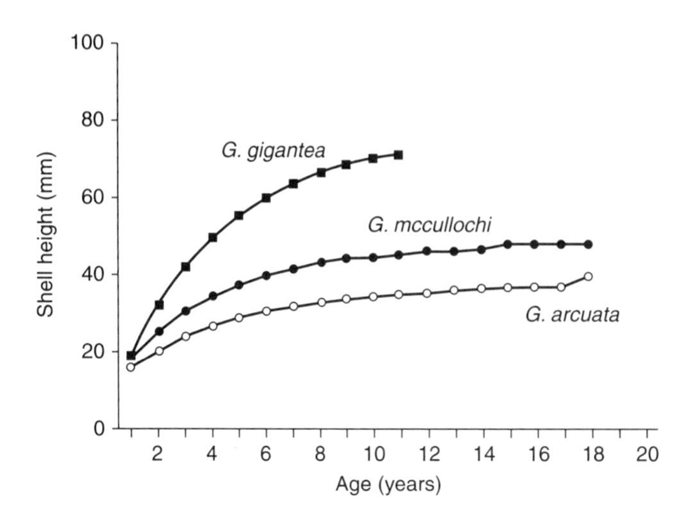


**FIGURE 2.35 Average growth curves showing shell height against time for three species of *Gryphaea*.** This shows that *G. gigantea* individuals grew for a shorter period of time, but at a higher rate, than those of *G. arcuata*. *(From Jones & Gould, 1999)*

ancestral adults.) In the *Gryphaea* lineage, the transition from *Gryphaea arcuata* to *G. mccullochi* to *G. gigantea* (Figure 2.33) involves striking changes in shape. The adults of the descendant *G. gigantea* resemble juveniles of the ancestral *G. arcuata* in being broad and flat in profile and less tightly coiled (Figure 2.33). These evolutionary changes appear to have a functional basis. Experimental flow-tank studies have verified that larger shells, with the shape of *G. arcuata* juveniles, are more stable on soft substrates such as those on which *Gryphaea* lived.

That the *Gryphaea* lineage underwent paedomorphic heterochrony seems clear, but, developmentally, how was the juvenile-like morphology produced in the ontogeny of *G. gigantea?* Did individuals develop over the same or a longer growth period than individuals of the ancestral species, but at a slower rate of shape change? Alternatively, did *G. gigantea* individuals grow more rapidly for a shorter interval of time? Clearly, either possibility could yield a paedomorphic adult, one with a large size but still retaining the juvenile form.

To distinguish these very different possibilities, it is necessary to analyze absolute age and growth rates. Fortunately, the age of individuals can be determined by chemical analysis of shells. It is well known from observations on a wide range of living aquatic organisms that the two most abundant **isotopes** of oxygen, $^{16}O$ and $^{18}O$, are incorporated into the carbonate of shells in relative proportions that depend in part on ambient water temperature, with heavier isotopes preferentially incorporated at lower temperatures [SEE SECTION 9.5]. This effect is dictated by basic thermodynamics. Because of seasonal variation in water temperature, there is a regular variation in the ratio of $^{18}O$ to $^{16}O$ incorporated into the shell, as illustrated in Figure 2.34 for a specimen of one of the *Gryphaea* species. It is therefore possible to determine the number of seasonal cycles through which a shell has passed, and thus its age. Doing this for the three species of *Gryphaea* shows that *G. gigantea* grew for fewer years than did its ancestors, but at a substantially higher rate of size increase, before attaining its adult form (Figure 2.35). Thus, the precise nature of the evolutionary change via heterochrony was identified in this instance: The descendant form grew more rapidly for a shorter period of time.

## 2.4 CONCLUDING REMARKS

Although paleontologists have a wide range of tools readily available for describing and measuring form and ontogeny, these methods are still being developed at a rapid pace. It will be interesting to see in the coming years which approaches prove most useful and for which questions.

Continued improvements in the measurement of absolute growth rates will enable the evolution of development to be studied more precisely, as in the *Gryphaea* example. Note that in this example we have accepted the ancestral–descendant relationships as given in Figure 2.33 and have bypassed the important question of how these relationships are known. In fact, how to determine the genealogy of evolving lineages is one of the most important issues in biology and paleontology. This subject will be taken up in Chapter 4.

The *Gryphaea* study is a success insofar as it clarifies the nature of evolutionary change by reconstructing growth relative to absolute age. Like most studies of heterochrony, however, it is limited to just a few traits. We would like to know whether heterochrony has molded a broad range of features in evolving lineages. Studies that consider a large suite of traits are only now becoming common. In some cases, it has been found that many traits simultaneously evolve by heterochrony; thus, the idea of achieving a large evolutionary change from a small developmental change is tenable. In other cases, however, the evolution of form is far more complex, with different parts of the organism evolving in different ways, some heterochronic, some not. For example, structures may shift in position in ways that cannot easily be explained by simple changes in the timing of development. In light of these studies, an important question for the future is just how important heterochrony really is in the evolution of form.

## SUPPLEMENTARY READING

Feldmann, R. M., Chapman, R. E., and Hannibal, J. T. (1989) *Paleotechniques* (Paleontological Society Special Publication Number 4). Knoxville, Tenn., The University of Tennessee, 358 pp. [A series of papers on preparation and illustration of fossil specimens.]

Gould, S. J. (1977) *Ontogeny and Phylogeny*. Cambridge, Mass., Harvard University Press, 501 pp. [Authoritative historical review of the study of allometry, heterochrony, and related topics.]

Kummel, B. H., and Raup, D. M. (eds.) (1965) *Handbook of Paleontological Techniques*. San Francisco, W. H. Freeman and Company, 852 pp. [A collection of specialized articles on a range of techniques for collecting, preparing, and illustrating paleontological material.]

McKinney, M. L. (ed.) (1988) *Heterochrony in Evolution: A Multidisciplinary Approach*. New York, Plenum, 348 pp. [A series of papers on general methods for studying development and evolution, with applications to a range of organisms.]

Rohlf, F. J., and Bookstein, F. L. (eds.) (1990) *Proceedings of the Michigan Morphometrics Workshop*. (University of Michigan Museum of Zoology Special Publication Number 2.) Ann Arbor, Mich., University of Michigan, 380 pp. [Pragmatic overview of methods for measuring and analyzing form.]

Schmidt-Nielsen, K. (1984) *Scaling: Why is Animal Size So Important?* Cambridge, U.K., Cambridge University Press, 241 pp. [Important overview of the effects of body size on function and physiology of animals.]

Thompson, D'A. W. (1942) *On Growth and Form*. Cambridge, U.K., Cambridge University Press, 1116 pp. [Classic work on ontogeny and other aspects of organic form. Essential reading for all interested in the interpretation of form.]

Wolpert, L., Beddington, R., Jessell, T., Lawrence, P., Meyerowitz, E., and Smith, J. (2001) *Principles of Development,* 2nd ed. Oxford, U.K., Oxford University Press, 568 pp. [Comprehensive text on developmental biology.]

Zelditch, M. L. (ed.) (2001) *Beyond Heterochrony: The Evolution of Development*. New York, Wiley-Liss, 371 pp. [A series of case studies of developmental change in evolution.]

## SOFTWARE

National Institutes of Health. NIH Image, http://rsb.info.nih.gov/nih-image/ [Software for digitizing and analyzing images.]

Rohlf, F. J. tpsDIG, http://life.bio.sunysb.edu/morph/ [Software for digitizing images and analyzing outlines and landmarks.]

*Chapter 3*

# POPULATIONS AND SPECIES

The previous chapter emphasized the description and measurement of single individuals within a species. But no two individuals are exactly alike, and the variation between them is a central fact of biology. In this chapter, we consider the sources of variation, why it is important from an evolutionary perspective, and some of the special problems facing paleontology. Variation among individuals must be considered in light of the nature of populations, to which we turn now.

## 3.1 POPULATIONS IN BIOLOGY AND PALEONTOLOGY

The individual organisms whose morphology we treated in the previous chapter exist in the biological context of the **population,** which can be defined as a group of individuals of the same species that live close enough together that they have ample opportunity for interbreeding. This emphasis on breeding applies, of course, to sexually reproducing species. The population shares a single **gene pool.** The gene pools of adjacent populations of a species may be partially or completely isolated from one another. When two populations interbreed, there is said to be **gene flow** between them. Depending on the dispersal ability and behavior of organisms and on the fragmentation of the physical habitat, populations differ widely in how large a geographic area they occupy and in how isolated they are from neighboring populations.

The geographic structure of populations is commonly studied in the context of metapopulation theory. The larger population, or **metapopulation,** consists of a number of smaller subpopulations. Subpopulations that are in particularly favorable environments may produce many individuals that disperse to other areas; they are said to be *sources*. Other subpopulations, by contrast, may accumulate migrants; they are *sinks*. The factors that govern the dynamics of sources and sinks within metapopulations, as well as other fine-scale aspects of geographic structure, are quite important to ecologists. The spatial structure of populations is also important for paleontological questions, because it plays a role in the origin of new species and in the pattern of evolutionary change over time [SEE SECTIONS 3.3 AND 9.3].

### Variation among Individuals within Populations

Each individual within a population has a particular **genotype**—its genetic composition encoded in its DNA sequences—and a **phenotype**—its form, structure, physiology, biochemistry, and behavior. The ultimate sources of variation within a population are **genetic mutation** and **recombination** of existing genetic material into new genotypes, through the production of sex cells and through sexual reproduction.

***Importance of Variation*** Variation is not only a fundamental property of populations. It also underlies all evolutionary change. **Evolution** within populations generally occurs if two simple conditions are met:

1. There is a regular relationship between genotypic and phenotypic variation, so that the phenotypic variation is heritable from parents to offspring. It is common to think of heritability in a direct sense—for example, a genetically determined trait such as eye color being inherited from the mother and father. In practice, however, heritability is studied by statistical analysis of populations, as we will see later in this section.
2. There is a relationship between heritable phenotypic variation and variation in reproductive success, reflecting both survival and fecundity. For example, suppose that bill size is heritable and that birds with larger bills tend to leave more offspring because they are able to eat larger and more nutritious seeds and therefore invest more energy into reproduction. Then the mean bill size of the population in this hypothetical case would tend to increase over time unless it were offset by other factors.

We just illustrated these simple requirements for evolution using a hypothetical case in which there was a direct, cause-and-effect relationship between the phenotype and reproductive success. For morphological traits such as body size and shape, as well as for many other traits, direct effects of this kind are generally accepted as the prevailing reason for evolutionary change. Such relationships are examples of **natural selection.** Traits can also evolve without being directly selected if they are genetically correlated with other traits that are under selection. For example, extra digits in vertebrates occur more commonly in larger individuals. Selection for increased body size could therefore lead indirectly to an increase in the number of digits.

An alternative to natural selection is that the correlation between phenotype and reproductive success is a matter of chance. This is likely in only two situations:

1. The traits are truly neutral with respect to selection, meaning that individuals are equally well adapted regardless of the trait value that they have. True neutrality is thought to be uncommon except for certain cases involving alternative forms of proteins and other biomolecules.
2. The population is so small that chance fluctuations in reproductive success are not averaged out. For example, even if it is true on average that larger-billed birds feed more effectively and leave more offspring, the occasional smaller-billed bird will be lucky, finding a cache of seeds, for instance.

In small populations, chance events can be of significance. Evolutionary change that results from such chance fluctuations is known as **genetic drift.** In a broader sense, chance fluctuations can also occur in other ways, such as by extinction of local populations that may differ in genetic composition relative to the larger metapopulation.

***Heritability of Variation*** One often reads debates in the press on the subject of "nature versus nurture"—whether particular traits, such as aspects of human behavior, are genetic or environmental in origin. In fact, the entire phenotype arises through ontogeny from the interplay between an organism's genetic composition and its environment. For example, it is well known that growth in oysters and other animals that live on hard surfaces molds the organism to the substrate. There is clearly an environmental effect, yet the capacity to grow in such a malleable way has a genetic basis.

Many factors, including light, temperature, nutrition, water and soil chemistry, and substrate, can yield environmental variation in the phenotype. Phenotypic variation that is attributable to environmental variation is referred to as **ecophenotypic;** the tendency for a genotype to produce different phenotypes in different environments is known as **phenotypic plasticity.** The environmental effect on the phenotype may be adaptive. Many such cases of adaptive plasticity have been documented, including animals that detect the presence of predators via chemical cues and grow protective ornament in response (Figure 3.1).

Because each individual's form is both genetically and environmentally determined, evolutionary biologists study the sources of phenotypic variation among individuals rather than the phenotype of a particular individual. Figure 3.2 shows an analysis of bill size in a population of finches on an island in the Galápagos Archipelago. Each point compares the mean bill size of offspring produced by a pair of parents with the mean bill size of the parents. The open and closed symbols represent measurements that were taken in two different years. The positive relationship between parental size and offspring size, shown by the lines, indicates that the trait has a heritable component. The less scatter there is around this line, the higher the heritability.

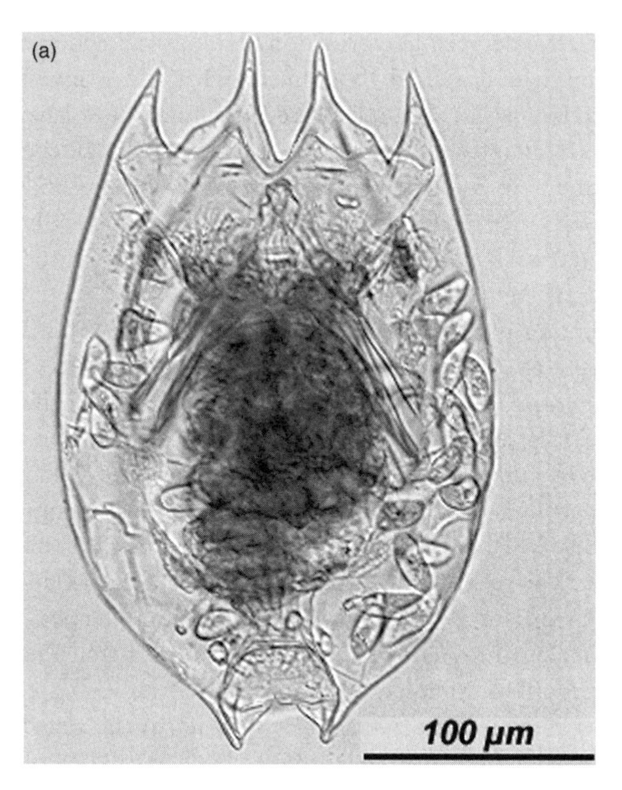


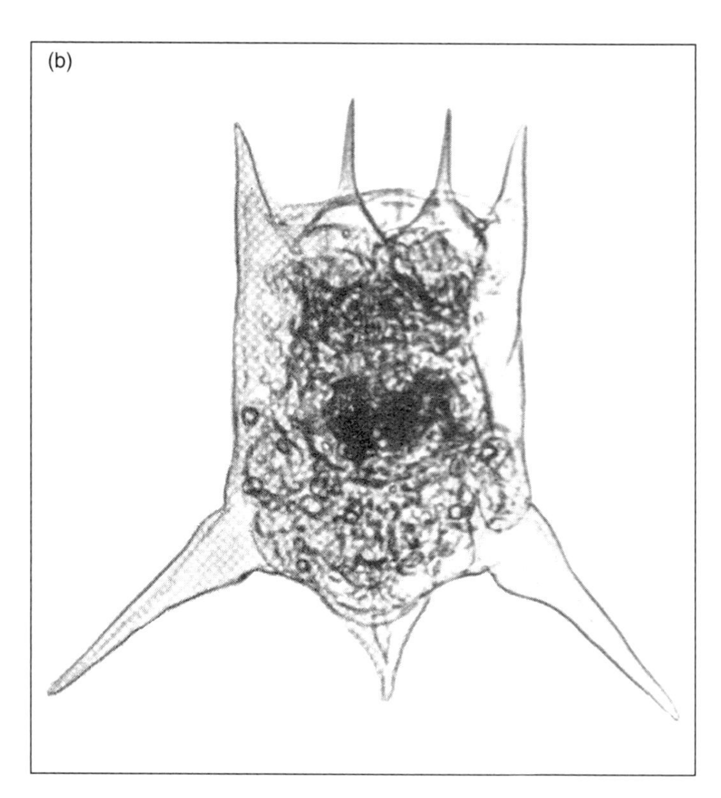


**FIGURE 3.1 Two specimens of the living rotifer *Brachionus calyciflorus*.** Individuals that grow in the presence of a predatory rotifer *Asplanchna*, or in chemical extracts derived from *Asplanchna*, develop elongate spines, as in the specimen on the right. The specimen in part (b) is approximately 150 microns across, excluding the spines. *(a: The Academy of Natural Sciences; b: From Gilbert, 1966, http://www.schweizerbart.de)*

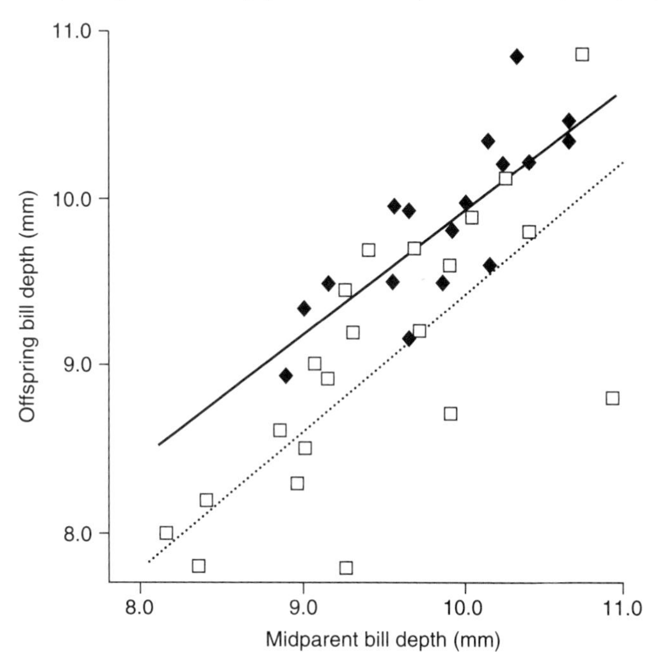


**FIGURE 3.2 Heritability of bill size in the ground finch *Geospiza fortis*.** Midparent value is the mean measurement of a pair of parents; offspring value is the mean of the offspring of these parents. The positive correlation between the bill size of the offspring and that of the parents indicates heritability of this trait. Open and closed symbols denote measurements taken in two different years. Lines are fitted to each set of points. *(From Boag, 1983)*

This scatter can be measured with standard statistical methods which, in this case, indicate that about 60 percent of the variation in bill size among the offspring is heritable. This is a statistical statement about the population as a whole in the given environment. It does not tell us to what extent the bill size of any individual bird is attributable to its genotype and how much to its environment.

Thus, variation among individuals of a given ontogenetic stage has both genetic and environmental components. In the population at large, change through ontogeny and differences between the sexes also contribute to the overall variation of the population. We typically try to factor out these last two sources of variation by studying the same sex at a comparable ontogenetic stage. There are still other sources of variation that affect fossil populations.

### *Additional Sources of Variation in Fossil Populations*

The populations of paleontology, consisting of individuals collected from a given locality, differ from living populations in some important ways. They have passed through various taphonomic filters such as postmortem distortion, and they may represent a time-averaged assemblage [SEE SECTION 1.2].

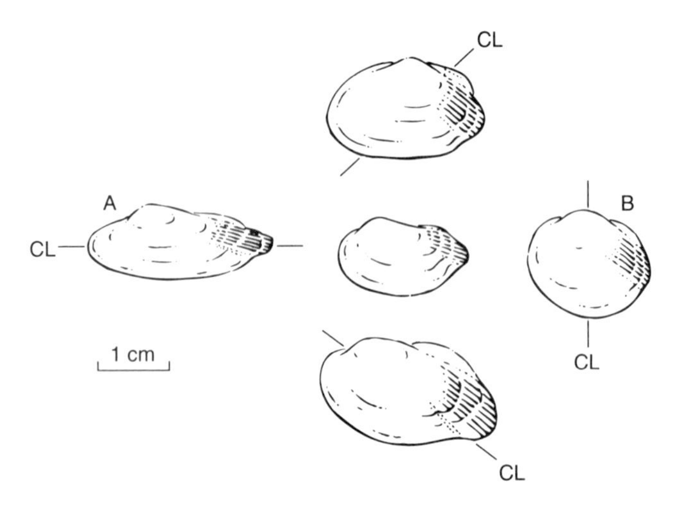


**FIGURE 3.3 Effects of rock deformation on fossil morphology.** The drawing in the center shows an undeformed specimen of *Arisaigia postornata*. The four drawings surrounding it show different patterns of deformation. In each case, the direction marked CL corresponds to rock cleavage and is perpendicular to the direction of maximum shortening. The specimens marked A and B were deformed in directions perpendicular to one another. *(From Bambach, 1973)*

One of the main taphonomic processes tending to increase apparent variation is distortion produced by compaction of sediments or deformation of sedimentary rocks. A particularly striking but by no means rare case is illustrated in Figure 3.3. Richard Bambach (1973) analyzed shape variation in a large sample of the infaunal bivalve *Arisaigia postornata* from Silurian rocks of Nova Scotia. For each specimen, the orientation of the rock's cleavage relative to the bivalve's morphology was noted. Because these specimens are actually two-dimensional molds of the original shells, the shapes taken by the fossils leave an accurate account of the deformational history of the rock, with the direction of maximal compression being perpendicular to the direction of cleavage.

Figure 3.3 shows drawings of four typical specimens covering the range of geometric relationships of cleavage direction to morphology. Despite the different appearance of these specimens, they can be assigned to the species in question because they possess characteristic surface ornament. The fifth drawing (center) is a reconstruction of an undeformed specimen. The reconstruction was aided by standard methods from structural geology. In essence, forms A and B are end-members that, based on orientations of rock cleavage, must have been deformed in perpendicular directions. The relative length:height ratios of A and B were used to estimate the relative degree of deformation in the two directions; this in turn was used to estimate what the original length:height ratio of undeformed specimens must have been. In Bambach's collections, we know that all the specimens were deformed because co-occurring brachiopods, which must originally have been bilaterally symmetrical, are also deformed. Thus, the form in the center of Figure 3.3 was not found.

Another potential source of added variation in fossil populations is the process of time averaging [SEE SECTION 1.2]. Figure 3.4 shows a hypothetical case in which the variation of a trait within a population is constant, as indicated by the width of the curve, while the average trait value changes over time. Typically, many successive populations will be averaged together into a single fossil sample. The resulting variation of the time-averaged sample depends on the amount of variation within the population at a moment in time and on how much the population's morphology shifts over time.

In principle, time averaging could act to the point where we could not obtain reasonable estimates of

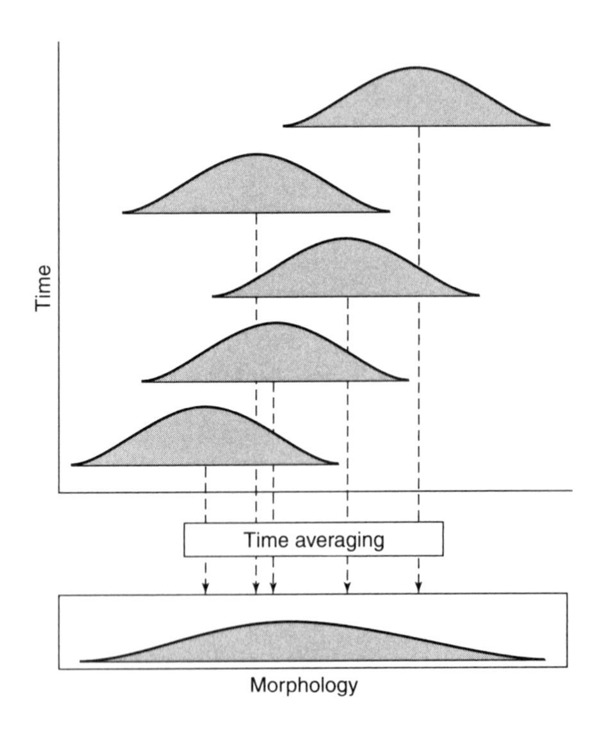


**FIGURE 3.4 Hypothetical effects of time averaging on variation within fossil populations.** Each curve in the upper part of the figure depicts variation within a population at a point in time, and the population average is shown by the position of the curve along the $x$ axis. If the population shifts over time, the resulting time-averaged sample, depicted by the bottom curve, will be more variable than the population. *(From Hunt, 2004a)*

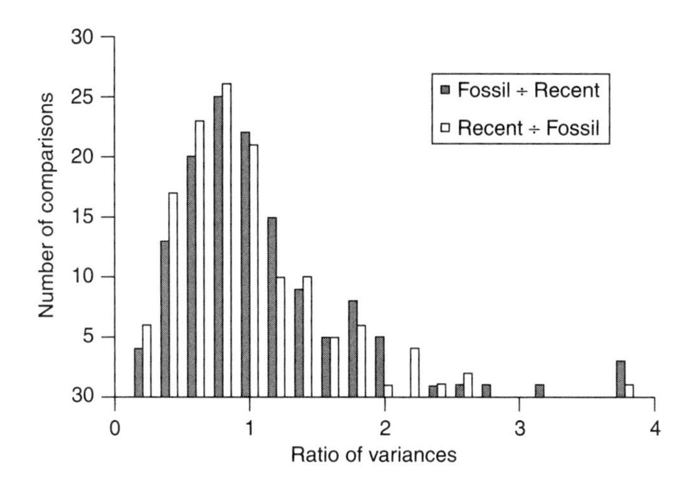


**FIGURE 3.5 Comparison of variance between living and fossil populations of the same species.** The solid bars show the frequency distribution of about 130 ratios between the variance of a fossil sample and the variance of a living population of the same species. The open bars depict the ratio of variances between living populations and fossil samples. The two distributions are indistinguishable, implying that these fossil samples are, in general, neither more nor less variable than their living counterparts. *(Data from Hunt, 2004b)*

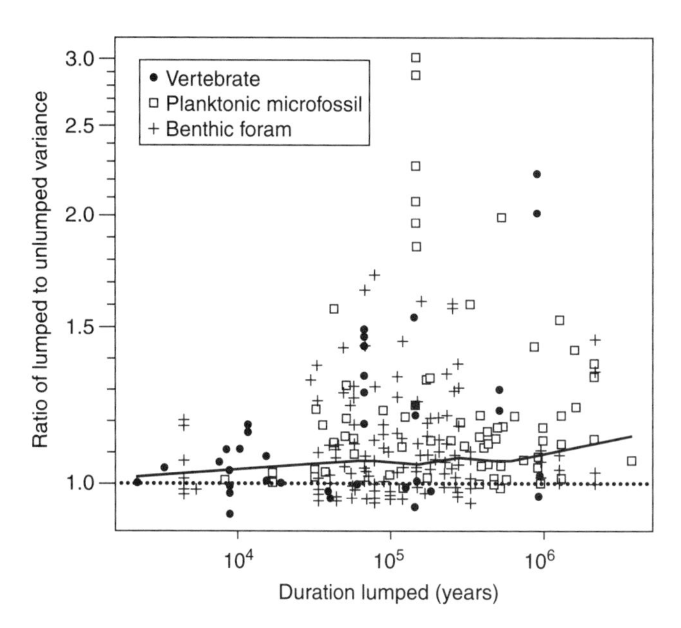


**FIGURE 3.6 Inflation of population variance relative to the extent of analytical time averaging.** Successive samples were lumped together, and the ratio of lumped to unlumped variance is plotted against the temporal extent of lumping. The thick line shows the median ratio. On average, there is less than a 10 percent inflation of variance due to time averaging. *(From Hunt, 2004b)*

variation within fossil populations. But how much does time averaging matter in practice? This can be assessed in two ways: (1) by comparing the variation of living populations with that of fossil samples of the same species; and (2) by comparing the variation of fossil samples with the duration of time averaging of those samples.

Figure 3.5 depicts comparisons among living populations of a number of species and fossil samples of the same species. Variation is expressed by the statistical measure known as the *variance* (see Box 3.1). For each comparison, the ratio between the variance in the fossil sample and that in the living population was calculated. These ratios are summarized by the solid bars in Figure 3.5. A ratio greater than 1 means that the fossil population has a greater variance than does the living one. Clearly, some of the fossil samples are more variable than their living counterparts and some are less so. The open bars in Figure 3.5 show the ratio of variance of living to fossil populations. Here, ratios greater than 1 mean that living populations have greater variance. There is no appreciable difference between the two sets of ratios. In other words, fossil samples are sometimes more variable than corresponding living populations and sometimes less so, but there is no predominant tendency one way or the other.

Paleontologists have no control over the amount of time represented by naturally occurring beds. The extent of time averaging can be varied artificially, however, by combining fewer or more beds together into a single sample. This practice, which results in **analytical time averaging,** allows one to explore how variance changes as more time is incorporated into a sample. Figure 3.6 summarizes data on variance from a number of fossil studies. Samples of the same species were analytically time averaged to determine how the variance of the combined samples is affected by time averaging. Each point in this figure compares the variance of a time-averaged sequence of populations with the duration of time averaging. The thick line shows the average trend through the points. Although there is an overall increase in variance with time averaging, it is generally rather small—on average, less than 10 percent even when millions of years are averaged together.

Thus, available evidence suggests that the variance within fossil samples is not dominated by time averaging. Variation within samples can be meaningfully studied, provided that gross distortion of the kind shown in Figure 3.3 is ruled out. That variance barely increases with time averaging in many cases implies that the population is morphologically relatively stable over time. We will return to this point in Chapter 7.

## 3.2 DESCRIBING VARIATION

In this section, we provide a brief treatment of some of the most important procedures for describing and analyzing variation. Our coverage is only introductory, and the sources listed at the end of the chapter should be consulted for additional details. The availability of high-speed computers and software makes it easy to perform a wide range of analyses, but it is essential to have a firm understanding of the goals, assumptions, and calculations underlying each analysis. We focus on variation among individuals within a population and variation among similar populations. Because of the importance of populations in evolution, these levels of analysis play a special role. Nevertheless, many of the same procedures we will describe can also be applied, often with only minor modification, to study other aspects of variation—within the growth of a single individual or among the distantly related species of a larger biologic group.

*Box 3.1*

### DESCRIPTIVE STATISTICS

A histogram shows the number or proportion of individuals having trait values falling within specified intervals (Figure 3.7); the histogram may be smoothed into an idealized frequency curve. The graphical summary of the histogram is often accompanied by other statistics, as outlined herein.

Paleontologists and biologists are generally interested in two main aspects of univariate data within a population: the central tendency and the dispersion or variation. Which statistics are appropriate to express central tendency and dispersion depends on the nature of the variables. There are three main kinds of biologic variables: (1) **nominal** or **categorical;** (2) **ordinal** or **ranked;** and (3) **quantitative.**

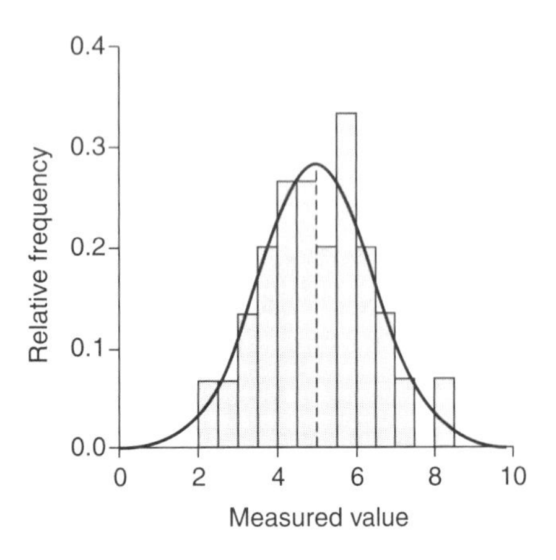


**FIGURE 3.7 Hypothetical histogram and idealized, smooth frequency curve.** Each bar shows the proportion of individuals having a trait value within the corresponding interval of values on the $x$ axis. The dashed line shows the position of the sample mean.

Nominal data can take on only particular, distinct values, and there is no natural ordering to the values; one value is not inherently greater or smaller than another. Examples include gender (male or female); presence or absence of a specified structure; and features such as surface ornament (none, spines, tubercles, and so on).

Ordinal data also take on only distinct values, but there is a natural ordering to them. Examples include small, medium, and large; absent, rare, common, and abundant; and compressed, equidimensional, and elongate. There is often an unmeasured continuum underlying the values. We can express size as small or large even though size can take on any number of values if measured more precisely. The differences between values on an ordinal scale generally have no consistent meaning. The difference between small and large, for example, is not twice as great as the difference between small and medium.

Examples of quantitative measures include length, width, area, volume, mass, and angle. The units on the scale have a consistent meaning. For example, the difference between a length of 10.0 and 10.2 mm is twice as great as the difference between 10.0 and 10.1 mm. Quantitative measures are the most common form of data in the study of fossil

## Describing Variation in One Dimension

Basic descriptive statistics are an important part of formal taxonomic work (Box 3.1). Summaries of measurements are typically univariate, involving only a single variable at a time. A graphical and tabular summary of measurements allows other workers to compare the data with similar measurements for other species [SEE SECTION 4.1]. These data are important, for example, in assessing whether two samples of specimens are likely to belong to the same species or to different species. Whether data are summarized in graphical or tabular form, it is important to make available to other workers the original measurements on which these summaries are based.

## Describing Variation in Two or More Dimensions

Most univariate analyses, such as the assessment of differences between two samples of specimens (Box 3.1),

populations. We therefore focus on such measures. Meristic counts [SEE SECTION 2.2] are, strictly speaking, ordinal. For many purposes, however, they are treated as quantitative, especially if the counts vary over a wide range of values. The statistics of nominal and ordinal data are discussed in the sources listed at the end of this chapter.

The **arithmetic mean** or **average** can be used to measure central tendency for quantitative data. If the measured values are denoted $x$, the mean of a sample, $\bar{x}$ (read "x bar"), is simply the sum of values divided by the number of individuals measured $(n)$: $\bar{x} = \sum x/n$.

If the distribution of data is highly asymmetric or if $n$ is small, the mean can be unduly influenced by a small number of high or low measurements. In such cases, the **median** often provides a more reliable estimate of central tendency. (The skewed distribution of wealth is one reason economic statistics commonly report median income and assets; the mean wealth of one billionaire and 1000 paupers would be roughly $1 million per person.) In the case of an asymmetric distribution, the median also better represents what we think of as the typical form. By definition, half of the observations are at or below the median, and half are at or above the median. For example, suppose we measure total length in centimeters in a small sample of seven specimens and obtain the following values, placed in increasing order: 12, 13, 13, 13, 14, 15, and 32. The median would be the fourth value, or 13 cm. This is a clear case in which one would want to report the median. The mean value of 16 cm is dominated by one large observation and is outside the range of the six remaining measurements. It does not adequately represent the typical size one is likely to encounter in the species from which this sample was drawn.

Dispersion for quantitative traits is typically measured by the average squared difference between observed values and the mean. The **variance** $s^2$ of a sample is defined as $\sum(x - \bar{x})^2/(n - 1)$. Variance has units that are the square of the original unit of measurement. To express dispersion in the same units as the original measurements, it is common to use the **standard deviation** $s$, equal to the positive square root of the variance. In the example of the previous paragraph, the variance is 51 cm$^2$ and the standard deviation is 7.1 cm.

One of the most common uses of univariate data is to interpret observed differences between two populations. Several samples, even if drawn from the very same population, will inevitably differ somewhat because of chance variation. Every sample statistic, such as the sample mean $\bar{x}$, has an associated **standard error,** which is a measure of the uncertainty in the statistic. If we took a very large number of samples from the same population and calculated $\bar{x}$ for each one, then the standard deviation of the values of $\bar{x}$ would be the standard error of the mean. The smaller the standard error relative

*continued on next page*

have bivariate and multivariate analogs. In addition, there are certain bivariate methods designed to analyze data when the focus is on the relationships between two measured variables. Such methods are useful in studying, for example, growth [SEE SECTION 2.3], function [SECTION 5.3], and heredity (Figure 3.2).

Two of the principal goals of bivariate analysis are to measure the strength of correlation between two variables and to describe the form of the relationship between them. For the first goal, a number of **correlation coefficients** are commonly employed (see Box 1.1). These typically vary between −1 and +1, with values closer to these extremes indicating stronger correlations. Negative values indicate that an increase in one variable tends to correspond with a decrease in the other, while positive values indicate that the two variables tend to increase together. Values near zero indicate that there is little relationship between the variables. Referring back to the example of bill size in Figure 3.2, the correlations between parent and offspring for the two sets of points

*Box 3.1 (continued)*

to the difference between two sample statistics, the smaller the probability that the observed difference reflects chance variation, and therefore the greater the chance that the two samples come from truly different underlying distributions. In general, the less variable the population and the more individuals in the sample, the smaller will be the standard error of a sample statistic. It is therefore necessary to compare any observed difference with the intrinsic variance of the population (Figure 3.8).

The samples of Figure 3.9 are quite different relative to their intrinsic variability. If the two histograms were superimposed on the same graph, they would barely overlap. Thus, it is unlikely that the difference between the samples is due to chance. The samples of Figure 3.10, by contrast, could easily have been drawn from the same statistical distribution. They have essentially the same mean and would fully overlap if plotted together. Whether a difference between samples is due to chance can be very unlikely (Figure 3.9), very likely (Figure 3.10), or anywhere in between—but one can never know with absolute certainty. Nonetheless, a low probability that the observed difference is due to sampling error provides reasonable, operational grounds for considering the difference to be meaningful unless proven otherwise. This probability is assessed with formal statistical tests described in several of the sources listed at the end of this chapter.

The pair of samples depicted in Figure 3.9 shows a large difference in mean values, whereas the pair in Figure 3.10 shows a small difference. In fact, each of these two comparisons involves the same pair of subspecies; the contrasting results reflect different traits that were analyzed. Because different traits can show different patterns of variation between populations, it may be necessary to measure and analyze several traits—hence, the need for multivariate analysis.

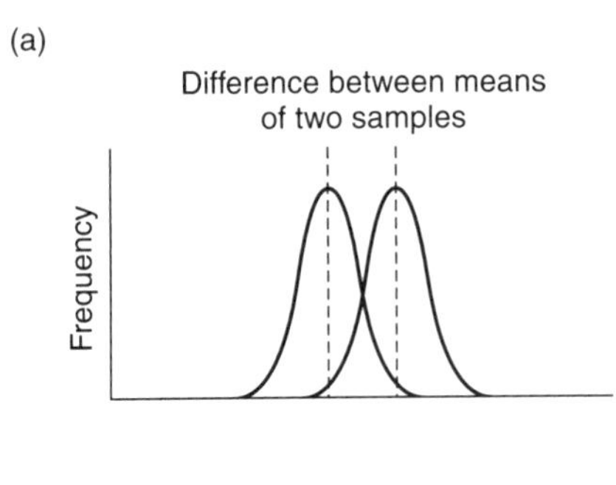


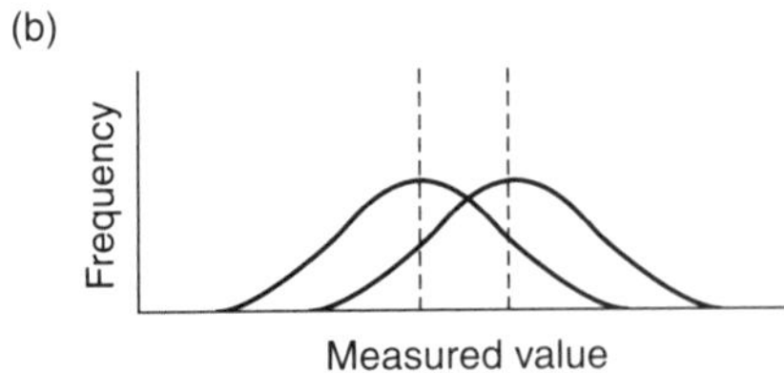


**FIGURE 3.8 Pairs of hypothetical frequency distributions illustrating the importance of variation in assessing the significance of an observed difference between mean values.** It is much more likely that the observed difference occurred by chance in part (b) than in part (a).

are between 0.75 and 0.80. These are relatively high values, in agreement with our earlier statement that there is clear evidence for heritability.

The form of the relationship between two variables expresses *how much* of a change in one is seen with respect to a change in the other. This is typically studied with a linear model of the form $Y = aX + b$. The slope $a$ estimates how much of a change in $Y$ there is for a given change in $X$. The intercept $b$ is the value that the variable $Y$ takes on when the variable $X$ has a value of zero [SEE SECTION 2.3]. Referring again to Figure 3.2, the slopes of the lines fitted to the data are approximately 0.8. This means that for every millimeter difference in parental bill size, there is, on average, a difference of 0.8 mm between the corresponding offspring.

There are two main reasons to fit a line to data. The first is to describe a mutual relationship between two variables without giving primacy to one or the other. This use is common in the study of allometry, as in Figure 2.29. Second, a fitted line can be used for predictive

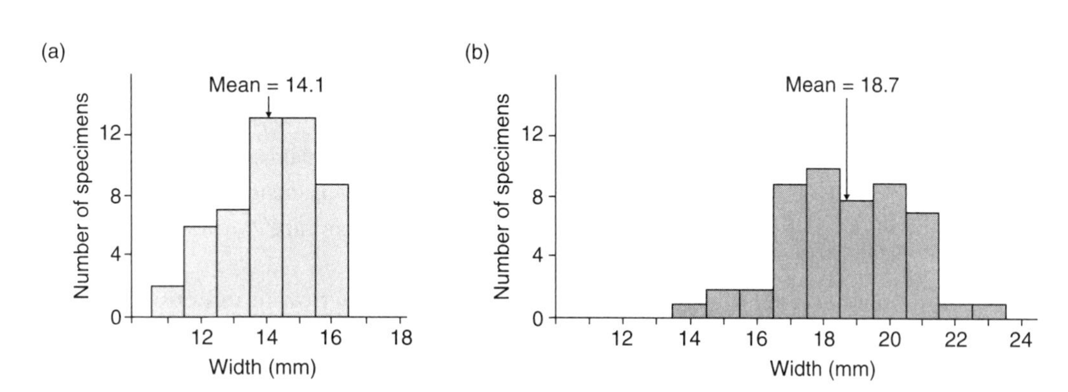


**FIGURE 3.9 Frequency distributions of shell width in two subspecies of the Devonian brachiopod *Pholidostrophia*.** (a) *Pholidostrophia gracilis nanus.* (b) *Pholidostrophia gracilis gracilis.* The mean values are different enough, relative to variation about the mean, that the difference is not likely to be due to chance in sampling alone. *(Data from Imbrie, 1956)*

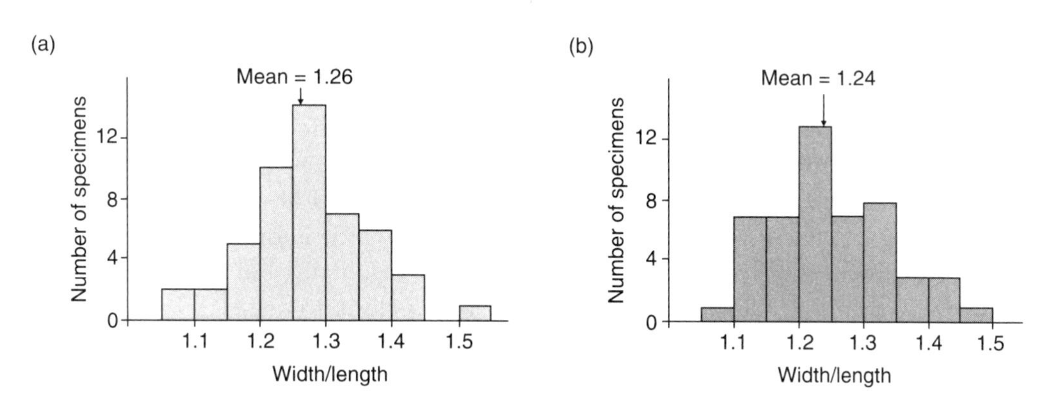


**FIGURE 3.10 Frequency distributions of width-to-length ratio in two subspecies of the Devonian brachiopod *Pholidostrophia*.** (a) *Pholidostrophia gracilis nanus.* (b) *Pholidostrophia gracilis gracilis.* The slight difference between the mean values could easily be due to chance errors in sampling. *(Data from Imbrie, 1956)*

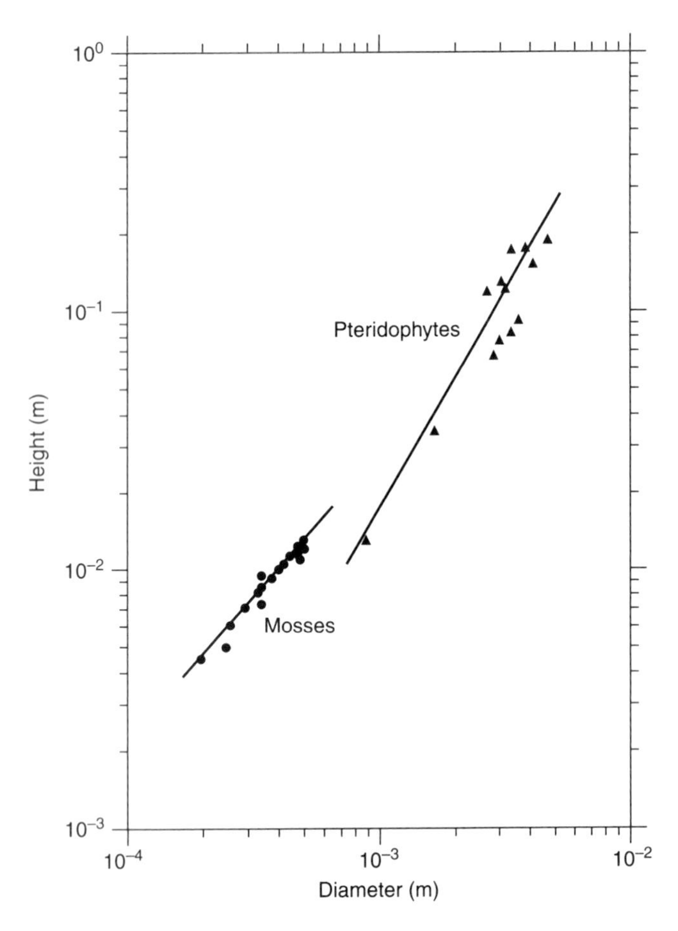


**FIGURE 3.11 Predictive regression lines that can be used to estimate stem height from stem diameter in two groups of land plants.** Both variables are measured on a logarithmic scale. Pteridophytes are an informal grouping of primitive vascular plants. *(From Niklas, 1994a)*

purposes. For example, Figure 3.11 shows the relationship between stem diameter and height in a number of species of living mosses and primitive vascular plants. The strength of the correlations means that stem diameter could reasonably be used to predict stem height—or vice versa. The prediction of stem height from stem diameter is actually much more useful in paleontology. Because material is often fragmentary, it is unlikely that the entire height of the stem will be preserved. Of course, this approach depends on having living or completely preserved representatives with which to establish the predictive relationship. Descriptive and predictive line fitting involve slightly different assumptions and procedures, which are covered in any elementary text on statistics.

It is important to bear in mind that our treatment of bivariate data with the equation $Y = aX + b$ assumes that the relationship between the variables is linear. Two traits may in fact be nonlinearly correlated, as with the brachiopods of Figure 2.23. In such a case, the correlation coefficient can greatly underestimate the strength of association between the traits, and a straight line fitted to the data is all but meaningless. A nonlinear relationship such as that of Figure 2.23 can sometimes be made linear by measuring the variables on a logarithmic scale. This, of course, is what the allometric equation [SEE SECTION 2.3] accomplishes. Other transformations can often be used to linearize the data.

Rarely is organic form sufficiently well represented by one or two features. It is often necessary to take measurements on many traits to gain a more complete picture of form. Doing so leads to problems, however, for the human mind cannot so easily visualize all the mutual relationships among numerous variables the way it can grasp a simple bivariate relationship. Therefore, a large class of approaches has been developed, collectively referred to as **multivariate analysis.** These approaches share the common goal of data reduction, in other words, summarizing, in a small number of dimensions, data that represent a large number of variables. The dimensions used are often synthetic in the sense that they are combinations of the original variables.

Any such reduction in dimensionality in effect represents a projection of the original data, just as a map is a projection of the globe into two dimensions. A projection generally produces distortion, and most methods have associated with them some means for assessing this distortion. Figure 3.12 shows hypothetical cases in which there are two original variables. Although there is variation in both dimensions, the strong correlations among the variables in Figure 3.12a imply that most of the variation can be summarized by the major axis running through the points from the lower left to the upper right. That is to say, if we were to treat this axis as a single, synthetic variable, and represent each point by a single number—its projected position along this axis—there would be little distortion and we would lose relatively little information. We may have measured two traits originally, but the number of meaningful variables is closer to one.

A contrasting case is shown in Figure 3.12b. Here, the variables are more weakly correlated, so there is more dispersion around the major axis. This means that the number of meaningful variables is much closer to two than to one, and we would lose a great deal of information by considering only the position of the points along

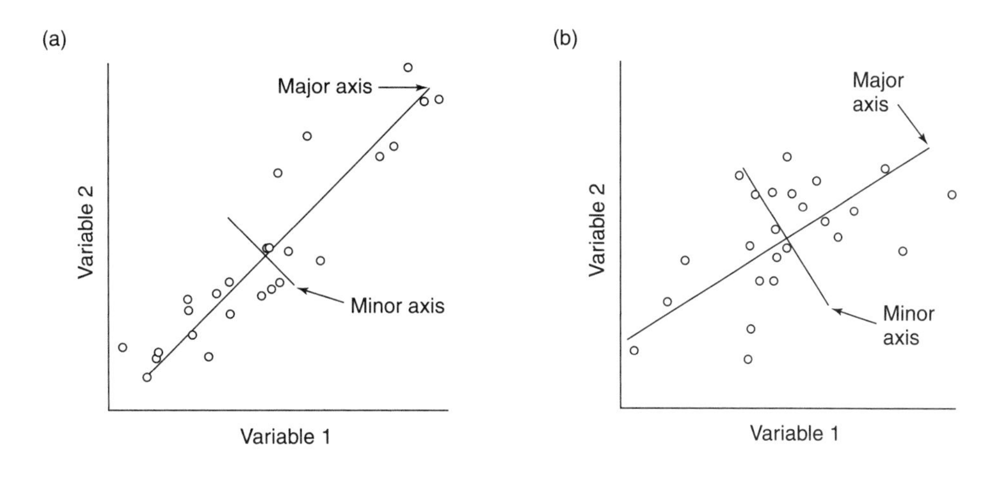


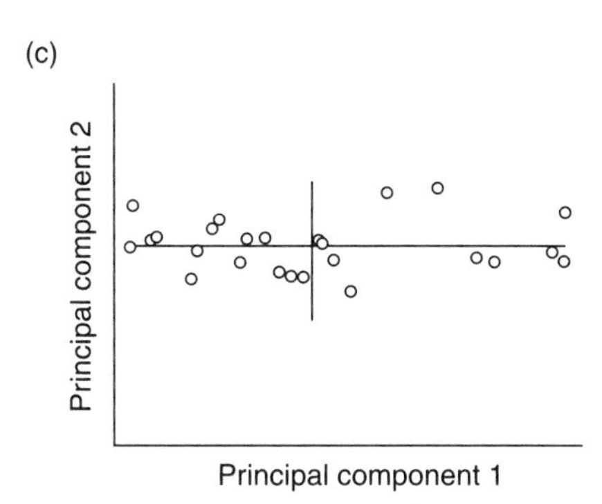


**FIGURE 3.12 Illustration of the rationale behind multivariate analysis.** (a) Here the two variables are highly correlated. In this case, projecting all points onto the major axis would result in little loss of information. (b) Here the variables are more weakly correlated, and therefore much information would be lost by reducing the two-dimensional data to a single dimension. (c) Shown here are the same data as in part (a). The major and minor axes are the first and second principal components, and the projections of the data onto these axes are the principal-component scores.

the major axis. Because of mutual intercorrelations among anatomical traits, biometric data are typically closer to the graph in Figure 3.12a. For example, if we measured the lengths of two limb bones in a sample of **TETRAPOD** vertebrates, we would find that the larger species or individuals tend to have greater lengths with respect to both measures.

In this section, we illustrate multivariate analysis with several different methods. The goal of data reduction pervades all of them, but each one focuses on a different kind of question. There are numerous other methods that are similar in spirit to the ones we present here, while differing in the particulars. Some of these will be covered later in this book.

***Ordination of Specimens*** One of the main uses of multivariate analysis is to facilitate visual inspection of data. In a bivariate plot, it is easy to see which specimens are most similar, how specimens differ, how the data trend, and so on. To do the same with multivariate data requires an **ordination**—a representation of the positions of the specimens relative to one another. One of the most widely employed methods to achieve this goal is **principal-component analysis.** Figure 3.12c shows the same hypothetical data as Figure 3.12a. The points have simply been rotated so that the major and minor axes running through the data in Figure 3.12a are now in the same direction as the new the $x$ and $y$ axes of Figure 3.12c. The direction of the major axis is the direction of maximal dispersion in the data and defines the first principal component. There is still residual variation around this axis, indicated by the minor axis that is perpendicular to the first axis. This minor axis defines the second principal component.

(a)

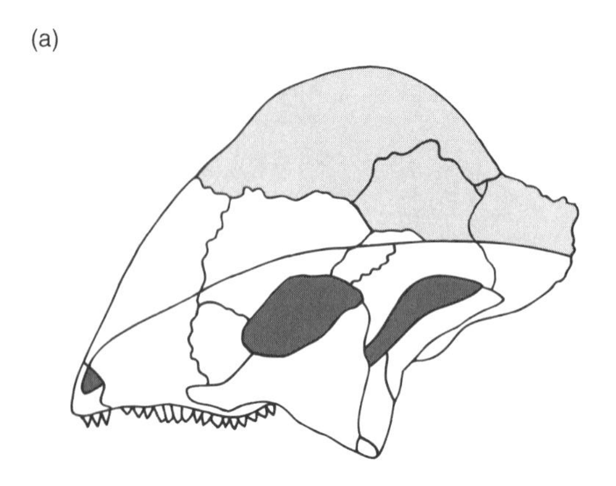

(b)

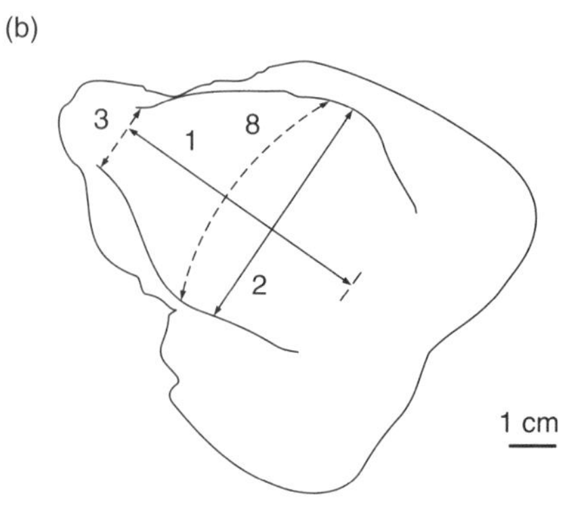


(c)

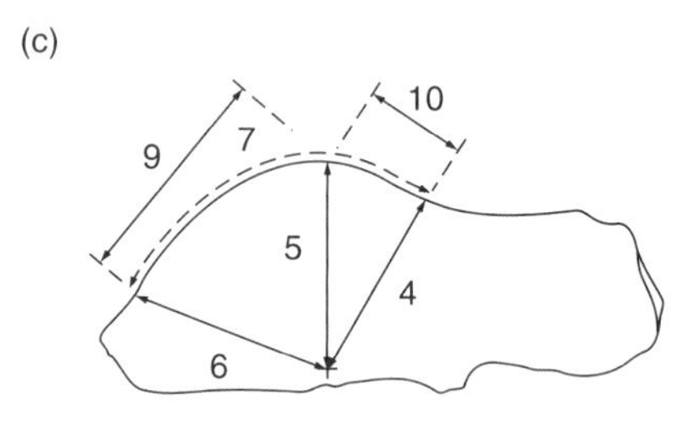


(d)

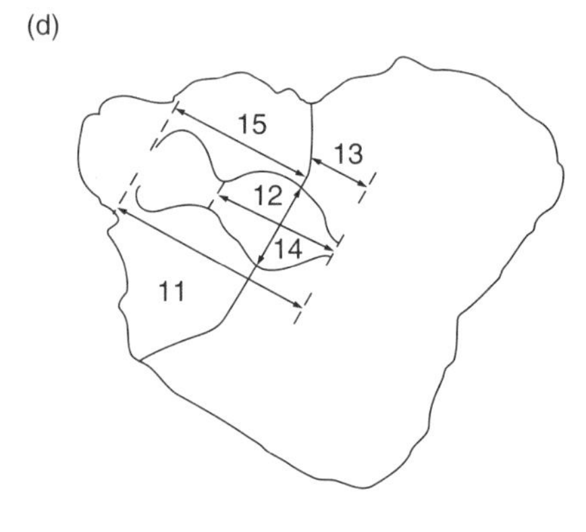


**FIGURE 3.13 Reconstructed skull of the dinosaur *Stegoceras*.** (a) The shaded part is the cranium. (b–d) Measurements taken on the skull in (b) dorsal, (c) lateral, and (d) ventral views. The region of the skull illustrated in parts (b) through (d) is roughly the shaded portion in part (a). *(From Chapman et al., 1981)*

The method of principal components extends to any number of dimensions. Each successive axis is always perpendicular to all the previous ones, and it runs in the direction of maximal remaining dispersion around the previous axes. The position of each specimen along a particular principal-component axis is referred to as its **score** on that axis. The length of each axis tells how much of the variance in the data is accounted for by the corresponding principal component; it is expressed by a number called the **eigenvalue** (see Table 3.1).

Let us consider a paleontological example of principal-component analysis. Figure 3.13a shows a reconstruction of the skull of the dinosaur *Stegoceras*. The shaded portion is the cranium, which includes the braincase and a prominent dome. A number of crania were measured on specimens from the Upper Cretaceous of western North America; the measurements are shown schematically in Figures 3.13b–d. There are ten measurements of the dome and five of the braincase. Figure 3.14 portrays the scores of about 30 specimens on the first two axes that result from a principal-component analysis. The specimens of the species *Stegoceras validus* appear to sort rather naturally into two groups, indicated by the closed and open circles.

***Structural Relationships among Variables*** If our only goal were to ordinate specimens to determine how much they differ from one another or whether they seem to sort into different groups, the analysis shown in Figure 3.14 would be sufficient. If we want to understand the nature of the differences, however, then it is essential that we know something about how the original variables are combined to produce the synthetic principal components. Groups of variables that are mutually correlated will tend to be represented in similar ways in the synthetic variables. Thus, in a general way, principal-component analysis allows us to explore structural relationships among variables.

This aspect of principal-component analysis is illustrated for the *Stegoceras* data in Table 3.1, which lists the correlations between the original variables (arrayed in rows) and the principal components (arrayed in columns). Each entry in this table is termed a **loading.**

TABLE 3.1

Summary of Principal-Component Analysis of *Stegoceras* Skull Measurements

| Variables | Principal Components 1 | 2 | 3 | 4 |
|---|---|---|---|---|
| 1. Dome length | 0.960 | −0.167 | −0.050 | −0.005 |
| 2. Dome width | 0.954 | −0.187 | −0.098 | 0.049 |
| 3. Anterior dome width | 0.918 | −0.040 | 0.090 | 0.187 |
| 4. Posterior dome thickness | 0.909 | −0.214 | −0.184 | −0.087 |
| 5. Dome thickness | 0.837 | −0.351 | 0.167 | −0.095 |
| 6. Anterior dome thickness | 0.947 | −0.086 | −0.097 | 0.051 |
| 7. Dome length (on curvature) | 0.945 | −0.166 | −0.071 | −0.115 |
| 8. Dome width (on curvature) | 0.946 | −0.110 | −0.138 | 0.028 |
| 9. Anterior dome length | 0.916 | −0.084 | −0.045 | 0.034 |
| 10. Posterior dome length | 0.918 | −0.108 | −0.143 | −0.170 |
| 11. Braincase length | 0.714 | 0.629 | 0.254 | −0.060 |
| 12. Length to braincase constriction | 0.595 | 0.695 | 0.248 | −0.206 |
| 13. Posterior braincase length | 0.220 | 0.824 | −0.311 | −0.363 |
| 14. Braincase width | 0.277 | 0.636 | −0.435 | 0.556 |
| 15. Anterior braincase length | 0.685 | 0.238 | 0.597 | 0.270 |
| Eigenvalue | 10.0 | 2.32 | 0.91 | 0.66 |

*SOURCE:* Chapman et al. (1981)

*NOTE:* The table shows the loadings of variables (rows) on principal components (columns). Refer to Figure 3.13 for the definition of the variables. Each eigenvalue is equal to the sum of the squared loadings for the corresponding principal component. The larger the eigenvalue, the greater the proportion of information summarized by the principal component. The first two eigenvalues are much larger than the remaining ones, indicating that most of the variation in the data is summarized by the first two principal components.

Relatively high loadings mean that the variable makes a substantial contribution to the principal component.

With biometric data of the kind represented here, it is common for most or all variables to have mutually high loadings on the first principal component. This component can then be interpreted, albeit only roughly, as a general measure of size. The second and higher principal components may have substantial loadings for just

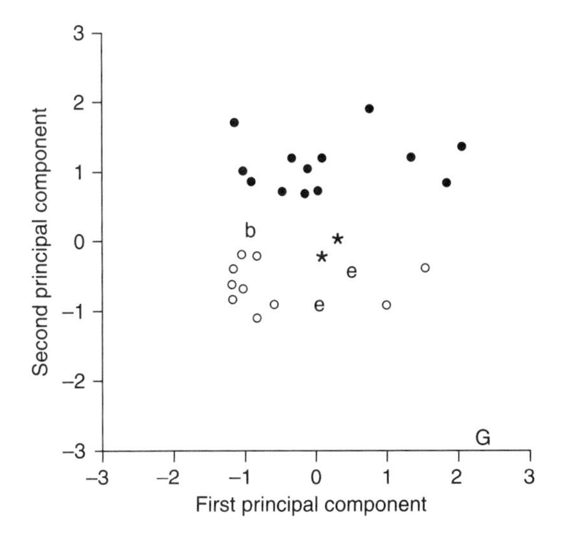


**FIGURE 3.14 Principal-component scores of *Stegoceras* specimens and the related species *Gravolithus albertae*.** Open and closed circles are specimens of *S. validus*. *S. browni* and *S. edmontonensis* are labeled *b* and *e*, the asterisks denote *Stegoceras* specimens of uncertain species affinity, and *Gravolithus* is labeled *G*. In contrast to the hypothetical case of Figure 3.12, this analysis has adopted a fairly common practice of standardizing the scores to have a mean of 0 and a variance of 1 on each principal component. This standardization is often done to portray each component as a biological factor of equal importance. It has little effect on the interpretation of results in this example. *(From Chapman et al., 1981)*

a few variables. The second principal component in this case has high loadings for braincase measurements only, suggesting that this component reflects the relative development of the braincase versus the dome.

The interpretation of the second principal component as a contrast between braincase and dome allows us to make some sense of the separation of individuals along this axis (Figure 3.14). Because the loadings for the dome characters are low and those for the braincase characters are high on this axis, a specimen with a high score will tend to have relatively low values of the dome variables and high values of the braincase variables. Thus, the upper group of specimens should have relatively larger braincases and relatively smaller domes.

This suggestion can be tested directly with bivariate analyses that compare dome and braincase measurements. A plot of dome length against braincase length (Figure 3.15) shows that the two sets of specimens do in fact differ in the relative development of the dome and braincase. A further question, one which

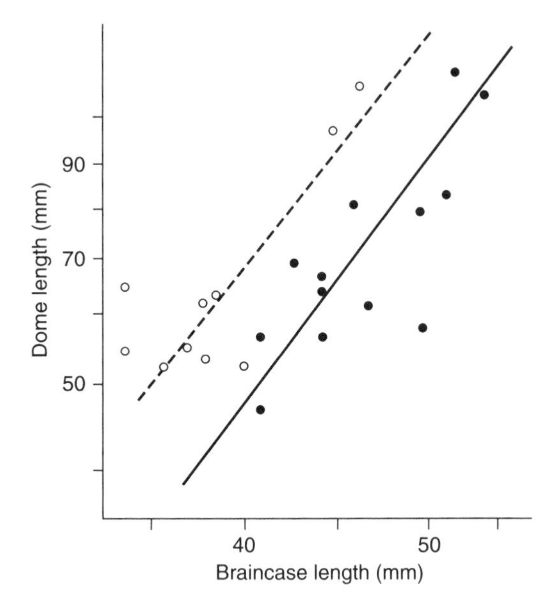


**FIGURE 3.15 Bivariate comparison of braincase length and dome length in specimens of *Stegoceras validus*.** Open and closed circles correspond to the two groups of specimens in Figure 3.14. Lines are fitted to each set of points. *(From Chapman et al., 1981)*

*Box 3.2*

## CLUSTER ANALYSIS

Cluster analysis begins with a matrix of similarities or dissimilarities between specimens. In this example, we focus on dissimilarity, which can be measured in numerous ways; here, it is simply calculated as the straight-line distance between two specimens in the complete, multivariate space representing the measurement data. Thus, if the number of variables measured is $m$ and if $x_{ik}$ represents the value of variable $k$ on specimen $i$, then the distance between any two specimens $a$ and $b$ is given by

$$d_{a,b} = \sqrt{\sum_{k=1}^{m} (x_{ak} - x_{bk})^2}$$

Table 3.2 shows the dissimilarities between a subset of the *Stegoceras* specimens. The **phenogram** or **dendrogram** (Figure 3.16) is constructed by finding those pairs of specimens that share mutually smallest dissimilarities; each is more similar to the other than to the rest of the specimens. In Table 3.2, these pairs are specimens 2 and 3, and specimens 6 and 7. Once these mutually most similar pairs have been found, the remaining specimens are joined with the existing clusters, and clusters are then joined together at several nested levels until all the clusters have been joined.

TABLE 3.2

Distances between a Subset of Specimens Used to Construct the Dendrogram of Figure 3.16

| | 1 | 2 | 3 | 4 | 5 | 6 | 7 |
|---|---|---|---|---|---|---|---|
| 1 | — | | | | | | |
| 2 | 1.9 | — | | | | | |
| 3 | 2.2 | 1.1* | — | | | | |
| 4 | 2.7 | 2.2 | 2.2 | — | | | |
| 5 | 2.1 | 2.1 | 2.1 | 2.1 | — | | |
| 6 | 1.7 | 1.8 | 1.6 | 1.3 | 1.0 | — | |
| 7 | 1.8 | 2.2 | 1.8 | 1.6 | 1.5 | 0.8* | — |

*NOTE:* These specimens are identified as cluster D in the dendrogram. Mutually most similar pairs are indicated by an asterisk.

cannot be answered by multivariate analysis alone, is *why* individuals differ in the degree of braincase development. One explanation that has been offered is that this difference reflects sexual dimorphism (Chapman et al., 1981).

We have seen that principal-component analysis can provide an ordination of specimens in a reduced number of dimensions and can facilitate the study of relationships among variables. In the example of *Stegoceras*, the ordination revealed what seem to be two distinct groups, which could be understood, by studying loadings, in terms of the original variables. Finding groups that are not known in advance is indeed one of the other major uses of multivariate analysis, to which we now turn.

***Classification of Specimens*** Paleontologists often start out studying a suite of specimens without knowing precisely how many natural groups are present. Determining the number of groups and the composition of each is the goal of **clustering** or **classification** techniques. One family of methods, collectively known as *cluster analysis*, is illustrated in Box 3.2, with the same *Stegoceras* measurements used in the principal-component analysis. The objective of cluster analysis is to summarize the morphological similarities and dissimilarities among specimens in the form of a **dendrogram.** This is a branching diagram that links similar specimens together into groups and separates them from other groups (Figure 3.16).

The *Stegoceras* specimens sort into about five clusters, labeled A through E in Figure 3.16. Comparing this dendrogram with the principal-component plot of Figure 3.14, we can see that specimens within each of the two groups identified on that plot tend to belong to the same clusters.

In contrast to the situation for which cluster analysis is used, we may want to determine whether two or more groups, designated in advance, differ appreciably in their measured traits. This is the problem of **discrimination,** as opposed to classification, and is discussed later in this

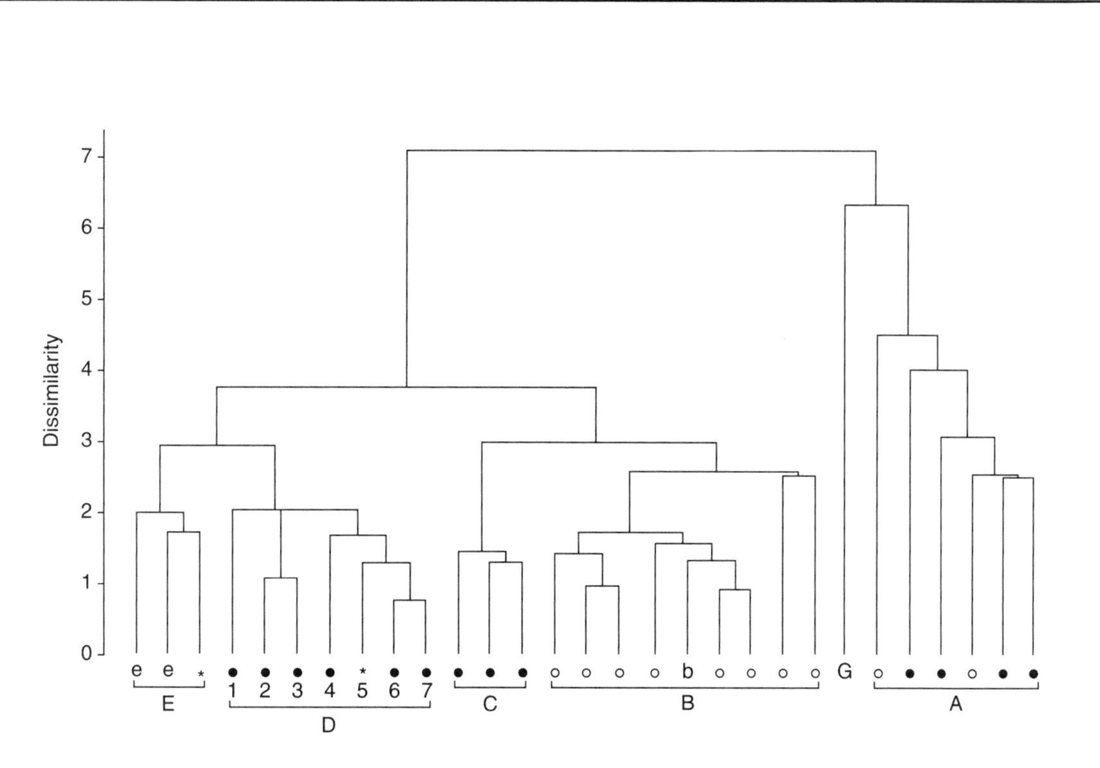


**FIGURE 3.16 Dendrogram depicting results of cluster analysis of *Stegoceras* specimens.** This analysis is intended to find groups based on overall morphological similarity. Symbols correspond to those in Figure 3.14. The numbered specimens in cluster D are discussed further in the text and in Tables 3.2, 3.3, and 3.4.

*continued on next page*

chapter. Discrimination often involves multivariate analogs of the analysis depicted in Figures 3.8 to 3.10.

## 3.3 THE BIOLOGICAL NATURE OF SPECIES

One of the most conspicuous and ancient observations, cutting across human cultures, is that organisms tend to sort out morphologically into relatively discrete clusters, the species of biology. Each species may have not only a distinctive form, but also physiology, behavior, trophic requirements, habitat, and so on. Species are, to varying degrees, ecologically as well as morphologically distinct. A major factor that maintains the distinctiveness of species is reproductive isolation, the evolution of which is at the core of the origin of new species from existing species. Likewise, the maintenance of reproductive isolation, once attained, is central to the maintenance of species distinctiveness.

### *Box 3.2 (continued)*

There is a large family of methods for determining how specimens join clusters and how clusters join one another. In this example, clusters of one or more specimens are linked if they have mutually smallest dissimilarity as measured by the average of all the pairwise dissimilarities between specimens in one cluster and specimens in the other.

Table 3.3 shows the dissimilarities that are relevant to the second round of clustering. Here, specimens 2 and 3 have been replaced by the cluster 2 + 3, and likewise for specimens 6 and 7. The dissimilarity between specimen 4 and cluster 2 + 3 is equal to the mean of $d_{2,4}$ and $d_{3,4}$ from Table 3.2. The remaining dissimilarities between specimens and clusters are calculated in the same way. The dissimilarity between clusters 2 + 3 and 6 + 7 is calculated as the mean of $d_{2,6}$, $d_{2,7}$, $d_{3,6}$, and $d_{3,7}$. There is a single mutually smallest distance in Table 3.2, namely, that between specimen 5 and cluster 6 + 7. Specimen 5 therefore joins this cluster. This procedure of recalculating the distance matrix and finding mutually closest pairs is repeated until all the specimens are joined.

A potential shortcoming of cluster analysis is that a nested structure is superimposed; all specimens eventually cluster together no matter how little they have in common. Moreover, multivariate data are compressed into a single dimension of overall morphological distance. It is therefore inevitable that there will be some distortion in the representation of dissimilarities. A simple and effective way to assess this distortion is to compare the dissimilarities implied by the dendrogram with the true, original dissimilarities based on all the variables.

When two clusters join in a dendrogram, every specimen in one cluster is represented as having the same dissimilarity vis-à-vis every specimen in the other cluster, even though the original pairwise

**TABLE 3.3**

Average Distances between Specimens and/or Clusters after One Round of Clustering of the Specimens in Table 3.2

| | 1 | 2 + 3 | 4 | 5 | 6 + 7 |
|---|---|---|---|---|---|
| 1 | — | | | | |
| 2+3 | 2.05 | — | | | |
| 4 | 2.7 | 2.2 | — | | |
| 5 | 2.1 | 2.25 | 2.1 | — | |
| 6+7 | 1.75 | 1.85 | 1.45 | 1.25* | — |

*NOTE:* The one mutually most similar pair is indicated by an asterisk.

**TABLE 3.4**

Implied Distances between Specimens of Cluster D, Based on the Dendrogram of Figure 3.16

| | 1 | 2 | 3 | 4 | 5 | 6 | 7 |
|---|---|---|---|---|---|---|---|
| 1 | — | | | | | | |
| 2 | 2.0 | — | | | | | |
| 3 | 2.1 | 1.1 | — | | | | |
| 4 | 2.0 | 2.0 | 2.0 | — | | | |
| 5 | 2.1 | 2.0 | 2.0 | 1.7 | — | | |
| 6 | 2.1 | 2.0 | 2.0 | 1.7 | 1.3 | — | |
| 7 | 2.1 | 2.0 | 2.0 | 1.7 | 1.3 | 0.8 | — |

## The Biologic Species Concept

The most widely accepted biologic definition of the species was formulated by Ernst Mayr (1942): "Species are groups of actually or potentially interbreeding natural populations, which are reproductively isolated from other such groups" (p.120). The species is referred to as a group of populations, emphasizing the fact that most species are divided geographically into subunits or breeding populations. It is explicit in the definition that such breeding populations are actually or potentially interbreeding with one another. Two populations are said to be reproductively isolated only if interbreeding would not occur if they both lived in the same area. Thus, "potentially" in the species definition is particularly critical. An important part of the species definition is that populations of different species are reproductively isolated from one another under natural conditions. There are many examples of species hybridizing readily in captivity or under domestication. This stems from the fact that

dissimilarities may vary quite a bit. This implied dissimilarity is equal to the height on the dendrogram at which the clusters join. Table 3.4 shows the implied dissimilarities for the specimens listed in Table 3.2, and Figure 3.17 compares the implied and original dissimilarities for all pairs of specimens used to construct the dendrogram. The correlation coefficient between original and implied dissimilarities measures how well the original data are represented by the dendrogram. In this case, it is equal to 0.83, a relatively high value, which suggests that the original dissimilarities are represented reasonably well by the dendrogram.

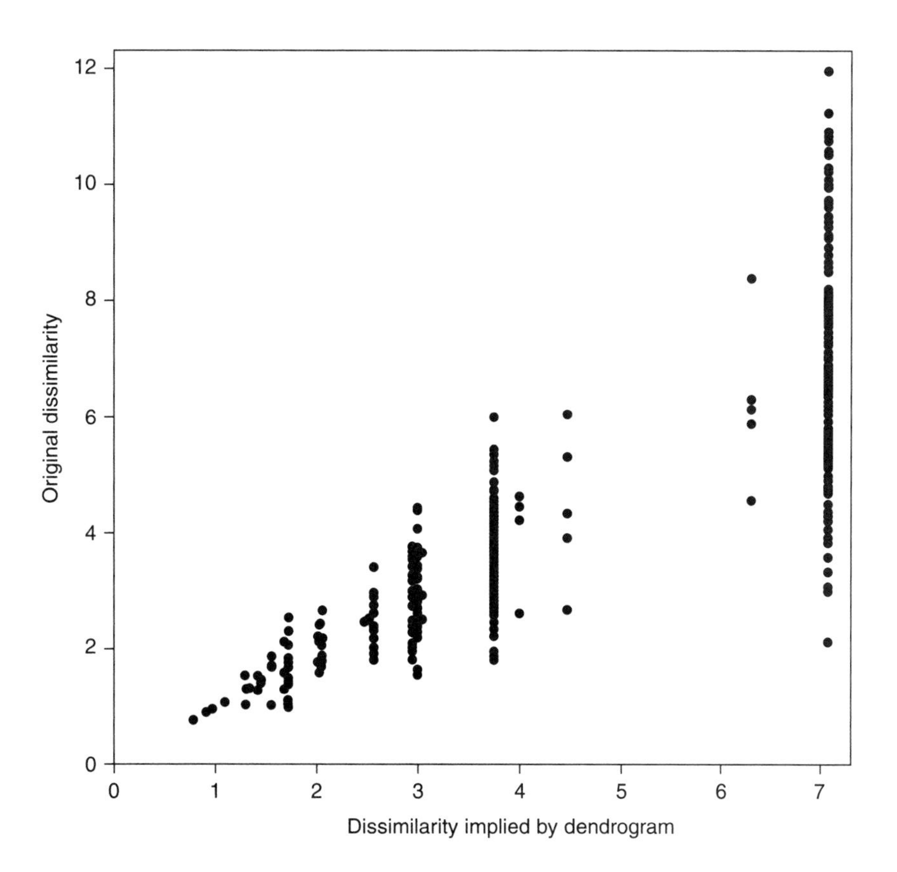


**FIGURE 3.17 Comparison of original dissimilarities between *Stegoceras* specimens and dissimilarities implied by the dendrogram of Figure 3.16.** Points line up vertically because all between-cluster specimen pairs have the same implied distance when two clusters join.

reproductive isolation often depends on ecologic or behavioral barriers that tend to break down in captivity.

The biologic species concept has some shortcomings. Chief among them are the occasional existence of evolutionary intermediates between species and the difficulty of applying the concept when reproduction is asexual. Biologists who study groups in which asexual reproduction is the rule sometimes adopt a species concept based on phenotypic attributes, such as biochemical properties in bacteria. We focus on the biologic concept because it is thought to apply reasonably well for many of the paleontologically important groups of organisms.

## The Origin of Species

If two or more populations of a species diverge to a sufficient extent genetically, they may become reproductively isolated and thus come to be distinct species. One of the principal questions in the study of the origin of species, or **speciation,** concerns the geographic relationships of the diverging populations. Do they have overlapping geographic ranges, in which case they are referred to as **sympatric,** or do they have disjunct ranges—that is, are they **allopatric?** Because gene flow can reduce distinctions between populations, and because populations living in the same broad area may be subject to largely the same forces of natural selection, it seems reasonable to presume that speciation should occur mainly between allopatric populations. In fact, this is the prevailing view among biologists, although there are many theoretical and empirical arguments in favor of sympatric speciation as well.

For allopatric speciation to take place, a population must first become geographically isolated from other populations of the species; then it must persist for some time; and finally it must attain reproductive isolation. Geographic isolates are forming all the time, as organisms disperse and found new populations geographically separated from parental populations, and as newly created geographic barriers, such as mountains, rivers, and emergent land, split populations. The resulting populations represent potential new species, but their fate is not at all assured. Many isolates become extinct, either because they start out with relatively few individuals and therefore are susceptible to fluctuations in population size, or because the environments they colonize may be unfavorable or ephemeral.

If a geographically isolated population does become established, even occasional migration of individuals between populations can lead to sufficient gene flow to prevent reproductive isolation from developing. Gene flow on a large scale is facilitated by the spatial shift of environments over time, which promotes migration as populations track the local conditions to which they are adapted. The probability that a geographically isolated population will actually become a new species is therefore generally quite low.

Our understanding of speciation comes mainly from biology rather than paleontology. Nonetheless, how species originate—that is to say, how populations become reproductively isolated and how evolutionary change is associated with this process—has important paleontological implications that we will pursue further in Chapter 7.

## Discrimination of Species

It is important to distinguish between how species are defined in principle and how they are recognized in practice. Biologists rarely perform breeding experiments to determine whether two populations are part of the same species, and of course paleontologists cannot do so with fossil populations. Except for the availability of behavioral data and the widespread analysis of genetic data in biology, the approaches of biologists and paleontologists are often rather similar: One typically starts by determining whether the phenotypic difference between two populations is large relative to the variation within the populations (see Figure 3.9).

Figure 3.18 shows an example of this approach with corals from the Silurian of Arctic Canada. Here there are three clear groups that do not overlap: *Heliolites* aff. *H. luxarboreus*, *H. diligens*, and *H. tchernyshevi*. These are accepted as distinct species on morphological grounds. A fourth form, *H.* sp., is rather similar to *H. tchernyshevi* with respect to the characters portrayed here, but it is not known from enough material to assess its variation in these characters. It is nevertheless accepted as a distinct species because it differs from the remaining species in other characters, such as the nature of the septa, or vertical plates within the corallites.

Genetic data, either in the direct form of DNA sequences or in the indirect form of proteins, have also proven invaluable in discriminating living species, and genetic analysis is now part of the standard toolkit of biologists. (See Box 3.3.) If two populations differ from each

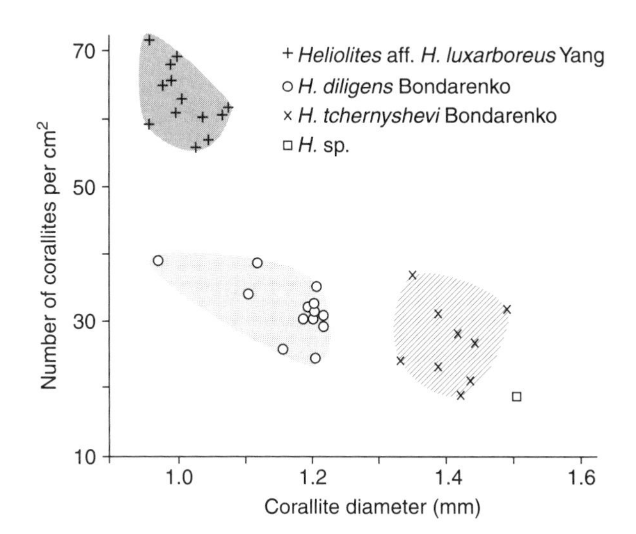


**FIGURE 3.18 Morphological discrimination of species of the coral *Heliolites* from the Silurian of Arctic Canada.** *(From Dixon, 1989)*

other by as much as two closely related species typically do, they are often regarded as belonging to distinct species. Genetic data can be used to great advantage when morphological differences are negligible or difficult to observe. As is true with morphological data, however, there is no formula that says how much genetic difference characterizes distinct species.

## Morphologic and Biologic Species

In practice, both biologists and paleontologists usually apply a morphologic species concept. There are several important problems that stem from this approach.

Failing to take variation into consideration can lead to biologically unrealistic results. Figure 3.19 shows an example involving the Triassic ammonoid genus *Parananmites* from the Great Basin of the western United States. This graph plots two separate characters, the whorl width ($W$) and the umbilical width ($U$), against the shell diameter. Each point is a single specimen and each field in the graph represents a separate bivariate comparison. Within each bivariate comparison, the points form a continuous distribution. There are no obvious divisions or clusters that would serve as evidence for multiple species. Partly on these grounds, Bernhard Kummel and Grant Steele (1962) concluded that the material represents a single species, *Parananmites aspenensis*.

Thirty years before Kummel and Steele performed this analysis, J. P. Smith (1932) studied a subset of this material. In addition to *P. aspenensis*, Smith erected three other species, based mainly on differences relative to *P. aspenensis* in overall size, whorl width, and umbilical diameter, as well as on details of sculpture. Given that Smith studied the same traits as Kummel and Steele, how

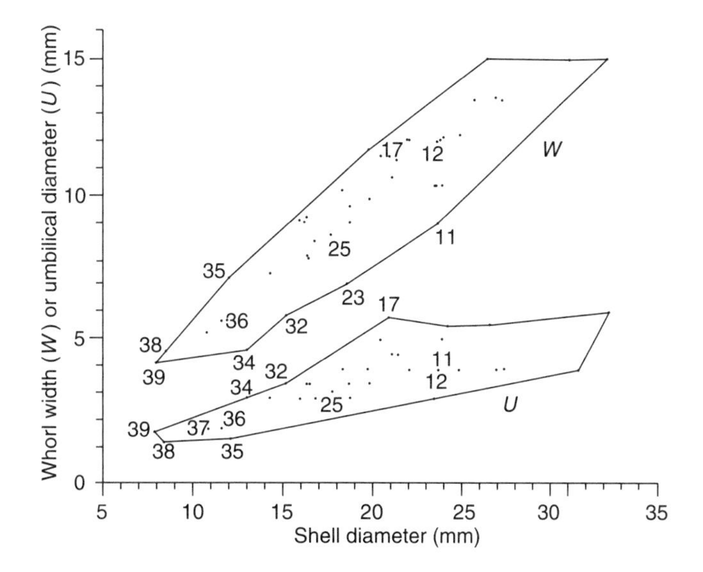


**FIGURE 3.19 Biometric analysis of the ammonite species *Parananmites aspenensis* from the Triassic of the Great Basin.** Two separate bivariate comparisons are shown here: whorl width ($W$) against shell diameter, and umbilical diameter ($U$) against shell diameter. Each point represents one specimen. The numbered points are type specimens that had previously been used to describe this species and three additional species. Because they show continuous variation, all the specimens are now considered to belong to a single species. The type specimens tend to fall near the extremes of the continuous distribution of form. *(From Kummel & Steele, 1962)*

can we account for the different numbers of species recognized by these authors? The numbered points on Figure 3.19 are Smith's type specimens—the exemplars he chose as representative of the species he described [SEE SECTION 4.1]. Most of these lie at the periphery of the scatter of points. Smith evidently focused on extreme forms and considered them to be representatives of separate species, rather than recognizing them as simply end-members of a continuum.

There are potential problems with the use of morphologic species, in both biology and paleontology, that cannot easily be overcome with more detailed assessment of morphological variation. First is the existence of **cryptic species,** also known as **sibling species.** Closely related species may be genetically and behaviorally distinct but may lack clear morphological differences. Second, species may contain numerous distinct morphological types, or **polymorphs.** The different forms within a polymorphic species are under genetic control, but they are not reproductively isolated and the genetic differences involved are generally small. Nonetheless, polymorphs are sometimes sufficiently different in form that they might be mistaken for distinct species on the basis of morphology alone. Finally, as we discussed earlier, some of the variation within species is ecophenotypic rather than heritable. Thus, two populations that belong to the same species could be mistaken for different species if they lived in environments that induced substantially different phenotypes.

There is no question that these problems exist in principle, but it is important to determine how common they are in reality. One study that explores this question involves living species of the cheilostome bryozoan genera *Steginoporella*, *Stylopoma*, and *Parasmittina* from the Caribbean Sea.

Using multivariate morphometric techniques similar to those we discussed earlier, Jeremy Jackson and Alan Cheetham (1990, 1994) analyzed a variety of skeletal measurements and found morphological clusters of specimens that were defined operationally as **morphospecies.** Once the morphospecies were established, Jackson and Cheetham sought to assess the importance of ecophenotypic variation. Embryos of known parentage were raised in environments different from those in which their parents had been raised. After rearing, the offspring were measured and assigned to prospective parents on the basis of morphological similarity. That is, each of the offspring was assigned to the parental colony with which it was morphologically most similar. For all seven species studied, these assignments were found to be correct—matching true parentage—99 to 100 percent of the time, despite the fact that parents and offspring did not share the same environment. On the whole, morphological variation was much more strongly affected by heritability than by variation in the environment in which the embryos grew.

Jackson and Cheetham then tested for polymorphism by asking whether morphologically distinct species have consistent genetic differences. To identify genetic differences, they used the standard technique of electrophoresis, which identifies alternative forms of proteins having different mass and electrical properties. Because proteins are coded by DNA, the alternative forms of protein are used as evidence for differences in DNA sequence. In general, different forms of the same gene are referred to as **alleles.** Here the different proteins are inferred to represent different alleles. For a given kind of gene, each individual inherits one allele from its mother and one from its father. For that gene, the combination of two alleles is the individual's genotype.

Box 3.3 gives one example of how the genetic results are interpreted to test for differences between populations. When this approach was applied to the bryozoans, every pair of distinct morphospecies within a genus was found to have at least one diagnostic genetic difference. Thus, these morphospecies are likely to be true biological species rather than polymorphs within a single species. Moreover, if genetic and morphological dissimilarity between populations are compared, it is found that the magnitudes of morphological and genetic difference are well correlated (Figure 3.20). Pairs of populations that are more dissimilar morphologically also tend to be more dissimilar genetically.

Finally, Jackson and Cheetham tested for the existence of cryptic species by determining whether different populations of the same morphospecies have diagnostic genetic differences. The analysis found no cases in which two populations of the same morphospecies could be genetically distinguished with confidence. In other words, populations that could not be distinguished morphologically could not be distinguished genetically, either. Thus, there was no compelling evidence for the existence of cryptic species in these genera.

*Box 3.3*

## TESTING FOR DIAGNOSTIC GENETIC DIFFERENCES BETWEEN POPULATIONS

In the genus *Stylopoma*, there are four alternative forms of the protein GPI. By genetically assaying many individuals (about 40 on average) within each morphospecies, it was found that these four alleles, denoted *a* through *d*, are present in different frequencies in the two morphospecies. Given the standard assumption of random mating between individuals within a species, the allele frequencies allow the genotype frequencies to be estimated. For example, the frequencies of the *b*- and *c*-alleles in *S.* sp. 1 are $f_b = 0.139$ and $f_c = 0.583$. The frequency of the *bc* genotype is therefore inferred to be equal to $2f_b f_c$, or 0.162. (The multiplication by 2 reflects the fact that an individual can inherit the *b*-allele from either its mother or its father, and likewise for the *c*-allele.)

Once the genotype frequencies are determined, we see that most genotypes are unique to one species or the other. If an individual has the *aa* or *ab* genotype, it belongs to *S.* sp. 2. If it has the *bc*, *bd*, *cc*, *cd*, or *dd* genotype, it belongs to *S.* sp. 1. The only ambiguous genotype is *bb*. Because the vast majority of *bb* individuals are in *S.* sp. 2, our best guess would be to assign any such individual to that species. If we assume that the two species are represented by the same number of individuals, then the probability that a randomly sampled individual will belong to *S.* sp. 1 *and* will have the *bb* genotype is equal to $f_b^2 \div 2$, which in this case is only $0.019 \div 2$, or less than 1 percent. In other words, if we use the GPI genotypes to assign individuals to morphospecies, we will be wrong less than 1 percent of the time. Operationally, genetic markers with which the expected probability of misclassifying an individual is less than 1 percent are considered to be diagnostic.

**TABLE 3.5**

Allele and Genotype Frequencies of the Protein GPI in Two Species of the Cheilostome Bryozoan *Stylopoma*

| | **Symbol for Frequency** | **Frequency *Stylopoma* sp. 1** | **Frequency *Stylopoma* sp. 2** |
|---|---|---|---|
| **Allele** | | | |
| *a* | $f_a$ | — | 0.188 |
| *b* | $f_b$ | 0.139 | 0.812 |
| *c* | $f_c$ | 0.583 | — |
| *d* | $f_d$ | 0.278 | — |
| | **Formula for Frequency** | **Frequency *Stylopoma* sp. 1** | **Frequency *Stylopoma* sp. 2** |
| **Genotype** | | | |
| *aa* | $f_a^2$ | — | 0.035 |
| *ab* | $2f_a f_b$ | — | 0.305 |
| *bb* | $f_b^2$ | 0.019 | 0.659 |
| *bc* | $2f_b f_c$ | 0.162 | — |
| *bd* | $2f_b f_d$ | 0.077 | — |
| *cc* | $f_c^2$ | 0.340 | — |
| *cd* | $2f_c f_d$ | 0.324 | — |
| *dd* | $f_d^2$ | 0.077 | — |

*SOURCE:* Jackson & Cheetham (1990)

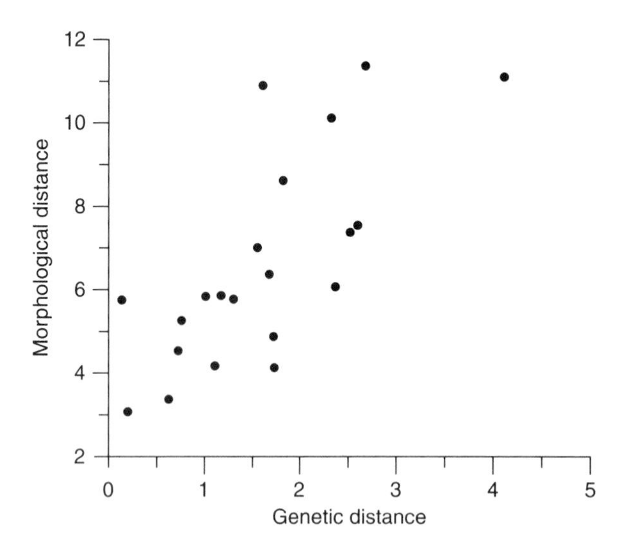


**FIGURE 3.20 Comparison of morphological and genetic dissimilarity between populations of the bryozoan *Stylopoma*.** Each point represents a comparison between two populations. Morphological distance is measured by a variant of the straight-line distance described in the discussion of cluster analysis in Box 3.2. Genetic distance is measured on the basis of differences in gene frequencies. (See Table 3.5 for examples of gene frequencies.) Morphological and genetic differences are positively correlated. *(From Jackson & Cheetham, 1994)*

Taken together, these results suggest that there is an excellent concordance between biological and morphological species in this sample of cheilostome bryozoans.

The question of concordance between morphological and biological species applies to biology as much as to paleontology. There is a special problem, however, that paleontologists must face because of the temporal dimension of the history of life. Our earlier discussion of speciation was restricted to the situation in which an evolving lineage splits into two distinct lineages. It sometimes happens that a single lineage may evolve over time to the point where it becomes morphologically quite distinct from earlier populations in the lineage, even though there has been no splitting (Figure 3.21). In cases like this, some paleontologists will divide the lineage into two or more named species. Because of the added time dimension, species such as A and B in Figure 3.21 may be referred to as **chronospecies.** Many workers today prefer, if possible, to place species boundaries at branching points and at true lineage terminations. It may be difficult to avoid erecting chronospecies, however, if the intermediate forms between A and B are not sampled.

## 3.4 CONCLUDING REMARKS

That there is a close correspondence between morphological and genetic species in a sample of bryozoans does not imply that the same is true for other groups of organisms, or even for other bryozoans. If these results prove to be general, however, then biologists and paleontologists are in a strong position to discriminate species on the basis of morphology. It is still too soon to assess fully the correspondence between morphospecies and biological species. Nonetheless, studies on many other groups of organisms have shown that, as in the bryozoans, morphologically defined species tend to be genetically distinct. At the same time, cryptic species are known to be common in some groups.

There is thus an asymmetry in the relationships between morphological and genetic species. If two populations are morphologically distinct, there is often a good chance that they belong to different species. But if they are morphologically indistinguishable, this need not imply that they belong to the same species. This asymmetry will be relevant when we consider the relationship between speciation and morphological evolution in Chapter 7.

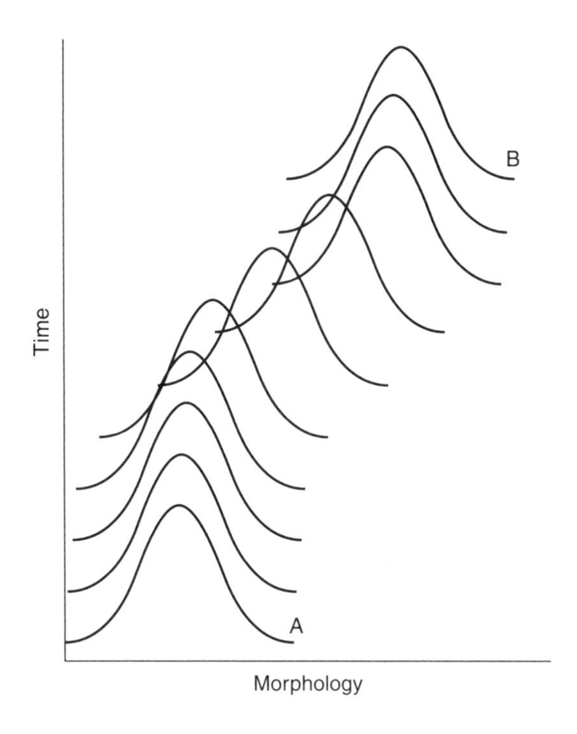


**FIGURE 3.21 The problem of species delimitation in an evolving lineage.** Each curve represents a frequency distribution for a trait. The lineage is a single, continuous succession of populations, yet populations at different points in time—for example, points A and B—may be so different from each other that they would be taken for different species if found together.

# SUPPLEMENTARY READING

Coyne, J. A., and Orr, H. A. (2004) *Speciation*. Sunderland, Mass., Sinauer Associates, 545 pp. [An in-depth review of ecological, genetic, and geographic aspects of speciation.]

Davis, J. C. (2002) *Statistics and Data Analysis in Geology*, 3rd ed. New York, John Wiley and Sons, 638 pp. [Includes intuitive and graphical explanations of many multivariate techniques.]

Futuyma, D. J. (1998) *Evolutionary Biology*, 3rd ed. Sunderland, Mass., Sinauer Associates, 763 pp. [Comprehensive text on genetics, evolution, and speciation.]

Hanski, I., and Gilpin, M. E. (eds.) (1996) *Metapopulation Biology: Ecology, Genetics, and Evolution*. San Diego, Academic Press, 512 pp. [A collection of theoretical papers and case studies.]

Knowlton, N. (1993) Sibling species in the sea. *Annual Review of Ecology and Systematics* **24:**189–216. [A broad survey of the problem of cryptic species.]

Reyment, R. A. (1991) *Multidimensional Palaeobiology*. Oxford, U.K., Pergamon Press, 377 pp. [Techniques of multivariate analysis applied to a range of paleontological problems.]

Siegel, S., and Castellan, N. J., Jr. (1988) *Nonparametric Statistics for the Behavioral Sciences*, 2nd ed. New York, McGraw-Hill, 399 pp. [Statistical analysis of univariate and bivariate data, with special emphasis on categorical and ordinal variables.]

Sneath, P. H. A., and Sokal, R. R. (1973) *Numerical Taxonomy*. New York, W. H. Freeman, 583 pp. [Includes important discussions of biological dissimilarity, ordination, clustering, and discrimination.]

Sokal, R. R., and Rohlf, F. J. (1995) *Biometry: The Principles and Practice of Statistics in Biological Research*, 3rd ed. New York, W. H. Freeman, 887 pp. [Comprehensive text on statistical analysis of biological data.]

# SOFTWARE

Hammer, Ø., Harper, D. A. T., and Ryan, P. D. (2001) PAST: Paleontological statistics software package for education and data analysis. *Palaeontologia Electronica,* volume 4, issue 1, article 4, 9 pp. [Noteworthy for worked examples of many methods of data analysis. Available for free download at http://folk.uio.no/ohammer/past/.]

McCune, B., and Mefford, M. J. (2002) *PC-ORD: Multivariate Analysis of Ecological Data, Version 4.* Glenden Beach, Oreg., MjM Software Design. [Software package with a variety of options for ordination and clustering of data. Available for purchase at www.ptinet.net/mjm.]

R Development Core Team (2004). *R: A Language and Environment for Statistical Computing.* Vienna, Austria, R Foundation for Statistical Computing. [Useful software for univariate, bivariate, and multivariate analysis, with powerful graphical capabilities. Available for free download at http://cran.r-project.org.]

# *Chapter 4*

# SYSTEMATICS

In the previous chapter, we discussed the nature and recognition of species. An important observation, predating the acceptance of biological evolution, is that all known species, despite their great numbers and enormous differences, appear to be naturally organized into more inclusive units. **Systematics,** broadly speaking, is the study of the diversity of organisms and the relationships among them. A major part of systematics is **taxonomy,** the theory and practice of describing and classifying organisms. Systematics accounts for a large part of all paleontological research, and the results of systematic studies form the foundation of many other areas of investigation. In this chapter, we focus on the procedures most important to practicing paleontologists: species description, inference of evolutionary relationships, and classification of species into more inclusive taxa or higher categories.

## 4.1 FORMAL NAMING AND DESCRIPTION OF SPECIES

A new species can be erected in biology or paleontology either because previously unnamed specimens have become available or because a previously recognized species is judged actually to be two or more species. An individual worker's views on the breadth of species and on the division of genera into species are guided by judgment that reflects the accumulated experience with a group of organisms (Chapter 3). In contrast, the formal naming of species is governed by widely accepted systems of rules and procedures. One of the most important of these systems is the *International Code of Zoological Nomenclature.* The Code applies to taxonomic categories from the subspecies to the superfamily. Emphasis in this chapter will be on its application to the species category.

A comparable set of procedures for plants is known as the *International Rules of Botanical Nomenclature.* In addition, other rules are commonly used for groups like the bacteria. Because these codes are independent, it is possible for the same formal name to be applied both to a plant and an animal, although in practice this is uncommon. For our purposes, the differences between these codes are minor. (For example, the Botanical Code requires a new species name to be accompanied by a description or diagnosis in Latin—a rule that, perhaps fortunately for the paleontologist, does not apply to fossils.) We will therefore focus on the Zoological Code. The possibility of adopting a single set of standard rules for plants and animals has been seriously discussed but not yet implemented.

For a species to be officially recognized, it must be given a name in binomial form; that is, the name must consist of two words. The official name for the human genus is *Homo*, and *Homo sapiens* is the species name. A species name like *sapiens* (sometimes called the *trivial name* or *epithet*) is meaningless unless associated with a genus name. In practice, most newly discovered species can be assigned readily to an existing genus and thus the act of describing a new species involves the invention of only one name. If the new species cannot be accommodated within an established genus, a new genus must be erected and named at the same time, and the new species is assigned to it.

Except for the genus assignment just mentioned, the Code does not insist on the complete classification of a new species at all levels from the family up to the kingdom, recognizing that complete classification is often difficult or impossible, particularly if the new species is quite distinct from all other known species.

Another requirement of the new name is that it not already be in use (**occupied**). This restriction refers to the combination of genus and species names. By convention, repeating trivial names in closely related genera is also avoided because genus affiliations may change as knowledge of the evolutionary relationships between species changes.

The names for species and genera must be Latin words or words that have been latinized. There is considerable latitude in the choice of words to be used as names—latinized place names, names of people, and descriptive words are all used.

For the name of a new species to be recognized, it must be published in an approved and widely accessible medium. A new name is not officially recognized if it has been used only in the labeling of a museum specimen or described orally before a scientific meeting. Nor is anonymous publication recognized. If a new species name is erected according to the rules of nomenclature and is validly published, it is said to be **available.** The nature of publication itself has evolved over the history of biological nomenclature, and the Code has changed to reflect this. Until recently, a new species name had to be published in ink on paper. The latest edition of the Code also recognizes publication in media such as read-only optical disks, although conventional print is still recommended. Dissemination over the Internet is not recognized, although this may soon change. To simplify bibliographic work, the Code recommends that publication be in French, German, English, or Russian.

The Code specifies that each newly described species be accompanied by the designation of a type specimen or set of type specimens. Type specimens must be clearly labeled, and suitable measures must be taken for their preservation and accessibility, which means that type specimens are usually deposited in a major museum where curatorial facilities are available. Illustration of specimens is strongly recommended and may be considered mandatory.

The type specimens do not in fact *define* a species; rather, they are the name bearers for that species. When a species is named, the name is formally attached only to the one or more specimens that are designated as type specimens. In practice, type specimens are often somewhat unusual representatives of the species (see Figure 3.19). The most common biases are toward large size and good preservation.

If a single specimen is designated, it is called the **holotype.** If several specimens serve this purpose, they are called **syntypes.** Both alternatives are officially acceptable, although the Code urges the use of a holotype rather than a series of syntypes because it is always possible that the series will be judged by later workers to contain representatives of more than one species.

Several other kinds of type specimens figure prominently in taxonomic work. A **paratype** is a specimen other than the holotype, which is formally designated by the author of a species as having been used in the description of the species. The designation of a single holotype and a series of paratypes thus contains some of the advantages of both the holotype system and the syntype system. The holotype remains the name bearer but the paratypes, which may be numerous, serve to express more fully the author's concept of the species. Some of the more important kinds of types are summarized in Table 4.1.

**TABLE 4.1**
Kinds of Type Specimens

| Name | Usage | Etymology |
|---|---|---|
| Holotype | Single specimen designated as name bearer | *holo-*, complete |
| Syntypes | Several specimens designated as name bearers | *syn-*, together |
| Paratype | Specimen, other than holotype, used in species description | *para-*, side by side |
| Lectotype | Syntype later chosen as definitive type | *lecto-*, chosen |
| Neotype | Replacement for lost or destroyed type | *neo-*, new |
| Plesiotype | Specimen used in redescription of existing species | *plesio-*, near |

A description of a new species typically contains a number of elements:

1. *Headings* include the name and author of the genus and other higher taxa to which the species is assigned.
2. The *scientific name* is in binomial form.
3. *Figure numbers* indicate where in the publication the species is illustrated.
4. The *diagnosis* is a listing of characteristics by which the new species can be distinguished from other species.
5. *Type material* is explicitly listed.
6. The *etymology* explains the derivation of the name.
7. A *description* is given; in this context, a description is a full assessment of characteristics without particular reference to similarities and differences relative to other recognized species.
8. A *discussion* section may include information about nongenetic variation, ontogenetic stages, evolutionary affinities with other species, and, for fossils, state of preservation.
9. A section on *occurrence* lists information on habitat and, for fossils, stratigraphic horizon.
10. The *distribution* section may be a list of places at which the new species has been found (in addition to the type locality). With regard to habitat, paleontologists are most concerned with the geologic setting (rock type, for example).

In addition to these standard parts of a description, it is also common to include a list of material examined, with reference to museum repositories of specimens other than those formally designated as types, and a section with biometric data. At a minimum, the major dimensions of the type specimen or specimens should be included. For a description of a new species based on previously known material, or for a redescription of a known species, the species description typically also includes a section on taxonomic references and a history of nomenclature for the species.

Three actual examples of species description are given in Boxes 4.1, 4.2, and 4.3, with illustrations of type specimens.

Because one of the prime objectives of diagnosis and description is communication of information, there is a premium attached to consistency. This means, for example, that standardized morphologic terminology is used wherever possible. The description, as distinct from the diagnosis, serves several purposes—not the least important of which is to provide an assessment of attributes that may at some future time be critical in diagnosis. If the species is part of a well-known group and is similar in most regards to other species, much of the description may be neglected in deference to the existing descriptions of closely related species, and a simple diagnosis may suffice.

For a species belonging to a relatively unknown group, the description must be more comprehensive so that relevant comparisons can be made if related species are discovered subsequently. When a new species is assigned to a new genus and family containing only that species, the diagnoses of the species, genus, and family will generally be quite similar. In such cases, the genus and family diagnoses are likely to require revision if closely related species are discovered in the future.

If possible, a description should include discussion of ontogenetic development, particularly if the organism's ontogeny is accompanied by a substantial change in form. Also important is an assessment of variation encountered within and between populations of the species.

How much specimen material is necessary to establish a new species? No unequivocal answers can be given to this question because what is necessary and possible in a particular description depends on the amount of difference between related species and on the quantity and quality of preserved material. Often a single specimen demonstrates that the organism is different from all other known organisms. On the other hand, if a new species belongs to a well-known group in which differences between species tend to be rather subtle, a large amount of material must be accumulated to make the description complete and effective and to establish that the new species is truly distinct from other, related species [SEE SECTION 3.2].

There are some noteworthy cases in which different parts of a fossil organism were initially described as separate species because they were found in isolation. Only later were the different "species" found in association in such a way that they clearly belong to the same organism [SEE SECTION 10.2]. The Upper Devonian **PROGYMNOSPERM** *Archaeopteris* represents a striking but by no means unique case. The leafy branches of *Archaeopteris* were first described in the 1870s, and the genus *Callixylon* was erected in 1911 for various woody stems. Although both genera are very common and were sometimes found in the same sedimentary

*Box 4.1*

## *STENOSCISMA PYRAUSTOIDES* COOPER AND GRANT, N. SP.

The following refers to Figure 4.1 and Table 4.2.

Large for genus; outline broadly subelliptical to subtrigonal, sides diverging between 80° and 125°, normally over 100° in adults, maximum width near midlength, normally slightly farther toward the anterior; profile strongly biconvex to subtrigonal; commissure uniplicate, fold moderately high, standing increasingly high anteriorly, beginning 1–5 mm anterior to brachial beak; sulcus rather shallow, but dipping steeply at anterior, extending forward as broad tongue, producing emargination of anterior. Costae strong and sharp crested on fold and in sulcus, lower, broader, and rounder on flanks, beginning at beaks, frequently bifurcated, especially on fold and sulcus, numbering 6–10 on fold (normally 9), one less in sulcus, 4–9 on each flank, number not necessarily equal on both sides; stolidium better developed on brachial valve, varying from broad and fanlike to nearly absent.

Pedicle valve flatly convex transversely and from beak to flanks, strongly convex longitudinally through sulcus; beak short, only moderately thick, suberect to erect but not hooked; beak ridges gently curved, ill-defined; lateral pseudointerareas elongate, narrow, normally covered by edge of brachial valve; delthyrium moderately large, sides only slightly constricted by small, normally widely disjunct deltidial plates; foramen large for genus, nevertheless small, opening ventrally.

Brachial valve strongly convex transversely, only moderately convex along crest of fold owing to anterior increase in height of fold, convexity uniform without swelling in umbonal region; beak bluntly pointed, apex only slightly inside pedicle valve.

Pedicle valve interior with small teeth, continuous with dental plates that form short, boat-shaped spondylium just above floor of valve; median septum low, extending slightly forward of spondylium; troughs of vascula media diverging from midline of valve just anterior to median septum, extending directly across floor of valve; muscle marks in spondylium faint and undifferentiated.

Brachial valve interior with short, broad hinge plate, semicircular to crescentic; cardinal process at apex of hinge plate, located just beneath apex of valve, low or rather high, knoblike, normally not polylobate, shallowly striate for muscle attachment; hinge sockets short, narrow, at lateral extremes of hinge plate, finely corrugated; crural bases slightly diverging anterior to cardinal process, space between filled by narrow crural plates dipping along center line attaching crural bases to top of intercamarophorial plate; brachial processes not observed, presumed to be normal for

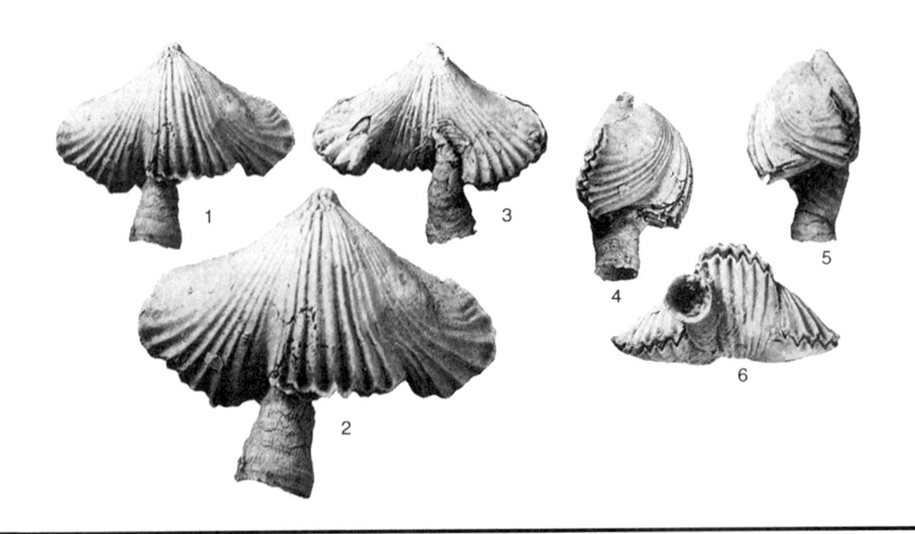


**FIGURE 4.1 Permian brachiopod *Stenoscisma pyraustoides* Cooper and Grant.** The original species description is printed in this box. The photographs are of the holotype (with a coral cemented to it). *(From Cooper & Grant, 1976)*

genus; median septum high, thin, exceptionally short, length increasing greatly with height; camarophorium narrow, relatively short, anteriorly widening; intercamarophorial plate low, thick, relatively long; muscle marks not observed.

STRATIGRAPHIC OCCURRENCE. Skinner Ranch Formation (base); Hess Formation (Taylor Ranch Member); Cibolo Formation.

LOCALITIES. Skinner Ranch: USNM 705a, 705b, ?709a, 711o, 711z, 715c, 716p, 720e, 726j, 729j. Taylor Ranch: USNM 716o. Cibolo: USNM 739-1.

DIAGNOSIS. Exceptionally large and wide *Stenoscisma* with numerous bifurcations of costae on posterior of fold and flanks.

TYPES. Holotype: USNM 152220i. Figured paratypes: USNM 152219a-d; 152220b,c,k; 152221a,b; 152225. Measured paratypes: USNM 152220a-h,j; 152225. Unfigured paratypes: USNM 152220a,d-h,j.

COMPARISONS. *Stenocisma pyraustoides* is characterized by its exceptional width, large maximum size, numerous and frequently bifurcating costae on flanks, short beak with small disjunct deltidial plates, relatively short spondylium and camarophorium.

The only known species that is closely related to *S. pyraustoides* is *S. multicostum* Stehli (1954, cited in Cooper & Grant, 1976) from the Sierra Diablo. *Stenoscisma pyraustoides* is larger, wider, and less strongly costate, especially on the flanks where the costae are lower, broader, and fewer. The species bears superficial resemblance to *S. trabeatum*, new species, which is smaller, more triangular in outline, less strongly convex, has a longer beak, and a stolidium that is continuous from flanks to fold.

TABLE 4.2

Measurements (in mm; measurements exclude stolidium)

| Localities and Types | Length | Brachial Valve Length | Width | Thickness | Apical Angle (°) |
|---|---|---|---|---|---|
| **USNM 705a** | | | | | |
| 152220a | 13.0 | 10.7 | 14.5 | circa 6.0 | 95 |
| 152220b | 15.0? | 13.0 | 16.7 | 10.3 | 89 |
| 152220c | 13.5 | 12.8 | 18.4 | 11.0 | 104 |
| 152220d | 18.2 | 16.2 | 23.5 | 14.0 | 103 |
| 152220e | 19.0 | 16.8 | 26.0+ | 14.0 | 107 |
| 152220f | 23.7 | 22.4 | 28.0 | 16.0 | 93 |
| 152220g | 26.0 | 25.2 | 35.9 | 21.3 | 116 |
| 152220h | 28.3 | 26.6 | 45.1 | 22.7 | 104 |
| 152220i (holotype) | 32.5 | 30.5 | 50.0 | 26.6 | 114 |
| 152220j | 34.7 | 32.5 | 56.0? | 21.0? | 118 |
| **USNM 716o** | | | | | |
| 152225 | 35.5 | 33.5 | 50.5 | 23.2 | 109 |

Box 4.2

## *DIPLOCAULUS PARVUS* OLSON, N. SP.

HOLOTYPE. UCLA VP 3015, partial skull and skeleton including vertebrae, ribs, shoulder girdle, humerus, radius, and ulna. (See Figure 4.2 and Table 4.3.)

HORIZON AND LOCALITY. Chickasha Formation (Permian: Guadalupian, equivalent to the middle level of the Flowerpot Formation) about 2 miles east of Hitchcock, Blaine County, Oklahoma. Site BC- I (Olson, 1965; cited in Olson, 1972). SW 1/4 SW 1/4, sec.6, T. 17N., R. 10W., Blaine County, Oklahoma.

DIAGNOSIS. A small species of *Diplocaulus*, in which the adult ratio of skull length to skull width is attained when the skull length is approximately 60 mm, as contrasted to *D. magnicornis* and *D. recurvatus*, in which the adult ratio is reached at skull lengths of between 80 and 110 mm. Otherwise similar in all features to *D. recurvatus*. (See Figure 4.3.)

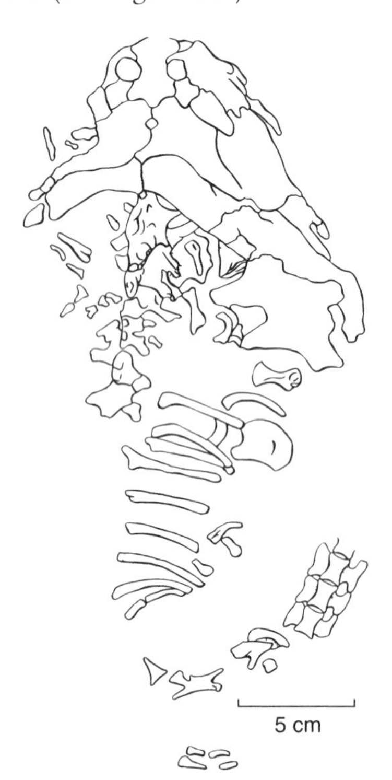


**FIGURE 4.2 Permian amphibian *Diplocaulus parvus* Olson.** The original species description is printed in this box. The drawing is a dorsal view of the holotype. *(From Olson, 1972)*

**TABLE 4.3**

Measurements of Skull Dimensions Based on UCLA VP 3015 (as described and figured in Olson, 1953; cited in Olson, 1972)

| | mm |
|---|---|
| Skull length | 63.0 |
| Skull width | 172.0 |
| Pineal-frontal length | 5.0 |
| Interparietal length | 14.1 |
| Parietal length | 19.6 |
| Frontal length | 25.2 |
| Orbito-snout length | 14.5 |
| Interorbital width | 10.8 |
| Orbital width | 10.2 |
| Orbital length | 9.9 |
| Parietal width* | 84.0 |
| Interparietal width | 94.0 |

* Based upon measurement of right-hand element and multiplied by 2 to give full width as used in various other papers.

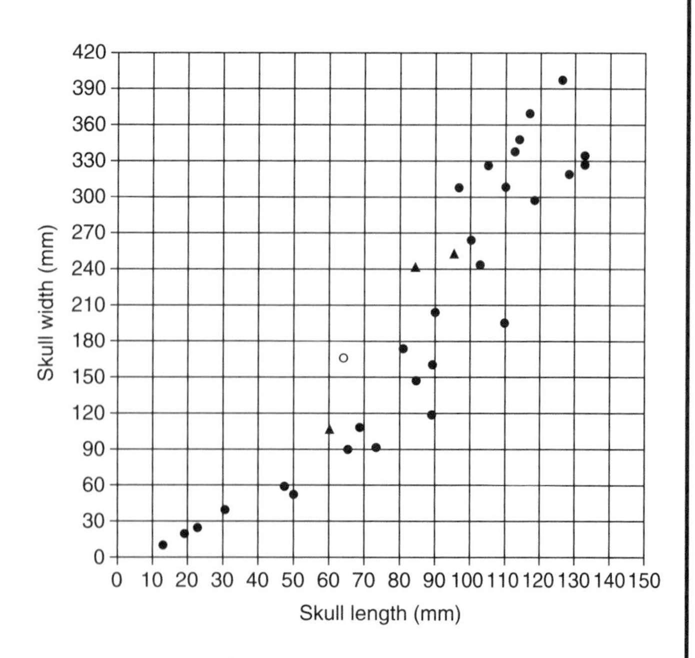


**FIGURE 4.3 The relationships of skull length and skull width in *Diplocaulus*.** Closed circles: *D. magnicornis*; open circles: *D. parvus*; triangles: *D. recurvatus*. *(From Olson, 1972)*

deposits, it was not until 1960 that a specimen was reported in which *Archaeopteris* was attached to *Callixylon*, making it clear that they were one and the same organism (Figure 4.6). This was important, as it demonstrated a previously unknown kind of plant—with wood like that known in seed plants but with branches that bear spores rather than the characteristic reproductive structures of seed plants.

Another case involves the CONODONTA—a primarily Paleozoic group of organisms known from characteristic toothlike elements made of calcium phosphate (Figure 4.7). It was generally thought, from shortly after the initial discovery of conodonts in the 1850s, that several different elements probably came from the same organism. Without clear evidence of this, however, a **form taxonomy,** which treats each element separately, was widely applied. Starting in the 1930s, the repeated discovery of particular assemblages of different conodonts, sometimes diagenetically fused, made it clear that several different elements must be part of the same organism. By the 1960s, therefore, form taxonomy had been largely abandoned in favor of a multielement taxonomy that tried to identify assemblages of conodonts belonging to the same species.

Finally, in the 1980s, several finds were reported of soft-bodied animals with bilaterally symmetrical assemblages of conodonts. Located in the pharyngeal region of the animal, the conodonts were evidently part of a feeding apparatus. This and many subsequent discoveries have revealed a number of characteristics of the soft anatomy that suggest the conodont-bearing animal was a CHORDATE.

Just as the question of what constitutes adequate material varies from case to case, there are differences from group to group in the attributes of organisms that are used in diagnosis. Such attributes are referred to as *taxonomic characters*. Despite these differences, any taxonomic character should satisfy several conditions. It must be a fairly obvious attribute, especially in fossil material in which shortcomings of preservation often yield an incomplete picture of the total organism. Nevertheless, in certain groups, sectioning or other procedures may be required to identify taxonomic characters. Ideally, a taxonomic character should be present and recognizable throughout an organism's ontogeny, and not just during certain growth stages. It should also show limited variation within a species. Characters should be avoided if they are known, from the study of living relatives, to show substantial ecophenotypic variation [SEE SECTION 3.1].

## Presentation of Taxonomic Names

Formal taxonomic names appear in print for a variety of reasons: description of new species, taxonomic revision, inventory of species found in a sample, labeling of museum specimens, and so on. In any case, accurate communication requires that the names be presented in a standard form. By convention, the genus and species names are italicized when printed and underlined when written or typed. The genus name is capitalized and the species name is not. Immediately following the species name, the name of the author of the species is given.

The following (from Farrell, 1992) is an example of a typical fossil list, which helps to illustrate the conventions that surround the presentation of taxonomic names. Notice that the names of some authors of species are enclosed within parentheses, which indicates that the genus assignment has been changed since the species was erected. Notice also that a genus name may be abbreviated to its initial letter in a list of two or more **congeneric** species. This is acceptable as long as it is unambiguous. In the text of a paleontologic paper, genus names are often abbreviated and the names of authors of species deleted (see Boxes 4.1, 4.2, and 4.3).

**Selected Fossils from the Garra Formation, Early Devonian, New South Wales**

*Dolerorthis angustimusculus* n. sp. Farrell
*Skenidioides* sp. cf. *S. robertsensis* Johnson, Boucot, and Murphy
*Muriferella* sp. cf. *M. punctata* (Talent)
*Iridistrophia mawsonae* n. sp. Farrell
*Eoschuchertella burrenensis* (Savage)
*Colletostracia roslynae* n. gen. n. sp. Farrell
*Gypidula pelagica austrelux* n. subsp. Farrell
*Grayina magnifica australis* (Savage)
*Machaeraria catombalensis* Strusz
*Atrypina* sp. cf. *A. erugata* Amsden
*Reticulatrypa fairhillensis* Savage
*Spirigerina (Spirigerina) supramarginalis* (Khalfin)
*S. (S.) marginaliformis* Alekseeva
*Megakozlowskiella* sp.
*Reticulariopsis* sp.
*Straparollus (Straparollus)* sp.
*Straparollus (Serpulospira)* sp.
Hyalospongea indet.
*Heliolites daintreei* Nicholson and Etheridge
*Pleurodictyum megastoma* M'Coy
Calymenina indet.

Box 4.3

## *NAMACALATHUS HERMANASTES* GROTZINGER, WATTERS, AND KNOLL, N. GEN. N. SP.

**Genus *Namacalathus* n. gen.**

TYPE SPECIES. *Namacalathus hermanastes* n. sp.

DIAGNOSIS. Centimeter-scale, chalice- or goblet-shaped fossils consisting of a calcareous wall less than 1 mm thick; a basal stem open at either end connects to a broadly spheroidal cup perforated by six or seven holes with slightly incurved margins distributed regularly around the cup periphery and separated by lateral walls; the cup contains an upper circular opening lined by an incurved lip.

ETYMOLOGY. From the Nama Group and the Greek *kalathos*, denoting a lily- or vase-shaped basket with a narrow base or, in latinized form, a wine goblet.

***Namacalathus hermanastes* n. sp.**

DIAGNOSIS. A species of *Namacalathus* distinguished by cups 2–25 mm in maximum dimension, with aspect ratio (maximum cup diameter/cup height) of 0.8–1.5.

DESCRIPTION. Goblet-shaped calcified fossils; walls flexible, ca. 100 μm thick (original wall dimensions commonly obscured by diagenetic cement growth); basal cylindrical stem, hollow and open at both ends, 1–2 mm wide and up to 30 mm long, attached to spheroidal cup; cup with maximum dimension 2–25 mm, broad circular opening at top with inward-curving lip, perforated by six to seven slightly incurved holes

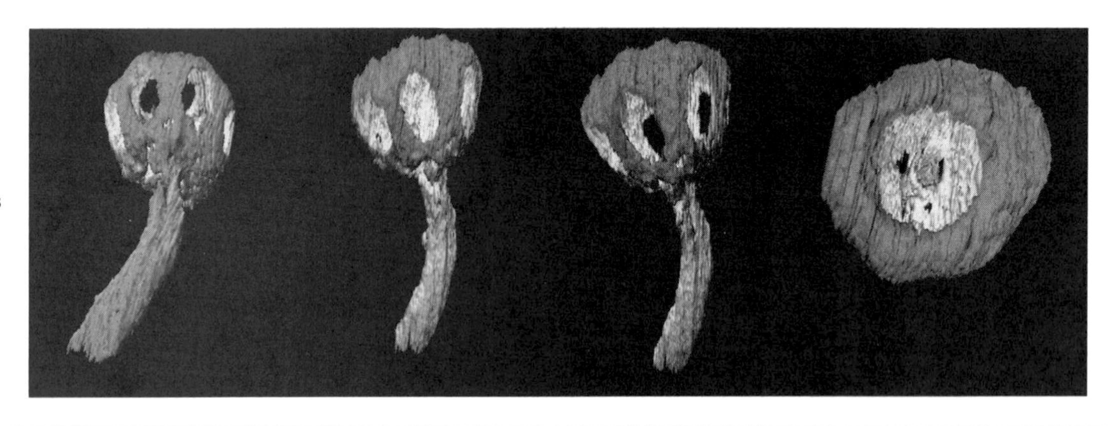

**FIGURE 4.4 Tomographic reconstructions of the calcified fossil *Namacalathus hermanastes*.** *(From Grotzinger et al., 2000)*

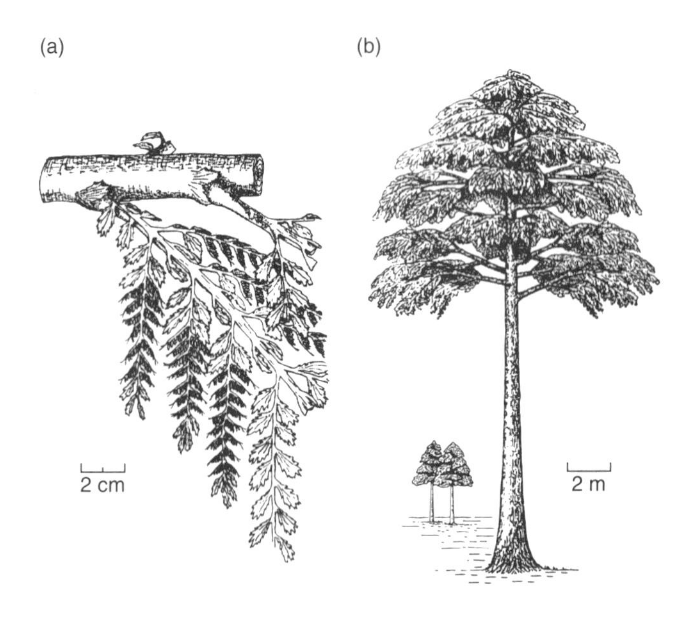


**FIGURE 4.6 Reconstruction of the Upper Devonian land plant *Archaeopteris*.** (a) A branch. (b) An entire tree. *(From Beck, 1962)*

of similar size and shape distributed regularly about cup periphery. Specimens preserved principally by void-filling calcite, with rare preservation of primary, organic-rich wall.

ETYMOLOGY. From the Greek *herma*, meaning "sunken rock or reef," and *nastes*, meaning "inhabitant."

MATERIAL. More than 1000 specimens from biohermal carbonates of the Kuibis and Schwarzrand Subgroups, terminal Proterozoic Nama Group, Namibia.

TYPE SPECIMEN. Our understanding of *Namacalathus hermanastes* derives principally from virtual fossils modeled from serially ground surfaces (see Figure 4.4). Systematic practice, however, requires that real fossils be designated as types. Accordingly, the specimen illustrated in the lower right corner of Figure 4.5 is designated as holotype for the species. The type specimen is to be reposited in the paleontological collections of the Museum of the Geological Survey of Namibia, as collection No. F314. Representative specimens are also housed in the Paleobotanical Collections of the Harvard University Herbaria (HUHPC No. 62989).

TYPE LOCALITY. Reefal biostrome developed at the top of the Omkyk Member, Zaris Formation, Kuibis Subgroup, exposed along the Zebra River near the boundary between Donkergange and Zebra River farms, south of Bullsport, Namibia.

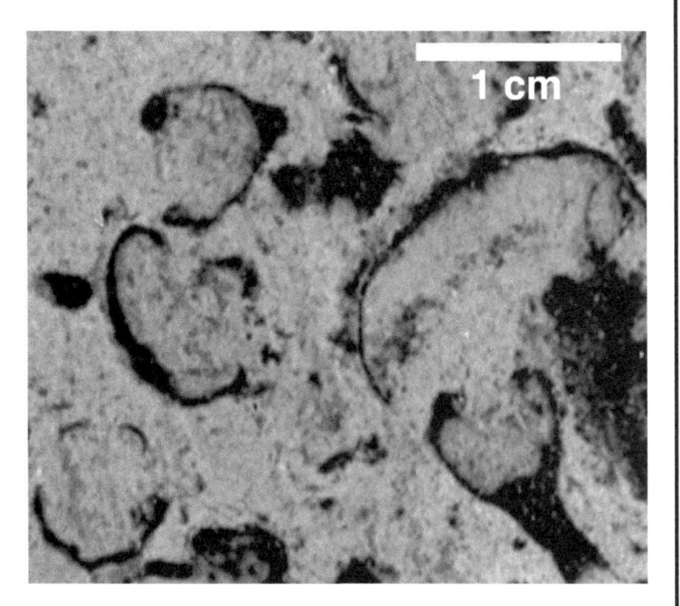


**FIGURE 4.5 Assemblage of calcified *Namacalathus* fossils within reefal and related thrombolitic horizons within the Nama Group.** Original goblet-shaped fossil described by Grotzinger et al. (1995; cited in Grotzinger et al., 2000). Note adjacent cup-shaped fossils, which also represent cross sections of *Namacalathus*.

In a number of entries in the list, the subgenus taxonomic rank is given, enclosed in parentheses and following the genus name (for example, *Serpulospira*). In a few cases, a subspecies name follows that of the species. An "sp.," rather than a species name, following a genus or subgenus name indicates that the species could not be identified with confidence. In a few cases, "sp." is followed by "cf." (for the Latin *confer*, compare) and a species name, indicating a questionable or doubtful species identification. A higher taxonomic name followed by "indet." indicates that identification could not be made below the level of the higher taxon. When "n. sp." follows a species name, this means that the author of the list is naming the species for the first time. A new genus is indicated by "n. gen."

## Changing Species Names

A worker may change the name of a taxon either because its use violates a rule of nomenclature or because it is judged to be improperly classified. We will restrict our discussion of name changes to species names.

**Homonyms** are identical names that denote different species. There are two varieties. Primary homonyms are identical names that were erected for different taxa (with different holotypes) belonging to the same genus. The author of the later-named homonym was in error, not knowing that the species name had been occupied. Once such an error is discovered, only the first published, or senior homonym,

(a)

(b)

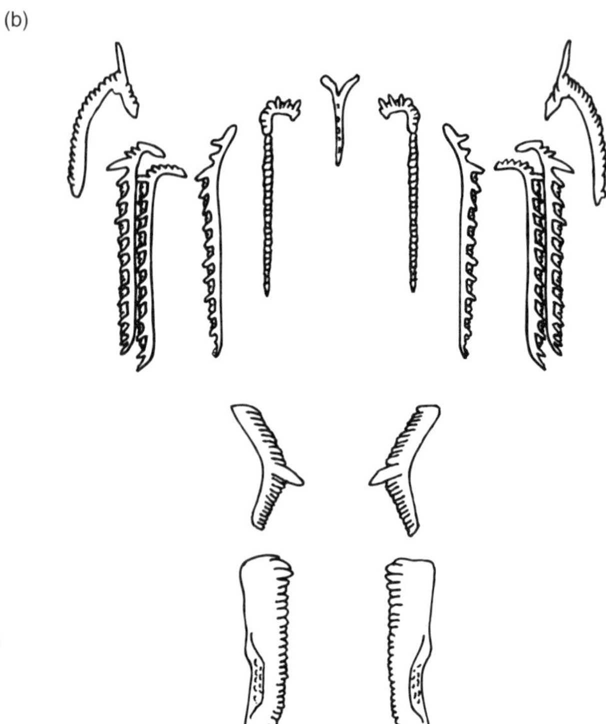

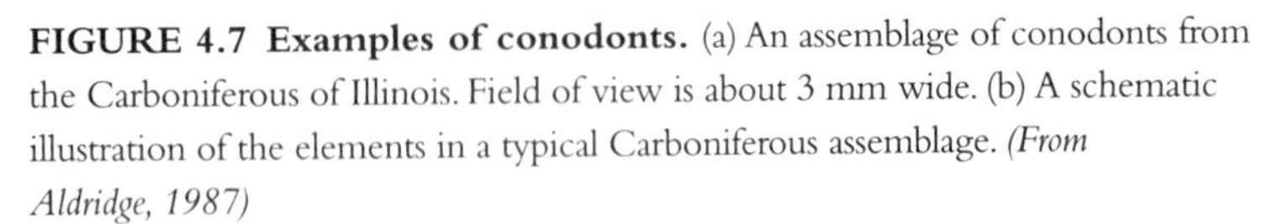

**FIGURE 4.7 Examples of conodonts.** (a) An assemblage of conodonts from the Carboniferous of Illinois. Field of view is about 3 mm wide. (b) A schematic illustration of the elements in a typical Carboniferous assemblage. *(From Aldridge, 1987)*

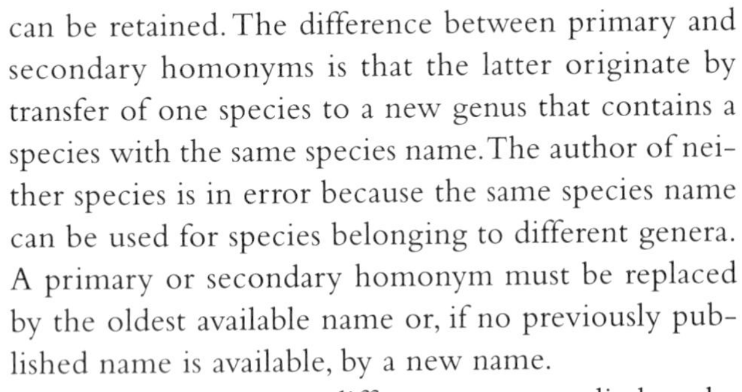

can be retained. The difference between primary and secondary homonyms is that the latter originate by transfer of one species to a new genus that contains a species with the same species name. The author of neither species is in error because the same species name can be used for species belonging to different genera. A primary or secondary homonym must be replaced by the oldest available name or, if no previously published name is available, by a new name.

**Synonyms** are two different names applied to the same taxon. There are two varieties. Objective synonyms are different names that are based on the same type specimen or specimens. Here there is no question of taxonomic opinion; the senior (first published) synonym must generally be retained, and the junior synonym must be permanently rejected. This is the law of **priority.** Subjective synonyms are names that were established for different type specimens that are later judged by a worker to belong to one species. Another worker, however, may judge that the type specimens belong to separate species; this worker will not consider the names to be synonyms and will retain both. In other words, while a junior objective synonym is eliminated automatically, a junior subjective synonym remains available as a name, its use depending entirely upon taxonomic opinion.

Rejection of names on the basis of priority is sometimes unfortunate because it eliminates familiar names and may make it difficult for future workers to trace older literature. The formal change in genus name of the familiar Eocene "dawn horse" from *Eohippus* to *Hyracotherium* was unpleasant to many workers. The Commission on Zoological Nomenclature, which administers the Code, is empowered to suspend the rule of priority at special request. Many familiar names found to be junior homonyms or synonyms have been retained by this procedure. Partly to relieve the Commission of the burden of taking numerous actions of this sort, the most recent edition of the Code empowers authors to choose common usage over priority in specific circumstances.

Box 4.4 gives an example of a list of taxonomic names known as a *synonymy*. A synonymy is a brief history of the taxonomic treatment of a species, with bibliographic citations to important works. Synonymies are important parts of new species descriptions as well as systematic revisions of higher taxa. This example is somewhat complicated but by no means unusually so. Because a synonymy is in part an historical record and in part an interpretation of a taxonomic situation, it is common that synonymies written by different specialists for a single species name do not agree.

Box 4.4

**SYNONYMY OF *ARCHAEOCIDARIS ROSSICA***

The following synonymy was written by Robert Tracy Jackson (1912) for a Lower Carboniferous echinoid species. The heading indicates Jackson's opinion as to the valid species name and the genus to which it belongs: *Archaeocidaris rossica* (von Buch). (Note that Jackson refers to L. von Buch as *Buch*; current convention is to consider *von Buch* the surname.)

This species was first described by E. Eichwald in 1841 under the name of *Cidaris deucalionis*, but the name is disallowed by Jackson because Eichwald's description was too vague. The next entry is to von Buch's description of the species as *Cidaris rossicus*. As the first valid description of the species, the name *rossicus* has priority over all names subsequently applied to the species (although the spelling has been altered to conform grammatically to a change in genus affiliation).

The third entry in the synonymy records the assignment of the species to the genus *Cidarites* (meaning "fossil *Cidaris*") by R. I. Murchison, E. Verneuil, and A. Keyserling. Several subsequent entries record similar shifts in genus affiliation, most reflecting changes or differences of opinion regarding the taxonomic relationships of the species. One entry in the synonymy stands out from the others: *Echinocrinus deucalionis*. This is credited to Eichwald (1860), who evidently recognized as valid his 1841 publication of the name *Cidaris deucalionis*.

The use of the genus name *Echinocrinus* raises another nomenclatural problem. This name was proposed (quite validly) in 1841 by L. Agassiz. *Archaeocidaris* was proposed independently for the same group of echinoids three years later by F. McCoy (1844). Technically, the name *Echinocrinus* is the correct name because it was proposed first. A special exception was made in 1955, however, by the International Commission on Zoological Nomenclature partly because *Echinocrinus* had rarely been used by echinoid specialists and partly because it was misleading in being very similar to genus names common in nonechinoid echinoderms (particularly crinoids).

***Archaeocidaris rossica* (Buch)**

(?) *Cidaris deucalionis* Eichwald, 1841, p. 88. [Description is unrecognizable so the name cannot hold.]
*Cidaris rossicus* Buch, 1842, p. 323.
*Cidarites rossicus* Murchison, Verneuil, and Keyserling, 1845, p. 17, Plate 1, figs. 2a–2e.
*Palaeocidaris rossica* L. Agassiz and Desor, 1846–1847, p. 367.
*Echinocrinus rossica* d'Orbigny, 1850, p. 154.
*Palaeocidaris (Echinocrinus) rossica* Vogt, 1854, p. 314.
*Eocidaris rossica* Desor, 1858, p. 156, Plate 21, figs. 3–6.
*Echinocrinus deucalionis* Eichwald, 1860, p. 652.
*Eocidaris rossicus* Geinitz, 1866, p. 61.
*Archaeocidaris rossicus* Trautschold, 1868, Plate 9, figs. 1–10b; 1879, p. 6, Plate 2, figs. 1a–1f, 1h, 1i, 1k, 1l; Quenstedt, 1875, p. 373, Plate 75, fig. 12; Klem, 1904, p. 55.
*Archaeocidaris rossica* Lovén, 1874, p. 43; Tornquist, 1896, text fig. p. 27, Plate 4, figs. 1–5, 7, 8.
*Archaeocidaris rossica* var. *schellwieni* Tornquist, 1897, p. 781, Plate 22, fig. 12.
*Cidarotropus rossica* Lambert and Thiéry, 1910, p. 125.

## Importance of Taxonomic Procedure

The importance of proper taxonomic procedure cannot be overstated. Adequate description and designation of types ensure that other workers know what an author had in mind when erecting a new species. Strict adherence to rules of nomenclature and to accepted conventions of reporting taxonomic names is necessary for communication. It is as important for paleontologists to use the name *Tyrannosaurus rex* consistently as it is for chemists to apply the name *hydrogen* to one and only one element, or for mathematicians to recognize that the number $\pi$ is a constant. The name in a sense takes on a life of its own.

Proper and consistent reporting is especially important when large inventories of fossil species are compiled for studies of ecology, evolution, and geology. Earlier in this section, we presented a partial list of taxa identified from a Lower Devonian formation in Australia. This is but one of the tens of thousands of similar lists that have been collated into electronic databases for purposes of paleontological analysis. We discuss such databases elsewhere in this book [SEE SECTION 8.7]. For now, it is important to note two points. First, keeping proper track of species names and their authors potentially saves great effort by allowing future taxonomic revisions to be automatically applied to a compilation. Second, the user of such a compilation is unlikely to go back to each original list and verify that the list as represented in the database is faithful to the author's intentions.

Consider the brachiopod identified as *Skenidioides* sp. cf. *S. robertsensis* on the list of Devonian fossils. The species *S. robertsensis* itself is known from deposits of roughly the same age in Nevada and Arctic Canada. If the Australian occurrence of *S.* sp. cf. *S. robertsensis* had been inadvertently entered into the database as *S. robertsensis*, this could easily mislead the user into thinking that this species is known with confidence from a much broader geographic distribution than is in fact the case.

The example of *Skenidioides* is but a minor instance of the kinds of problems that can arise if names are not reported accurately. In one sense, this situation is no different from the publication of a synonymy or taxonomic revision—where we are likely to accept an author's summary of how previous authors used a name. But in another sense, there is a fundamental difference. One could in principle verify every entry in a typical synonymy, such as that in Box 4.4. Given the scale of many secondary taxonomic compilations, however, such verification could not even be contemplated. It is therefore necessary to take steps to minimize errors when compiling taxonomic names.

## 4.2 PHYLOGENETICS

Many areas of paleontology depend on knowing the evolutionary or genealogical relationships among species and among more inclusive groups. These include heterochrony [SEE SECTION 2.3], rates of evolution [SECTION 7.1], and evolutionary trends [SECTION 7.4]. **Phylogenetics** is the enterprise that attempts to deduce evolutionary relationships. This field is distinct from **classification,** the organization of species into a hierarchical system of named categories. Phylogenetics and classification are linked, because most workers prefer a system of classification that reflects inferred phylogeny in some way. Nonetheless, even if one knew the evolutionary relationships among species, there could be numerous alternative classifications that legitimately reflect these relationships.

In this section, we present some of the simplest methods that illustrate the logic by which evolutionary relationships are estimated. We do not treat the analysis of DNA sequences and other molecular data that are generally available only for living organisms. Such data are essential for biologists, however, and students should consult the references listed at the end of this chapter to become familiar with the analysis of molecular data.

### Cladograms and Trees

Because all species living today and all those known from the fossil record are descended from a single origin of life, all species are related to some extent. For many purposes, it is useful to focus on the most proximal relationships between species and to distinguish two fundamental patterns of relationship. First, one species may be ancestral to another, descendant species, either directly or via intermediates. Second, two species may merely share a common ancestor.

Figure 4.8 portrays relationships for a small group of hypothetical species in the form of a **cladogram** (from the Greek *klados*, branch). This is a branching diagram that portrays proximity of relationship without a temporal dimension. The points where branches join are **nodes.** The information in this cladogram can be portrayed in other ways, such as by nested parentheses, as follows: ((AB)C)(DE). Although a cladogram does not

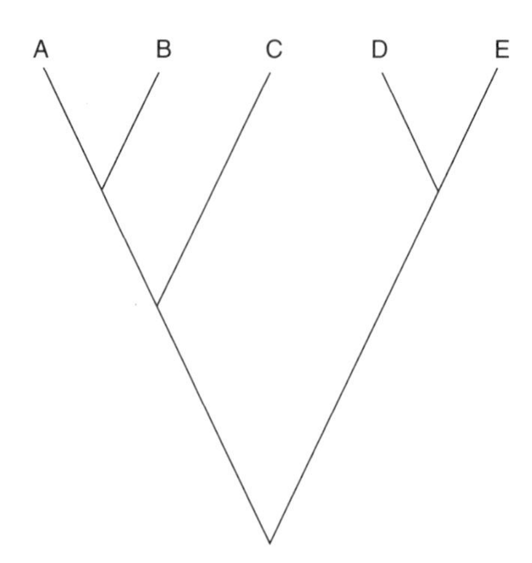


**FIGURE 4.8 Cladogram showing relationships among five hypothetical species.** Species A and B are mutually most closely related, as are D and E. Species C is more closely related to A and B than to D and E.

explicitly portray time, the association of species A and B implies that they share a common ancestor more recently in time than either does with any of the remaining species. In this sense, A and B are mutually most closely related. The same is true of D and E. Together, A and B share a more recent common ancestor with C than any one of A, B, or C does with either D or E. By definition, A and B are **sister species,** as are D and E. Two or more species that share a common ancestor, together with all the other descendants of that common ancestor, are said to form a **clade.** Species A and B form a clade, as do D and E. The groupings (AB)C and ((AB)C)(DE) are also clades. Just as A and B are sister species, (AB)C and DE are **sister clades.** The node linking two sister species or sister clades represents the common ancestor to them.

We must distinguish this cladogram from numerous **evolutionary trees** that are consistent with the pattern of relationships in the cladogram. Evolutionary trees portray ancestral–descendant relationships over actual time and can also depict other aspects of evolution, such as morphological change. Although we commonly speak of relationships among species and other taxa, in general only *samples* of lineages are available, sometimes referred to as **lineage segments.** For paleontologists, these are populations from particular localities and stratigraphic horizons. For biologists, they are populations that happen to be alive today.

Figure 4.9 shows the temporal positions of the five hypothetical species of Figure 4.8, along with two of many possible evolutionary trees consistent with the cladogram. In the first of these trees (Figure 4.9b), some sampled species occupy an ancestral position. Species C gave rise to the common ancestor of A and B, and D gave rise to E. The "species" D and E are in fact two samples or segments of the same lineage. In the other tree (Figure 4.9c), all the species are linked through unsampled common ancestors, and each sample represents a distinct lineage. A species or lineage segment that does not give rise to any descendants is said to be **terminal.** A, B, and E are terminal in Figure 4.9b, and all lineage segments are terminal in Figure 4.9c.

Our earlier discussion of the process of speciation [SEE SECTION 3.3] emphasized the splitting of an evolutionary lineage into two separate lineages. This still leaves the question of whether either of the two resulting lineages should be regarded as ancestral to the other. If speciation occurs via the attainment of reproductive isolation in a population that is separate from the main geographic

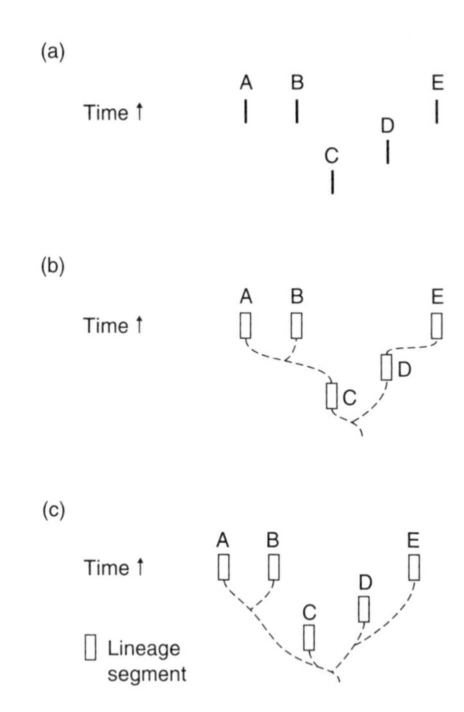


**FIGURE 4.9 Illustration of lineage segments.** (a) Temporal position of the five sampled parts of lineages (lineage segments) shown in the cladogram of Figure 4.8. (b, c) Two of many possible evolutionary trees corresponding to this cladogram. Dashed lines show lines of descent between lineage segments. In part (b), lineage segments C and D are ancestral to others. In part (c), all sampled lineages are terminal.

range of the species, then it is straightforward to consider the species in the main range as the ancestor. It may be that one of the two lineages exhibits less evolutionary change than the other [SEE SECTION 7.3]. In such a situation, it is conventional to regard the more static lineage as the ancestor and the more divergent one as the descendant.

We discuss the construction of cladograms and trees later in this chapter. For now, consider an actual example of each to see some of their main features. Figure 4.10 presents a highly simplified cladogram for selected tetrapod vertebrates. Each branch may represent numerous species. On this figure, the tick labeled *T* marks the evolution of a set of features that define tetrapods to the exclusion of other vertebrates; these include details of bones and their arrangement in the forelimb, hindlimb, and vertebral column. The ticks show the evolution of additional characters, such as the amniote egg (*a*), mammary glands (*m*), and wings (*w*). These are not necessarily taxonomically diagnostic characters.

An evolutionary tree of tetrapods, consistent with the cladogram of Figure 4.10 but even further simplified, is

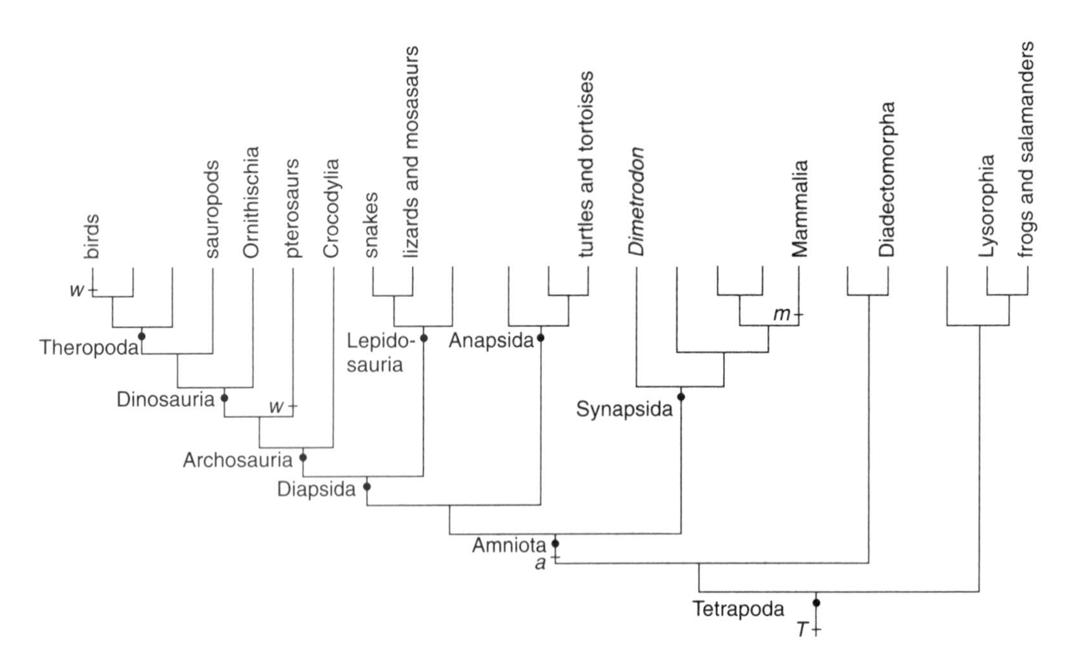


**FIGURE 4.10 Highly simplified cladogram of tetrapod vertebrates with the evolution of selected characters indicated.** Selected branches, mostly familiar groups, are labeled. Unlabeled branches represent less familiar groups. Names in uppercase letters are scientific names; others are common names. The *T* denotes a number of skeletal characters of the limbs and vertebral column that characterize tetrapods. Other characters shown are the amniote egg (*a*), mammary glands (*m*), and wings (*w*). The names of selected clades appear at nodes. All the taxa that diverge from a node are in the named clade.

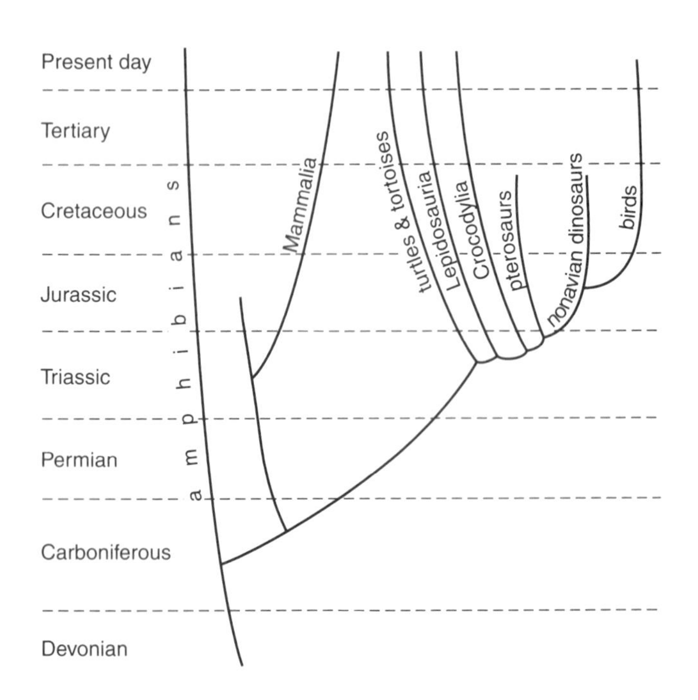


**FIGURE 4.11 Highly simplified evolutionary tree of tetrapods, showing some of the main lineages in the cladogram of Figure 4.10.**

given in Figure 4.11. The time of origin of groups is based both on their time of appearance in the fossil record and on the branching sequence of the cladogram. For example, lepidosaurs, crocodylians, pterosaurs, and dinosaurs first appear in the Triassic, but the order in which they appear is not the same as the order of branching of the cladogram. For Figure 4.11, the cladogram is used to establish the relative time of origin of these groups.

## Shared Novelties and Evolutionary Relationships

To understand phylogenetic inference, it is helpful to start with a case in which we think we know the true evolutionary relationships. This will enable us to discern what kind of information would be useful if we had to reconstruct the cladogram, or tree, given only this information.

Assume for the sake of argument that the cladogram of Figure 4.10 is correct. We first distinguish between a **homologous** trait (one that is shared, with possible modification, in two or more species because they inherited it from a common ancestor) and an **analogous** or **convergent** trait (one that is shared in two or more species whose lineages evolved it independently). According to the cladogram of Figure 4.10, the amniote egg is homologous in all the amniote tetrapods, and lactation is homologous in mammals. The wing, by contrast, evolved independently in the pterosaur and bird lineages, so it is convergent in those two groups. Thus, by their very nature, homologous characters carry information on proximity of evolutionary relationship, whereas convergent characters do not.

Determining whether similar traits are homologous has long been one of the most important and difficult tasks in biology. Detailed structural similarity, similarity of embryological unfolding of the trait, and tracing of evolutionary transitions through well-documented fossil sequences are among the most important clues to homology. Whether a trait is homologous depends to some extent on the scale of analysis. The bird and pterosaur forelimbs are not homologous as wings—the common ancestor of these two groups did not have wings—but they are homologous as forelimbs.

In the case of the pterosaur and bird wings, the modification of the forelimb is radically different in the two groups. The digits represent a good example. In pterosaurs, the first three digits are of normal size, the fourth digit is greatly elongated and supports the membranous wing, and the fifth digit is absent (Figure 4.12). In the oldest known bird, *Archaeopteryx*, there are three fully and essentially equally developed digits—the condition inherited from dinosaurs (Figure 4.13). (It should be noted here that modern birds possess numerous skeletal specializations not found in *Archaeopteryx*.)

Whether a homologous character is informative about evolutionary relationships depends on the scale of analysis. Consider the amniote egg. It is a **derived character** or **novelty** in vertebrate evolution that apparently evolved in the common ancestor of mammals, birds, and reptiles. Thus, when we consider the relationships among major vertebrate groups, the amniote egg is useful in uniting these three into a natural group, the **Amniota,** to the exclusion of amphibians and other vertebrates. Within the Amniota, however, this character is primitive; all lineages share this character by inheritance from their common ancestor. Although the character is homologous, it gives no information that would allow us to determine that any pair among, say, birds, lizards, crocodiles, and mammals is most closely related to the exclusion of the other groups.

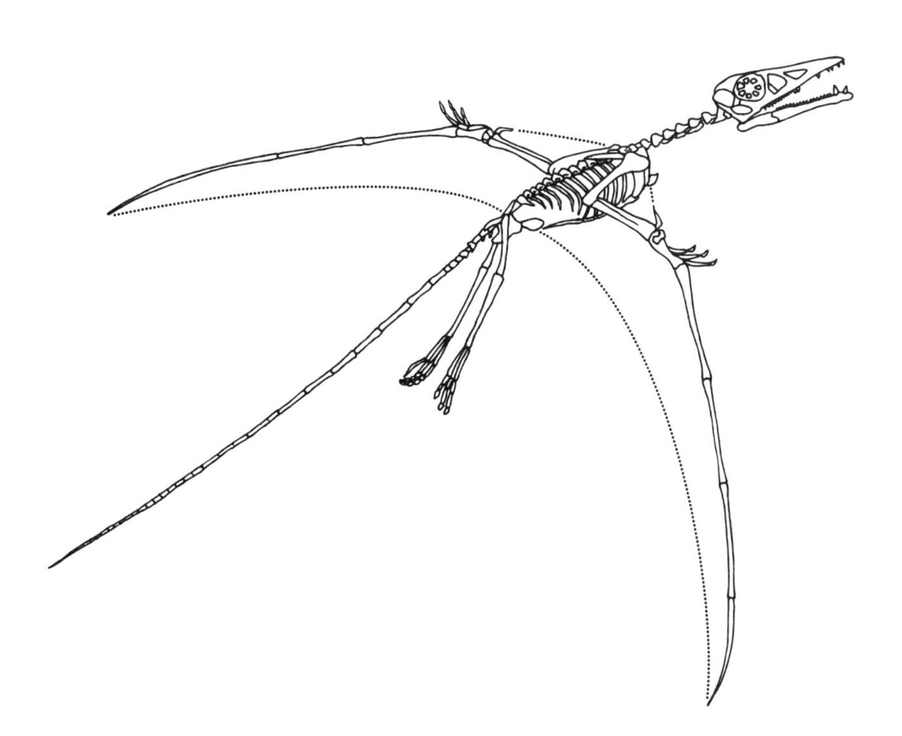

**FIGURE 4.12 Skeletal reconstruction of the primitive pterosaur *Eudimorphodon* from the Late Triassic of Italy, showing the modified forelimb characterized by a very elongate fourth digit.** *(From Carroll, 1988)*

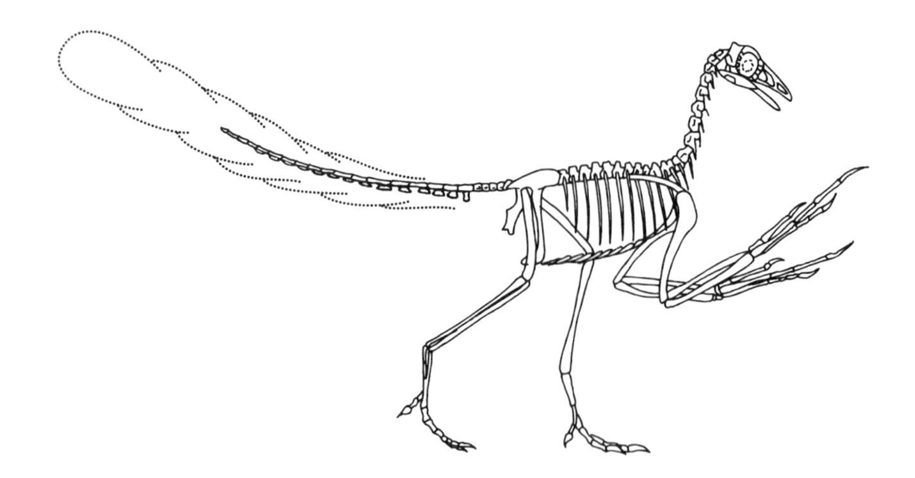

**FIGURE 4.13 Skeletal reconstruction of *Archaeopteryx*, the oldest known bird, from the Late Jurassic of Germany, showing three elongate digits.** *(From Carroll, 1988)*

As this case illustrates, the key to phylogenetic inference is to find characters that are homologous and novel at the appropriate scale of analysis. Homology tells us that a group of lineages shares a character by evolutionary descent. Novelty tells us that the lineages within the group share this character to the exclusion of some other lineages and are thus most closely related. Novel traits are referred to as **apomorphies** (*apo-*, away from), and primitive traits are **plesiomorphies** (*plesio-*, near). A novel character that is shared by a group of lineages is a **synapomorphy,** while one that is unique to a particular lineage is an **autapomorphy.**

In summary, synapomorphies are the key characters that allow us to deduce that groups of lineages are most closely related in the sense of sharing a common ancestor to the exclusion of other lineages.

***Deep Homology*** The example of convergence between the bird and pterosaur wings is one in which the pathways of evolution and resulting structures are profoundly different. There are other cases, however, in which convergently evolved traits represent repeated evolution along similar pathways from a similar starting condition. An example is found in bivalve molluscs, such as mussels, that are attached as adults via a specialized tuft of filaments called the *byssus* (Figure 4.14). Several bivalve lineages have independently evolved this style of attachment. As it happens, many

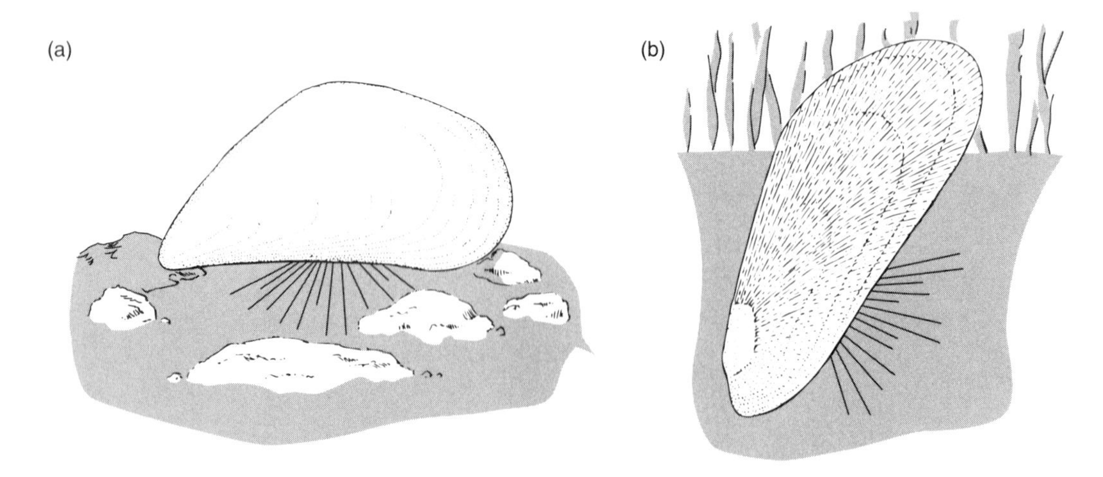


**FIGURE 4.14 Examples of living bivalves that are byssally attached as adults.** (a) *Mytilus*, which attaches to rocks and other firm substrates. (b) *Modiolus*, which attaches to debris within the sediment and is mostly buried within it. (*From Stanley, 1972*)

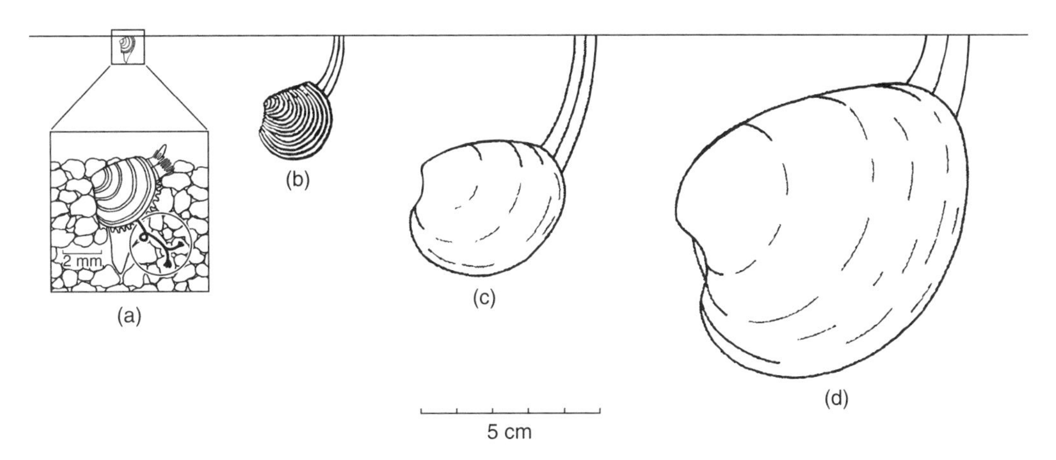


**FIGURE 4.15 Ontogenetic stages in the living bivalve *Mercenaria mercenaria*, which lives within the sediment.** (a) The very young juvenile is stabilized by byssal threads (circled). Subsequent stages lack byssal threads and are stabilized by (b) sharp ornamentation or by (c, d) sheer size. The organ extending to the sediment–water interface is the siphon, through which water is drawn and expelled for feeding and respiration. *(From Stanley, 1972)*

bivalve taxa that are not attached as adults, such as the common quahog *Mercenaria mercenaria* (Figure 4.15), have an early stage that remains stable in the sediment by using byssal threads. It is thought that the independent evolution of the adult byssus in various lineages may represent repeated instances of the retention of this juvenile condition via paedomorphosis [SEE SECTION 2.3].

In a similar vein, developmental genetics has revealed many cases of convergently evolved features deriving in part from similar genetic pathways in different lineages. To take just one such feature, the limbs of arthropods and vertebrates are clearly independently derived; the common ancestor to arthropods and vertebrates did not even have limbs. Yet both groups utilize a set of homologous genes, known as *Hox* genes, that set up the patterning of the body, including the limbs, during ontogeny. The ability of certain genes to establish patterning is apparently a homologous feature, but exploitation of these genes to produce limbs occurred independently in the arthropod and vertebrate lineages.

This phenomenon—in which structures have evolved independently in different lineages but nonetheless arise in ontogeny through the action of homologous genes—has been referred to as **deep homology.** Another example concerns eye development. Eyes have evolved numerous times independently in animals, yet many groups that have been studied use some of the same critical genes to produce eyes during ontogeny. In these examples, the shared genes are homologous characters, but the phenotypic structures—the limbs and eyes—are convergent.

## Inferring Relationships from Morphological Characters

We just showed, with reference to an assumed evolutionary tree, that synapomorphies are the key to inferring evolutionary relationships. In reality, of course, we are not given the true evolutionary tree. We have species on which we can make anatomical observations, allowing us to specify a number of traits or characters, and that we may be able to date stratigraphically. Given such information, how does phylogenetic inference proceed?

One way to approach this problem is to start by establishing the **polarity** of characters, whether they are primitive or derived. Several criteria can be used to determine polarity. None is foolproof, and it is best to use several of them together.

1. *Outgroup comparison.* If a character varies within a group of interest, the state of the character in a related group is generally assumed to be primitive. The group of interest is referred to as the **ingroup** and the related group is called the **outgroup.** Referring back to our tetrapod example, mammals vary in the mode of birth, some being born live and some hatching from eggs. By comparing mammals with other tetrapods, we infer that egg-laying is primitive for mammals, because the primitive members of other tetrapod groups lay eggs. Outgroup comparison is the most common method of polarizing characters. It is best used when the outgroup is known to be reasonably close to the ingroup, so that one can have

confidence in homology of characters. Thus, some degree of prior phylogenetic analysis underlies the designation of outgroups.

2. *Stratigraphic position.* We expect traits appearing earlier in history to be primitive, on average, relative to later-appearing traits. Of course, the fossil record is incomplete, so this is a statistical rather than an absolute statement. It is especially important to be cautious with this criterion in cases where the acquisition of a character greatly affects preservation potential. Consider vascular plants, for example. Vascular plants are thought on good grounds to be derived relative to nonvascular plants, yet they have a richer early fossil record. This is evidently because the evolution of vascular tissue enhanced the preservability of land plants. As a consequence, a number of novelties within land plants appear in the fossil record before the primitive states.
3. *Developmental biology.* Traits that appear earlier in ontogeny are often interpreted as primitive relative to traits that appear later. For example, sharks and their relatives have a cartilaginous skeleton throughout life. Bony fishes, on the other hand, have a cartilaginous skeleton early in ontogeny, and this later ossifies. The cartilaginous skeleton would therefore be interpreted as the primitive state. It should be kept in mind, however, that the order of appearance of traits can be reversed, and that developmental stages can be lost altogether.

Let us suppose that we have a set of polarized characters for a number of species and that we wish to infer the evolutionary relationships among these species, expressed as a cladogram. Consider the simple data matrix of Figure 4.16a, which shows five species and five polarized characters, with zero being the primitive state in each case. Let us propose some alternative cladograms and see what they imply about the evolution of certain traits.

One of many possible cladograms is shown in Figure 4.16b. This one implies five evolutionary transitions: the acquisition of the novel state of character 1 in the common ancestor to A through D; the acquisition of novelty 2 in the common ancestor to A and B; the acquisition of novelty 3 in the common ancestor to C and D; and the acquisition of novelties 4 and 5 in the lineages leading up to B and D, respectively. Note that the number of evolutionary transitions implied by this cladogram is equal to the number of derived character states.

Figure 4.16c shows another possible cladogram. Compared with the first cladogram, there are two main differences in implied character evolution: (1) Novelty 3 evolved twice—once in the lineage leading up to C and once in the lineage leading up to D; and (2) novelty 2 evolved in the common ancestor to A, B, and C, but this character subsequently reverted to the primitive state in the lineage leading up to C. Thus, there are seven rather than five implied evolutionary steps.

Given that both of these cladograms (and numerous others that could be constructed) are perfectly consistent with the character data, which should we prefer? One reasonable way to think of this problem is as follows: We have not in fact *observed* evolutionary transitions directly. Rather, we are *postulating* them as a means of explaining the character data. That is, we are suggesting each transition as a hypothesis. Each hypothesis is proposed to explain a particular subset of data that results from an unobserved evolutionary process. There is a strong intuitive appeal to favoring simpler explanations

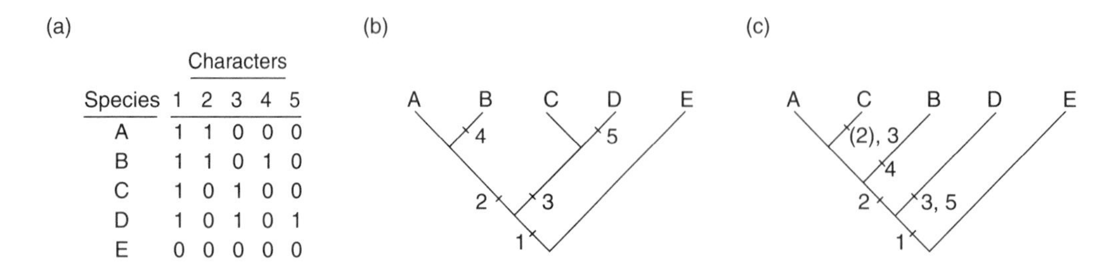


**FIGURE 4.16 Example of cladogram construction.** (a) Character data: 0 is the primitive state and 1 the derived state for each character. (b) Cladogram preferred on the basis of parsimony. This implies that each derived character state, denoted by the tick marks, evolved exactly once. (c) According to this alternative cladogram, derived character 3 evolved twice and character 2 reverted to the primitive state, as denoted by the character number in parentheses.

for data when more complex explanations are not needed. Thus, all else being equal, the first cladogram, with only five postulated evolutionary transitions, is preferred to the second one, with seven postulated transitions.

The cladogram that requires the fewest postulated character transitions to explain the observed character data is said to be most **parsimonious.** The number of transitions is commonly referred to as the *length* of the cladogram. Thus, the shortest cladogram is preferred by parsimony. Parsimony and cladogram length are discussed in more detail in Box 4.5.

It is important to bear in mind that parsimony is generally used as a scientific operating principle; we are not assuming that evolution in fact proceeds in such a way that the number of character transitions is minimized. Strict considerations of parsimony can and should be overturned when there are compelling reasons to do so. For example, sheetlike, or laminate leaves—as opposed to needles, scales, and other structures—appear in many groups of land plants. These include the progymnosperms (Figure 4.6), **SEED PLANTS** (Figure 2.6), **SPHENOPSIDS,** and **FERNS.** Although the relationships among these groups are not fully understood, enough is known about relationships *within* the groups to say with confidence that species with laminate leaves are nested within clades, the primitive members of which did not have laminate leaves. Therefore, laminate leaves must have evolved independently several times.

In light of this, a proposed genealogy that invokes several acquisitions of laminate leaves is not so unparsimonious as might be suggested by character data that simply scored such leaves as "absent" versus "present." In practice, the various laminate leaves could be coded as different characters to indicate the prior evidence that they are not in fact homologous, provided that one has reasonable confidence in this prior evidence.

Finding the most parsimonious cladogram is a computational problem in which every possibility must be explicitly evaluated before one can be sure that a more parsimonious solution does not exist. The examples treated here are simple enough that they can be solved by inspection. Most real studies, however, involve many species and characters, so that an exact, manual evaluation of all possible cladograms is not feasible. Many computer programs have therefore been written to facilitate the job of finding one or more cladograms that are maximally parsimonious. In fact, for more than a handful of taxa, the number of cladograms that must be considered is so large that they cannot all be evaluated even by a computer. Many algorithms have therefore been developed to find approximate solutions.

For simplicity, we have outlined the inference of evolutionary relationships starting with polarized characters. In fact, polarity is often determined after the phylogenetic analysis is done, as outlined in Box 4.6. For paleontologists and biologists interested in understanding evolutionary events as they actually occurred, polarity is of the utmost importance. If one is interested only in the topology of relationships (see Figure 4.18c in Box 4.6), however, it is not necessary to know character polarity.

Our treatment of phylogenetic inference implicitly assigns equal weight to all characters. The number of characters supporting each possible union of species is simply tabulated (Box 4.5), irrespective of anything we might know of these characters, such as their functional significance, or their tendency to revert from the derived to primitive state or to be attained convergently multiple times. In principle, a character that is thought to be less prone to evolve convergently or to revert to the primitive state would be given more weight. Rational weighting schemes, while quite desirable, have been elusive. Phylogenetic analysis therefore usually proceeds, at least in the initial stages, with equal weighting of characters.

The problem of character weighting is inherent in one of the fundamental assumptions underlying much of phylogenetic analysis. By assigning equal weight to every character, we are tacitly assuming that characters evolve independently. But this assumption is unlikely to be met in reality, if for no other reason than that many traits are under common genetic control [SEE SECTIONS 2.3 AND 3.2]. Moreover, several traits may evolve in concert because of an evolutionary modification in the way of life.

Earlier we mentioned the case of byssal attachment in bivalves. Within this style of attachment are two major categories: endobyssate (Figure 4.14b), in which most of the organism is within the sediment, and epibyssate (Figure 4.14a), in which the organism is attached on the surface of the substrate. The epibyssate condition, exemplified by the living mussel *Mytilus*, is more derived. For functional reasons, this condition is correlated with other modifications. These include the flattened ventral side of the shell, which confers stability by allowing close attachment to the substrate, and details of the musculature, which allow the byssus to be pulled tightly in a direction at right angles to the substrate. Thus, at least three traits—the adult byssus,

*Box 4.5*

## CHARACTER DISTRIBUTION AND INCONSISTENCIES IN DATA

Another way to illustrate phylogenetic inference is to consider patterns of **character distribution** and what they tell us about relationships among species. In the example of Figure 4.16, character 1 has the pattern 1 1 1 1 0, meaning that it has state 1 in species A through D and state 0 in species E. The remaining characters have the patterns 1 1 0 0 0, 0 0 1 1 0, 0 1 0 0 0, and 0 0 0 1 0, respectively. Each cluster of 1s denotes a shared derived character state, which, as we have already stated, supports the association of one or more species to the exclusion of the others. Thus, character 1 implies that species A through D are mutually more closely related to each other than any of them is to species E, although this character permits no finer resolution of the relationships among A through D. Similarly, character 2 unites A and B, and character 3 unites C and D.

Characters 4 and 5, while they allow diagnosis of species B and D, respectively, are present in the derived state in only one species each, and therefore do not unite any species as most closely related. The constraints implied by these character distributions, when taken as a whole, yield the exact pattern of relationships expressed in the preferred cladogram of Figure 4.16b. We see that finding the arrangement of species that is supported by more character distributions than any other arrangement is exactly equivalent to finding the cladogram with the fewest implied evolutionary steps.

In the example of Figure 4.16, the smallest number of steps needed to explain the character data is exactly equal to the number of derived character states in the data. In general, the number of derived states is the same as the theoretical minimum length that the cladogram could have. A cladogram with this minimum length is one in which none of the associations of species implied by a character distribution conflicts with any of the others. For instance, character 2 in Figure 4.16 leads us to associate species A and B, and there are no traits that would lead us to associate either of these species with any of the remaining ones. In other words, stating that there are no conflicting interpretations of how character data lead to associations of species is equivalent to stating that the cladogram has the theoretically smallest number of steps.

In reality, paleontologists study tens to hundreds of characters, each with a different evolutionary history. It is therefore all but inevitable that they will show some conflict or **incongruence.** The most parsimonious cladogram is the one with a grouping of species supported by more characters than any other grouping. For example, Figure 4.17a shows a case in which there are five species and seven polarized characters. Character 1 unites A through D, and no other characters are in conflict with this union (Figure 4.17b). Characters 2 and 3 unite A and B to the exclusion of all other species, and characters 5 and 7 unite C and D to the exclusion of all others. Yet two conflicts affect the strength of support for the union of A with B and that of C with D. Character 4 unites A and C, and character 6 unites B and D. In this case, the unions of A with B and of C with D are each supported by two characters. Therefore, the conflicting arrangements (A + C and B + D), each supported by only one character, are overruled by the available evidence.

Figure 4.17c shows the most parsimonious cladogram consistent with these character distributions. Note that, because of the two character conflicts, there are nine steps on the cladogram—the evolution of characters 1, 2, 3, 5, and 7, and the evolution of characters 4 and 6 twice each—rather than the theoretical minimum of seven that would be expected if there were no conflicts among different characters.

In real examples, we are generally not so fortunate as to have character data that unambiguously support one cladogram over all others. Suppose, for example, that characters 2 and 7 were absent from the foregoing example. Then there would be exactly one character supporting each of the unions A + B, A + C, B + D, and C + D. As a result, there would be two rather different, but equally parsimonious, cladograms, each with seven evolutionary steps (Figure 4.17d).

A standard approach to the problem of multiple, equally parsimonious cladograms is to enumerate them, if possible, and extract the features that they have in common. The character data in this example would not enable us to say with confidence which pairs of species are most closely related, but these data would argue that neither B + C nor A + D is a good candidate for a pair of sister species.

(a)

| Species | Characters 1 | 2 | 3 | 4 | 5 | 6 | 7 |
|---|---|---|---|---|---|---|---|
| A | 1 | 1 | 1 | 1 | 0 | 0 | 0 |
| B | 1 | 1 | 1 | 0 | 0 | 1 | 0 |
| C | 1 | 0 | 0 | 1 | 1 | 0 | 1 |
| D | 1 | 0 | 0 | 0 | 1 | 1 | 1 |
| E | 0 | 0 | 0 | 0 | 0 | 0 | 0 |

(b)

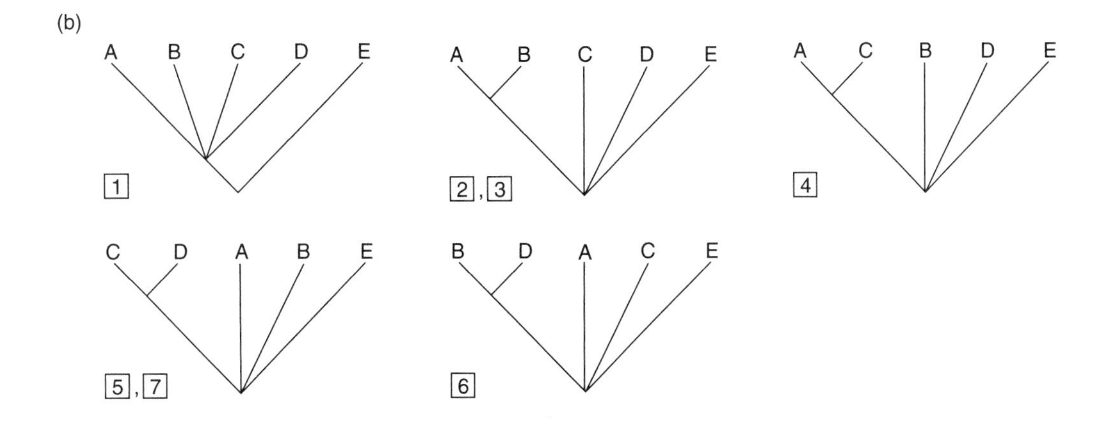


(c)

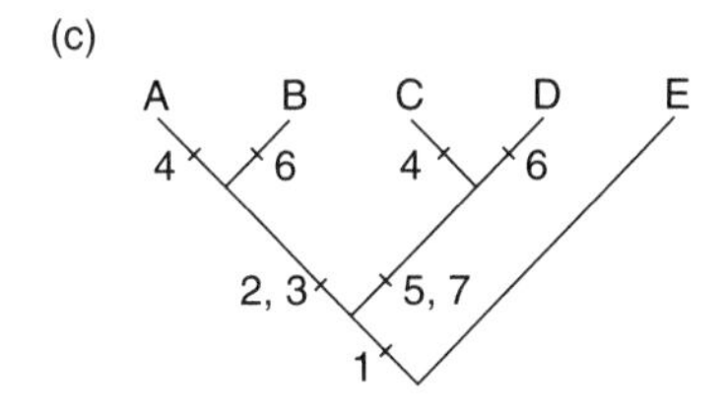


(d)

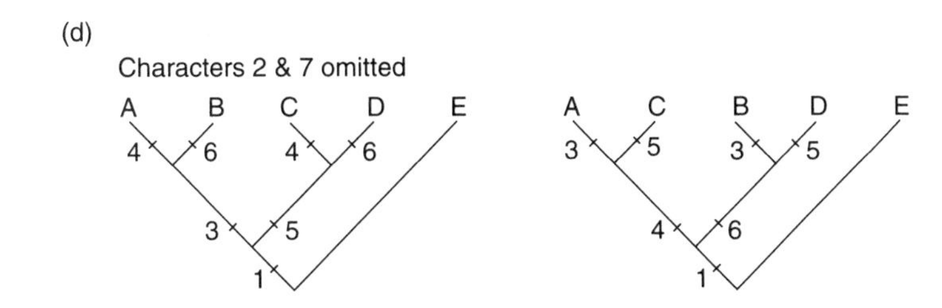


**FIGURE 4.17 Example of cladogram construction by finding associations of species supported by the maximal number of characters.** (a) Character data: 0 is primitive and 1 is derived. (b) Cladograms implied by each of the seven characters. For example, character 1 unites A through D to the exclusion of C but allows no finer resolution, and characters 2 and 3 unite A and B to the exclusion of C, D, and E. (c) Cladogram for all characters. (d) Two equally parsimonious cladograms that would result if characters 2 and 7 were omitted.

*Box 4.6*

**PHYLOGENETIC INFERENCE WITH UNPOLARIZED CHARACTERS**

Consider the hypothetical example of Figure 4.18, which shows four species and five traits, each of which exists in two alternative states. Let us suppose that we do not know which states are primitive and which are derived. Because each of the two states (0 and 1) could be either primitive or derived, and because there are five characters, there are $2^5$ or 32 possibilities for which states are primitive and which are derived.

Let us simplify by focusing on four possibilities, each corresponding to the case in which one of the observed species shows the primitive state for all characters. The four resulting cladograms with their character-state changes are shown in Figure 4.18b. These look quite different from one another, but in fact they have some features in common. In all four cases (as well as the 28 other cases not shown here), tracing the path between A and B, or between C and D, involves two character transitions on the cladogram; tracing between A or B on the one hand and C or D on the other hand involves three transitions.

These common features can be summarized in an unpolarized or **unrooted network,** in which the characters that must be changed to transform one set of character states to another are indicated, and in which character change can be traced in either direction. Regardless of character polarity, there will always be one node between A and B, one between C and D, and two between A or B on the one hand and C or D on the other hand.

In practice, computer programs generally proceed by constructing unrooted networks because, with fewer unrooted than rooted networks, the computational load is substantially reduced. The network is then converted to a **rooted** cladogram, typically by choosing a set of taxa as the ingroup and one or more taxa as outgroups, which polarizes the characters.

the shape of the shell, and the modified musculature—constitute a correlated character complex rather than three independent characters.

There is as yet no general solution to the problems of character correlation and weighting, and this represents an important area for future work. One should ideally avoid treating characters as independent if they are known to be strongly correlated, but the presence of character complexes is not always so clear as in the case of epibyssate bivalves.

## Accuracy of Estimated Phylogenies

An obvious question will have occurred to the reader by now: How well in fact does the most parsimonious cladogram estimate the true evolutionary relationships that it attempts to portray? We have little reason to think that evolution acts parsimoniously (minimizing convergence and reversal), so why should we expect parsimonious cladograms to be accurate?

Two general approaches have been taken to assess the accuracy of parsimony and other phylogenetic methods. First, methods have been tested in groups that have been studied for so long and in such detail that we think we know something about their evolutionary relationships, as with parts of the tetrapod cladogram (Figure 4.10). Second, they have been tested with artificial evolutionary trees, generated by computer simulation of evolution, controlled laboratory breeding, and other procedures.

Although the detailed results of such investigations are beyond the scope of this book, one generalization

(a)

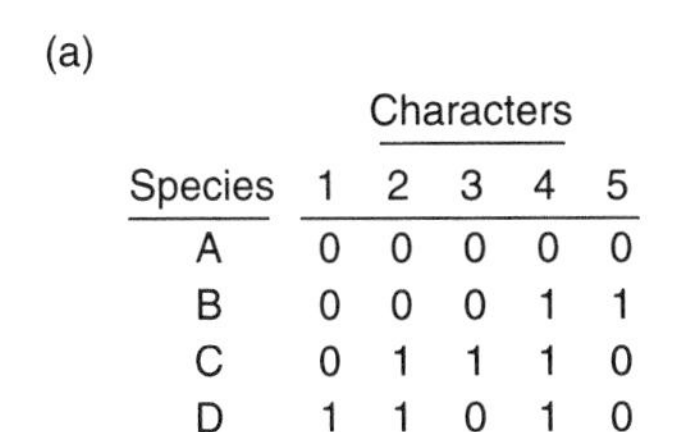

| Species | Characters 1 | 2 | 3 | 4 | 5 |
|---|---|---|---|---|---|
| A | 0 | 0 | 0 | 0 | 0 |
| B | 0 | 0 | 0 | 1 | 1 |
| C | 0 | 1 | 1 | 1 | 0 |
| D | 1 | 1 | 0 | 1 | 0 |

(b)

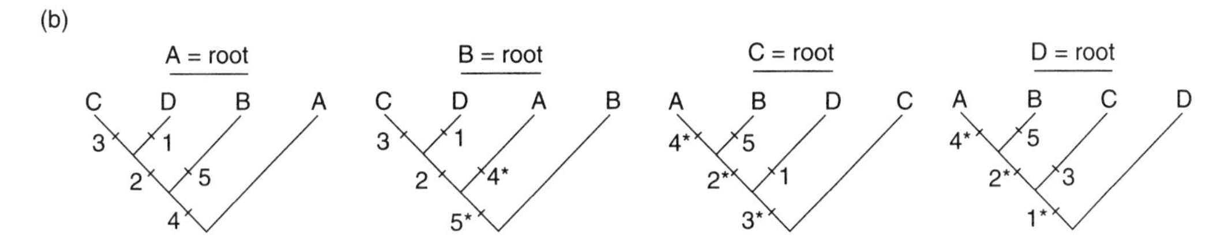


(c)

**FIGURE 4.18 Example of cladogram construction when character polarity is unknown.** (a) Character data. (b) Four possible cladograms, each corresponding to the assumption that one of the sampled species shows the primitive state in all characters. For example, if B is assumed primitive, then 0 is the primitive state for characters 1, 2, and 3, while 1 is the primitive state for characters 4 and 5. Each character number on the cladogram denotes a character transition. Those marked with an asterisk indicate transition from state 1 to state 0; others indicate transition from state 0 to state 1. (c) Unrooted network showing the topology of relationships among the four species. Arrows indicate the direction of change from state 0 to 1. Regardless of the polarity of characters, A and B are separated by two transitions (4 and 5); C and D are separated by two transitions (1 and 3); AB is separated from CD by one transition (2); and either of the pair AB is separated from either of the pair CD by three transitions (4 or 5; 2; and 1 or 3).

that emerges is that parsimony seems to work well when the rate of evolution is relatively low (so that the chances of reversal or convergence are low), and when characters evolve independently (as required by the practice of equal weighting). These conditions make obvious sense, because they should lead evolution to act parsimoniously. That parsimony works well when these conditions hold does not mean that it absolutely requires them, however. In fact, the set of evolutionary assumptions underlying parsimony has not yet been fully specified (Sober, 2004).

Cladistic parsimony will also tend to be more accurate to the extent that all lineages have the same rate of character evolution. The violation of this condition is of special interest in paleontology and evolutionary biology. If lineages differ substantially in evolutionary rate [SEE SECTION 7.1], and if the characters studied have a limited number of alternative states, then there will be a tendency for those lineages that evolve more rapidly to group together even if they are not most closely related. Although the issue is complex, it can be seen intuitively that the artificial grouping of distantly related lineages reflects their increased probability of attaining the derived character state independently, because of the high rate of evolution in those lineages to the exclusion of others. This problem is generally referred to as **long-branch attraction** (the length of a branch referring to the amount of evolutionary change along that branch of the evolutionary tree).

## Other Approaches

The approach to phylogenetic inference we have described here, cladistic parsimony, is by far the most commonly used method when the data at hand are morphological, as is nearly always the case in paleontology. Many other methods exist, however. What most of them have in common is that they evaluate numerous alternative cladograms or trees and find the set of these that optimize some criterion—just as parsimony minimizes the number of evolutionary steps implied by a cladogram.

One such optimality criterion that has received much attention is the probability that a hypothesized cladogram, *under an assumed model of evolution*, would yield the observed character data. In this context, the corresponding probability is proportional to what is known as the **likelihood;** the cladogram that maximizes this quantity is considered the **maximum-likelihood estimate** of the phylogeny (see Box 4.7). The evolutionary model

*Box 4.7*

### MAXIMUM-LIKELIHOOD ESTIMATION OF PHYLOGENY

Figure 4.19 illustrates the likelihood approach with a very simple example involving three hypothetical species. In actual cases with more realistic assumptions, the steps will differ from those shown here, but the ultimate criterion for choosing among cladograms is the same. Figure 4.19a shows the data matrix, and Figure 4.19b gives the three alternative cladograms whose likelihoods are to be evaluated.

The following assumptions are made for the sake of this example: (1) The characters are polarized, so that 0 is the primitive state for each. The basal node, marked X on the cladograms, is primitive in all characters. (2) For each of the cladograms there are four opportunities for character change, corresponding in three instances to the transition from a node (common ancestor) to a terminal taxon, and in one instance to the transition between the two nodes, X and Y. For each of these four opportunities, the probability of change from 0 to 1 is assumed to be the same. This probability is denoted $P$. The probability that character state 0 will not change to state 1, given that is has the opportunity to do so, is equal to $1 - P$. (3) There is no reversal from the derived state to the primitive state (from 1 to 0).

The distributions of character states among the three species are given by 0 0 1 for character 1, by 0 1 1 for character 2, and by 1 0 0 for character 3. The essence of the analysis is to compute the probability of attaining these character distributions under each alternative phylogenetic hypothesis. Consider character 1, and look first at cladogram I. To be present in the derived state only in species C requires that the character changed once, from node X to C, and that it failed to change three times, from X to Y, from Y to A, and from Y to B. The net probability is thus $P(1 - P)^3$ (see Figure 4.19c).

Similar reasoning for the other two cladograms shows that in each case the probability of attaining the character distribution 0 0 1 is equal to $P(1 - P)^3$. Because the probability of observing the distribution 0 0 1 is the same regardless of the cladogram, it is not possible to distinguish the likelihood of the three cladograms on the basis of this character. The same is true of character 3, with distribution 1 0 0, which also has a probability of $P(1 - P)^3$ for all three cladograms. These results should not be surprising because the characters in question are present as unique rather than shared novelties.

Let us turn now to character 2, which has distribution 0 1 1. In cladogram I, the derived state must have evolved twice independently, once from X to C and once from Y to B. Keeping in mind that we have assumed no reversal from 1 to 0, character 2 must also have failed to evolve from X to Y and from Y to A. Thus, the overall probability of distribution 0 1 1, given cladogram I, is $P^2(1 - P)^2$. The same result holds for cladogram II.

In cladogram III, there are two possible ways that the character distribution may have resulted. First, the derived state may have evolved twice independently, from Y to B and from Y to C, and failed to evolve twice, from X to Y and from X to A. The corresponding probability is $P^2(1 - P)^2$. Second, the derived state may have evolved once, from X to Y, after

generally contains elements such as the probabilities of transition between alternative character states and the number of different evolutionary rates—ranging from the simplest model in which all lineages and characters evolve at the same rate to more complex models in which each lineage is characterized by a unique rate.

Likelihood methods have several potential advantages. In contrast to cladistic parsimony, the underlying model of evolution is made explicit and can therefore be assessed directly. Moreover, the statistical foundations of likelihood analysis are well developed, and the relative strength of support for alternative cladograms can be evaluated rigorously. Perhaps most important, likelihood may yield more accurate cladograms than parsimony in certain cases, notably that of unequal rates, where long-branch attraction is relevant. Unfortunately for paleontologists, likelihood analysis has been developed mainly for DNA sequence data, where models of evolution are easiest to define. It is nonetheless applicable in principle to morphological data, and the

which it was passed on to both B and C, and failed to evolve once, from X to A. The corresponding probability is $P(1 - P)$. Thus, the overall probability of character distribution 0 1 1, given cladogram III, is equal to $P^2(1 - P)^2 + P(1 - P)$. This is obviously greater than the probability of the same character distribution given either cladogram I or II, namely, $P^2(1 - P)^2$. Because the probability of the observed data is highest with cladogram III, this cladogram is considered the maximum-likelihood estimate. In light of the assumptions that were made for this exercise, it should come as no surprise that this is the same cladogram that would be preferred by cladistic parsimony. In general, however, this need not be the case.

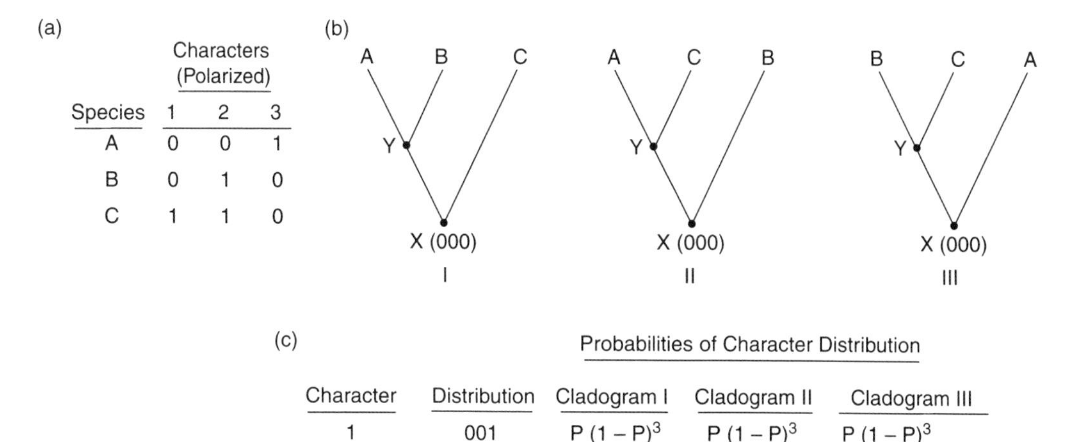
(a)

| Species | Characters (Polarized) 1 | 2 | 3 |
|---|---|---|---|
| A | 0 | 0 | 1 |
| B | 0 | 1 | 0 |
| C | 1 | 1 | 0 |



(c)

| Character | Distribution | Cladogram I | Cladogram II | Cladogram III |
|---|---|---|---|---|
| | | Probabilities of Character Distribution | | |
| 1 | 001 | $P(1-P)^3$ | $P(1-P)^3$ | $P(1-P)^3$ |
| 2 | 011 | $P^2(1-P)^2$ | $P^2(1-P)^2$ | $P^2(1-P)^2 + P(1-P)$ |
| 3 | 100 | $P(1-P)^3$ | $P(1-P)^3$ | $P(1-P)^3$ |

**FIGURE 4.19 Example of cladogram assessment by the principle of likelihood, which evaluates the probability of observing the given data if the postulated cladogram is correct.** (a) Character matrix. (b) Three postulated cladograms. The probability that any character will evolve from 0 to 1 along a segment of the cladogram is *P*; the probability that it will not do so is $1-P$. The relevant cladogram segments are between one node and the other (X to Y) or between a node and a terminal species. It is assumed that characters do not revert from 1 to 0. (c) Characters 1 and 3 yield equal likelihoods for all cladograms and so are not informative. Character 2 yields the highest likelihood for cladogram III, which is therefore preferred.

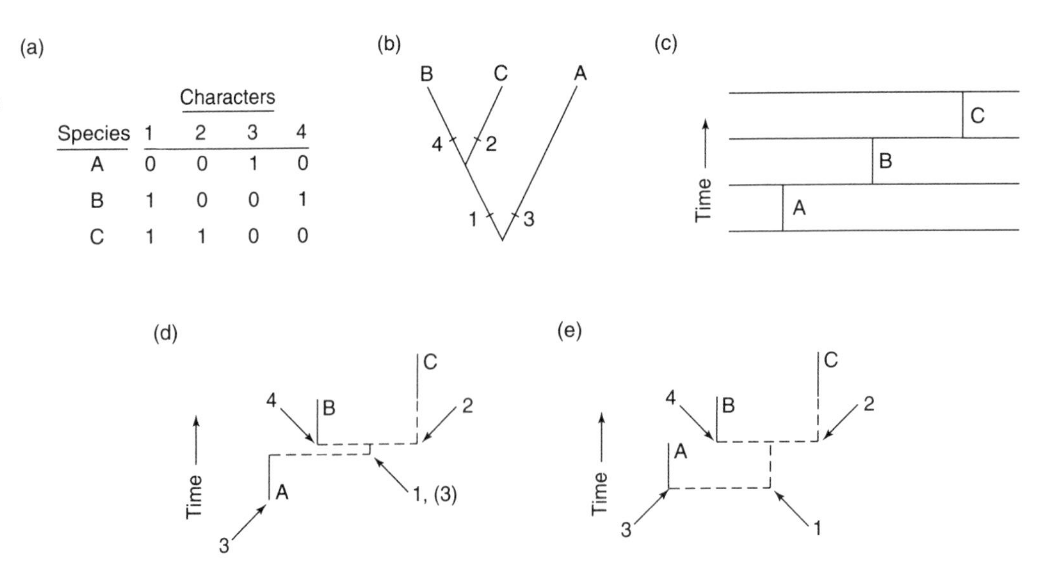

| Species | 1 | 2 | 3 | 4 |
|---|---|---|---|---|
| A | 0 | 0 | 1 | 0 |
| B | 1 | 0 | 0 | 1 |
| C | 1 | 1 | 0 | 0 |



**FIGURE 4.20 One way of incorporating stratigraphic position after a cladogram is established.** (a) Character data: 0 is primitive and 1 is derived. (b) The most parsimonious cladogram for these data. (c) Stratigraphic position of sampled species. (d) Postulated tree with A ancestral to BC. Because A is autapomorphic in character 3, this tree requires secondary loss of this character. (e) Postulated tree with A and BC sharing unsampled common ancestor. No character reversal is required.

continued refinement of likelihood methods is expected to be an important development for systematic biology and paleontology.

## The Temporal Dimension in Genealogy

We have so far focused on the reconstruction of genealogical relationships in the form of a cladogram, but for many paleontological questions it is important to have evolutionary trees. For all practical purposes, the construction of evolutionary trees is equivalent to the incorporation of stratigraphic or temporal data [SEE SECTION 6.1] into the process of phylogenetic inference.

There are two principal ways of incorporating stratigraphic data into phylogenetic analysis. The first uses morphological data to reconstruct cladistic relationships and only subsequently adds stratigraphic position to form evolutionary trees. The second incorporates data on stratigraphic occurrences as a fundamental part of the reconstruction of evolutionary relationships. Here we present just one variant of each general approach.

Consider the character data, cladogram, and temporal occurrences for three hypothetical lineage segments (Figure 4.20). The character data clearly support the grouping A(BC); B and C share the derived state of character 1 to the exclusion of A. The fact that A occurs stratigraphically below B and C might seem to suggest that it is ancestral to these two (Figure 4.20d). However, A possesses a novelty that should be present in B and C if A were ancestral. To maintain that A is ancestral requires the loss in the (BC) lineage of that trait unique to A. This is less parsimonious than supposing that A and (BC) share a common ancestor (Figure 4.20e), with A's novelty evolving after the divergence between the lineage leading to A and that leading to (BC).

A contrasting case occurs when A is fully primitive relative to B and C and it contains no novelties of its own (Figure 4.21). Here it is plausible and parsimonious to reconstruct A as ancestral to (BC). Thus, the general guideline is that a taxon that is stratigraphically lower and fully plesiomorphic relative to its cladistic sister taxon may be interpreted as potentially ancestral to it. An actual example of this procedure is shown in Box 4.8.

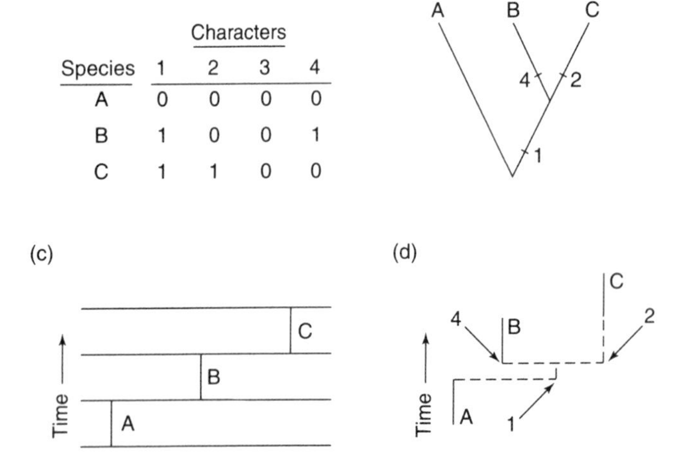

| Species | 1 | 2 | 3 | 4 |
|---|---|---|---|---|
| A | 0 | 0 | 0 | 0 |
| B | 1 | 0 | 0 | 1 |
| C | 1 | 1 | 0 | 0 |



**FIGURE 4.21 One way of incorporating stratigraphic position when a species is fully plesiomorphic.** (a) Character data: 0 is primitive and 1 is derived. (b) The most parsimonious cladogram for these data. (c) Stratigraphic position of sampled species. (d) Evolutionary tree with A ancestral to BC. Because A has no apomorphies, postulating that it is ancestral to BC requires no evolutionary reversals.

Many paleontologists have recognized that stratigraphic data have the potential to do more than distinguish between the terminal and ancestral placement of a taxon after the cladogram is derived from morphological data. One method that has been developed to incorporate stratigraphic position into the reconstruction of

*Box 4.8*

## TREE CONSTRUCTION IN A SAMPLE OF ARBACIOID ECHINOIDS

Construction of the evolutionary tree in this example follows the principles illustrated in Figures 4.20 and 4.21. Figure 4.22a shows the cladogram resulting from parsimony analysis of character data. Names of outgroups are in boldface. The asterisk indicates an extinct genus, the (−) indicates a genus that has no apomorphic character states relative to its sister taxon, and the (+) indicates one or more apomorphic states. The (?) indicates that it is uncertain which of the sister genera *Coelopleurus* and *Murravechinus* is primitive relative to the other—an uncertainty that stems from the presence of both primitive and derived states in *Coelopleurus*.

Figure 4.22b shows the stratigraphic ranges of sampled genera and the evolutionary tree consistent with the cladogram. In constructing this tree, the stratigraphically earlier member of a pair of sister taxa is placed in an ancestral position if it lacks apomorphic states, as in Figure 4.21. An example is that of *Acropeltis*, which gives rise to *Goniopygus*. By contrast, the stratigraphically earlier member of a pair of sister taxa is placed in a terminal position if it has apomorphic states. An example is that of *Glyphopneustes*, which does not give rise to *Arbia*. The placement of *Coelopleurus* as ancestral to *Murravechinus* is based on stratigraphic position. The cladistic relationships of *Hemicidaris*, *Hypodiadema*, and *Gymnocidaris* are based on a number of equally parsimonious cladograms that, unlike Figure 4.22a, place *Hemicidaris* and *Hypodiadema* as sister taxa.

(a)

**Hemicidaris*** (–)
*Asterocidaris** (–)
*Glypticus** (+)
*Codiopsis** (–)
*Dialuthocidaris* (+)
*Pygmaeocidaris* (+)
*Arbacia* (–)
*Tetrapygus* (+)
*Coelopleurus* (?)
*Murravechinus** (?)
*Acropeltis** (–)
*Goniopygus** (+)
*Glyphopneustes** (+)
*Arbia** (+)
**Hypodiadema*** (–)
*Gymnocidaris** (–)

(b)

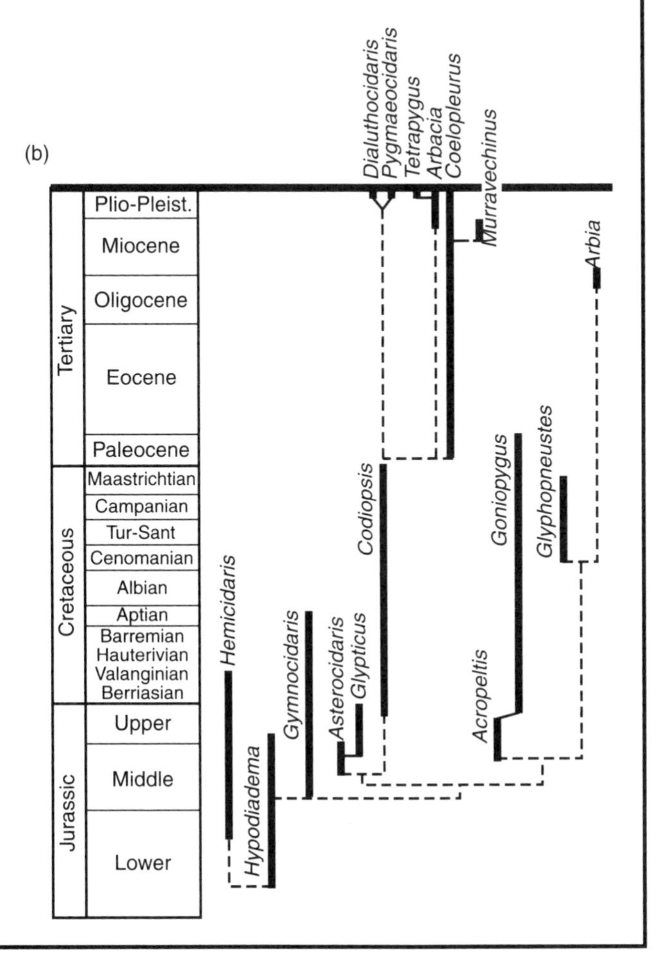


**FIGURE 4.22 Tree construction in a sample of arbacioid echinoids.** (a) Cladogram. (b) Stratigraphic ranges of sampled genera and evolutionary tree. Observed ranges are shown by thick vertical bars. Dashed lines show inferred evolutionary relationships. *(a: Courtesy of Andrew B. Smith; b: From Smith, 1994)*

the cladogram itself is **stratocladistics.** The essence of stratocladistics is to treat morphological and stratigraphic data as logically equivalent classes of information. This means that the same criterion—parsimony—is used to evaluate the consistency between a postulated evolutionary tree and observed data, whether the data are morphological or stratigraphic (Box 4.9).

## 4.3 CLASSIFICATION

One of the main purposes of a classification, whether of species or of inanimate objects, is to summarize and retrieve information efficiently. This aids the memory and also facilitates communication. The ideal classification of species summarizes information on morphological attributes while at the same time reflecting what is known about evolutionary relationships. For example, if a paleontologist reports a new fossil arthropod, nearly any scientist will immediately infer that the organism is characterized by those traits typical of the phylum Arthropoda, such as a chitinous exoskeleton and jointed limbs. If the paleontologist goes on to report that the specimen is in the class Trilobita, the order Asaphida, and the superfamily Asaphoidea, then the typical trilobite features—such as a three-lobed body, fused tail region or pygidium, and two-branched

*Box 4.9*

### STRATOCLADISTICS

We have already seen what parsimony means for morphological data. Figure 4.23 illustrates the concept of stratigraphic parsimony. An evolutionary tree under evaluation is considered unparsimonious to the extent that it postulates the existence of unobserved lineages at a time and place where we would expect to observe them if they had in fact existed—that is, in deposits suitable for their preservation. The echinoid tree of Figure 4.22, for example, implies many unobserved lineages (dashed vertical lines). The suitability of preservation can be judged on the basis of facies characteristics or the presence of taphonomic control taxa [SEE SECTION 1.1].

The morphological data in Figure 4.23a support the grouping A(BC) (Figure 4.23c). One possible evolutionary tree that is consistent with this grouping (Figure 4.23d) has A ancestral to C, which in turn is ancestral to B. This tree requires the existence of C, or a lineage leading to it, throughout two stratigraphic intervals in which it was not in fact sampled. Each such unsampled lineage segment is said to contribute a unit of **stratigraphic parsimony debt.** Because the number of evolutionary steps on this tree is the minimal number that there can be for two derived character states, there is no **morphological parsimony debt.**

Another possible tree (Figure 4.23e) has A giving rise to B, persisting for some time, and giving rise to C as well. This tree is not consistent morphologically with the grouping A(BC) because the tree implies that A is as close to B as it is to C. The tree involves one unit of stratigraphic debt, because there is a postulated lineage segment leading from A to C that is not preserved, and one unit of morphological debt, because the derived state in character 3 must have evolved twice. If stratigraphic and morphological debt are given equal weight, this tree is overall just as parsimonious as the first, even though it is less parsimonious morphologically.

A third tree, of many that could be postulated, has A ancestral to B, which in turn is ancestral to C (Figure 4.23f ). This tree requires no unsampled lineage segments, so there is no stratigraphic debt. It does, however, imply one unit of morphological debt because it requires the secondary loss in C of the derived state of character 2. Of the trees considered here, this one involves the lowest combined morphological and stratigraphic parsimony debt. It is therefore preferred by the method of stratocladistics.

In the hypothetical example of Figure 4.23, we assumed that units of morphological and stratigraphic debt carry the same weight. Nevertheless, just as cladistics allows for the differential weighting of characters if some are believed to be more or less susceptible to convergent evolution, stratocladistics can give different weight to certain unobserved lineage segments if there is evidence that the fossil record is more or less complete in the corresponding stratigraphic intervals. In other words, if we know that the record is fairly complete, we impose a large penalty for pos-

limbs—will come to mind for nearly any paleontologist or biologist. The typical asaphide features, such as the specialized larva, will also be suggested to the trilobite specialist.

Thus, the classification allows detailed information on anatomy to be conveyed with just a few words. In the trilobite example, the phylum, class, and order are also clades. Therefore, the placement of the specimen in this classification allows one to infer that it is genealogically more closely related to other arthropods, such as crustaceans, than it is to molluscs; that it is closer to other trilobites, such as phacopides, than it is to crustaceans; and that it is closer to other asaphide trilobites, such as trinucleoids, than it is to phacopides.

## Nature of Higher Categories

The classification of species into higher categories that has developed over the years, deriving from the standardization of Carl Linnaeus (1758), is a nested hierarchy. The kingdom contains one or more phyla, the phylum contains one or more classes, and so on down to species within genera. At the same time, it is a nonoverlapping hierarchy. A species belongs to only one genus, a genus to only one family, and so on. The most commonly used

tulating lineages that are not actually observed. And if we know that geologically appropriate deposits are absent or scarce, we forgive the stratigraphic debt implied by missing lineage segments because it is quite plausible that they existed but simply were not preserved. Stratocladistics can also give less weight to stratigraphic data as a whole than to morphological data if the taphonomic controls are not strong.

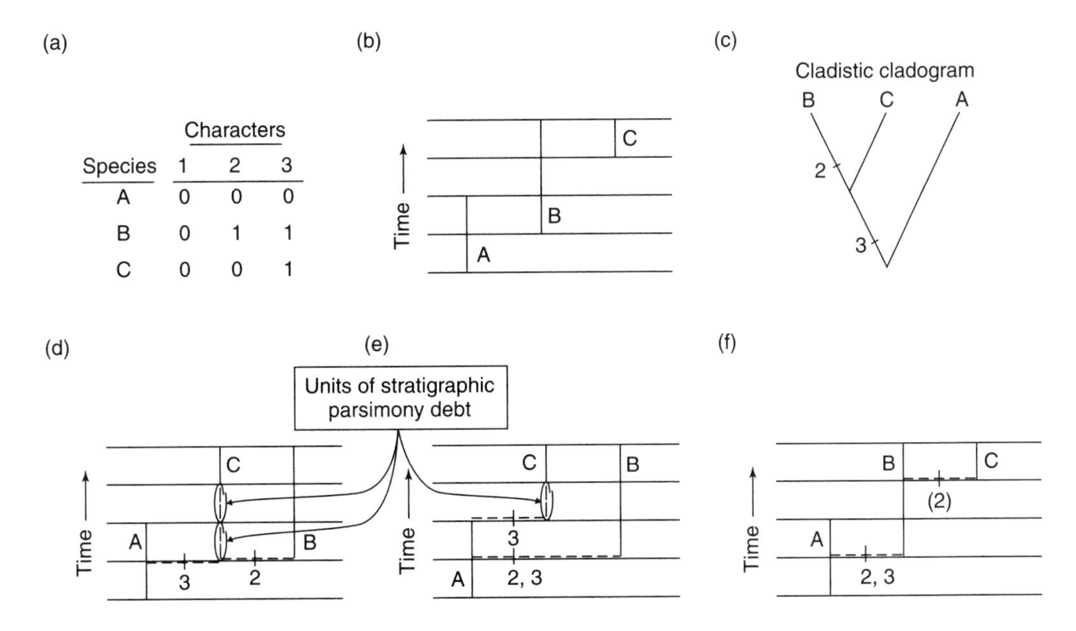


**FIGURE 4.23 Stratocladistic analysis of hypothetical data.** (a) Character data: 0 is primitive and 1 is derived. (b) Stratigraphic ranges of sampled taxa. (c) The most parsimonious cladogram corresponding to character data. (d–f) Three possible evolutionary trees corresponding to character data and stratigraphic ranges. Tree (d) requires neither character reversal nor repeated evolution of any character, but it does require that the lineage leading to species C existed for 2 time units without being sampled. Tree (e) requires that character 3 evolved twice and that the lineage leading to species C went for 1 time unit without being sampled. Tree (f) requires that character 2 reverted to the primitive state, but it does not require any unsampled lineages. Tree (d) is preferred by cladistics, while tree (f) is preferred by stratocladistics.

taxonomic categories between the kingdom and species levels are the phylum (or division) class, order, family, and genus, although it is customary, especially in groups with many species, to use numerous intermediate levels such as the subphylum, subclass, superorder, suborder, superfamily, subfamily, and subgenus.

The *International Code of Zoological Nomenclature* includes rules and recommendations for formation of higher categories. The rules are approximately parallel to those for species. It is generally assumed that all genera can be assigned to families, orders, and so on, although allowance is given for the possibility that the evolutionary affinities of a genus may be unknown. In such a case, the genus may be defined in isolation or in **open nomenclature** ("incertae sedis"). The type concept also extends to the definition of higher categories. When a new genus is proposed, a **type species** designation must accompany the original description. The type species thus becomes the name bearer for the genus. Similarly, a family must have a **type genus,** and so on.

A grouping of species may be **monophyletic,** meaning that it consists of a common ancestor and all its descendants (Figure 4.24)—in other words, a clade, as defined earlier. It is a discrete branch of an evolutionary tree. Familiar examples of monophyletic higher taxa include the phylum **CHORDATA,** the class Mammalia, and the order Primates. Similar in some ways to monophyletic groupings are those that are **paraphyletic,** consisting of a common ancestor and some but not all of its descendants. A paraphyletic group is, broadly speaking, ancestral to one or more groups, meaning that the lineage giving rise to the descendant group is part of the paraphyletic group.

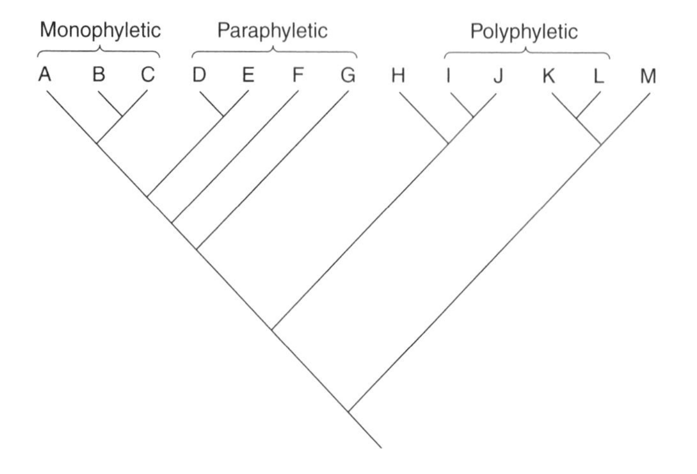


**FIGURE 4.24 Cladogram showing examples of monophyletic, paraphyletic, and polyphyletic groups.**

Referring back to the tetrapod cladogram of Figure 4.10, the groups traditionally referred to as amphibians and reptiles are paraphyletic. Roughly speaking, amphibians are tetrapods that are not amniotes, and reptiles are amniotes that are neither birds nor mammals. A **polyphyletic** grouping of organisms is one whose members do not derive from a single common ancestor within the group. Winged vertebrates (including pterosaurs, birds, and bats) would be an example of a polyphyletic group. In light of the dual goals of a classification—summarizing evolutionary relationships as well as morphological traits—polyphyletic groups are generally undesirable, and we will not discuss them further.

This threefold terminology is the most widely used, and we will adopt it for the sake of discussion. Some systematists use the term *monophyletic* to incorporate both paraphyletic and monophyletic groups as defined here, referring to the latter as **holophyletic** or **strictly monophyletic.**

## Inclusiveness and Rank

In erecting a category above the species level, two quite distinct decisions need to be made: Which lower-level taxa will be included within it, and which level or **rank** (genus, family, order, and so on) will be designated for it? The answer to these questions in any given case is subject to considerable influence of judgment and experience of practicing systematists, and such issues are not legislated by the Code. Unlike phylogenetic inference, to which algorithmic approaches have been applied quite successfully, attempts to erect classifications by strict rules have generally failed.

Given an accurate evolutionary tree, there can be no question as to whether two species are on the same branch or not. This, however, does not answer the question of whether they should be classified together in a given higher taxon. A classic example of this problem is seen in alternative classifications of humans and closely related primates. A large body of genetic data supports the union of humans and *Pan* (chimpanzee and bonobo) as most closely related among living primates, to the exclusion of gorillas and other great apes (Figure 4.25). There have nonetheless been conflicting classifications of this group, some emphasizing the apparent morphological divergence of humans and placing them in a distinct family (with *Pan* and *Gorilla* in a separate, paraphyletic family). A generally accepted approach today unites *Homo*, extinct humans, *Pan*, *Gorilla*, and *Pongo* into a single, monophyletic family, Hominidae.

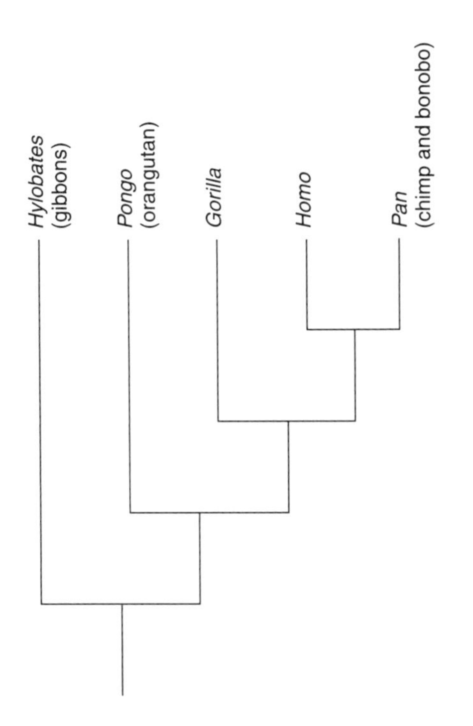


**FIGURE 4.25 Cladogram showing relationships among living humans and great apes.** This represents a composite of numerous studies, based on both morphological data and DNA sequences. *(After Purvis, 1995)*

Just as the inclusiveness of a higher taxon cannot be determined by rigid rules, the rank is also subject to considerable judgment even if taxonomic composition is not at issue. Should humans and chimpanzees be classified as different species within the same genus, or as different genera within the same family? The answer is largely subjective. Many systematists maintain the ideal that rank should be proportional to the magnitude of morphological differences among taxa. Species are separated by small differences, genera by larger differences, and so on.

Although morphological difference is undoubtedly important in assigning rank, other criteria have been used as well. Perhaps most common is evolutionary success, or the accumulation of diversity over time. It has often been said that if birds had not survived past the Jurassic, the class Aves would not have been erected, and *Archaeopteryx* and its relatives would instead have been classified as perhaps a single family within theropod dinosaurs (Figure 4.10).

Higher rank has also been applied to primitive, paraphyletic groups from which arose one or more groups that are given high rank. For example, within the echinoderm subphylum **BLASTOZOA,** the primitive group referred to as **EOCRINOIDS** evidently gave rise to numerous class-level taxa (Figure 4.26). Many of these—for instance, the Blastoidea and Coronata—are clearly monophyletic. In most classifications, the paraphyletic eocrinoids are also formally named as a class.

If a particular higher rank is used for some members of a phylum or class, it may be used for all or most

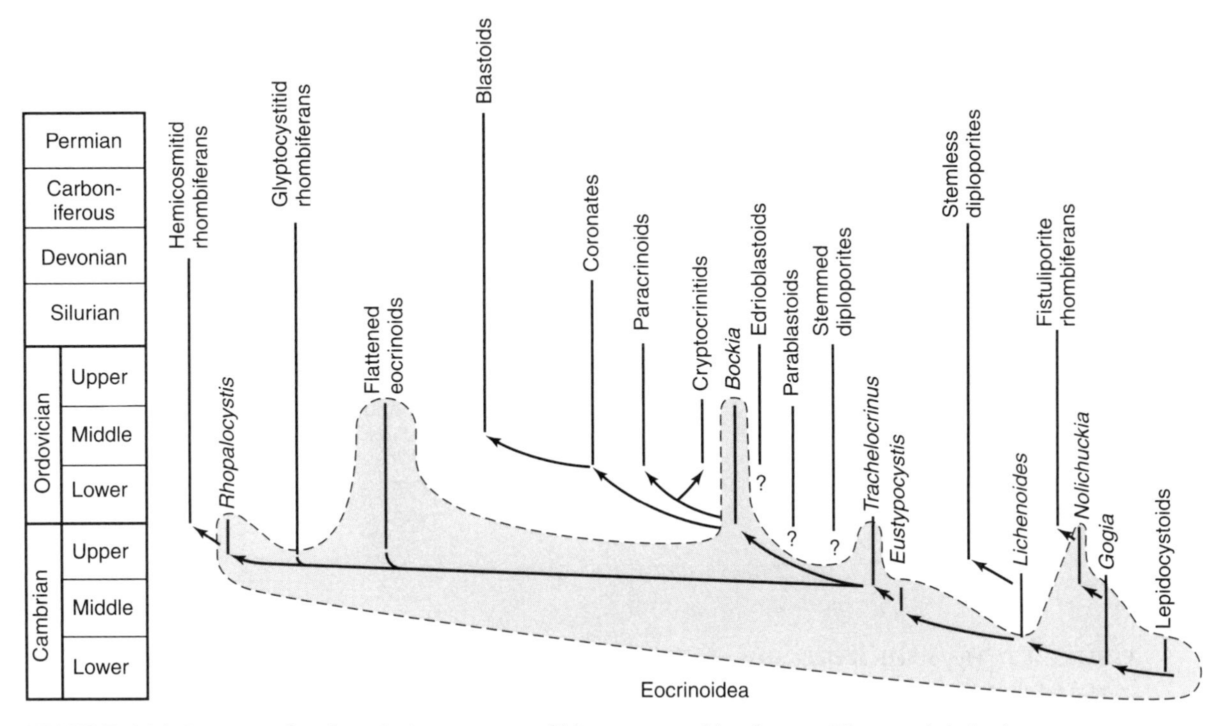


**FIGURE 4.26 One postulated evolutionary tree of blastozoan echinoderms.** The paraphyletic class Eocrinoidea is indicated by the shading. *(From Paul & Smith, 1984)*

members, even though this is not mandatory. For example, the subclass rank is customarily used in classifying crinoids. The Middle Cambrian *Echmatocrinus brachiatus*, initially described in 1973, is thought by some workers to be a crinoid, but its affinities with other crinoids are uncertain. It has been assigned to a **monotypic** genus, family, order, and subclass. Similar to this practice is that of assigning identical rank to sister clades.

*Box 4.10*

## PARAPHYLETIC TAXA IN PALEONTOLOGY

There have been several arguments against the use of paraphyletic taxa:

- Drawing the line between a paraphyletic group and a monophyletic group derived from it is often seen as arbitrary, as this line could be drawn at many different places on the evolutionary tree equally well. Of course, the resulting monophyletic group is no less arbitrary in this regard than is the paraphyletic group. Phylogenetic taxonomy only dictates that groups be defined on the basis of shared novelties; it does not give a formula for which particular novelties to use and therefore which set of species to include in a given taxon. The only completely nonarbitrary system would be to define every single monophyletic grouping as a higher taxon (Figure 4.27). This, however, would involve such a proliferation of taxonomic names as to make the resulting classification, in many cases, cumbersome and inconvenient.
- Paraphyletic taxa are defined in part by the traits they lack rather than the traits they possess. Referring back to Figure 4.26, the eocrinoids possess the feeding structures called *brachioles* that characterize blastozoan echinoderms, but they are distinguished from other groups of blastozoans mainly in lacking novel traits that evolved in other lineages. From the perspective of phylogenetic systematics, the retention of primitive characters is of no particular significance, and there is little that unites eocrinoids; they are simply the residue left after recognition and extraction of many monophyletic groups.
- The first and last appearances in time of genera and families, many of which are paraphyletic, are often used to document patterns of taxonomic origination and extinction in the fossil record [SEE SECTIONS 7.2, 8.5, and 8.6]. Times of elevated turnover of higher taxa are commonly thought to mark elevated turnover of species as well. Some workers have questioned this, however, suggesting that paraphyletic taxa tend to distort underlying species-level patterns.

This last problem is illustrated in Figure 4.28. Part (a) shows a hypothetical evolutionary tree with a concentration of species extinctions at the end of stratigraphic interval 5. Figure 4.28b presents one

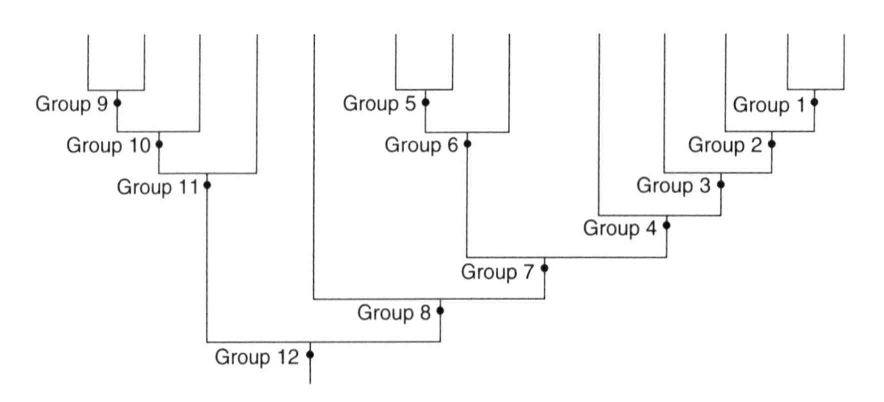


**FIGURE 4.27 Hypothetical cladogram showing many nested, monophyletic groups.** A worker constructing a cladistic classification must decide which of these groups to describe as formal higher taxa, and what taxonomic rank to apply to them.

This has an obvious logic, but, like other approaches, it leaves open the question of what the rank should be.

One question about taxonomic inclusiveness has assumed much importance in paleontology: Is it good practice to erect paraphyletic higher taxa? An essential component of what has been called **phylogenetic classification** is that only monophyletic higher taxa should be permitted. For more detail, see Box 4.10.

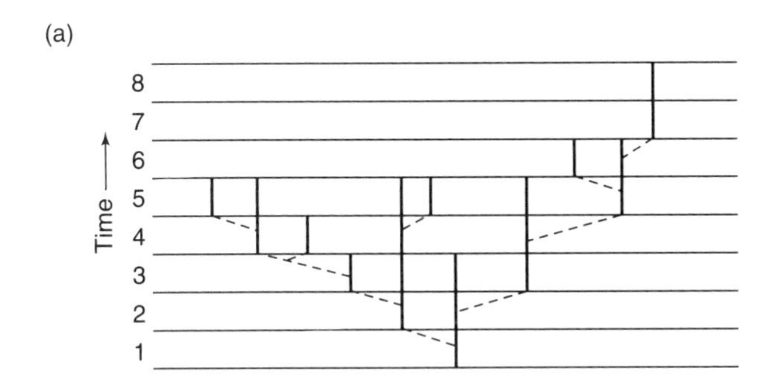


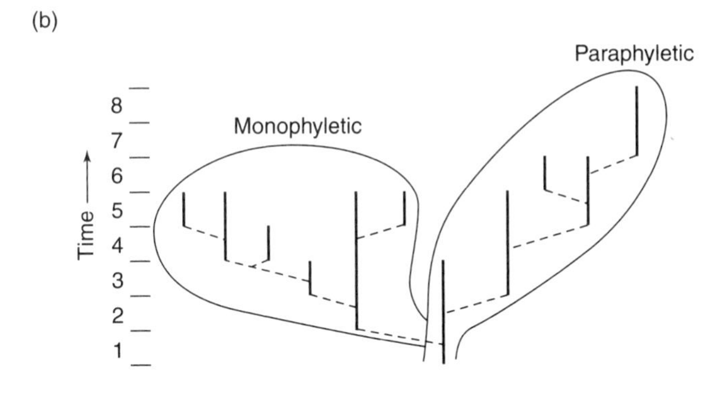


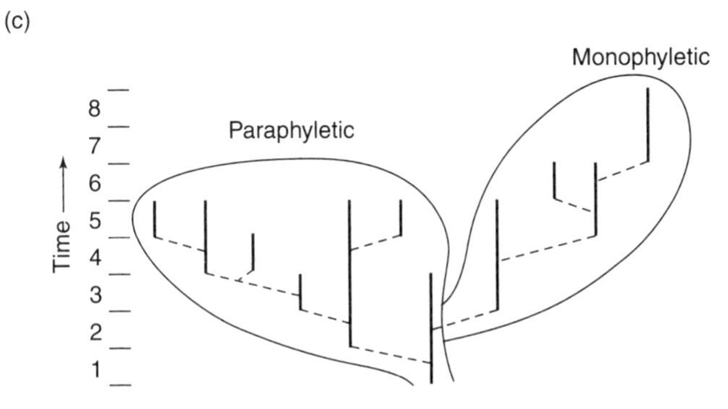


**FIGURE 4.28 Species extinction and higher taxonomic extinction.** (a) Hypothetical evolutionary tree divided into higher taxa in two different ways (b, c), each of which yields one monophyletic group and one paraphyletic group. Many species become extinct in interval 5. In part (b), this event is marked by the last appearance of the monophyletic taxon; in part (c), it is marked by the last appearance of the paraphyletic taxon. *(After Fisher, 1991)*

possible way of dividing this tree into two higher taxa, one monophyletic and one paraphyletic. According to this classification, the last appearance of the paraphyletic taxon does not coincide with the concentration of species extinctions, but that of the monophyletic taxon does. Thus, it would be misleading to use the extinction of the paraphyletic taxon as a surrogate for an underlying species extinction event. There is nothing inherent in paraphyletic taxa, however, that requires them to be misleading in this way. According to an alternative classification (Figure 4.28c), it is in fact the paraphyletic taxon whose last appearance marks the extinction of numerous species, whereas the last appearance of the monophyletic taxon coincides with little in the way of species-level turnover.

In point of fact, it is as yet unknown whether the situation in Figure 4.28b or Figure 4.28c is more common, and actual examples of each have been documented. In other words, it is not known just how serious this particular concern about paraphyletic taxa will turn out to be.

Paraphyletic taxa can be quite useful in reflecting important morphological, functional, and ecological distinctions. Birds are phylogenetically nested within dinosaurs, but maintaining a separate, paraphyletic taxon for dinosaurs is nonetheless thought by many to summarize substantial anatomical and physiological distinctions.

From the paleontological standpoint, perhaps the most compelling argument in favor of permitting paraphyletic higher taxa is that evolutionary trees reconstructed from the fossil record incorporate not only sister-taxon relationships but also ancestor–descendant relationships. Every ancestor is by its very nature paraphyletic. In groups with a sparse fossil record, ancestor–descendant pairs are relatively unlikely to be discovered; most sampled taxa will be terminal, and a classification free of paraphyletic groups

*continued on next page*

## 4.4 CONCLUDING REMARKS

We have discussed phylogenetic inference as if it were mainly a matter of how to process data. In fact, the most crucial step in phylogenetics is careful morphological analysis, with a detailed understanding of ontogeny, sources of variation, correlations among traits that would affect the assumption of character independence, and prior knowledge of evolutionary transitions. Experience has shown that those paleontologists who have produced the most compelling and believable phylogenies are not necessarily those with the best computational skills, but rather those with the detailed knowledge of a group of organisms that is needed for sound morphological analysis.

Even when two paleontologists agree completely on the subject of evolutionary relationships, they may disagree on the ideal manner in which to summarize these relationships, as well as other information, in a classification. It is our feeling that the information and convenience sacrificed by a classification system that allows only monophyletic taxa outweighs the advantages of such a system (see Box 4.10). Nevertheless, a number of paleontologists have adopted fully phylogenetic classifications of the groups they study. We expect to see extensive developments in the area of phylogenetic classification, as evolutionary

*Box 4.10 (continued)*

may be feasible. In those groups that make up the majority of the fossil record, however, it is quite likely that ancestor–descendant pairs are preserved and that paraphyletic taxa will therefore be a practical necessity.

One proposed system for circumventing the conflict between phylogenetic classification and paleontological data involves dispensing with the need to place every species in a taxon of every rank from genus up through phylum. This system distinguishes the **crown group,** the monophyletic group containing living species and extinct species that nest among them, from the **stem group,** the paraphyletic remainder that consists only of extinct species (Figure 4.29). The stem group is not formally named as a higher taxon, and monophyletic groups within the stem group are named as **plesions.** These plesions may be given an optional rank, but they are not nested within successively higher taxa.

Table 4.4 shows such a classification for the echinoids of Figure 4.22a. Note that informal taxonomic categories are used (such as "Unnamed subfamily 1" and "Group 1") and that the genus *Hemicidaris* is interpreted not to be monophyletic but is divided

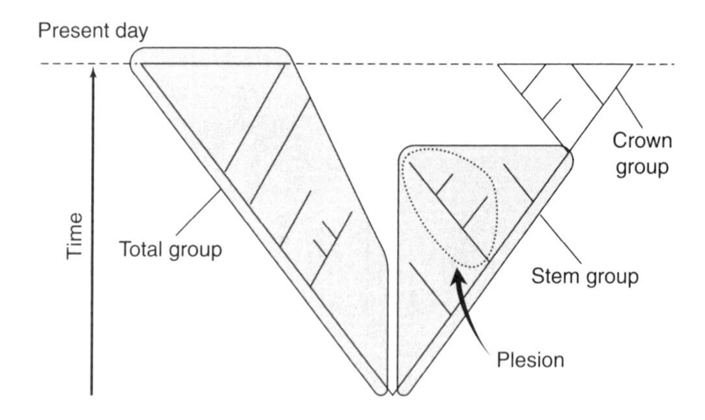


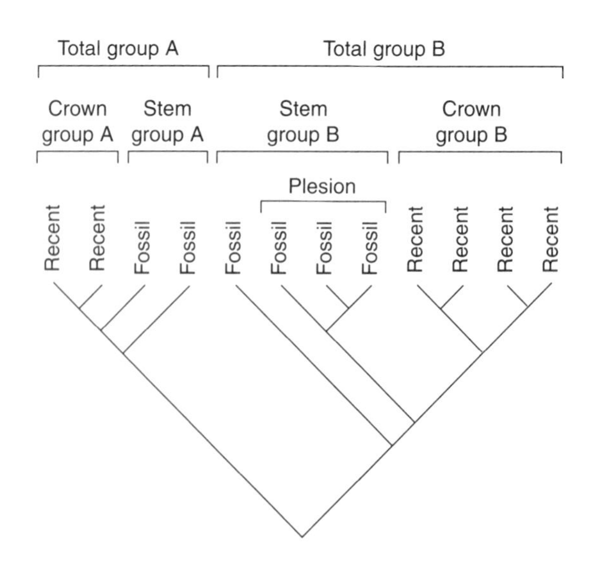


**FIGURE 4.29 Illustration of crown groups, stem groups, and plesions.** (a) Hypothetical evolutionary tree. (b) Cladogram. Each part of the figure shows one plesion of many that could be identified. *(From Smith, 1994)*

relationships are progressively resolved and as electronic storage and retrieval of classification schemes make the dense nesting of exclusively monophyletic groups and the crown–stem–plesion system (Box 4.10) easier to apply.

One such development that is currently under way is the writing of a **phylogenetic code** for species nomenclature. One version of this code would dispense with assigning species to genera and other higher taxa unless these higher taxa are demonstrably monophyletic. This would be a radical departure from the traditional, binomial nomenclature that has evolved since the time of Linnaeus. Related to this development is the consideration of an alternative to the biological species concept [SEE SECTION 3.3]. The proposed **phylogenetic species concept** defines a species as having a shared evolutionary history independent of other species. This history should lead the populations within a species, just like the species within a clade, to be united by synapomorphies. Operationally, this concept therefore considers a species to be "the smallest aggregation of populations . . . diagnosable by a unique combination of character states" (Nixon & Wheeler, 1991). To what extent a formal phylogenetic code of nomenclature will prove useful to working biologists and paleontologists, and will therefore be adopted, remains to be seen.

between two subfamilies. Note also that three plesions (*Hypodiadema*, *Gymnocidaris*, and *Codiopsis*) may be ancestral, and thus paraphyletic, according to the evolutionary tree of Figure 4.22b. Unnamed subfamily 2 and the sister genus pair *Dialuthocidaris* + *Pygmaeocidaris* constitute the two crown groups in this classification.

Although the crown–stem–plesion system enables phylogenetic systematics to cope with paleontological data to some extent, it is not applicable to the many fossil groups, such as trilobites, that are completely extinct. Moreover, by not mandating the ranking of extinct taxa, it may lead to classifications that fall short in their potential for information storage and retrieval, especially for groups such as brachiopods, cephalopods, and crinoids, which contain vastly more extinct than living forms.

**TABLE 4.4**

Classification of the Echinoids of Figure 4.22, Illustrating the Use of Plesions

- Order Arbacioida Gregory
- Plesion (Genus) *Hypodiadema* Desor
- Plesion (Family) Hemicidaridae Wright
  - Subfamily Hemicidarinae Smith and Wright
    - Genus *Hemicidaris* Agassiz (in part)
  - Subfamily Pseudocidarinae Smith and Wright
    - *Hemicidaris termieri* Lambert
- Plesion (Genus) *Gymnocidaris* (Agassiz)
- Unnamed plesion 1
  - Family Acropeltidae Lambert
    - Genus *Acropeltis* Agassiz
    - Genus *Goniopygus* Agassiz
  - Family Glyphopneustidae Smith and Wright
    - Genus *Glyphopneustes* Pomel
    - Genus *Arbia* Cooke
- Unnamed plesion 2
  - Genus *Glypticus* Agassiz
  - Genus *Asterocidaris* Cotteau
- Family Arbaciidae Gray
  - Unnamed subfamily 1
    - Plesion (Genus) *Codiopsis* Agassiz
    - Genus *Dialuthocidaris* Agassiz
    - Genus *Pygmaeocidaris* Doderlein
  - Unnamed subfamily 2
    - Group 1
      - Genus *Arbacia* Gray
      - Genus *Tetrapygus* Agassiz
    - Group 2
      - Genus *Coelopleurus* Agassiz
      - Genus *Murravechinus* Philip

*SOURCE:* Smith (1994)

# SUPPLEMENTARY READING

Bodenbender, B. E., and Fisher, D. C. (2001) Stratocladistic analysis of blastoid phylogeny. *Journal of Paleontology* **75:**351–369. [Explanation and application of stratocladistic methods.]

Cantino, P. D., Bryant, H. N., De Queiroz, D., Donoghue, M. J., Eriksson, T., Hillis, D. M., and Lee, M. S. Y. (1999) Species names in phylogenetic nomenclature. *Systematic Biology* **48:**790–807. [A proposal for a code of nomenclature based on phylogenetic principles.]

Felsenstein, J. (2004) *Inferring Phylogenies*. Sunderland, Mass., Sinauer Associates, 664 pp. [Comprehensive treatment of phylogeny estimation and applications of phylogenetic analysis to evolutionary questions.]

Hillis, D. M, Moritz, C., and Mable, B. K. (1996) *Molecular Systematics*, 2nd ed. Sunderland, Mass., Sinauer Associates, 655 pp. [Comprehensive treatment of the collection and analysis of data on DNA and other biomolecules in phylogenetics.]

International Botanical Congress. (2000) *International Code of Botanical Nomenclature (Saint Louis Code).* Königstein, Germany, Koeltz Scientific Books, 474 pp. [Internationally accepted rules and recommendations for naming species and other categories of plants.]

International Commission on Zoological Nomenclature. (1999) *International Code of Zoological Nomenclature*, 4th ed. London, International Trust for Zoological Nomenclature, 306 pp. [Rules of nomenclature for animals.]

Kitching, I. J., Forey, P. L., Humphries, C. H., and Williams, D. M. (1998) *Cladistics: The Theory and Practice of Parsimony Analysis,* 2nd ed. Oxford, U.K., Oxford University Press, 228 pp. [A useful primer on cladistic analysis.]

Lewis, P. O. (2001) A likelihood approach to estimating phylogeny from discrete morphological character data. *Systematic Biology* **50:**912–924. [A pioneering attempt to apply likelihood methods to morphological rather than genetic data.]

Mayr, E., and Ashlock, P. D. (1991) *Principles of Systematic Zoology,* 2nd ed. New York, McGraw-Hill, 475 pp. [Discussion of scientific foundations of systematics, with reference to noncladistic approaches.]

Smith, A. B. (1994) *Systematics and the Fossil Record: Documenting Evolutionary Patterns*. Oxford, U.K., Blackwell Scientific, 223 pp. [Treatment of phylogenetic analysis with particular reference to paleontological problems.]

Winston, J. E. (1999) *Describing Species: Practical Taxonomic Procedure for Biologists*. New York, Columbia University Press, 518 pp. [Guide to taxonomic procedure, with many actual examples.]

# SOFTWARE

Huelsenbeck, J., and Ronquist, F. (2001) MRBAYES: Bayesian Inference of Phylogeny. *Bioinformatics* **17:** 754–755. [Description of software for phylogenetic analysis using likelihood and related methods. Available for download at http://morphbank.ebc.uu.se/mrbayes.]

Maddison, W. P., and Maddison, D. R. (2000) *MacClade: Analysis of Phylogeny and Character Evolution, Version 4.0*. Sunderland, Mass., Sinauer Associates, 398 pp. [Phylogenetic analysis software for Macintosh operating systems; includes capabilities for stratocladistic analysis. Available for purchase at www.sinauer.com.]

Swofford, D. L. (2002) *PAUP: Phylogenetic Analysis Using Parsimony (and Other Methods), Version 4.0*. Sunderland, Mass., Sinauer Associates. [Popular software for phylogenetic analysis. Available for purchase at www.sinauer.com.]

*Chapter 5*

# EVOLUTIONARY MORPHOLOGY

The subject of biological diversity [SEE SECTION 8.1] is often conveyed with the question: Why are there so many kinds of organisms? We may just as well turn this question around, however, and ask why there are so *few* kinds. In discussing the nature of populations and species in Chapter 3, we saw that form is not randomly or uniformly distributed, but rather that organisms form more or less discrete units. Form is also nonrandomly distributed at higher taxonomic levels. The species that have lived on the earth represent a very small subset of all imaginable forms. In other words, most forms that are conceivable have not in fact evolved. By contrast, some aspects of form have evolved numerous times convergently [SEE SECTION 4.2]. Given that life has been evolving on earth for well over three billion years, why is the spectrum of biologic form so limited?

Broadly speaking, **evolutionary morphology** is concerned with understanding the diversity and the nonrandomness of form. This is obviously an enormous subject. We will emphasize two main aspects of this area of research: **functional morphology,** which interprets the function of organisms in relation to their form, and **theoretical morphology,** which compares the spectrum of conceivable forms to those that have actually evolved.

## 5.1 ADAPTATION AND OTHER UNDERLYING ASSUMPTIONS

We usually start with the working assumption that the distribution of form can largely be explained by **adaptation.** The distinction can be made between adaptation as a state (the fit between an organism's phenotype and its environment and way of life) and adaptation as a process (the evolutionary mechanisms and pathways that produce adaptive traits in a lineage). This distinction is most relevant when there have been evolutionary shifts in function. Natural selection may have produced a structure to perform a particular function in a particular environment, and the structure may have been subsequently co-opted and modified, over evolutionary time, to suit a new functional need.

A persuasive example of such a functional shift is found in the wings of insects. Insect wings must exceed a critical size to generate flight. Because it is practically impossible that fully developed wings were produced by genetic mutation in a single step, the earliest stages in the evolution of the wing must have been small organs that could not have been used for flight. In other words, it seems unlikely that the wing initially evolved by natural selection for the function of flight.

This does not mean that a small, winglike structure would have been useless, however. Functional modeling of the kind we discuss later in this chapter has shown that small, winglike appendages can be useful in regulating body temperature by absorbing solar radiation. In a series of experiments carried out by biologists J. Kingsolver and M. Koehl (1985), wings attached to model insects became more effective at thermoregulation as they were made larger, but only up to a certain size. Above that size, the wings began to generate appreciable lift and to confer other aerodynamic benefits to the models. This suggests that natural selection for the function of thermoregulation could have produced a wing sufficiently large that selection for the new function of flight could have taken over.

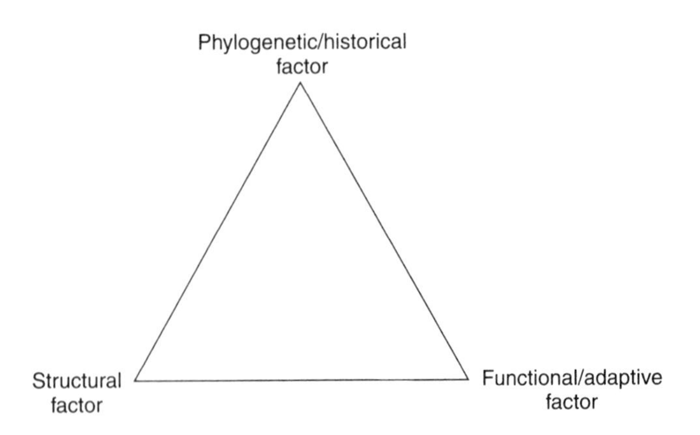


**FIGURE 5.1 Schematic diagram depicting principal factors that contribute to biologic form, using the model of a ternary diagram familiar to geologists.** Every form represents an interplay between immediate adaptation (functional factor), phylogenetic history, and constraints imposed by physical law and the properties of materials (structural factor). *(After Seilacher, 1970)*

In interpreting form, it is useful to consider a framework that distinguishes two major determinants of form in addition to adaptation (Figure 5.1). This framework, which has been developed extensively by the paleontologist Adolf Seilacher and his colleagues, is commonly referred to as **constructional morphology.** The **historical** or **phylogenetic factor** reflects those aspects of form that tend to be fixed within a biologic group because of their shared ancestry. For example, in interpreting the form of a specialized bivalve such as a scallop, we do not ask why it has two valves. This is a fundamental part of the bivalve body plan, one that did not vary in the history of the scallop lineage. Whether we interpret an aspect of form to reflect phylogenetic inheritance depends on the scale of analysis. In studying the *origin* of bivalve molluscs, we might well consider the adaptive value of having two valves; the **functional factor** might then be quite prominent.

The **structural factor** may be the least familiar, although it was discussed at great length by D'Arcy Thompson in his book *On Growth and Form* (1942) [SEE SECTION 2.3]. This factor pertains to consequences of physical law and the properties of materials rather than direct selection. For example, a number of natural structures such as honeycombs, coral colonies, arthropod eyes, and many echinoderm skeletons show a regular arrangement of hexagonal units. In the case of the honeycomb, individual bees are not genetically programmed to produce hexagonal cells in the beehive. An isolated bee would produce a circular cell; it is the simultaneous action of many individual bees, each one pushing outward as it constructs a single cell, that results in the geometrically close-packed structure. Similarly, comparison between the roughly circular perimeters of solitary corals and the hexagonal perimeters of the corallites of many colonial corals suggests a consequence of close packing.

Natural selection can act only upon the variation present in populations [SEE SECTION 3.1]. If there is no genetic variation for a trait, that trait cannot evolve, even if some modification would be advantageous to the organism. Thus, the absence of certain forms in the history of life need not imply that they would have been maladaptive. Likewise, some genetic variants may be generated by spontaneous mutation more often than others. The term **developmental constraint** is sometimes used to describe the nonrandomness of variation that results from the interaction between the genome and developmental processes. We saw in Chapter 4, for example, that the repeated evolution of byssal attachment in adult bivalve molluscs was probably facilitated by the presence of the byssus in the juvenile stage, a feature that could be retained by a simple modification of developmental timing. Thus, the variation on which selection can act is not strictly random. This reflects a combination of the historical and structural factors of Figure 5.1.

## 5.2 FUNCTIONAL MORPHOLOGY

The existence of a correlation between form and function is one of the oldest observations in biology. In some cases, the reason for the correlation may be at least partly phylogenetic. For example, living mammals that chew their cud have an even number of toes, but the number of toes clearly represents deep phylogenetic inheritance rather than an adaptation for digestion. In other cases, however, the form–function correlation clearly has an adaptive basis. Quadrupeds that run fast also have long limbs, for instance, and this can be understood from the mechanics of muscles and levers. The causal understanding of form–function correlation lies at the heart of functional morphology.

Given the assumption that adaptation is one of the main determinants of form, it is essential to identify those aspects of form that were the targets of selection in specific cases. Here we outline the basic ways in which function is inferred for extinct organisms, and we follow these with examples that illustrate the main approaches.

Throughout any functional analysis, it is essential to keep in mind that a structure may be involved in multiple functions. If these functions make conflicting demands on the structure, then not every function can be optimally performed. The result is a compromise, or **trade-off.**

## Approaches to Functional Morphological Analysis

***Inference from Homology*** The simplest way to infer function in extinct organisms is to consider homologous structures in living relatives. For example, we infer that most extinct birds used their wings to fly, unless the details of form suggest otherwise. Homology as a key to function is limited, because it is of no help in the many fossil species that have no close living relatives. It is also an inexact guide to function. *Archaeopteryx* and other primitive birds may have used their wings for flight. This does not imply that the style of flight was like that in any group of extant birds, however, for there has been extensive modification of the skeletal, muscular, and respiratory systems since the initial evolution of birds. Despite such limitations, evidence from homology remains an important component of many studies of functional morphology.

***Inference from Analogy*** The function of skeletal structures may also be inferred from their close physical resemblance to convergent structures in distantly related species. This is what we do, for example, when we interpret the wings of pterosaurs as an apparatus for flight or gliding and when we interpret the streamlined form of ichthyosaurs as an adaptation for swimming. Because analogies can be inexact, analogous structures require thorough analysis before the precise details of their function can be understood.

***Biomechanical Analysis*** Nearly all studies of functional morphology today involve **biomechanical analysis.** The function of problematic structures is deduced in light of the physical properties of biologic materials; mechanics of beams, levers, joints, and other structures; and aero- and hydrodynamics. Biomechanical analysis can be broadly categorized into two general approaches, the **paradigm approach** and the **experimental approach,** although the two are not completely distinct.

The concept of the paradigm in functional studies was introduced to paleontology in the 1960s by M. J. S. Rudwick, who was especially concerned with inferring function when homology and analogy could not be easily recognized. He defined the paradigm as "the structure that can fulfill the function with maximal efficiency under the limitations imposed by the nature of the materials" (Rudwick, 1961, p. 450). The paradigm approach typically involves three steps:

1. One or more potential functions are postulated for a problematic structure.
2. For each potential function, engineering principles are used to design the hypothetical structure optimally suited for carrying out that function. This structure is referred to as a *paradigm*. Because of trade-offs and limitations in the inherited body plan and materials, the optimal structures are not the best conceivable designs, but the best ones possible in light of these constraints.
3. The resemblance between the actual structure and the set of paradigms is assessed, and the paradigm that most closely resembles the actual structure is identified. The function corresponding to the closest paradigm is inferred to be the one that the actual structure most likely performed.

Assessing the resemblance between a paradigm and an actual structure involves some degree of subjectivity. For example, how closely must an elevated region on a bryozoan colony resemble a chimney for us to be confident that it indeed functioned, like a chimney, to facilitate fluid flow away from the colony surface? This uncertainty has contributed to a general preference for the experimental approach, which allows functional performance to be measured and verified. The experimental approach to biomechanics also involves three steps:

1. As with the paradigm approach, several potential functions are postulated for an unknown structure.
2. A model of the organism or structure is made. This model can be physical or numerical, and it can be highly simplified or a nearly exact replica.
3. The capacity of the structure to perform the function is assessed experimentally. Experimentation often involves manipulations such as removing a structure of interest from the model organism to determine whether its presence makes an appreciable difference to mechanical properties and function. For a simple physical or numerical model, there may be exact equations to determine its performance.

The paradigm approach and the experimental approach to biomechanical analysis share the advantage that they rely on universal physical laws and properties of materials. Both have the disadvantage of being limited by the range of postulated functions, and therefore by the imagination of the investigator. It is always possible that a structure may be best suited to a function that has not even been considered. Biomechanical analysis can only tell us whether an organism was capable of functioning in a specified way, not that it actually did so. Biomechanics has nonetheless been of great use in understanding the relationship between form and function in living as well as fossil organisms.

## Examples of Biomechanical Analysis of Extinct Organisms

***Vision in Trilobites*** The eyes of trilobites are similar in many regards to those of living arthropods. Therefore, much can be learned about the functional morphology of the trilobite eye by analogy with living forms. But there are important structural differences that suggest that the optical systems used by trilobites were significantly different. Much of our understanding of trilobite vision derives from an unusual collaboration between paleontologist Euan Clarkson and physicist Riccardo Levi-Setti. When the two met at a conference in Oslo in 1973, both had for several years been active students of trilobite morphology; Clarkson had done considerable work on trilobite vision, and Levi-Setti had a physicist's knowledge and understanding of optical systems.

Trilobites possessed a compound eye, consisting of numerous lenses arranged in rows (Figure 5.2). The lenses were usually deployed in a geometrically closely packed configuration. The lenses themselves were made of calcium carbonate in the form of the mineral calcite and are sometimes preserved. It has been possible experimentally to produce focused images through individual lenses. It is not known whether the animals could perceive a clear image, because this depends on

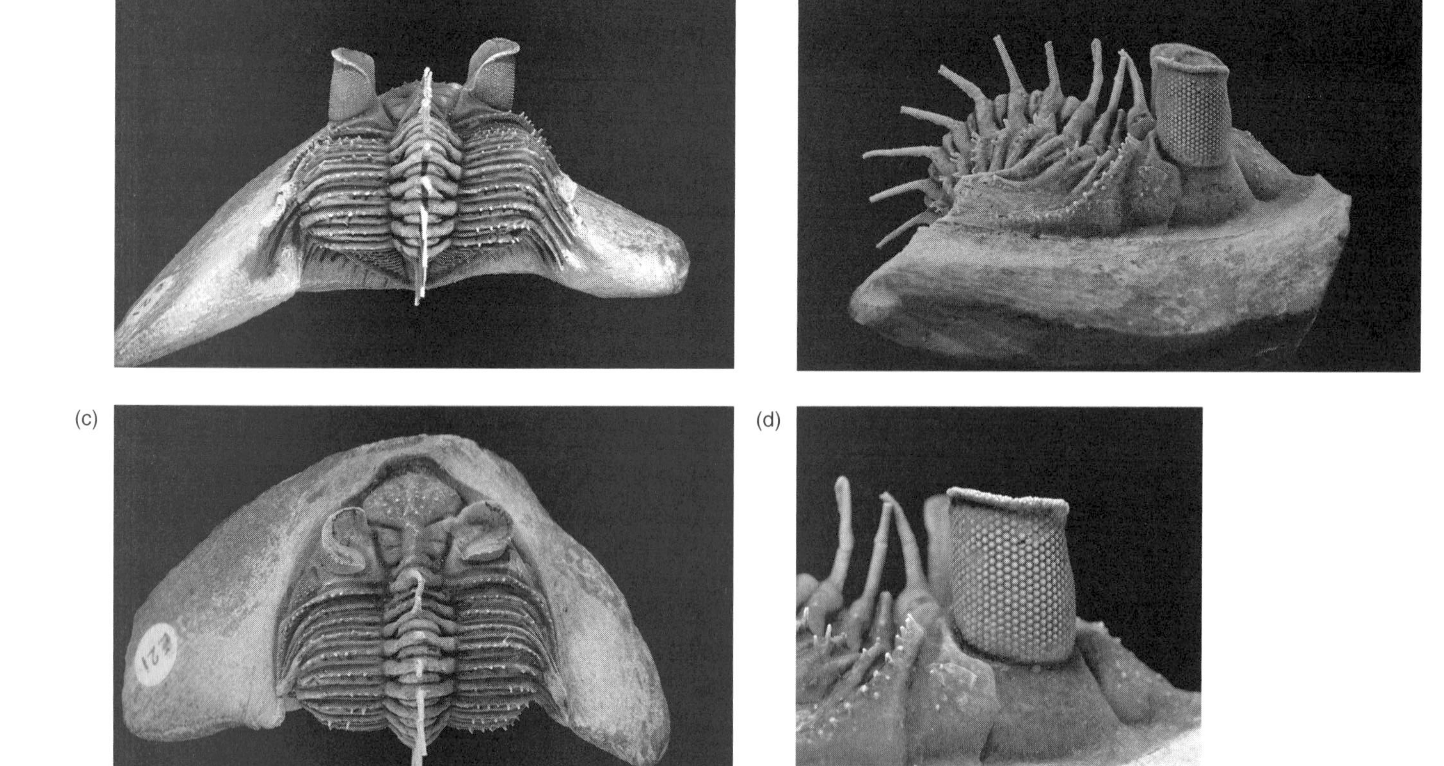


**FIGURE 5.2 A trilobite with well-developed compound eyes, *Erbenochile* from the Lower Devonian of Morocco.** (a–c) Posterior, lateral, and dorsal views. (d) Detail of eye, showing arrangement of individual lenses. Width of the head is 32 mm. *(From Fortey & Chatterton, 2003)*

the nervous system and unpreserved details of the eyes. But they could, at the very least, recognize movements of an object and estimate its size.

Lens morphology and the arrangement of lenses vary considerably from one group of trilobites to another. A particularly interesting lens shape is illustrated in Figure 5.3. It is a doublet consisting of an upper unit that is convex on its upper surface but has a more complex shape on its lower surface. Two variants of the shape of the lower surface are shown at the center in the illustration. In both variants, the lower part of the doublet has an upper surface that fits the shape of the upper lens and a lower surface that is simply convex. The two lenses together thus make a biconvex compound lens.

Upon examination of Clarkson's reconstruction of trilobite eyes, Levi-Setti noticed that the upper lenses just described are very close approximations of lens designs published by René Descartes and Christiaan Huygens in the seventeenth century. The Descartes and Huygens drawings are reproduced for comparison in Figure 5.3, left and right. The purpose of both designs was to produce what is known as an *aplanatic lens*—one that avoids certain kinds of distortion. The similarity between the shapes of the upper trilobite lens and the lenses designed by Descartes and Huygens is remarkable. Indeed, the lenses differ little, other than in the presence of the lower lens in the trilobite, an element that does not appear in the designs of either Descartes or Huygens. But this is understandable when it is noted that the aplanatic lens was designed to operate in air. Calculations have shown that in the trilobite's aqueous environment, the lower lens would be necessary to compensate for the relatively high refractive index of seawater. Thus, the trilobite lens doublet appears to be an optimal modification of basic designs that became a part of human technology only as recently as the seventeenth century. Similar correcting lenses have since been recognized in some living insects, ostracodes, and even scallops.

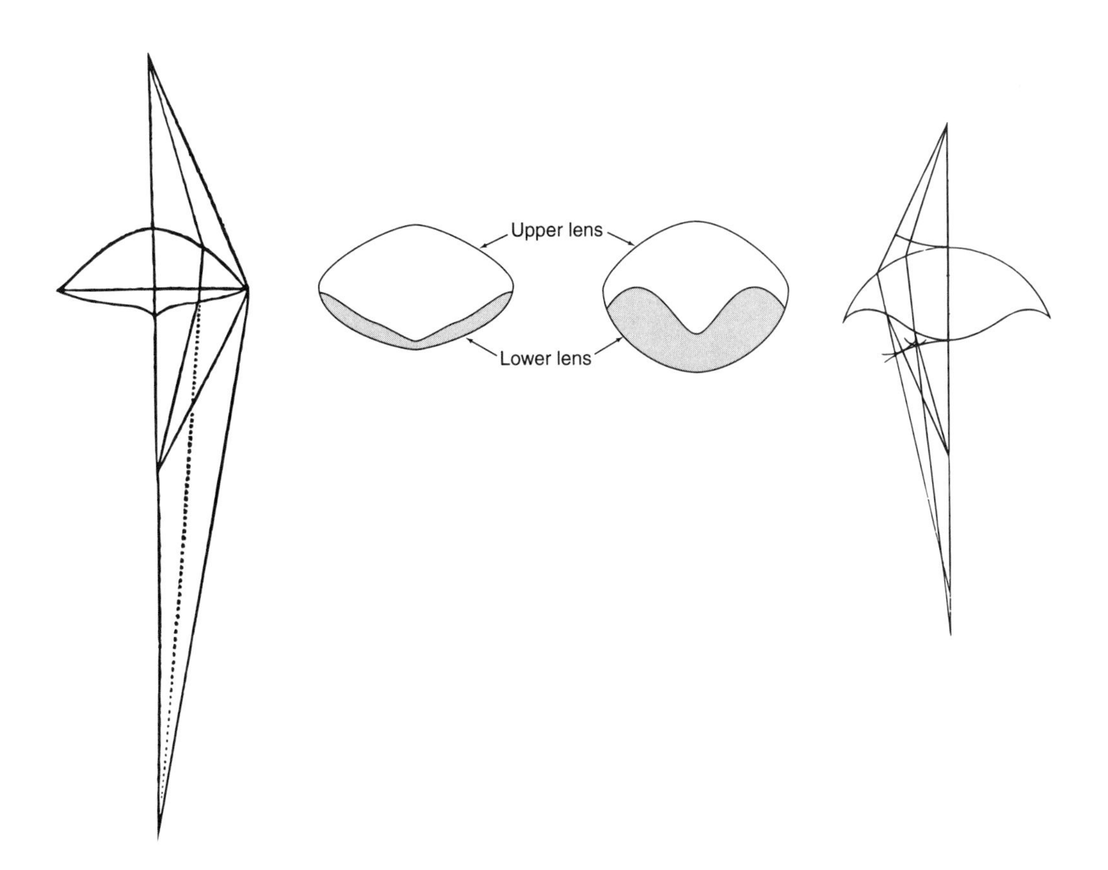


**FIGURE 5.3 Lens morphology of two trilobites (*Dalmanitina socialis*, center left; *Crozonaspis struvei*, center right) compared with the original drawings for aplanatic lenses published by Descartes (left) and Huygens (right).** Lenses are drawn in cross section; the vertical axis of each trilobite lens is normal to the surface of the compound eye. Both trilobites are of Ordovician age. *(Based on Clarkson & Levi-Setti, 1975)*

The trilobite lens is optimal in yet another way. Light was transmitted through the calcite lenses to photoreceptive cells within the eyes. The properties of calcite are such that light impinging on a crystal from virtually any angle is refracted in two directions, leading to a double image. However, if the crystal is oriented so that the light is moving parallel to its principal optical axis, the *c* axis (see Figure 2.14), the light will travel through the crystal as if it were glass. And this is precisely the orientation observed in trilobite lenses. The individual eye cells are oriented with respect to the curved eye surface in such a way that the *c* axes of the calcite lenses are normal to the eye surface.

To summarize, it appears from the work of Clarkson and Levi-Setti (1975) that trilobites evolved a remarkably sophisticated optical system. For an engineer to develop such a system would require considerable knowledge of optics and quite a bit of ingenuity. As an application of the paradigm approach to problems of functional morphology, the example provided by the trilobite lens is nearly unsurpassed.

As with many classic case studies that illustrate a principle unusually clearly, the interpretation of trilobite lenses has been scrutinized and challenged. First, it has been suggested that the doublet structure of Figure 5.3 may be a preservational artifact (Bruton & Haas, 2003). This possibility has not been completely evaluated, however, and the general consensus at the moment is that the trilobites in question had genuine lens doublets. Second, and more interestingly, some calculations have shown that the Descartes lens, which involved some mathematical approximations in its design, may not actually be well suited for minimizing optical distortion (Gál et al., 2000). Yet some trilobite lenses have this shape. Why? Although we still do not know with certainty, some workers have suggested that the lenses may have functioned as bifocals, allowing focused images of near and far objects through different parts of the lens (Gál et al., 2000). We cannot predict how these questions will ultimately be resolved, but we can be sure that the function of trilobite eyes will continue to be a fascinating area of research.

***Ventral Wing Plates in Crinoids*** Living stalked crinoids are erect suspension-feeders. They use their arms to capture suspended organic particles, which are then passed along an ambulacrum, or food groove, that runs the length of the arms toward the mouth, located centrally on the ventral side of the calyx (Figure 5.4). The feeding posture of living stalked crinoids is shown in Figure 5.5. The arms are recurved into the current, which flows from left to right in the photograph.

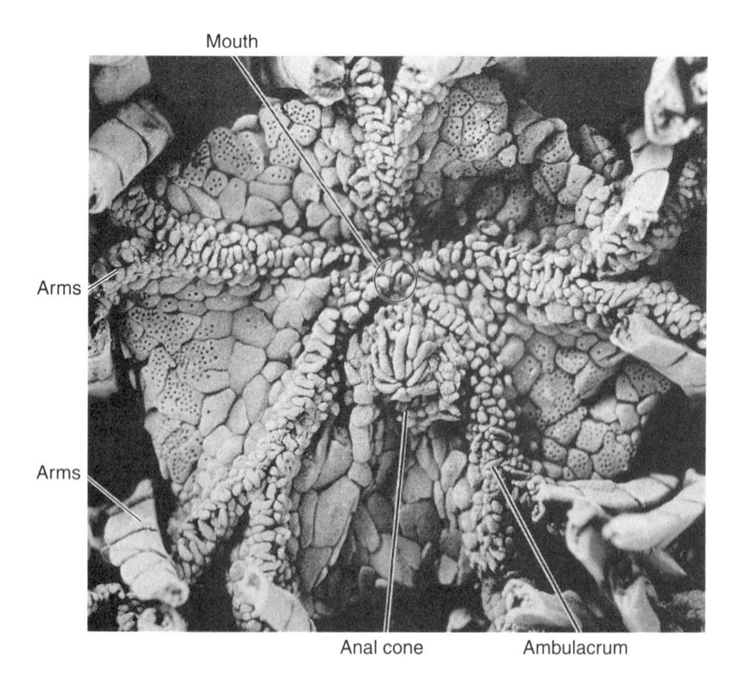


**FIGURE 5.4 Ventral view of the living crinoid *Neocrinus decorus*, showing the central mouth, plated ambulacra, and bases of arms.** Field of view is roughly 1 cm. *(From Moore & Teichert, 1978)*

**FIGURE 5.5 Feeding posture of the living stalked crinoid *Cenocrinus*.** The current flows from left to right, and the mouth is on the downstream side of the calyx. This crinoid is approximately 1 m tall. The photo was taken at between 200 and 300 m depth off the coast of Jamaica. *(Courtesy David L. Meyer)*

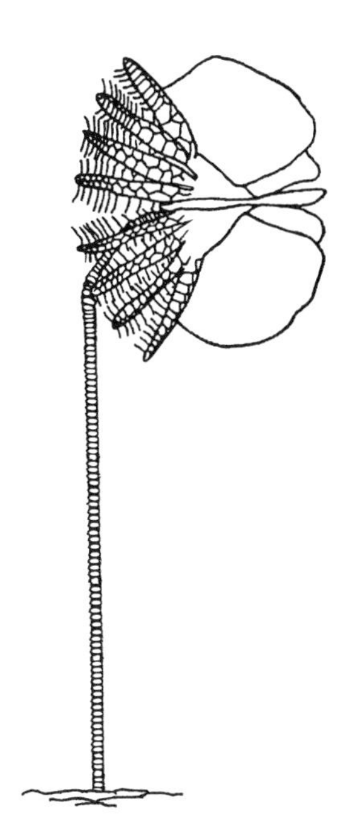

**FIGURE 5.6 Reconstruction of the Early Carboniferous crinoid *Pterotocrinus depressus* in feeding posture.** Posture is based on analogy to living stalked crinoids (Figure 5.5). The current flows from the left. Note the pronounced wing plates on the ventral side (facing to the right in this figure). *(From Baumiller & Plotnick, 1989)*

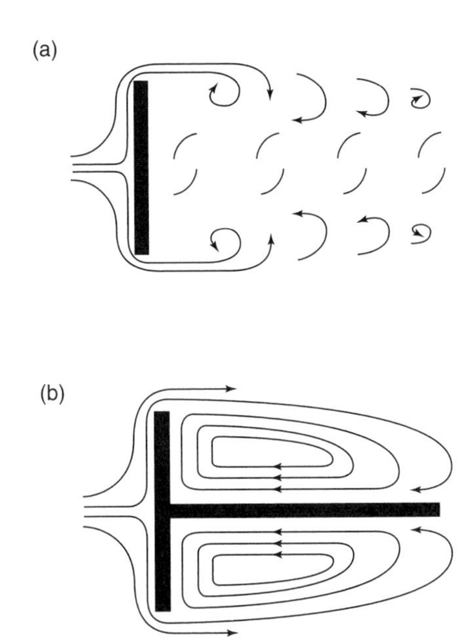


**FIGURE 5.7 Splitter-plate effect, with flow from left to right.** (a) Cross section of a blunt body in a flow; arrows indicate turbulent flow in the wake. The separation of flow induces a low-pressure region in the wake and thus increases drag. (b) Similar body with a splitter plate; arrows indicate laminar flow. Separation of flow is delayed and drag is reduced. *(From Baumiller & Plotnick, 1989)*

The current goes around the arms and through the openings between arms, and the food particles are captured on the downstream side of the arms. By homology, most extinct stalked crinoids are thought to have functioned in the same way.

Certain Carboniferous crinoids, most notably the genus *Pterotocrinus*, are unusual in possessing large, winglike plates that protrude from the ventral surface of the calyx (Figure 5.6). In an experimental study of *Pterotocrinus*, Tomasz Baumiller and Roy Plotnick (1989) postulated two potential functions for the wing plates.

First, they may have served as "splitter plates." It is well known from hydrodynamics that the flow around a blunt body separates, producing a low-pressure region in the wake and increasing drag on the body (Figure 5.7a). Adding a long plate to the object in the downstream direction helps to reduce drag by delaying the separation of flow, thus reducing the diameter of the wake (Figure 5.7b). Drag reduction could be beneficial to the crinoid by enabling it to maintain the appropriate feeding posture and by reducing stress on the ligaments of the stalk.

Second, the plates may have served as rudders, enabling the crinoid to maintain its feeding posture by reorienting passively when the current direction changed, much as the tail of a weather vane keeps it pointed into the wind.

To explore these two possibilities, Baumiller and Plotnick constructed an idealized physical model of a crinoid feeding apparatus: a fine steel screen formed into a hemispherical bowl (Figure 5.8). This model crinoid was attached via rigid rods to ball bearings so that the model could turn, and the apparatus was attached to a strain gauge so that the forces on the model could be measured. Experiments were conducted by placing the model in a flume—the hydrodynamic equivalent of a wind tunnel—and varying the speed of the current and its direction relative to the models. Models with and without wing plates were tested. To ensure that results did not depend critically on the particular experimental conditions, the experiments spanned a wide range in current speed; the angle between the current and the model; the coarseness of the wire mesh; and other aspects of the model, such as the distance between the ball-bearing pivot and the "calyx." Current speeds were also kept within a biologically realistic range.

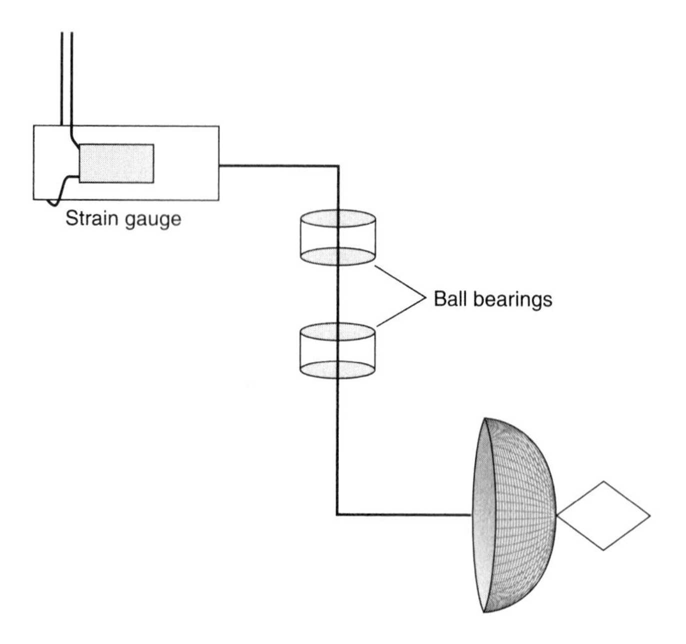


**FIGURE 5.8 Experimental design for measuring forces on crinoid models.** The wire-mesh hemisphere simulates the crinoid in feeding posture, and the diamond-shaped plate simulates the wing plate. Ball bearings allow the model to swivel passively, and the strain gauge measures the forces on the model. The current flows from left to right. *(From Baumiller & Plotnick, 1989)*

To test the hypothesis that the wing plates in *Pterotocrinus* may have functioned as splitter plates, the drag force on models with plates was compared with the force on models without plates. Models were oriented with the concave side of the wire-mesh hemisphere pointing upstream, consistent with the feeding posture of living crinoids. The splitter-plate hypothesis predicts that models with plates should experience lower drag forces. In fact, the drag forces on the two models were found to be indistinguishable (Figure 5.9a). This suggests that the wing plates were unlikely to have functioned to reduce drag.

To test the rudder hypothesis, the models were turned away from the concave-upstream posture by specified angles. If the wing plates functioned effectively as rudders, then the models with plates should experience greater rotational forces than the models without plates. This is exactly what happened (Figure 5.9b)—which suggests that the rudder hypothesis is plausible; crinoids with wing plates could have used them to reorient themselves passively.

In summary, a simplified but hydrodynamically relevant model of an erect crinoid shows that specialized structures—the wing plates—probably did not function to reduce drag but may well have enabled crinoids to maintain the proper feeding posture without expending energy to turn into the current. This represents an exemplary case of the use and experimental manipulation of physical models, combined with knowledge of living representatives, to deduce the function of extinct organisms. The next example illustrates these same themes, but differs in using replicas of actual specimens.

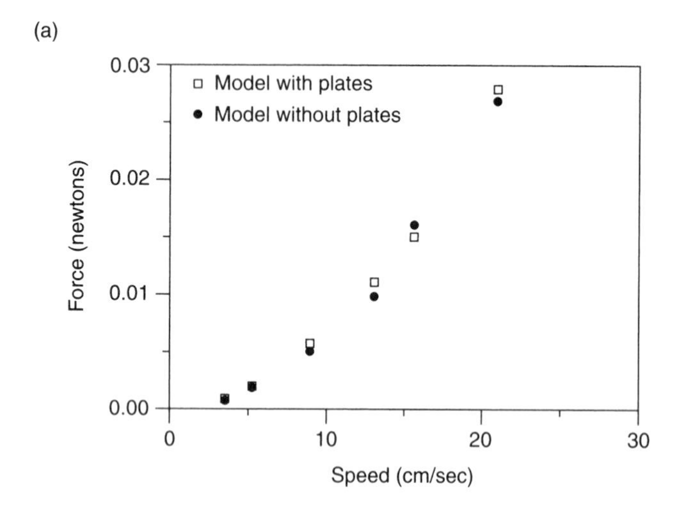


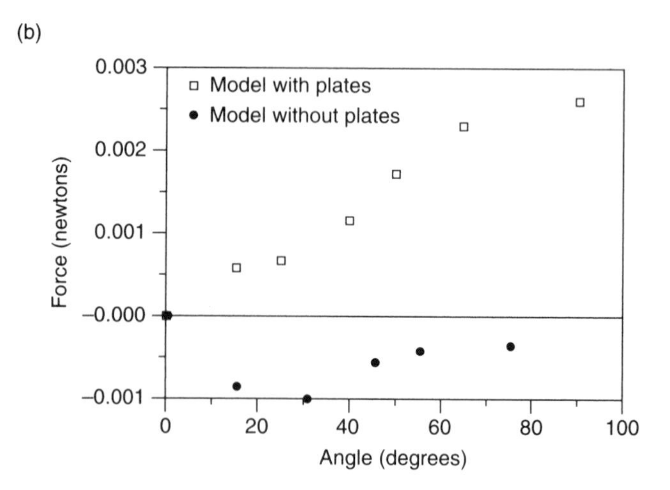


**FIGURE 5.9 Results of experiments on crinoid models of Figure 5.8.** (a) Drag force on the models versus current speed. There is essentially no difference in drag between models with and without wing plates. (b) Rotational force on the model versus angle between the model and the current. Forces with a positive sign are those that cause the wire-mesh bowl to turn into the current. The models with plates are able to reorient passively into the current, whereas the models without plates are not. *(From Baumiller & Plotnick, 1989)*

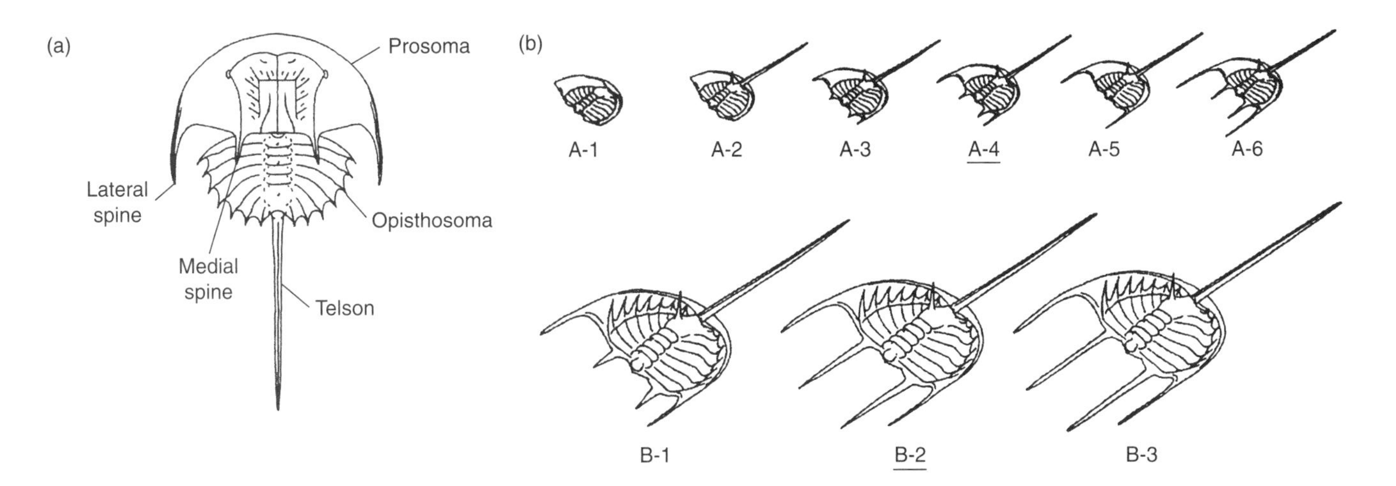


**FIGURE 5.10 Late Carboniferous horseshoe crab *Euproops danae*.** (a) Reconstruction. (b) Fabricated models. Models are in an enrolled posture, with the posterior tucked up under the anterior and the telson projecting forward. The A and B models are juveniles and adults. A-4 and B-2 are actual forms; the others have had the spines artificially lengthened or shortened. *(From Fisher, 1977)*

***Spines in Horseshoe Crabs*** Spines and other projections are common in a wide range of organisms. Their function is often regarded, quite reasonably, as protective. The adaptive value of the precise morphological details of spines—their size and shape—is usually less clear, however. *Euproops*, a Late Carboniferous arthropod related to living horseshoe crabs, possessed two pairs of spines on the anterior body region, or prosoma (Figure 5.10). One intriguing property of these spines is that, in juveniles, the lateral pair is longer than the medial pair. The medial spines grow faster, however, so in adults they are longer than the lateral spines. The function of spines was explored experimentally by Daniel Fisher (1977), who considered both the general role of spines and the reason for their relative sizes.

Living horseshoe crabs are known to burrow for protection. Because previous functional studies had shown that *Euproops* was probably a capable swimmer, Fisher reasoned that individuals were likely to encounter predators well above the substrate, where burrowing would not be an option. Numerous arthropods, living and extinct, are known to enroll, evidently in response to disturbance. There is anatomical evidence that *Euproops* was capable of enrollment—such as the fit in shape between the prosoma and rear body region, or opisthosoma. Moreover, specimens are commonly preserved in an enrolled position.

If an individual enrolled upon encountering a predator, it would settle toward the substrate and potentially escape predation. The predators of *Euproops* would have included fishes and amphibians with basic sensory systems similar to those of living forms. Observations of many modern fishes demonstrate that they are highly sensitive to horizontal motion in their prey. A smooth path of settling would therefore make the horseshoe crab less conspicuous to a predator than would an oscillatory or irregular path in which horizontal movements interrupt the general vertical descent.

Given this background, Fisher explored the role of spines in settling. He constructed models of horseshoe crabs, allowed the models to fall freely in sea water, and filmed their settling behavior. Figure 5.10b shows two sets of models, one of juveniles (A) and one of adults (B). For each set, one model has spines with realistic lengths while the other models have spines that have been made longer or shorter compared with real forms or have had their relative lengths changed. The models are reconstructed in an enrolled posture, with the opisthosoma tucked under the prosoma and the telson (tail spine) extending beyond the head.

Representative results of settling experiments are shown in Figure 5.11. Some of the models oscillate as they descend (A-6), others make abrupt horizontal shifts (A-2), and some attain a smooth and stable descent (A-4). This last type of settling is expected to be least conspicuous to a predator, and therefore to be most advantageous for predator avoidance. The models that settle in this way are in fact the ones corresponding to the observed forms. For juveniles, relatively long lateral spines and relatively short medial spines, both of

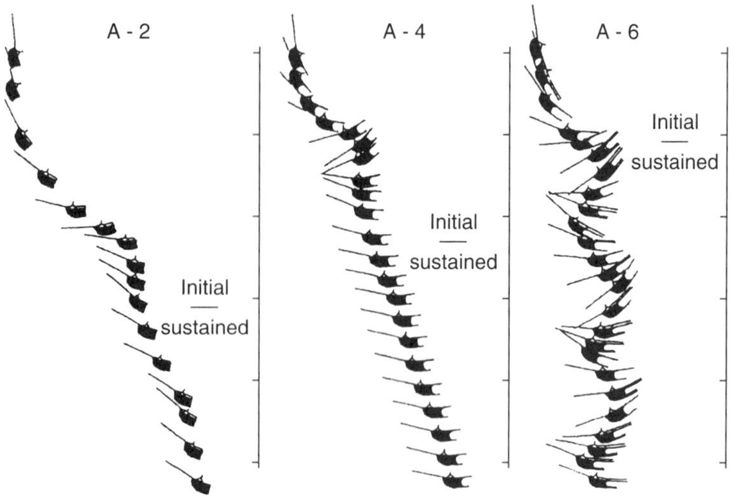

**FIGURE 5.11 Examples of settling behavior of juvenile *Euproops* models, drawn from time-lapse photographs.** Models were released in water with tail spine pointing up. They initially moved from this orientation, then achieved a sustained pattern of descent. A-4 shows the steady descent of a realistic form. A-2 and A-6 show examples of unsteady descent in forms that have had spines changed from their true lengths. *(From Fisher, 1977)*

Box 5.1

## LOCOMOTION IN NONAVIAN DINOSAURS

In any long-lived and diverse biologic group, it is unlikely that any aspect of function will be completely uniform throughout the group. Nevertheless, it may be possible to characterize the general functional style of a higher taxon and to contrast it with that of other taxa. The case of dinosaurs is especially interesting because there is a living group that is phylogenetically close to them—the birds, which are generally recognized to be an offshoot of theropod dinosaurs.

Despite their evolutionary descent from dinosaurs, living birds are highly derived in terms of physiology, behavior, feeding, and skeletal anatomy. Basal phylogenetic relationships within archosaurs (Figure 4.10) might also seem to suggest crocodilians as a possible living analog for some dinosaurs. At the same time, the wide range of ways of life apparently exploited by dinosaurs as a group, many of them similar to those of living mammals, suggest the possibility of mammals as living analogs. The following study focuses on a particular aspect of function—namely, the posture adopted in walking—to determine which living group is likely to represent the best analog.

Many aspects of skeletal morphology in dinosaurs suggest an upright posture, so the sprawling gait of crocodiles and other primitive archosaurs would seem to be ruled out. Yet within this upright posture, there are two principal styles of locomotion, broadly characteristic of birds and of mammals, respectively (Figures 5.12 and 5.13). The orientation of the femur changes throughout the step cycle in both groups, generally being relatively more vertical at the point of foot lift-off and more horizontal at the point of foot contact (Figure 5.13). The forces on the femur also vary regularly within the step cycle, being dominated by torsion and compression when the femur is horizontal, and bending and compression when it is vertical (Figure 5.12d and 5.12e). Regardless of the point in the step cycle, however, birds tend to have the femur at a position much closer to horizontal than do mammals (Figure 5.13).

The relatively horizontal posture of the femur has important consequences for avian skeletal structure. Consider the total length of the hindlimb and the proportion of this length made up by the femur, the tibia, and the metatarsal (Figures 5.14 and 5.15). The femur of a bird typically accounts for 20 to 40 percent of the total limb length. Bone is weaker in the face of torsional as opposed to bending forces. This, coupled with the horizontal attitude of the bird femur, implies that the torsional forces on this bone would be excessive if it were much longer than 40 percent of the limb length. With a more vertical femur, mammals experience a lower torsional force and can therefore achieve longer femoral lengths, up to 60 percent or more of the total limb length.

which are short relative to the prosoma, are required to yield a smooth descent. For adult models to settle smoothly, Fisher found that spines must be about the same length as the prosoma, and the medial spines must be longer than the lateral spines. This again is what is seen in actual specimens. These results strongly suggest that the right balance of spine sizes is an adaptation for smooth settling. Because there are nonlinear relationships between spine size and body size on the one hand and drag and other hydrodynamic forces on the other hand, the same spine sizes are not equally effective at all body sizes. There has evidently been natural selection for a particular pattern of anisometric growth [SEE SECTION 2.3] in order to accommodate this fact.

Although the spines of *Euproops* clearly have functional value, the settling behavior of this horseshoe crab is not perfectly ideal. Rapid, smooth settling would be better attained by a spherical object. Yet the organism had functional demands other than settling. For example, its overall form was elongate rather than spherical in order to facilitate swimming, and the spines projected posteriorly to facilitate movement through the sediment. *Euproops* could settle remarkably smoothly; it is as optimal as can reasonably be expected, given the constraints of competing functions and phylogenetic inheritance (Figure 5.1).

The additional example in Box 5.1 combines biomechanics with statistical analysis of anatomical measurements.

Thus, we have a form–function correlation that can be understood in terms of biomechanics. Can we use this to deduce the style of dinosaur locomotion? The argument for doing so is statistical in nature. When dinosaur limbs are measured, they largely overlap the mammalian field in the

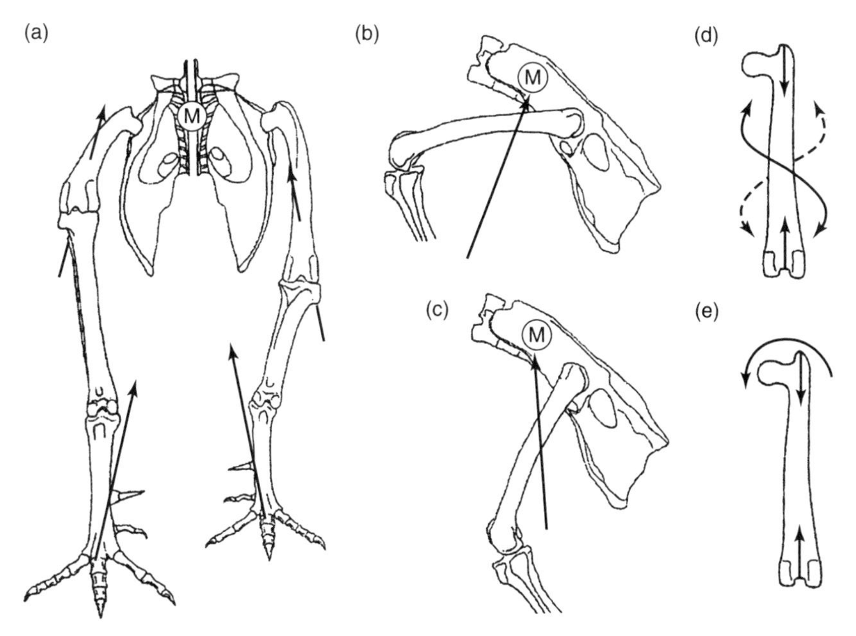


**FIGURE 5.12 Posture and biomechanics of a representative terrestrial vertebrate (a chicken).** (a) The pelvis and hindlimbs. The lower arrows indicate the direction of force between the ground and the center of mass, M. The upper arrows indicate forces through the femur that result from the offset between the femur and the center of mass. (b, c) Two parts of the stride in which the femur is (b) relatively more horizontal and (c) more vertical. (d, e) The forces on the femur, indicated diagramatically. When the femur is more horizontal (d), there are compressive forces (straight arrows) and twisting forces (curved arrows). When the femur is more vertical (e), there are compressive forces (straight arrows) and bending forces (curved arrow). *(From Carrano, 1998a)*

*continued on next page*

## Other Lines of Evidence in Functional Interpretation

Numerous other lines of supplementary evidence are often invoked in functional studies. As discussed in Chapter 1, trace fossils may reveal patterns of behavior that give clues to function and aspects of life habit. Studies of growth are also important in functional analysis. In the example of respiration in rhombiferan echinoderms that we discussed in Chapter 2 (Box 2.3), the details of allometric scaling were used to infer that some factor must have limited the efficiency of respiratory structures. In the example that follows, additional geologic and paleogeographic data are used to help infer the functional ecology of certain specialized trilobites.

***Life Habits in Pelagic Trilobites*** The range of individual lens orientations in the trilobite eye can be used to infer the size and shape of the visual field of the trilobite. In most trilobites, the field of view is lateral, over the surface of the sediment. A number of trilobite lineages independently evolved large eyes that, in the most extreme cases, gave them 360° vision in all directions, including downward (Figure 5.16). The specialized eyes of the forms in Figure 5.16 are accompanied by other unusual features that are not generally found in trilobites, the majority of which

*Box 5.1 (continued)*

femur–tibia–metatarsal ternary diagram (Figure 5.15), while they show only slight overlap with the avian field. Note that the oldest known bird, *Archaeopteryx*, falls within this small region of overlap. Bipedal and quadrupedal dinosaurs occupy nearly separate fields, but both are mainly coincident with the mammalian field.

Overall, the structure of dinosaur limbs suggests a style of locomotion more similar to that of living mammals than to that of living birds. This result, of course, does not mean that other aspects of dinosaur function and physiology are more mammalian than avian, but it is important in providing an analog for future studies of dinosaur locomotion.

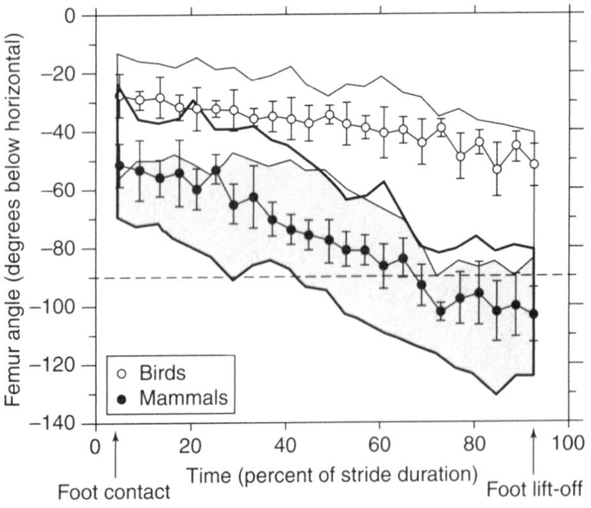

**FIGURE 5.13 Posture of the femur during the stride of birds (open circles) and mammals (closed circles).** The angle of the femur relative to the horizontal varies predictably during the stride, and this angle is consistently lower (more horizontal) for birds. Points show the mean ±1 standard error (see Box 3.1), based on four bird species and eight mammal species. *(From Carrano, 1998a)*

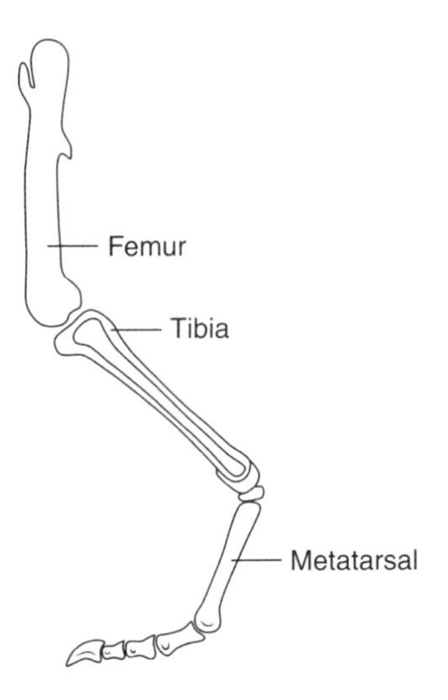


**FIGURE 5.14 Sketch of a hindlimb showing femur, tibia, and metatarsal that were measured for comparison among birds, mammals, and dinosaurs.** *(From Carrano, 1998b)*

were benthic (bottom-dwelling) and had the ability to walk and swim to a limited extent. The pleural (lateral) regions of the thorax in these specialized forms are greatly reduced, which would have contributed to flexibility and reduced the bulk of the trilobite. This would seem to serve as an adaptation for swimming. The reduction in pleural regions may also have facilitated backward vision.

The head is large and has genal spines that project downward. This is different from the majority of trilobites, whose spines project horizontally, and it would not have been conducive to a benthic existence. Moreover, the axial region is highly vaulted, suggesting well-developed musculature, like that of a shrimp, that would have been useful for active swimming. This combination of features suggests that these trilobites were pelagic (open-ocean) rather than benthic.

The well-developed eyes are similar to those seen in a number of specialized living species of amphipod and isopod crustaceans. These groups are mostly benthic, but specialized forms that live in the open ocean have evolved large eyes like those of the trilobites in question. By analogy, this suggests that these trilobites were also pelagic. As stated earlier, arguments from analogy can be rather inexact. But in this example, there are two additional lines of evidence that support the inferences drawn from functional arguments and analogy.

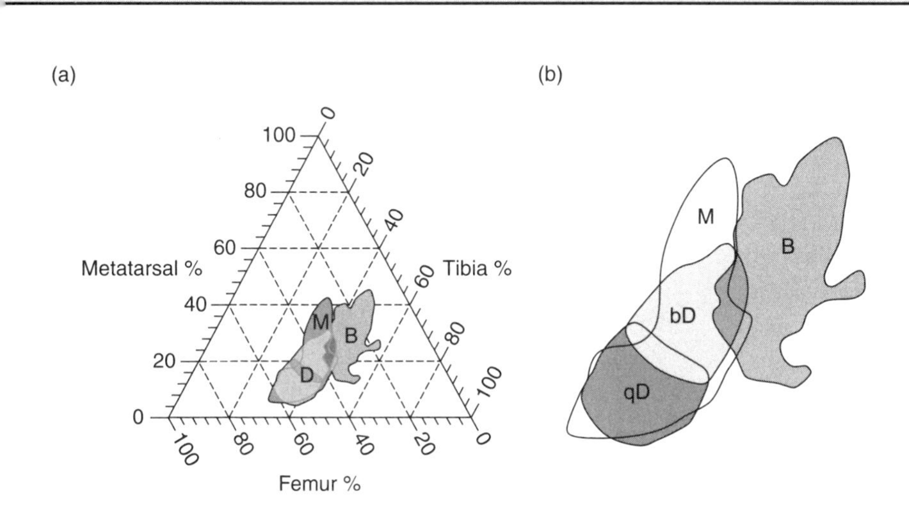


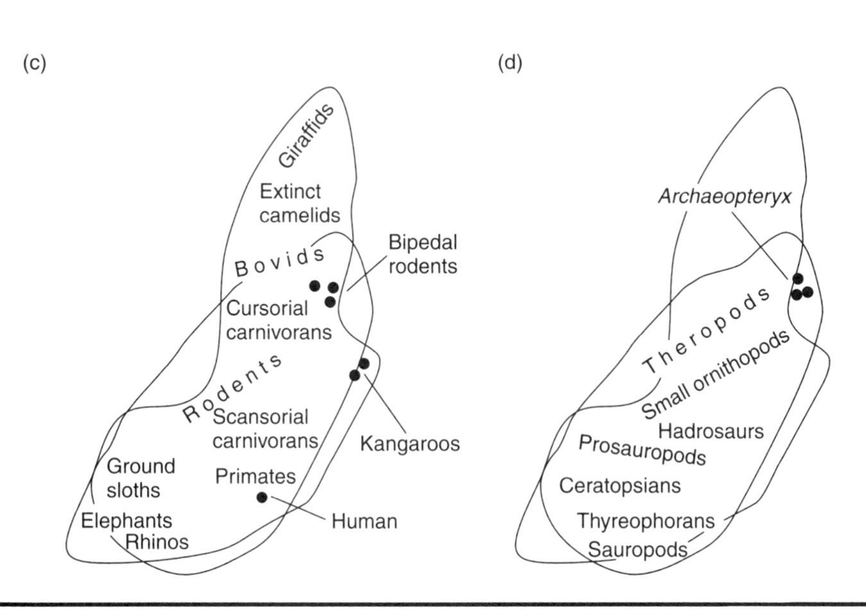


**FIGURE 5.15 Hindlimb measurements for birds (B), mammals (M), and nonavian dinosaurs (D).** (a) The percent of the total hindlimb length accounted for by the femur, tibia, and metatarsal is graphed in the ternary diagram. Lines circumscribe the entire field of values for each group. (b) Details of the fields occupied by the three groups. Bipedal dinosaurs (bD) overlap somewhat with birds, but both bipedal and quadrupedal (qD) dinosaurs overlap mainly with mammals. (c) The positions of some groups of mammals. Note that bipedal mammals (indicated by the dots) largely overlap with the field of bipedal dinosaurs. (d) The positions of some groups of dinosaurs. Note that three specimens of the oldest known bird *Archaeopteryx* are within the field of bipedal dinosaurs. *(From Carrano, 1998a)*

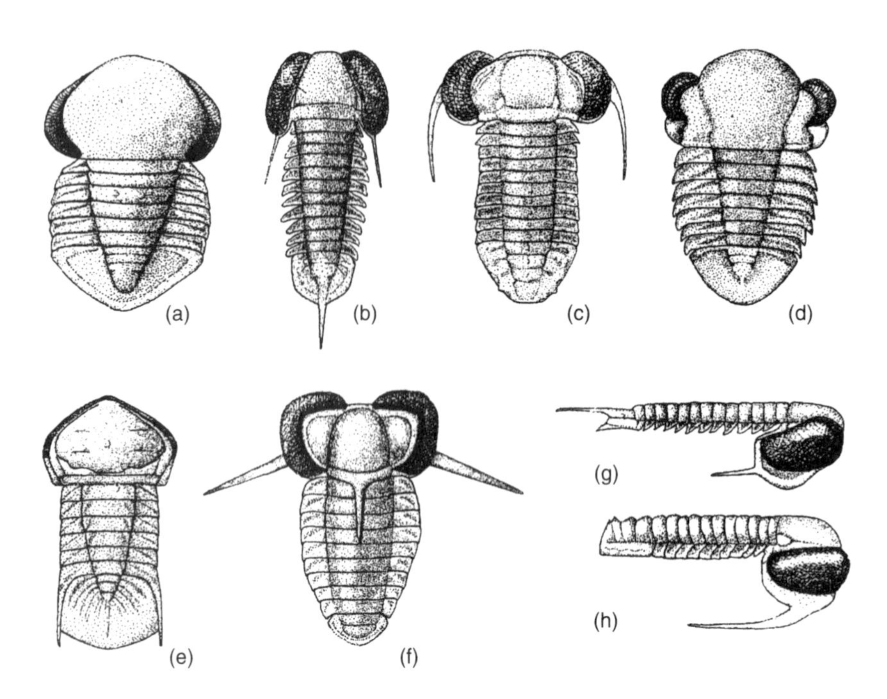


**FIGURE 5.16 Examples of trilobites with well-developed eyes, reduced pleural regions, and ventrally projecting heads.** (a–f) Dorsal views of *Pricyclopyge*, *Opipeuterella*, *Carolinites*, *Prospectatrix*, *Girvanopyge*, and *Telephina*. (g, h) Lateral views of *Opipeuterella* (b) and *Carolinites* (c). *(From Fortey, 1985a)*

First, individual taxa of these trilobites have very broad geographic ranges but are mostly restricted to near the paleoequator (Figure 5.17). This suggests that ocean conditions rather than dispersal ability and the arrangement of continents [SEE SECTION 9.6] limited their distribution. Second, the trilobites are found in a wide range of sediment types, ranging from the kind of shallow-water deposits in which benthic trilobites are usually found to deeper water sediments. It seems implausible that a species would occupy such a wide range of benthic environments without showing any anatomical modifications to suit the different habitats. In fact, other lineages of trilobites that live on the sediment surface in very deep waters, where little light penetrates, often have reduced or absent eyes.

The natural interpretation of this combination of geographic and geologic occurrence is that these trilobites, like the amphipods and isopods mentioned earlier, lived in the open ocean and that their molts and carcasses settled to the ocean floor to occupy a range of sedimentary environments. Further refinement of this interpretation is possible. Different genera of pelagic trilobites are found in a somewhat different range of sediment types. Molts and carcasses settled to the ocean floor, and the shallower the pelagic habitat of the taxon, the broader the range of depths to which it could settle (Figure 5.18).

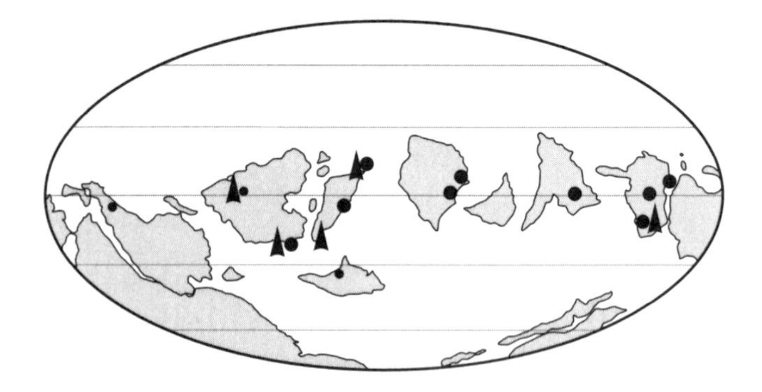

**FIGURE 5.17 Paleogeographic reconstruction showing the arrangement of continents in the Early Ordovician.** Circles show occurrences of the genus *Carolinites* (Figures 5.16c and 5.16h), and triangles show occurrences of *Opipeuterella* (Figures 5.16b and 5.16g). Both are geographically widespread but mostly near the paleoequator, suggesting that they were limited by ocean conditions rather than dispersal ability. *(From Fortey, 1985a)*

Genera such as *Carolinites* (Figures 5.16c and 5.16h) and *Opipeuterella* (Figures 5.16b and 5.16g), which are found in the full range of environments from shallowest to deepest, must have inhabited the surface waters. Others, such as *Pricyclopyge* (Figure 5.16a) and *Girvanopyge* (Figure 5.16e), which are absent from the shallowest sediments, must have lived within the deeper parts of the ocean.

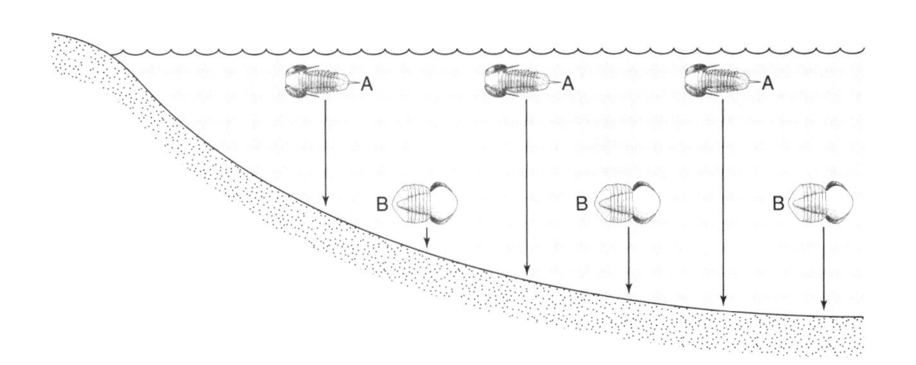


**FIGURE 5.18 The distribution of sedimentary environments in which pelagic trilobites are found depends on the depth at which they lived.** Arrows indicate settling to the sea floor after death or molting. A surface-dwelling form (A) can be found in sediments representing a wider range of water depths, whereas a form that lives at depth (B) will be absent from deposits representing shallower water. *(Based on Fortey, 1985a)*

## 5.3 THEORETICAL MORPHOLOGY

The foregoing examples of functional morphologic analysis illustrate adaptation and trade-offs in individual organisms and structures. These same factors are important in shaping the distribution of form within larger biologic groups. If some aspect of form is highly adaptive, we should expect it to be quite common, subject to the limitations imposed by history, structure, and competing functional demands.

There are generally three main features in a study of theoretical morphology: (1) A formal model of morphology is set forth [SEE SECTION 2.2]; (2) this model is used to generate the spectrum of possible forms that adhere to the assumptions of the model; (3) the distribution of known forms is compared with the theoretically possible spectrum. Differences between the possible and the actual, such as preferred modes and gaps in the actual distribution, are explored using functional morphology and other lines of reasoning.

In the sections that follow, we outline some major themes of research in theoretical morphology, illustrating each with a different model of form. Certain models, such as the harmonic analysis of curves (see Box 2.2), involve a large number of parameters and are therefore of limited practical use in exploring the relationship between conceivable and actual distributions of form. For our purposes, it is convenient to restrict discussion of theoretical morphology to models with relatively few parameters.

### Exploring Alternative Modes of Life

***Geometric Analysis of Shell Coiling*** A wide range of organisms produce coiled skeletons and skeletal parts that are mathematically well characterized. The growth of mollusc and brachiopod shells, for example, can be modeled as the movement of a generating curve around an axis of coiling, sweeping out a three-dimensional solid of revolution (Figure 5.19). The generating curve may approximate the aperture or opening of the shell, but the generating curve and coiling axis are mathematical constructs rather than biological structures.

The generating curve changes in size as it revolves about the axis. The whorl expansion rate $W$ expresses this change in size as the ratio of sizes of the generating curve separated by a full revolution, or $2\pi$ radians (360°). Because we are concerned with biological traits, which

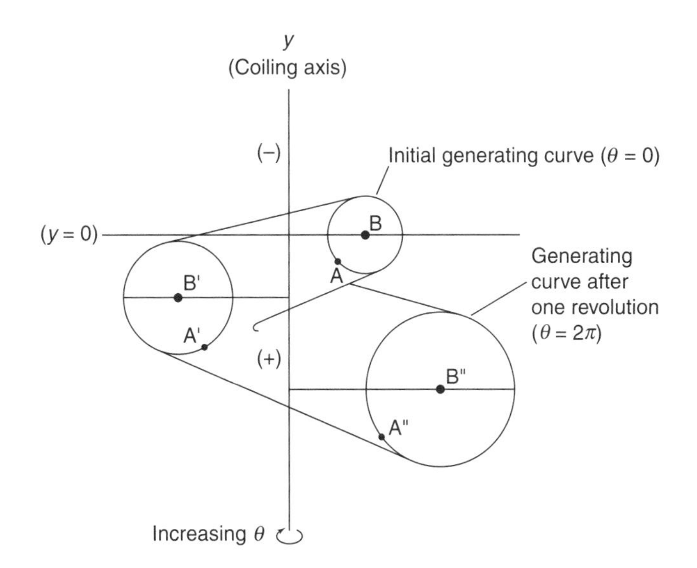


**FIGURE 5.19 Geometric model of shell coiling, depicted in cylindrical coordinates.** The movement of the generating curve about the coiling axis sweeps out a three-dimensional solid of revolution. $\Theta$ gives the angular revolution of the generating curve about the coiling axis, and $y$ represents the distance along the coiling axis. A, A′, and A″ represent a point on the generating curve at 0, $\pi$, and $2\pi$ radians of revolution. B, B′, and B″ represent the center of the generating curve. *(From Raup, 1966)*

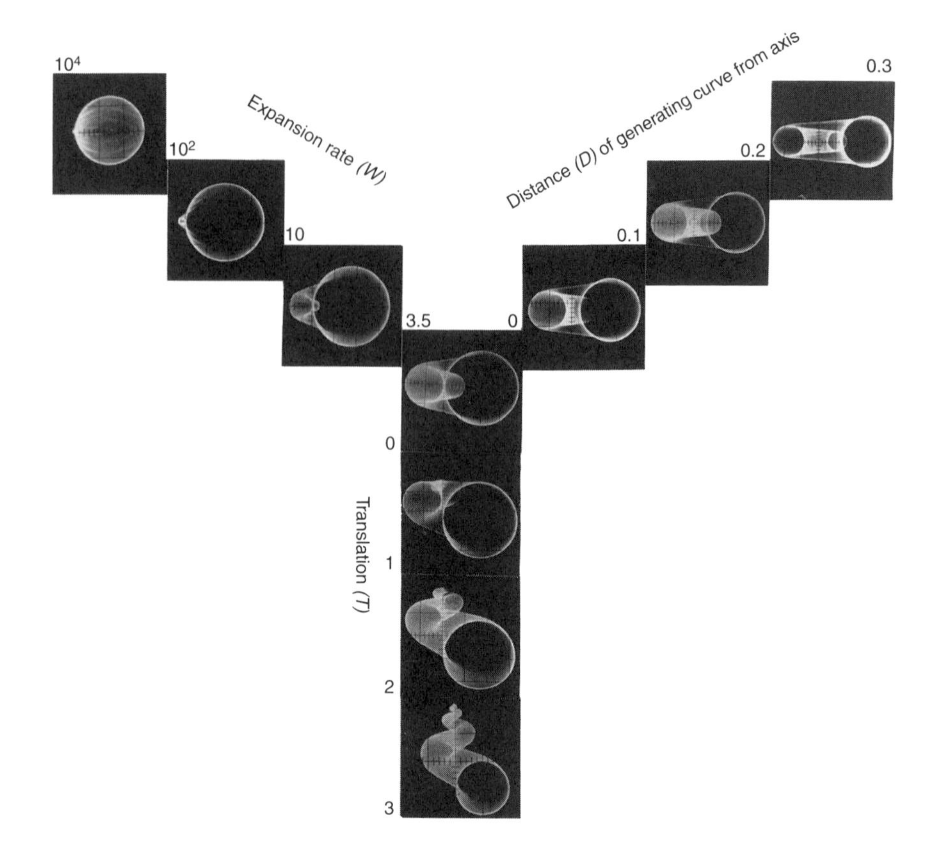


**FIGURE 5.20 Some hypothetical shells generated by computer using the model of Figure 5.19.** This figure shows the effect of varying each of the three parameters of the coiling model while keeping the others constant at the values indicated for the shell in the center. *(From Raup, 1966)*

generally increase in size as the organism grows, the theoretical lower limit on $W$ is 1. $W$ has no theoretical upper limit.

The generating curve may also move along the axis as it revolves and expands. The rate at which it does so is the translation rate $T$, expressed as a ratio of how far along the $y$ axis a point on the generating curve moves relative to how far away from the $y$ axis it moves. If $T$ is equal to zero, the shell coils in a plane (Figure 5.20, center), as is the case for most cephalopods. The model shell in Figure 5.19 is dextral; that is, it turns in the sense of a right-hand screw. For reasons that are still not fully understood, the vast majority of gastropods that have ever lived are dextral. For dextral shells, $T$ is positive—that is, the curve translates down the axis. For sinistral or left-hand shells, $T$ is negative and the curve translates up the axis. In either case, the magnitude of $T$ has no theoretical upper limit. All else being equal, the higher the translation rate, the greater the height-to-width ratio of the shell (Figure 5.20, bottom series).

The final parameter of the coiling model is the relative distance $D$ of the generating curve from the coiling axis. This is defined as the distance from the coiling axis to the inner margin of the generating curve, divided by the distance to the outer margin. The generating curve is assumed not to overlap the axis of coiling; thus $D \geq 0$. If the generating curve just touches the coiling axis, then $D = 0$. Distance $D$ has no theoretical upper limit.

From the definition of $W$, it is clear that this geometric model assumes multiplicative growth in size. If the coiling parameters are constant, the shape of the shell remains constant as size increases. Figure 5.20 shows a range of coiled shells that can be simulated with this simple model. In many ways, they succeed in mimicking real shells, although they also fail in some respects. For example, it is common for molluscs and other coiled organisms to change their coiling geometry as they grow, sometimes gradually and sometimes abruptly, as at the transition from larva to juvenile (see Figure 7.29), but this is not taken into account in the simple model of Figure 5.19. Also, the generating curve is assumed to be circular for simplicity, while the cross sections of real whorls vary enormously in shape.

Ontogenetic changes in coiling parameters could be incorporated into the model, and the shape of the generating curve could itself be modeled with one or more parameters to make the resulting forms more realistic. However, the goal of morphological modeling is generally not to produce exact replicas of organisms. To do so would involve a complex description with so many parameters as to make the model practically useless.

The model tells us how to simulate ideal shells. If we are to compare these with actual forms, it is necessary to estimate the coiling parameters from real shells. One way to do this is described in Box 5.2. Analogous operational procedures must be devised for other models considered later in this chapter. Although mathematical models of form may seem highly abstract, it is often easier to work with models than to measure actual specimens!

*Box 5.2*

## ESTIMATING COILING PARAMETERS

To calculate the values of $W$, $T$, and $D$ for a coiled shell, it is necessary to estimate the position of the coiling axis and to identify the generating curve. This is commonly done by cutting a cross section of the shell or by taking an X-ray.

Figure 5.21a shows the adult shell of the extant land snail *Theba pisana*. A radiograph of another specimen of this species, printed as a negative, is shown in Figure 5.21b. This simulates sectioning of the shell without actually damaging it [SEE SECTION 2.2]. Here we can make out the outline of the coiled tube at successive whorls. These outlines are assumed to represent the ideal generating curve.

A line drawing of the generating curve at increments of $\pi$ radians is shown in Figure 5.21c. Superimposed on these is an estimate of the position of the coiling axis. In this case, the coiling axis was fitted by eye, but more exact statistical approaches can be used to find the optimal position of the axis.

Assuming the shell fits the coiling model, $D$ can be estimated from the generating curve at any point, and $W$ and $T$ can be estimated from the generating curve

(a)

(b)

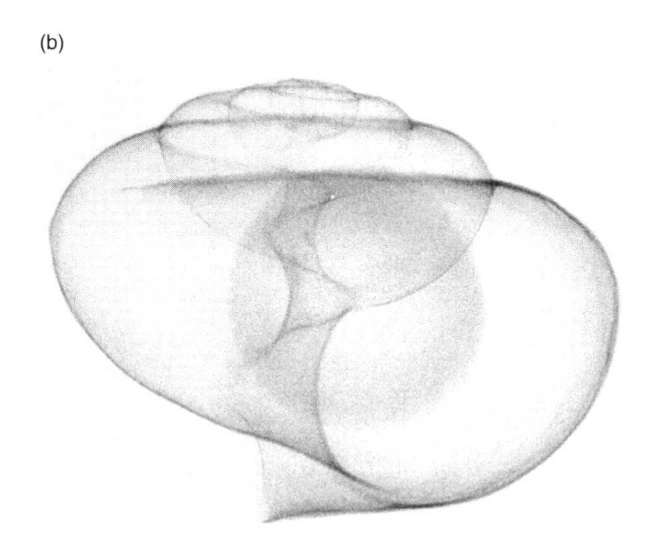

**FIGURE 5.21 Estimation of coiling parameters.** (a) Photograph of the living land snail *Theba pisana*. The width of the shell is about 1.5 cm. (b) Enlarged radiograph of a shell, printed as a negative, used to obtain the image of the cross section. The large, subcircular feature near the center of the image is a ball of plasticene used to hold the shell in place for radiography. *(Michael Foote)*

*continued on next page*

*Box 5.2 (continued)*

at any two points separated by $2\pi$ radians, as shown in Figure 5.21c. Here $D$ is calculated as $d_{in}/d_{out}$ on the final whorl. The heavy points show the position of the geometric centroid of the generating curve. The $x$- and $y$-coordinates of these points are used to calculate $W$ and $T$, as follows:

$$W = x_{\Theta+2\pi}/x_{\Theta}$$

and

$$T = (y_{\Theta+2\pi} - y_{\Theta})/(x_{\Theta+2\pi} - x_{\Theta})$$

That these expressions are appropriate can be verified by comparing them with the model in Figure 5.19. Other approaches are also commonly used. For example, if $A$ is the measured area of the generating curve, $W$ can be calculated as $\sqrt{(A_{\Theta+2\pi}/A_{\Theta})}$. The square root is taken because $W$ is defined as the rate of increase of a linear feature.

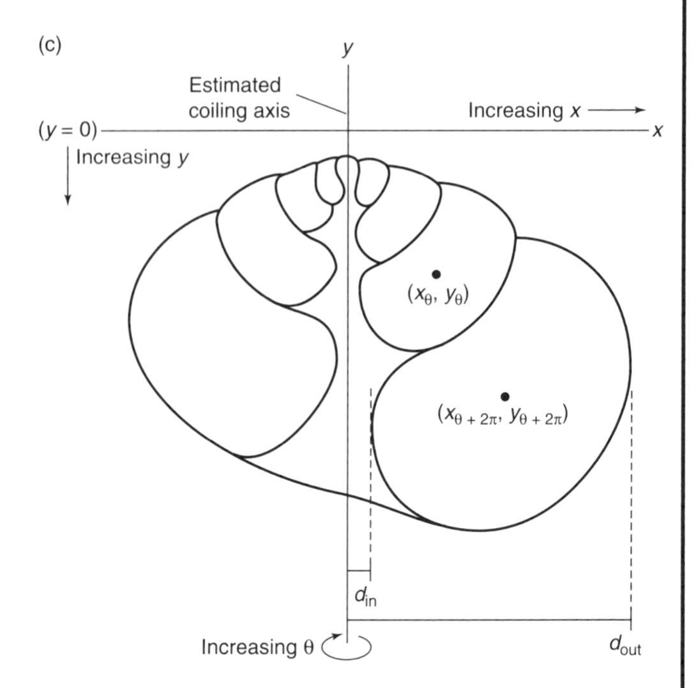


**FIGURE 5.21 (*cont.*)** (c) Line drawing of the whorl outlines from the radiograph, used to estimate the position and size of the generating curve. The vertical line is the estimated coiling axis. Heavy points denote the geometric centroid of the generating curve. $D = d_{in}/d_{out}$, $W = x_{\Theta+2\pi}/x_{\Theta}$, and $T = (y_{\Theta+2\pi} - y_{\Theta})/(x_{\Theta+2\pi} - x_{\Theta})$.

The three parameters $W$, $T$, and $D$ define an enormous spectrum of possible shell forms, yet real shells are confined to a relatively small region of the parameter space (Figure 5.22). This concentration of observed forms can be understood to a large extent by considering some of the functional demands of bivalved and univalved organisms and their different modes of life.

In order for bivalved shells to articulate effectively, it is important to have a high expansion rate and low value of $D$ (Figure 5.23). Deviations from this ideal, as in the upper part of Figure 5.23, would lead to extensive whorl overlap and therefore to interference between the two shells. The dashed lines in Figure 5.22 show the surface in the $W$–$T$–$D$ space that separates shells with overlapping whorls from shells with open coiling. Bivalved shells, with nonoverlapping whorls, are confined to below this surface.

In fact, even having the appropriate values of $W$ and $D$ does not completely eliminate the problem of shell interference for bivalves. Figure 5.24a shows two shells with high $W$ and low $D$ superimposed. The umbonal regions of the two valves clearly interfere with each other in these model shells. There are at least three ways that bivalves can avoid shell interference. The first is to deviate from the ideal model by depositing extra shell material between the umbones, in effect to have a biological generating curve that is distinct from the geometric generating curve (Figures 5.24b and 5.24c). The second is to have valves that are distinctly unequal in size (Figure 5.24d). The third is to have equal valves with positive allometry of expansion rate, that is, a value of $W$ that increases progressively with size (Figure 5.24e). The first and second strategies are widely exploited by both brachiopods and bivalve molluscs, while the third is most common in bivalve molluscs. The problem of valve interference is not always perfectly solved, however. A number of bivalve molluscs show beveled umbones caused by grinding together of the two valves.

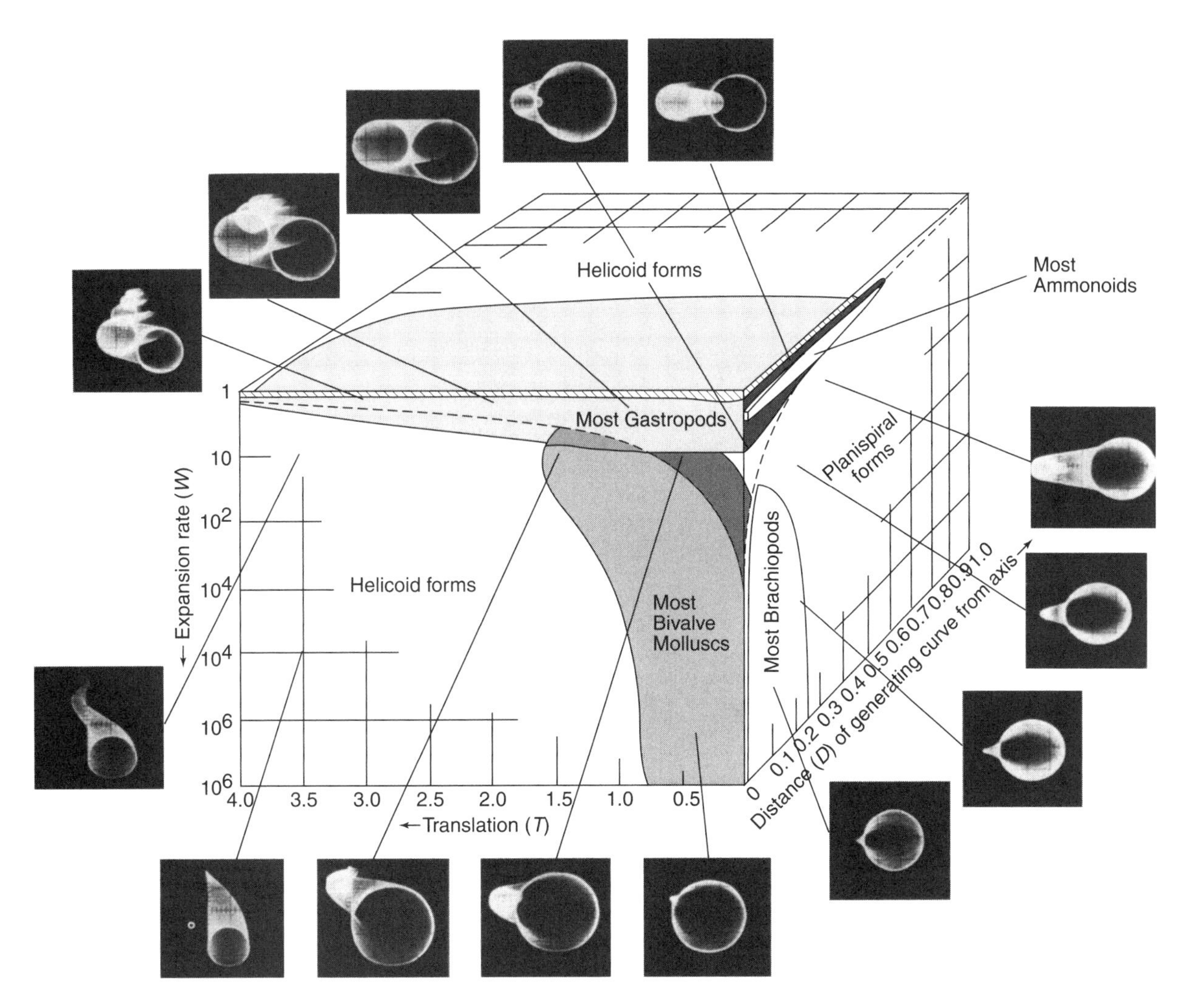


**FIGURE 5.22 General distribution of observed forms within the theoretically possible coiling space.** Representative computer-generated forms correspond to particular combinations of coiling parameters. Shaded areas show the combinations of coiling parameters typical of a few major taxonomic groups. Most of the theoretically possible space is not occupied by actual shells. The surface shown by the dashed lines separates shells with open coiling (below) from shells whose successive whorls overlap (above). *(From Raup, 1966)*

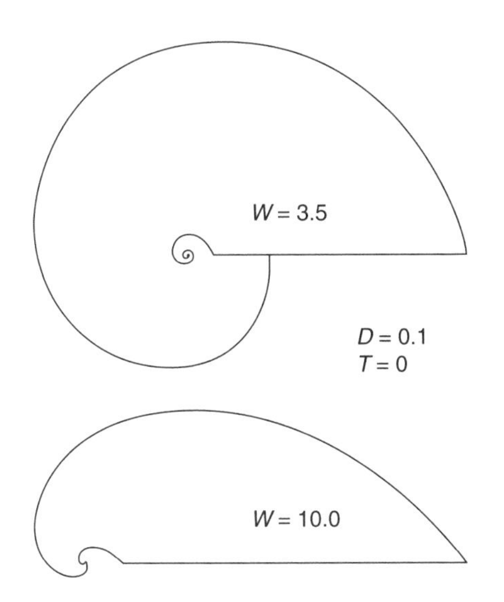


**FIGURE 5.23 The effect of expansion rate on whorl overlap.** A relatively low expansion rate yields whorl overlap, which is typical of univalves (top). A higher expansion rate produces no overlap, which is typical of each of the valves in a bivalve (bottom). *(From Raup, 1966)*

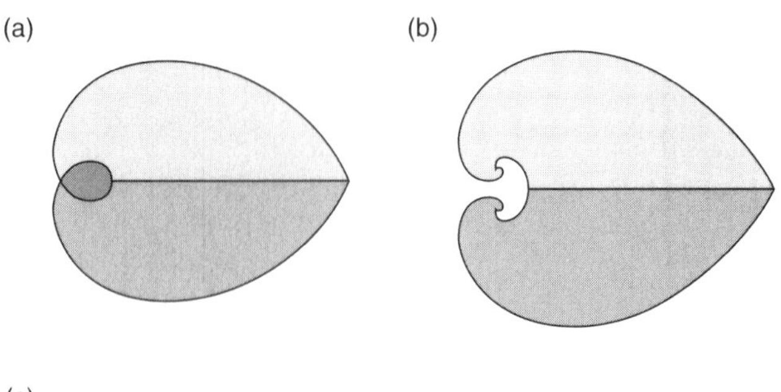


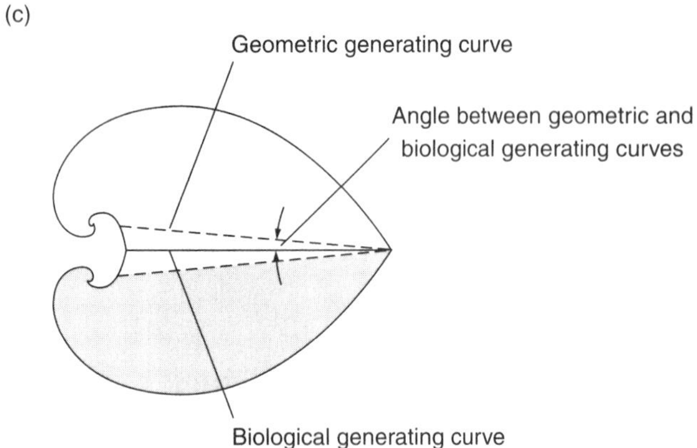


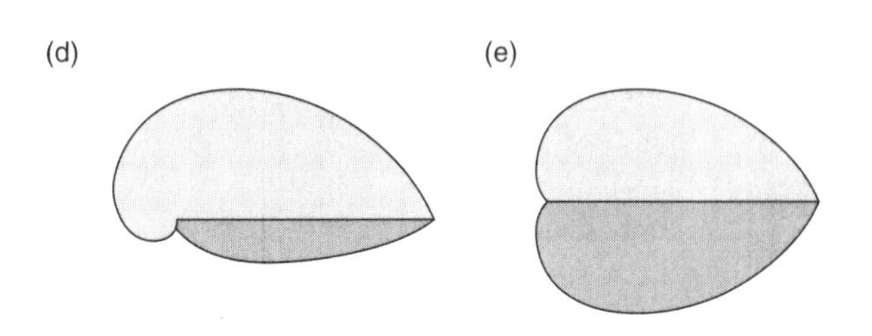


**FIGURE 5.24 The problem of valve interference in bivalved shells.** (a) The overlap in the umbonal region if both valves follow the ideal geometric model of coiling. (b) Interference is avoided by depositing additional shell material between the umbones. (c) Geometrically, this corresponds to having a biological generating curve, or actual growing shell margin, that is not the same as the geometric generating curve. The greater the angle between the biological and geometirc generating curves, the less the valves will interfere. Shell interference can also be avoided if the shells are highly unequal (d) or if whorl expansion rate increases during growth (e). *(a, b, d, e: From Ubukata, 2000; c: From Raup, 1966)*

The large apertures that characterize bivalved shells would be maladaptive for many univalves—such as snails—because this would make them highly susceptible to predation [SEE SECTION 9.4]. Univalves therefore tend to have relatively low values of $W$ (Figure 5.22). In general, only univalves that gain protection by adhering completely to the substrate, such as abalones and limpets, can cope with a high expansion rate.

In addition, shells are stronger as successive whorls overlap more. Univalves tend to sit above the dashed surface of Figure 5.22. This means, as we saw in the discussion of bivalves, that their whorls overlap.

The projection of this surface onto the $W$–$D$ plane, where $T = 0$, is a curve with the equation $W = 1/D$. Figure 5.25a depicts this curve along with a series of computer-generated shells that show closed coiling above the curve and open coiling below. For comparison, a sample of some 400 cephalopod genera is portrayed in Figure 5.25b. Estimates of $W$ and $D$ were obtained by measuring drawings and photographs of shells (see Box 5.2). The contour lines depict the density of occupation of the $W$–$D$ space, with the concentration of points increasing toward the inner contour. These contours are thus two-dimensional analogs of frequency curves (see Box 3.1).

Almost all the observed shells fall in the region of closed coiling. Even the species that fall on the other side of the curve in reality have closed coiling, as determined by inspection of the actual shells. As with any deviation between model and data, there are two possible reasons for the discrepancy. Either there is a problem with the data—namely, measurement error—or there is a problem with the model—for example, failure to take into consideration ontogenetic change in coiling parameters. In this case, it is likely that both factors play a role.

In summary, the coiling model can simulate the principal features of a wide range of shells with just three simple geometric parameters. Certain aspects of the nonrandom occupation of the parameter space can be understood in part by considering different ways of life and functional needs of organisms with coiled shells.

## Trade-Offs and Limits to Optimality

***A Model of Branching and Spiral Growth in Bryozoans*** Another large class of biological structures can be represented as a growing system of branches. These include circulatory systems, bacterial filaments, trees, antlers, crinoid arms, and the skeletons of some colonial invertebrates.

Suspension-feeding bryozoans have repeatedly evolved a form that combines helical growth along a main colony axis with the proliferation of lateral, branched extensions that contain feeding zooids (Figure 5.26). Such colonies can be modeled with a few simple

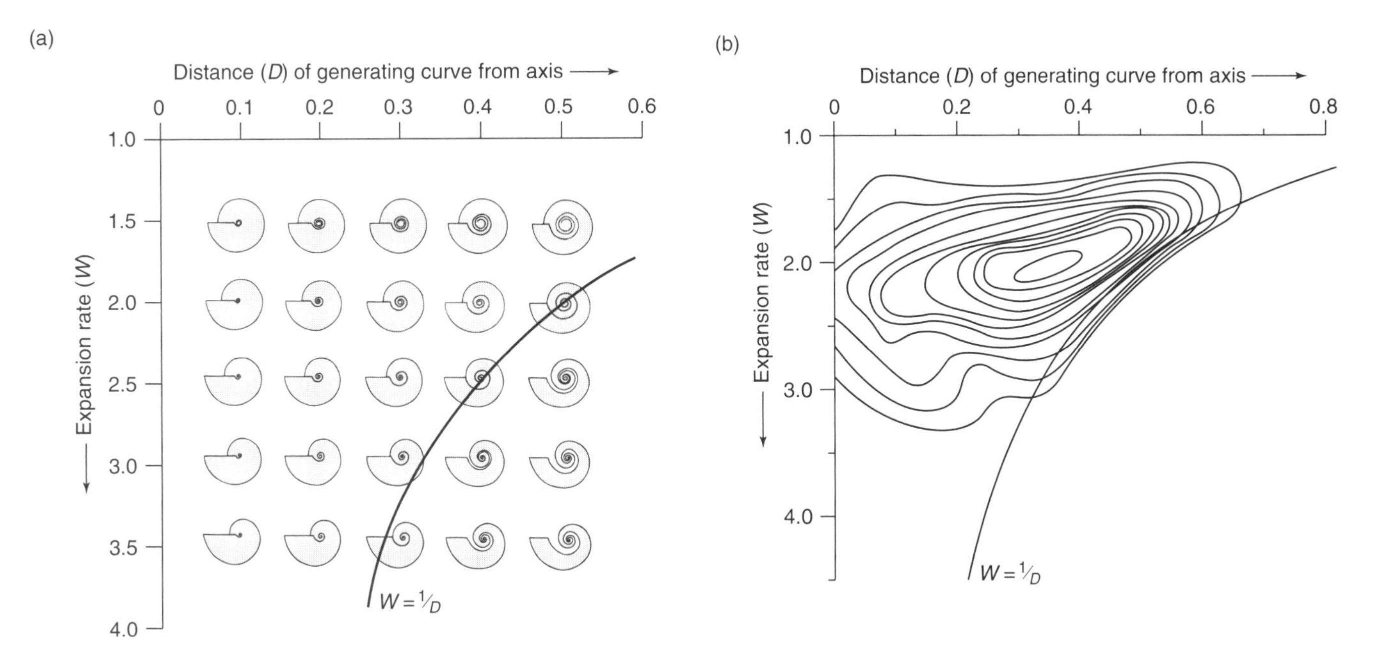


**FIGURE 5.25 Hypothetical planispiral shells and measured ammonoids in $W$–$D$ coiling space.** (a) Hypothetical shells. $T = 0$ for shells that coil in a plane. (b) Measured shells. The contour lines show density of occupation of coiling space of a sample of about 400 ammonoid genera. The curve $W = 1/D$ separates openly coiled forms (below) from those with whorl overlap (above). *(From Raup, 1967)*

**FIGURE 5.26 Examples of helically coiled, branching bryozoans.** (a) *Archimedes* from the Carboniferous. (b) *Crisidmonea* from the Eocene. (c) *Retiflustra* from the present day. (d) *Bugula* from the present day. Scale bars are 10 mm in parts (a) through (c) and 1 mm in part (d). Part (c) is an axial view (see Figure 5.27a). Other views are lateral (see Figure 5.27b). *(From McGhee & McKinney, 2003)*

parameters (Figure 5.27): the radial distance between the coiling axis and the inner colony margin (*RAD*); the angular separation between innermost branches (*ANG*); the minimum distance between lateral branches (*XMIN*); the difference in elevation between lateral branches separated by $2\pi$ radians of revolution (*ELEV*); and the angle between a lateral branch and the coiling axis (*BWANG*). A wide spectrum of theoretically possible colony forms can be generated just by varying these last two parameters (Figure 5.28).

The mathematical model of colony form can be used to calculate the surface area of lateral branch systems corresponding to each combination of parameters. Contours in Figure 5.28 show values of surface area, increasing toward the lower left. The highest surface areas correspond to low values of both *ELEV* and *BWANG*. Because higher surface area of branches that possess feeding zooids would seem to allow greater food uptake, one

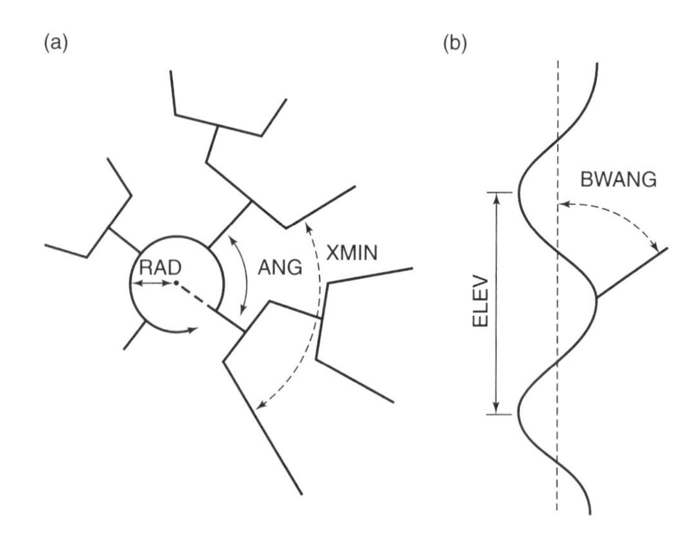


**FIGURE 5.27 Geometric model of helical, branched bryozoans, showing five growth parameters.** (a) View down the coiling axis. (b) Lateral view. *(From McGhee & McKinney, 2000)*

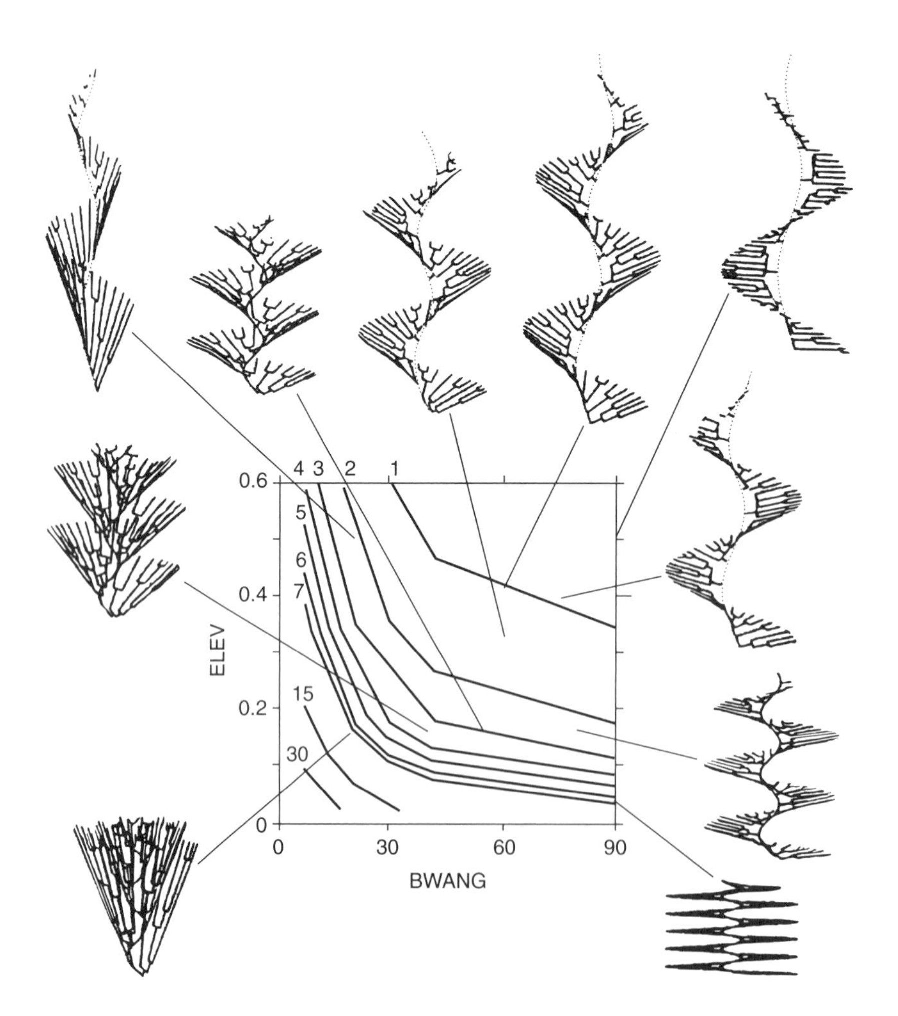


**FIGURE 5.28 Effect of varying two of the parameters of the bryozoan model.** Computer-generated forms correspond to particular parameter combinations. The lines inside the plot show contours of branch surface area, indicated by the numbers and increasing toward the lower left. *(From McGhee & McKinney, 2000)*

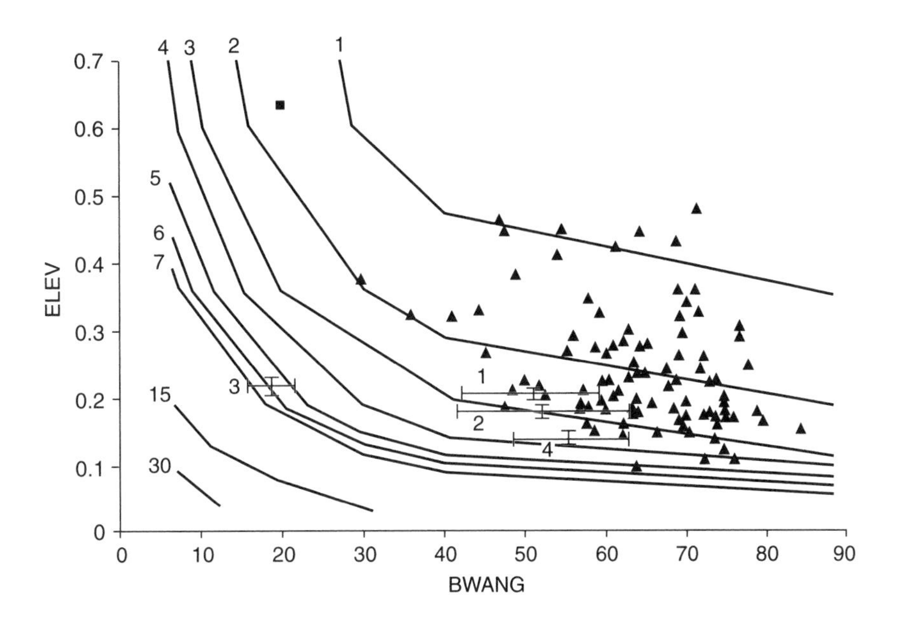


**FIGURE 5.29 Distribution of observed forms in bryozoan model space.** The axes and contour lines are as in Figure 5.28. The square shows a single specimen of the Devonian genus *Helicopora*. The triangles show multiple specimens of *Archimedes*. The numbered points with error bars show the mean and its standard error for three species of *Bugula* (1–3) and *Crisidmonea* (4). *(From McGhee & McKinney, 2000)*

might predict a concentration of bryozoan species in the lower left of this diagram. In fact, when real colonies are measured, they are found to be concentrated well away from this region (Figure 5.29).

The reason for the observed distribution of species can be understood by identifying an important trade-off. Based on observation of living helical bryozoans, it is known that water currents move through the colony from top to bottom, aided by the beating of cilia on bryozoan lophophores (feeding tentacles). As the water moves through the branches, it inevitably encounters resistance and slows down. Eventually it ceases to flow, producing a zone of stagnant water from which food cannot be extracted (Figure 5.30). Colonies with very low values of *ELEV* and *BWANG* would have deeply nested branches (Figure 5.28) and a correspondingly large stagnant zone. Thus, maximizing surface area makes feeding less efficient by increasing the size of the stagnant zone. The common colony forms represent a compromise that balances the need for feeding area with the need for fluid flow between the branches.

In the preceding example, the trade-off results, in essence, from a single functional demand—feeding—which is frustrated because the form that is optimal in terms of surface area is suboptimal in terms of fluid flow. We can gain further insight into the consequences of trade-offs by considering multiple functional demands that must be satisfied simultaneously (see Box 5.3). As the example of Box 5.3 shows, trade-offs between

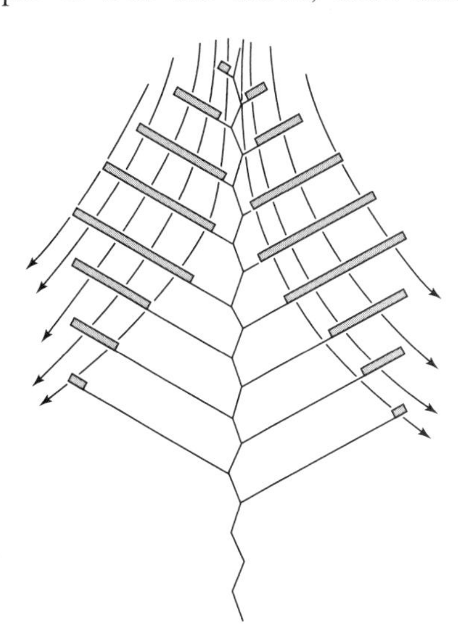

**FIGURE 5.30 Schematic diagram showing feeding in the living bryozoan *Bugula*.** The central line is the colony axis, and the radiating, thin lines show the branches. The thicker bars show the location of feeding zooids. The arrows indicate water flow; the area to the interior of these is a zone of relatively stagnant water. *(From McGhee & McKinney, 2000)*

Box 5.3

## TRADE-OFFS AS A SOURCE OF MULTIPLE ADAPTIVE MODES

Figure 5.31 depicts a simple model of branching growth in land plants. There are three parameters: $P$, the probability of branching per unit branch length; $\phi$, the angle between the two branches that diverge from a bifurcation; and $\gamma$, the angle of rotation between a new pair of branches and the branch from which it arises. Higher values of $P$ yield more bifurcations and thus lead to more densely branched model plants. Higher values of $\phi$ give a wider divergence between branches. And higher values of $\gamma$ produce new branches that are rotated more relative to their parent branch. The special case where $\gamma = 0$ would result in a plant restricted to a vertical plane.

There is thus a three-dimensional parameter space for branched plants in which the parameters have a definite theoretical range of values: $P$ from 0 to 1; and $\phi$ and $\gamma$ from 0° to 360°. Each point in this space specifies a different possible form. A very large but finite sample of the full spectrum of forms can be generated by varying each of the three parameters, in fine increments, over its entire range.

For certain functional demands, the performance of a theoretical form can be quantified. Therefore, all of the forms can be compared and the optimal ones can be identified. In this context, an optimum is a form that is functionally superior to all its neighbors in the parameter space. There can be multiple optima, and not all optima need to be equal in their functional performance. We confine this discussion to the biology that is relevant to the earliest vascular plants.

Consider first the single functional demand of reproduction. A spore can fall or be blown farther from the plant if it begins its descent from a greater height. Therefore, dispersal of spores will be maximized if plant height is maximized. At the same time, spores are produced at branch tips, so having more branches is advantageous in producing more spores. These two factors together lead to a single optimal phenotype, one that is very tall and has dense branching concentrated near the top (Figure 5.32a).

A second important function is the interception of light for photosynthesis. If performance is assessed with respect to this function alone, then there are three optima which differ in their details but share the property of numerous, broad, horizontal branches that enhance light capture (Figure 5.32b).

Finally, consider the function of mechanical stability. This correlates largely with the ability to resist bending. Resistance is highest when the plant is vertical, in other words, when $\phi$ is near zero (Figure

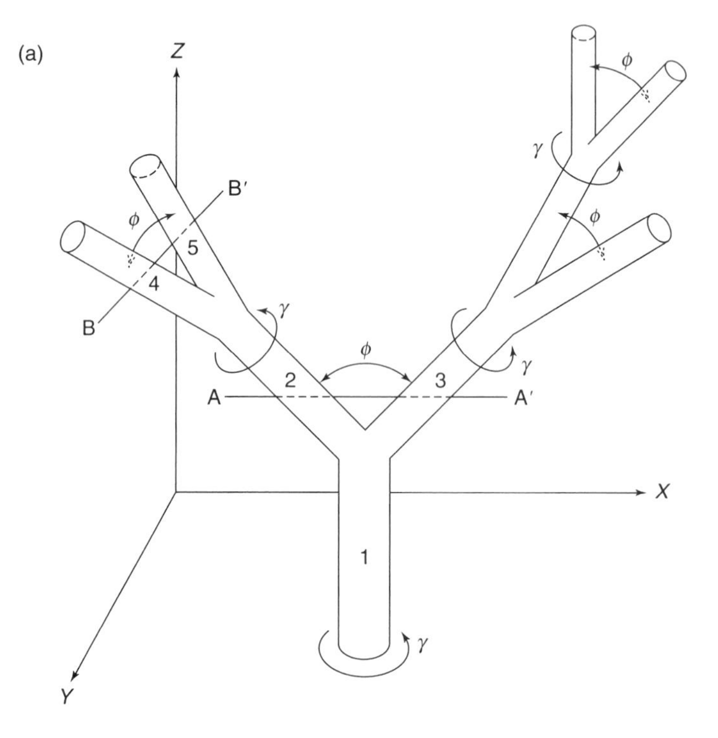


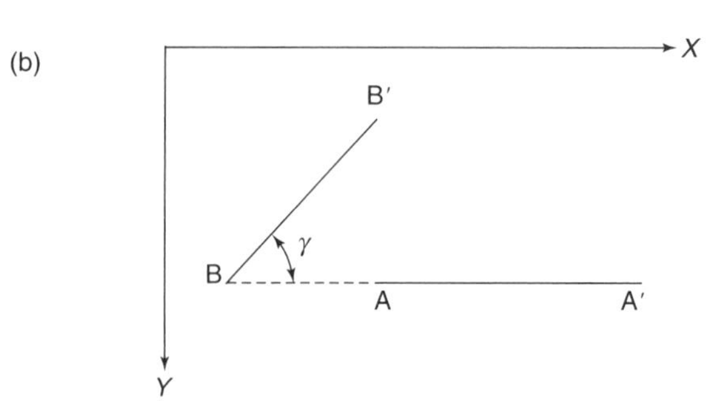


**FIGURE 5.31 Geometric model of branching growth in plants.** $P$ is the probability of branching per unit branch length; $\phi$ is the angle between branches that arise from the same bifurcation; $\gamma$ is the angle of rotation between the older and younger branches. (a) Branching in three dimensions. (b) The parameter $\gamma$. Line A–A′, between branches 2 and 3, and line B–B′, between branches 4 and 5, are projected into the $x$–$y$ plane; $\gamma$ is the angle between these projected lines. *(From Niklas & Kerchner, 1984)*

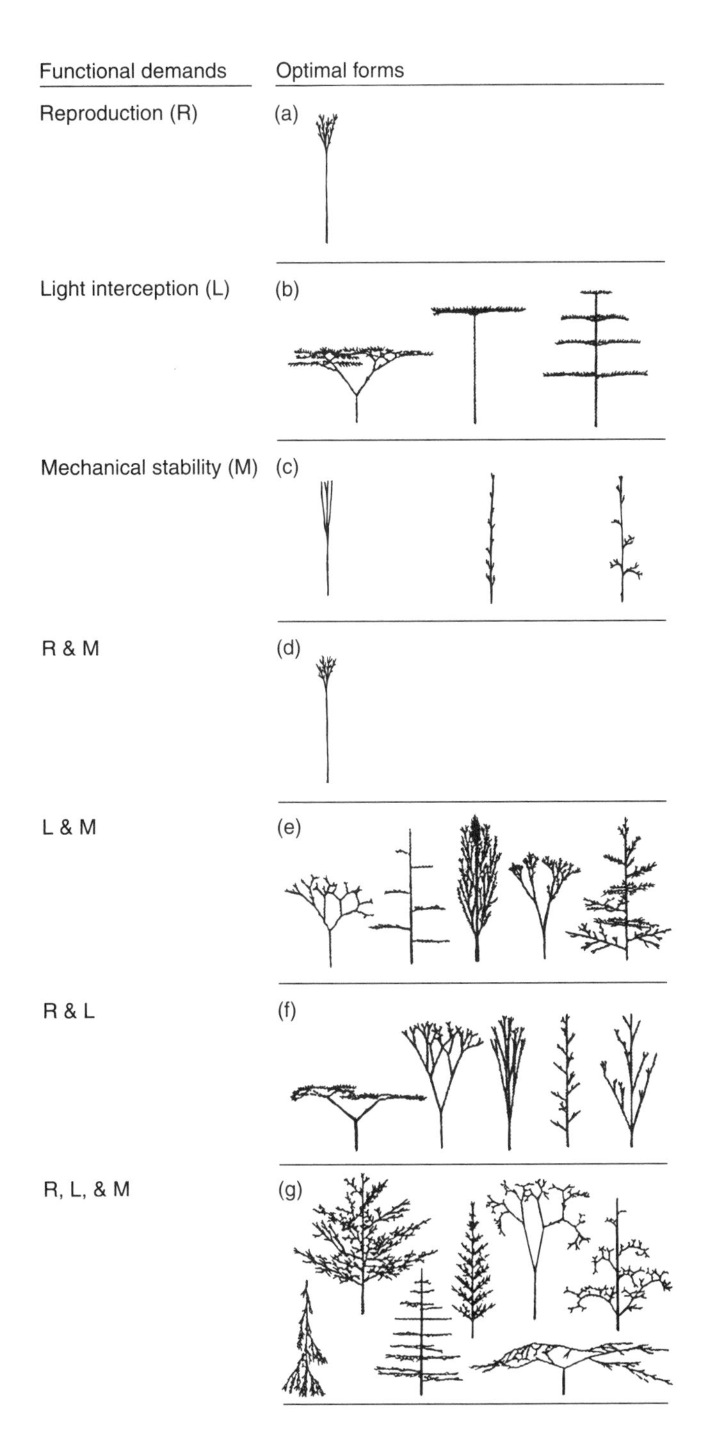


**FIGURE 5.32 Optimal forms of model plants that satisfy different functional demands and combinations of demands.** Because there are different ways to reach a compromise between conflicting demands, there are more distinct optima when more demands must be simultaneously satisfied. *(From Niklas, 1994b)*

5.31). If stability alone needs to be maximized, then there are three distinct optimal forms (Figure 5.32c).

Of course, it is unrealistic to suppose that a plant has only a single function, so let us see what happens when there are multiple functional demands. Figure 5.32d depicts the single optimum that results when both reproduction and mechanical stability must be satisfied. In this case, there is not a serious trade-off because both functions are served by a tall, vertical structure. The single optimum that satisfies both functions is essentially the same as the one that satisfies reproduction only (Figure 5.32a).

Other combinations of functions are not so compatibile, however. Light interception, which favors horizontal branches, conflicts with mechanical stability, which favors vertical branches. Thus, the combination of functional demands forces compromises, leading to the several optimal forms of Figure 5.32e. Light interception and reproduction conflict for similar reasons, with results evident in Figure 5.32f.

One obvious result of conflicting demands is that there are more distinct optima than when there is a single function or two compatible functions. When conflicts are inevitable, there are many different ways to compromise. This is seen even more strikingly when all three functional demands are combined (Figure 5.32g). The increase in the number of distinct optima results from the different ways of balancing functions against one another. The fact that many of the plant forms of Figure 5.32 appear biologically realistic suggests that such trade-offs may indeed have been important in plant evolution and that the three functions explored here are among the most important for real plants.

incompatible functions are limitations in the sense that they prevent all aspects of performance from being maximized simultaneously. But they are also likely to contribute to the diversity of form, as evolution produces a variety of compromise solutions.

## Phenotypic Change and Underlying Genetic Factors

***An Alternative Model of Shell Coiling*** To the extent that models of form approximate growth processes, it may be possible to compare the size of a genetic change and the size of the corresponding phenotypic change. Differences in adult form that seem to the eye to be large may prove in some cases to involve a genetic or developmental change that is relatively small, and vice versa. A slight modification of the coiling model has been used to illustrate this principle.

In this variant of the shell coiling model, the growing margin is characterized by a field of vectors around the aperture (Figure 5.33). The orientation and length of the vectors show direction and rate of growth at each point. The vectors define the "aperture map" in which the shape of the resulting shell is encoded. A few aperture maps and their corresponding shells are shown in Figure 5.34. The shell in Figure 5.34a is a helical spiral typical of many snails; Figure 5.34c is a coiled, limpet-like shell, similar to the living *Crepidula*; and Figure 5.34b is intermediate between shells (a) and (c). An interesting feature of shells (a) through (c) is that they all have aperture maps with the same *relative* vector lengths; they differ only in absolute vector lengths.

Figure 5.34c differs from the shell of true limpets—for example, of the genus *Patella*—which resemble Figure 5.34d. The patelliform shell is practically a straight cone with no coiling. Its aperture map is very different from that of the coiled, limpet-like form.

Thus, there are at least two ways for a limpet form to grow, and therefore at least two ways to derive a limpet from a coiled ancestor such as the shell in Figure 5.34a. An evolutionary change that reduced all growth vectors by the same proportion could produce the transition from shell (a) to the limpet-like shell (c), because the aperture maps differ only in scale. By contrast, the transition to a conical limpet (d), with its unusual aperture map, would require an evolutionary change that affected different growth vectors disproportionately.

It is commonly thought that evolutionary transitions involving uniform change across many features

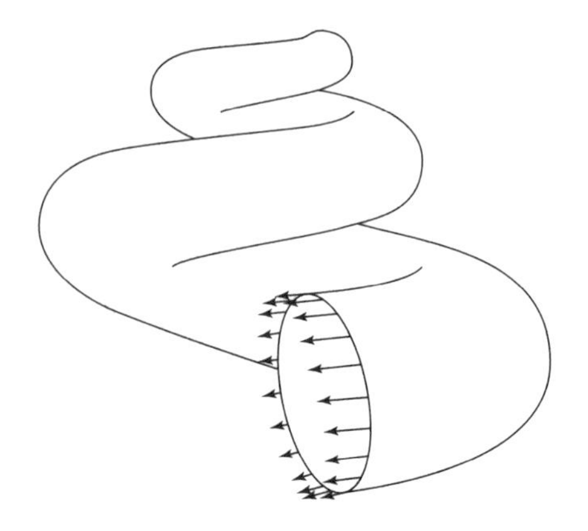

**FIGURE 5.33 Alternative geometric model of shell growth.** Each point on the margin has a rate and direction of growth, indicated by the length and orientation of the vectors (compare with Figure 5.19). *(From Rice, 1998)*

of growth are more likely to occur than those in which different aspects of growth change by different amounts. If this is true, then the transition to the limpet-like form (Figure 5.34c) represents a smaller genetic change than the transition to the conical limpet (Figure 5.34d). This leads to a testable (but not yet fully tested) prediction: The evolution of coiled, limpet-like forms should have occurred more frequently in the history of gastropods than did the evolution of the conical limpet form. In this context, it is interesting to note that there are more species with the conical limpet form than with the coiled limpet-like form. This does not tell us, however, which form arose independently a greater number of times.

# 5.4 CONCLUDING REMARKS

Despite the successes of theoretical morphology, the range of taxonomic groups to which formal models have been applied is still relatively small. There is an obvious need for new ways to look at particular groups of organisms. A more important and far more elusive goal is to generate the theoretically possible spectrum of form of even more inclusive groups such as the animal and plant kingdoms. The model of plant growth considered in Box 5.3 is certainly an important step in this direction.

The questions and approaches discussed in this chapter apply as much to biology as to paleontology, and indeed functional morphology is a vibrant area of biological research. At the same time, the subject of theoretical morphology has received more attention from

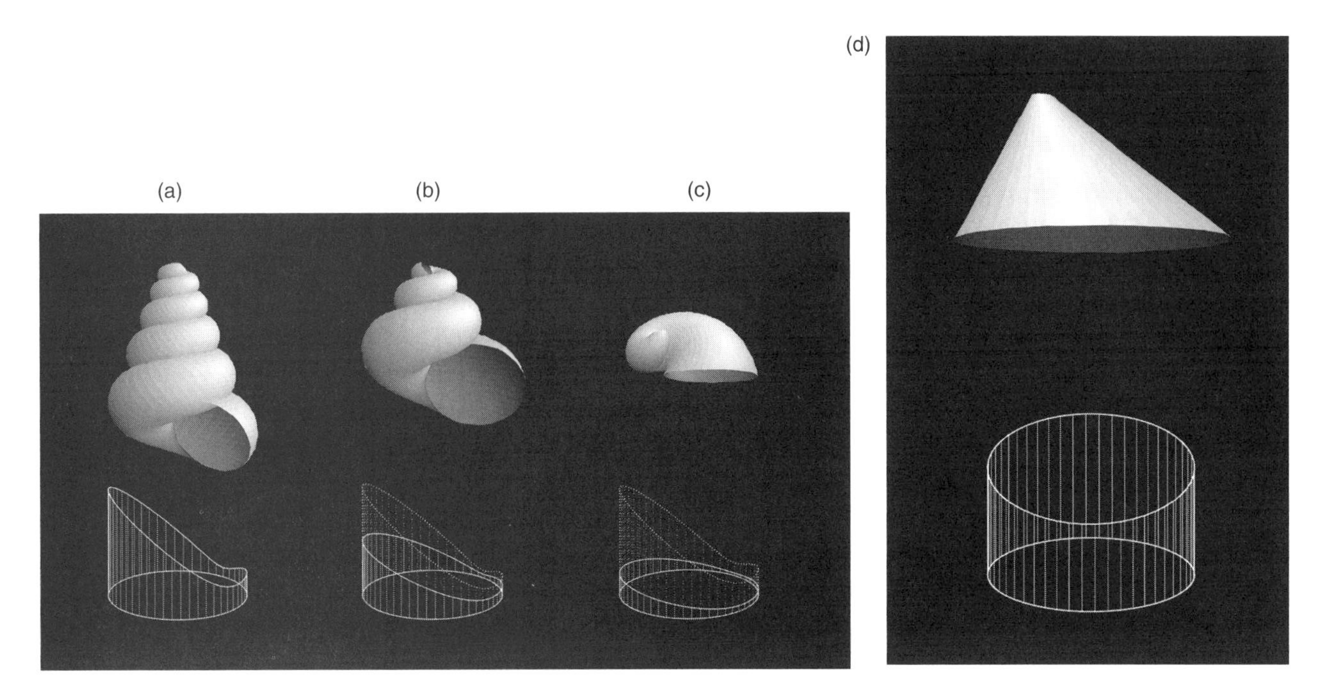


**FIGURE 5.34 Comparison between three coiled shells and a conical limpet.** (a) A high-spired, helically coiled form. (b) A form intermediate between parts (a) and (c). (c) A coiled limpet. (d) A conical limpet. The aperture maps below each computer-generated shell correspond to vector fields like the one in Figure 5.33; the length of each line segment is proportional to the rate of growth at a point on the margin. The dotted lines of maps (b) and (c) show the aperture maps of shells (b) and (c) magnified to a larger size. These magnified maps exactly match map (a). Therefore, it would be possible to derive (b) or (c) from (a) by a uniform scaling-down of the growth rates. The map of (d) is completely different from that of (a). Therefore, more substantial changes in the relative magnitudes of growth vectors would be required to derive (d) from (a). *(From Rice, 1998)*

paleontologists. There seem to be at least two reasons for this curious situation. First, paleontologists study fossil as well as living organisms, and therefore they must comprehend a broader diversity of form. Second, some of the first people to model morphology with the newly available computers in the late 1950s and early 1960s happened to be paleontologists, and their early work influenced later generations.

In this chapter, we have focused on adaptation in understanding individual forms. But this is only one of the three major determinants of form depicted in Figure 5.1. In fact, the relative importance of these three factors—that is, how much of the variance in form in the organic world can be attributed to each—is unknown. In a broader sense, our discussion of the overall distribution of form was also dominated by functional considerations. In Chapter 7, we consider rates of speciation and extinction as factors leading to the accumulation of species with particular morphologic features. These factors need not reflect adaptation.

# SUPPLEMENTARY READING

Fisher, D. C. (1985) Evolutionary morphology: Beyond the analogous, the anecdotal, and the ad hoc. *Paleobiology* **11:**120–138. [Discussion of the principles underlying evolutionary morphology.]

McGhee, G. R. (1999) *Theoretical Morphology: The Concept and Its Applications.* New York, Columbia University Press, 316 pp. [Principles of theoretical morphology, with emphasis on paleontological examples.]

Plotnick, R. E., and Baumiller, T. K. (2000) Invention by evolution: Functional analysis in paleobiology. *Paleobiology* **26** (Suppl. to No. 4):305–323. [An overview of functional morphology in paleontology.]

Prusinkiewicz, P., and Lindenmayer, A. (1990) *The Algorithmic Beauty of Plants*. Berlin, Springer, 228 pp. [One of a series of volumes exploring theoretical morphology through systems that generate form by replicating basic elements according to simple rules.]

Radinsky, L. (1987) *The Evolution of Vertebrate Design*. Chicago, University of Chicago Press, 188 pp. [Mechanistic description of vertebrate function.]

Rudwick, M. J. S. (1964) The inference of function from structure in fossils. *British Journal for the Philosophy of Science* **15:**27–40. [Landmark paper on the application of engineering principles to functional morphology in paleontology.]

Savazzi, E. (ed.) (1999) *Functional Morphology of the Invertebrate Skeleton*. Chichester, U.K., Wiley, 706 pp. [Case studies of functional morphology in living and fossil invertebrates.]

Thompson, D'A. W. (1942) *On Growth and Form*. Cambridge, U.K., Cambridge University Press, 1116 pp. [Includes consideration of structural factors that contribute to biologic form.]

Thompson, J. J. (ed.) (1995) *Functional Morphology in Vertebrate Paleontology*. Cambridge, U.K., Cambridge University Press, 293 pp. [Case studies of functional morphology in fossil vertebrates.]

Vogel, S. (1981) *Life in Moving Fluids*. Princeton, N.J., Princeton University Press, 352 pp. [Principles of fluid dynamics applied to fluid flow within organisms and to the function of organisms in water and air.]

Wainwright, S. A., Biggs, W. D., Currey, J. D., and Gosline, J. M. (1976) *Mechanical Design in Organisms*. Princeton, N.J., Princeton University Press, 423 pp. [Engineering principles applied to organisms, with particular emphasis on structural properties of biological materials.]

# *Chapter 6*

# BIOSTRATIGRAPHY

Because it is an historical science, a central goal of paleontology is to determine the relative timing of any given set of events, even if the events took place in geographic locations that were far apart. For example, if we want to know whether a time interval was characterized globally by a mass extinction [SEE SECTIONS 8.6 AND 10.3], we must first figure out whether sets of extinction events recognized in different places around the world actually occurred simultaneously. The assessment of age relationships such as these is a major goal of **biostratigraphy,** the study of the geometry, biotic composition, and time relations of fossiliferous rocks.

Since its inception, biostratigraphy has relied on a set of basic principles concerning the stratigraphic ranges of taxa preserved in the fossil record, and these have permitted the development of a global geologic timescale. More recently, however, numerical methods have been developed that seek to combine information on the stratigraphic ranges of taxa contained at several separate localities, providing correlations at much finer scales of resolution than is generally possible by traditional biostratigraphic means.

Biostratigraphy has also been transformed in recent years by the advent of **sequence stratigraphy,** which focuses on the processes that produced sets of strata. Sequence stratigraphers aim to recognize fundamental, repeated units of stratigraphic architecture, **parasequences** and **sequences,** that can be used as bases for correlation in their own right. It has also been recognized that fossil distributions tend to be affected predictably by the same depositional processes that produce the sequences. These patterns must be diagnosed and understood if paleontologists are to make full use of regional variations in the fossil content of rocks in studies of taxonomic origination and extinction.

We begin this chapter with a review of several fundamental principles of biostratigraphy and their utility for constructing a global geologic timescale. Then we turn to the palette of quantitative techniques that have greatly enhanced the ability of biostratigraphers to correlate fossiliferous rocks at high resolution. Finally, we consider the new generation of stratigraphic models that help to diagnose the fundamental architecture of the fossil record and its effects on regional biotic patterns in space and time.

## 6.1 THE NATURE OF BIOSTRATIGRAPHIC DATA AND CORRELATION

All biostratigraphic methods require a detailed accounting of the occurrences and, in some cases, abundances of taxa within a set of strata under investigation. At any given outcrop, a worker can seek to determine the stratigraphic limits to the distribution of any fossil taxon that is present. We can refer to this interval as the **stratigraphic range** of the taxon, bounded at the base by the taxon's **first appearance datum (FAD)** and at the top by its **last appearance datum (LAD).** The FAD and LAD constitute the fundamental data for many approaches to biostratigraphy. Of course, it is highly unlikely that this local stratigraphic range encompasses the entire global stratigraphic range of the taxon. In most cases, what we know about the

**FIGURE 6.1 Examples of common index fossils from throughout the Phanerozoic.** The sizes and nature of these examples are highly variable, but they were all associated with organisms that were capable of wide geographic dispersal. (a) A colony of the graptolite *Nemagraptus gracilis*, contained in a piece of the Ordovician Athens Shale, in Alabama (horizontal field of view is 12 cm). (b) An element of the conodont *Ozarkodina remscheidensis eosteinhornensis* (horizontal field of view is 1200 microns). (c) The ammonite cephalopod *Uptonia jamesoni*, from the Jurassic of France (approximate diameter is 9.6 cm). (d) The planktonic foraminferan *Gansserina gansseri*, from the Late Cretaceous (note 100 micron scale at bottom). *(a: From Prem Subrahmanyam's online fossil gallery, www.premdesign.com/fossil.html; b: Museum of Natural History, London; c: Courtesy Hervé Châtelier's Jurassic and Cretaceous ammonite database, http://perso.wanadoo.fr/herve.chatelier/; d: Smithsonian National Museum of Natural History)*

global stratigraphic ranges of taxa is based on composites of information from several localities. Moreover, the preserved global stratigraphic range of a taxon is unlikely to preserve its entire true temporal range. Because the majority of individuals belonging to any species are not likely to be preserved in the fossil record, the true time of origination of any taxon almost certainly predates its first documented appearance in the fossil record, and the true time of extinction almost certainly postdates its last documented appearance.

Barring the intervention of post-mortem processes that disturb its stratigraphic position, the presence of the same taxon at any two localities permits a paleontologist to make a simple but important statement of **correlation** about the strata at these localities: The strata must have been deposited during the evolutionary lifetime of the taxon. Of course, this statement is of real value only if the taxon in question is confined to a small global stratigraphic range. In general, for correlating strata in this way, biostratigraphers endeavor to use taxa (generally species) that have two important properties: (1) They are of limited stratigraphic duration; and (2) they are geographically and environmentally widespread, ideally occurring in a variety of rock types worldwide.

Taxa with these properties, known as **index, guide,** or **zone** fossils, have been used widely in the correlation of Phanerozoic strata, and global biostratigraphic inter-

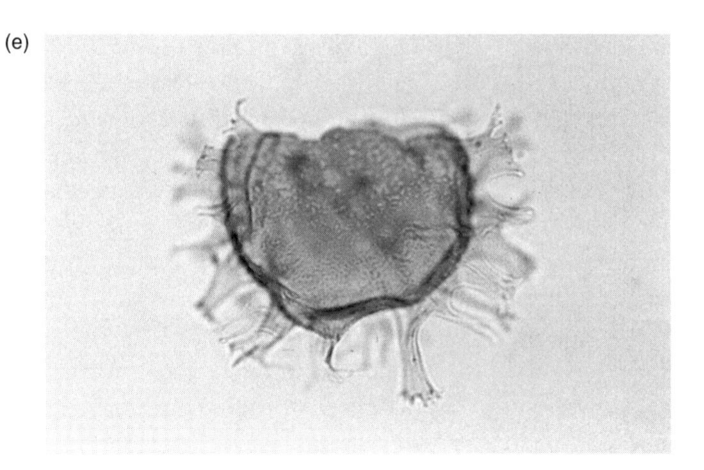


**FIGURE 6.1 (*cont.*)** (e) Walled cyst of the dinoflagellate *Chiropteridium galea,* from Oligocene sediments in a deep-sea core collected off the coast of Tasmania (approximate length from left to right of the main, dark shaded region is 60 microns). (*From Williams et al., 2003*)

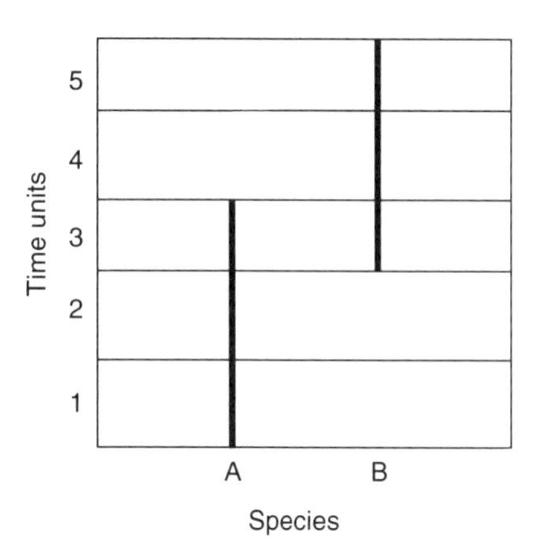


**FIGURE 6.2 A simplified, hypothetical example that illustrates the use of overlapping ranges as a means to refine relative age determinations.** In this **range chart,** the known stratigraphic ranges of two fossil species, A and B, are depicted, based on previous cataloguing of these species in strata from around the world. If a paleontologist found only individuals belonging to species A in a sedimentary stratum, she could assume that the stratum was deposited sometime in the range of time units 1 through 3 (barring post-mortem processes that affected the position of the fossil, relative to where it lived). If she found only species B in the stratum, then a time of deposition in the range of time units 3 through 5 would be implied. However, if she found that species A and B were *both* present, then the time range could be narrowed to unit 3 since this was the only interval during which species A and B are known to have coexisted (i.e., these species exhibited overlapping ranges encompassing only time unit 3).

vals have often been recognized on the basis of one or more pervasive, diagnostic fossil species. Formally, such an interval is referred to as a **zone** or **biozone.** In general, marine organisms that floated or swam in the water column, and terrestrial organisms that possessed components capable of windborne dispersal, are best suited as zone fossils because of the greater likelihood that they will be widespread (Figure 6.1).

In cases where the stratigraphic ranges of fossil taxa are too long for the taxa to be useful individually for correlation, they may nevertheless be useful *in combination* with the occurrences of other taxa. At its simplest level, this involves the diagnosis and use of **overlapping** (or **concurrent**) **ranges** among taxa (Figure 6.2). Even if two or more taxa exhibit long stratigraphic ranges, the interval of geologic time during which these taxa were extant *simultaneously* may have been far more limited. In such instances, the co-occurrence of these taxa permits correlations among strata in different venues that may be far more refined than what would be possible if only one of the taxa were present.

Primarily through the use of these biostratigraphic tools, a global geologic timescale has emerged (see inside front cover of this book). Whereas zones are demarcated based on the stratigraphic occurrences of individual fossil species or limited assemblages of species, more profound boundaries between **stages** and **series,** which tend to be local or regional in scope, as well as **systems** and **eras,** which are global in scope, are marked by the near-simultaneous loss of larger numbers of taxa and their replacements by other taxa. In fact, as we will see when we consider the history of global biodiversity (Chapter 8), biotic changes can be particularly profound at system and, especially, era boundaries.

The absolute ages of important boundaries have been subsequently determined using radiometric dating techniques, which permit the calibration of geologic time intervals that correspond to the aforementioned stratigraphic intervals. Samples appropriate for radiometric analysis, which are typically volcanic in origin, have been difficult to obtain for many important boundaries, and the absolute calibration of the timescale has therefore not been straightforward. As researchers discover new materials from key horizons that are suitable for absolute dating and continue to refine methodologies used for absolute age determination, we can expect significant improvements to previous estimates of absolute ages, as well as the addition of age determinations for key boundaries that have heretofore not been dated.

For example, analyses in the past decade and a half of a suite of samples obtained from lowermost Cambrian rocks in northeastern Siberia by Samuel Bowring and colleagues (1993) have led to a recalibration, to 544 million years ago (Ma), of the Proterozoic–Cambrian boundary. This stands in contrast to previous estimates that generally depicted the boundary as some 30 million years older. In this case, the reason for the refined estimates was twofold: (1) The samples used for dating were linked more definitively to the boundary than previous suites of samples; and (2) the method used for dating, which relied on the decay of two different isotopes of uranium that could serve as cross-checks on each other, were more reliable than previous methods based on other isotopes.

There are two operational limitations to the use of zone fossils or overlapping ranges as tools of correlation. First, it is appropriate to use biostratigraphic markers only in a positive way. If a stratum does not contain a fossil that is considered diagnostic of a particular stratigraphic interval, this does not necessarily imply that the stratum was not deposited during the interval in question. Even the most ubiquitous zone fossil is likely to be absent from a significant percentage of the strata deposited during the interval in which the taxon was extant, either because the taxon did not live at the locations represented by those strata or because it was removed after its death by taphonomic processes. In some important stratigraphic intervals (e.g., parts of the Ordovician System), it has been difficult to identify reliable and globally ubiquitous biostratigraphic markers; parts of the global timescale therefore remain in flux.

Second, there are limits to the degree of temporal resolution possible with these methods, owing to the simple fact that fossil species tended to persist for several million years [SEE SECTION 7.2]. For many paleobiological questions in which morphological and ecological patterns are assessed in space or through time, it would be useful to have a stratigraphic framework that permits much finer temporal acuity, at least on a regional basis.

## 6.2 COMPOSITE METHODS OF CORRELATION

Recognizing the need to develop timescales that are more highly resolved, biostratigraphers are adopting newer, more sophisticated methods that combine data on FADs and LADs from several outcrops into single, composite "sections." These composites can then be used as timescales that not only depict the order of all the events (i.e., the FADs and LADs) that they include, but can also serve as references for determining the probable stratigraphic placement of additional outcrops that are evaluated at some later date. In this respect, the methods we consider here have much in common, but they differ from one another in the ways that those data are assembled and treated to construct the composites. They are worth reviewing not only because of their growing usefulness, but because they illustrate well the jigsaw-puzzle-like challenge of piecing together information from a disparate set of localities.

### Graphic Correlation

First proposed by Alan B. Shaw (e.g., 1964), the fundamental principles of graphic correlation can be illustrated by first considering the case of two outcrops, A and B, presented as examples in Figure 6.3. Using conventional biostratigraphic methods, a paleontologist would seek to correlate directly the FADs and LADs of taxa present jointly at *both* outcrops (taxa 5 and 9 in Figure 6.3a). However, several additional taxa are present uniquely at each outcrop, and it would obviously be desirable to combine all of the stratigraphic range information from the two outcrops so that we could know the relative order of *all* FADs and LADs preserved at the two outcrops. This is accomplished with graphic correlation by first developing a **line of correlation (LOC)** between the two outcrops based on the events that they share, and then superimposing onto the LOC the additional biological events from each outcrop. The methodology for determining the LOC is described in Box 6.1.

Once the LOC is defined, events that are unique to outcrop A can be projected onto outcrop B (Figure 6.3e), providing a composite of all of the biological events preserved at both localities, thereby resulting in a **composite standard reference section** or, more simply, the **composite standard (CS).** At this point, the power of graphic correlation becomes evident because the procedure is not limited to the two initial sections. It can be repeated indefinitely by correlating additional outcrops with the CS, which continues to mature so long as taxa not recognized previously continue to be incorporated graphically into the CS.

(a)

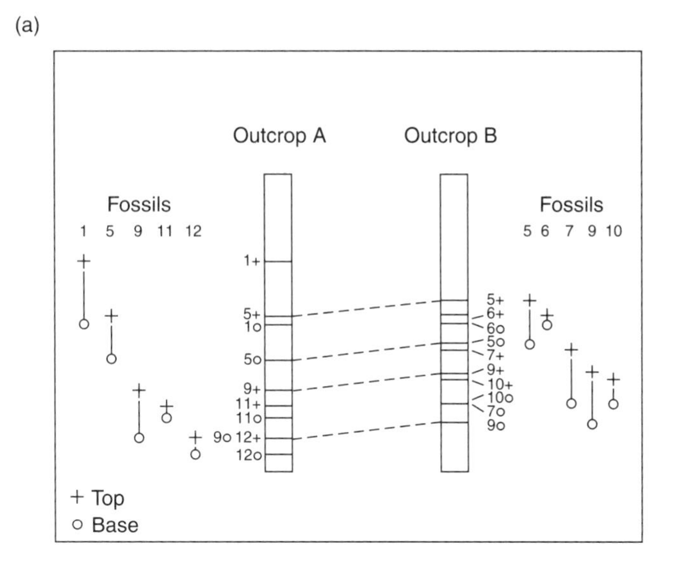


(b)

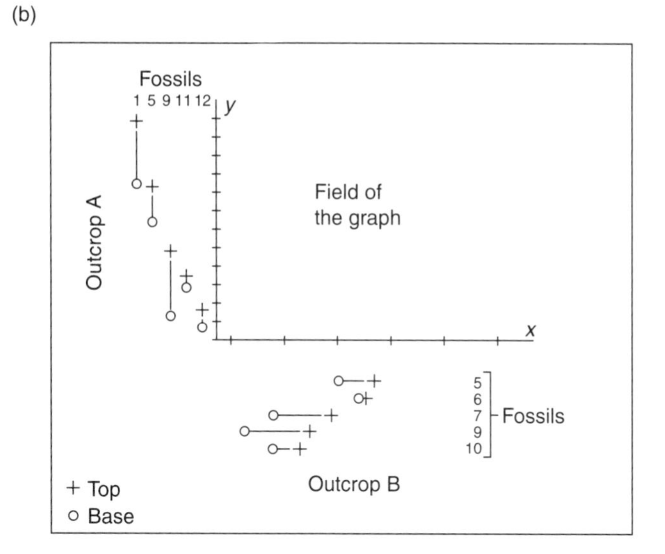


(c)

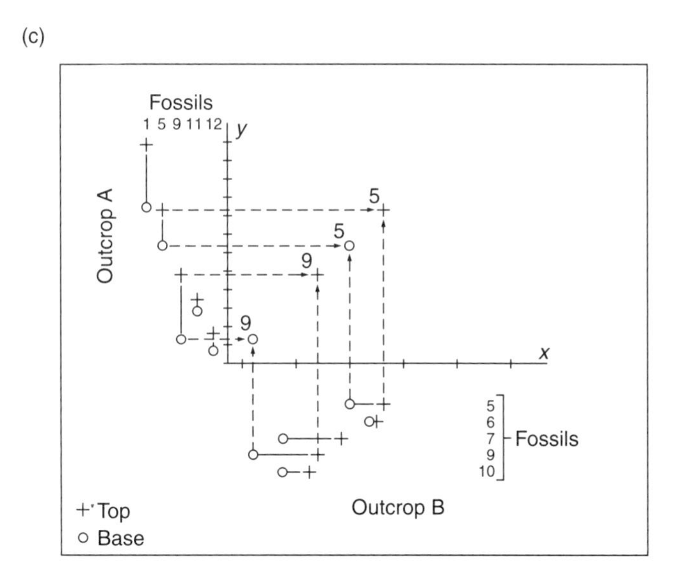


(d)

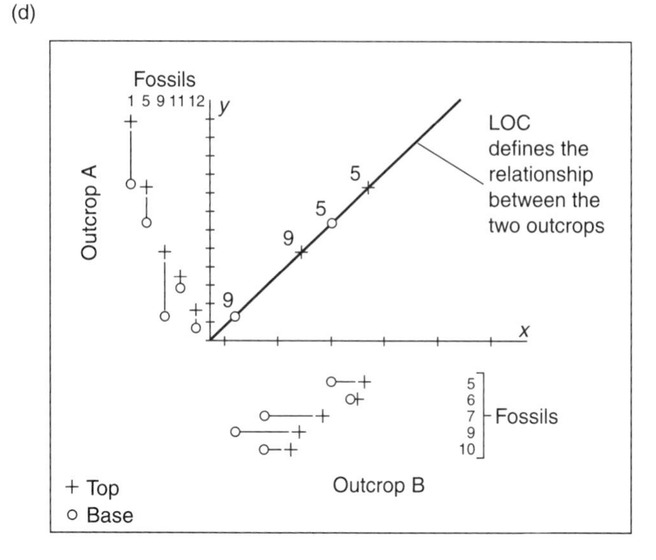


(e)

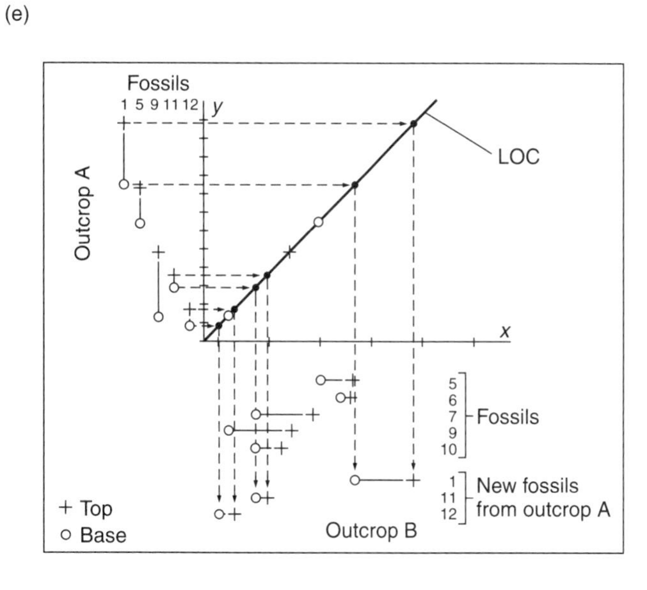


**FIGURE 6.3 Development of graphic correlation from hypothetical data at two outcrops.** (a) Correlation by conventional means of FADs and LADs for fossils shared jointly at both outcrops. (b) First step in graphic correlation: Range data for outcrops are placed as axes on a two-dimensional grid. (c) Second step in graphic correlation: The projection of shared events to the center of the graph. (d) Third step in graphic correlation: The line of correlation (LOC) is drawn. (e) Fourth step in graphic correlation: The projection of events unique to outcrop A onto the LOC, from which they can then be projected onto outcrop B, thereby compositing together all of the information on fossil ranges from both outcrops. *(From Carney & Pierce, 1995)*

*Box 6.1*

## GRAPHIC CORRELATION

In graphic correlation, the stratigraphic-range data for each section are placed at right angles to one another, as axes of a two-dimensional grid, with distances along each axis corresponding to distances above the base of each section. By convention, the section with the greater number of events is placed along the *x* axis (Figure 6.3b)—although, in this example, outcrops A and B contain equal numbers of taxa. As a next step, the events common to both sections (i.e., the FADs and LADs of taxa 5 and 9) are projected, horizontally in the case of the section on the *y* axis and vertically in the case of the section on the *x* axis, to the center of the graph (Figure 6.3c). The points where the projections from the two axes meet on the graph define the LOC between the two outcrops (Figure 6.3d).

In the idealized example of Figure 6.3, note that the two outcrops are of equal thickness; thus, the LOC resides at a 45° angle to each axis. This is typically not the case, however. If one outcrop exhibits a greater thickness than the other outcrop for the *same* stratigraphic interval, then the slope of the LOC will fall closer to the axis of the thicker outcrop than it will to that of the thinner one (Figure 6.4a). Moreover, in the real world, it is highly unlikely that all FAD and LAD events correlated among two outcrops will fall precisely on a straight line. For example, in cases where intermittent variations in stratigraphic thicknesses exhibited at one outcrop are not matched in corresponding intervals at the other outcrop, a connecting together of the "dots" representing joint FADs and LADs would result not in a straight line, but in a line that contains meanders (Figure 6.4b). Finally, because of preservational issues, the sequence of FADs and LADs for the same fossils (i.e., shared events) at two outcrops will not always be identical. The derived LOC might therefore have to accommodate contradictions among localities in the order of shared events. Numerous protocols have been proposed for numerically estimating the LOC in the event of these complications, some of which we discuss later.

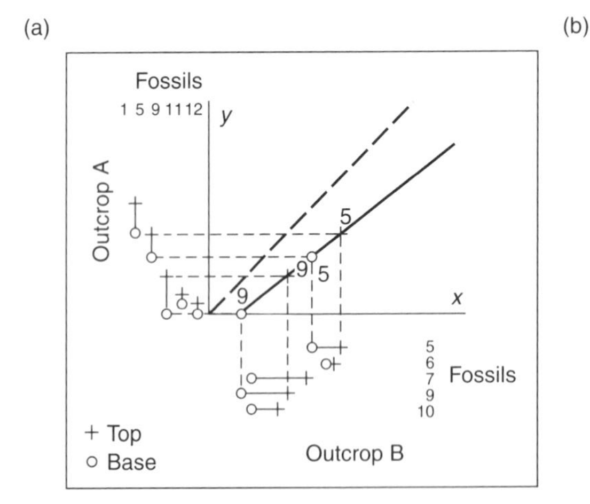


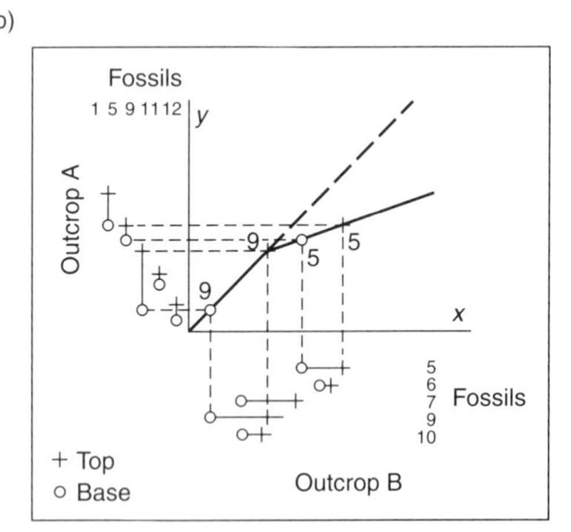


**FIGURE 6.4 Deviations from the straight line of correlation (LOC), with a slope of 1, illustrated in Figure 6.3.** (a) The stratigraphic thickness of outcrop A and the ranges of the fossil species that it contains are reduced relative to Figure 6.3. Outcrop B is unchanged. The resultant LOC (solid line) is shifted downward, and the slope decreases, relative to the original LOC (dashed line). (b) The lower portion of outcrop A remains the same as in Figure 6.3, but a localized decline in sediment accumulation rate coinciding with the FAD of species 5 results in a reduction of thickness thereafter. Outcrop B is unchanged. In this instance, the "true" LOC (solid line) deviates from the trajectory of the original LOC (dashed line) above the LAD of species 9.

This process is illustrated in Figure 6.5. By convention, the CS is placed along the $x$ axis of the two-dimensional Cartesian grid, and the additional outcrop to be correlated to the CS is placed along the $y$ axis. Note that the units designated along the $x$ axis in this example do not correspond to distances above the base of a single outcrop but are reported as **composite standard units (CSUs).** The values for the CSUs in this example are derived from a timescale for Paleocene through Oligocene planktonic **FORAMINIFERIDA** developed by the former Amoco Oil Company. The global ranges of these species in the composite standard had already been established firmly in CSUs based on previous biostratigraphic analyses.

As before, the LOC is drawn based on the FADs and LADs of taxa present jointly at the outcrop and in the CS (taxa 1 and 9, as shown in Figure 6.5a). The ranges of two additional taxa present uniquely at the outcrop (taxa 2 and 4) are then projected onto the LOC (Figure 6.5b). On this basis, the outcrop has been correlated to the CS and, at the same time, the CS has matured further, through the addition of the FADs and LADs of two additional taxa. Therefore, as the CS for a given region continues to mature, it not only provides a means of correlating outcrops throughout the region with increased resolution, but it also provides a regional timescale, recording the order of biological events represented by the FADs and LADs of taxa. In practice, the events correlated graphically need not be biological and can include discrete physical events. For example, volcanic ash falls are sometimes preserved in the stratigraphic record as **bentonites,** clay layers that can be recognized at numerous localities, sometimes over broad geographic ranges.

## Appearance Event Ordination

The procedure in graphic correlation of adding one locality at a time to the CS necessarily places greater weight on information added early in the process. If there were no contradictions in the information on stratigraphic ranges available from different localities, this would not be a problem. However, such contradictions are virtually inevitable because of preservational and lithological differences from locality to locality, and the CS becomes constrained increasingly as data from more and more localities are added to the composite, thereby favoring the information from localities added earlier. By contrast, the family of methods that we consider next all involve the simultaneous assessment of data from multiple localities.

Even in cases where there are no contradictions among a set of localities in the FADs and LADs of the

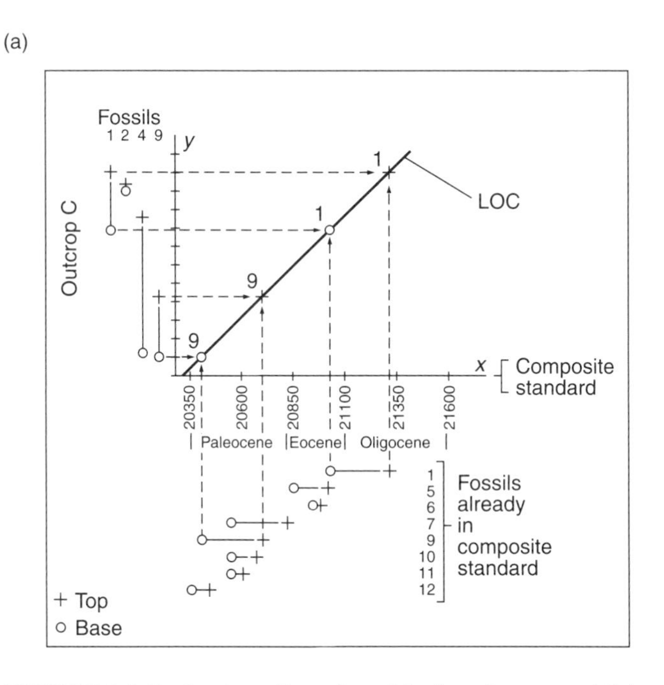


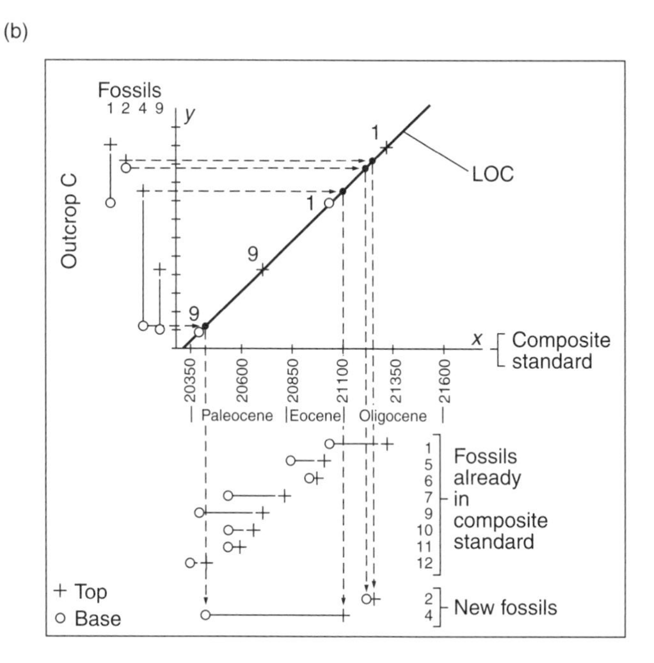


**FIGURE 6.5 Projection of stratigraphic data from an additional outcrop onto the composite standard (CS).** (a) Construction of the LOC between the new section (outcrop C) and the CS. (b) Projection of FADs and LADs for taxa unique to outcrop C (taxa 2 and 4) onto the CS. *(From Carney & Pierce, 1995)*

taxa that they share, the stratigraphic ordering depicted for some of these events in a composite section may nevertheless be incorrect. As a simple, if somewhat extreme, example, consider a case in which the evolutionary origination of some widespread taxon A preceded that of another widespread taxon B, but, simply by chance, depositional and taphonomic processes throughout an entire study area remove any evidence of taxon A in strata below those containing the FAD of taxon B. In such a case, the data from all localities would therefore be unanimous in depicting the FAD of taxon B in a position below that of taxon A, and a composite section would therefore depict incorrectly the first appearance of taxon B as preceding that of A.

More generally, if we consider the relative ordering of the FADs or that of the LADs for any pair of taxa preserved at an outcrop, it is always possible that additional sampling will reverse the ordering of either pair of events (Figure 6.6). However, if at some outcrop, we observe that taxon A occurs in strata that are below, or overlap with, strata that contain taxon B, there is one statement that will remain incontrovertible regardless of how much more sampling we conduct at the outcrop or anywhere else in the world (barring post-mortem reordering of taxa): The FAD and, indeed, the evolutionary first appearance of taxon A must predate the LAD and the extinction of taxon B (Figure 6.5). This must be so because, while additional sampling could revise the FAD of taxon A down or the LAD of taxon B up, it cannot negate the overlap already observed in the ranges of the two taxa. Similarly, if taxon A overlaps stratigraphically with taxon B, it must also be true that the FAD of taxon B predates the LAD of taxon A.

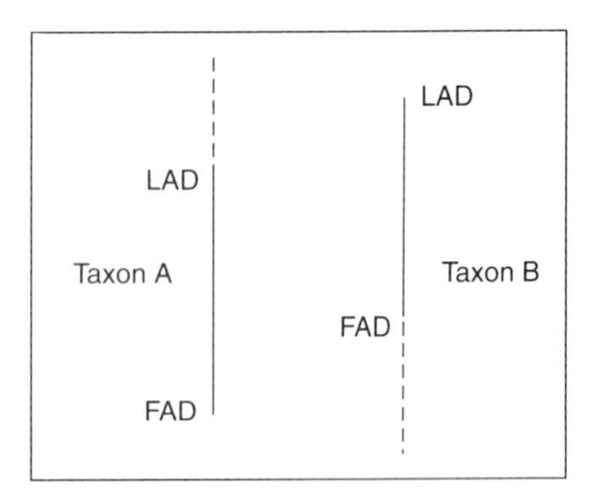


**FIGURE 6.6 Schematic representation of the stratigraphic ranges of two hypothetical taxa at an outcrop.** Solid lines depict the known ranges of each taxon. Based on these ranges, we can state that the FAD of taxon A precedes that of taxon B, and the LAD of taxon A precedes that of taxon B. However, if, with additional sampling, we extended the observed stratigraphic ranges of each taxon to include the dotted portions of each range, both of these statements would be overturned. By contrast, no amount of additional sampling at this locality or any other will overturn two additional statements that we can make: The FAD of taxon A precedes the LAD of taxon B, and the FAD of taxon B precedes the LAD of taxon A.

John Alroy's (1994a) **appearance event ordination (AEO)** makes exclusive use of these incontrovertible observations, called **F/L statements.** Not all F/L statements will be preserved at all localities in a region under investigation, and AEO is designed to piece together a CS from the combination of information preserved at all localities. The basic principles of AEO are described in Box 6.2.

Note that, in the example presented in Box 6.2, the positioning of events in the CS is relative; no absolute indication is provided for the timing of these events. However, as with graphic correlation and the other methods we will consider, a CS generated by AEO can be calibrated to an absolute timescale based on correlations of some of the events to horizons from which absolute dates have been collected.

## Constrained Optimization and Ranking and Scaling

While F/L statements are incontrovertible, this of course does not imply that other kinds of relationships observed among events are necessarily incorrect; the inclusion of these additional events, assuming that we can trust them, can obviously help to further refine a CS. **Constrained optimization (CONOP)** is a procedure that makes broader use than AEO of the stratigraphic events recognized collectively at a set of localities. As part of the procedure for CONOP, impossible solutions are first eliminated (constraint) and the best of all possible solutions is identified quantitatively (optimization).

At the outset, in constraining the roster of possible solutions, any solution can be viewed as impossible if it violates observations known to be incontrovertible. For example, it would obviously be inappropriate to establish a sequence in which the first appearance of any species is placed after its last appearance. Likewise, if the ranges of two species, A and B, are observed to overlap at any outcrop, we know from our discussion in the last section that the FAD of taxon A must precede the LAD of taxon B and the FAD of taxon B must precede the LAD of taxon A. Clearly, any solution that suggests otherwise would be incorrect.

Box 6.2

## APPEARANCE EVENT ORDINATION (AEO)

As a first step in appearance event ordination (AEO), consider the three hypothetical localities in Figure 6.7a, depicting the stratigraphic occurrences of five species. Not all species are preserved at all localities, and species may occur in multiple horizons at any given locality. A list of F/L statements derived from the information at each locality is provided in Figure 6.7b. By convention, these statements are presented with a syntax of $X < Y$, which means that the FAD of taxon $X$ predates the LAD of taxon $Y$.

For example, at section 1, we know that, because taxa 1 and 2 co-occur in the same horizon at locality 1, the following statements must be true: $1 < 2$ and $2 < 1$. Similarly, because taxon 3 occurs stratigraphically above taxon 1 at locality 1, we also know that $1 < 3$. However, it would not be appropriate to say that $3 < 1$ because taxon 3 does not occur at or below the horizon that contains taxon 1.

We can summarize the information from Figures 6.7a and 6.7b in the matrix presented in Figure 6.7c: Instances in which FADs (rows) predate LADs (columns) are designated with the symbol $<$. Although it is inevitable that the FAD of any taxon must predate its own LAD, we include these statements in the matrix because we will depict these events in our CS.

From the information in the F/L matrix, we construct a *composite event sequence* (really, the CS), as depicted in Figure 6.7d. We begin by first finding all taxa

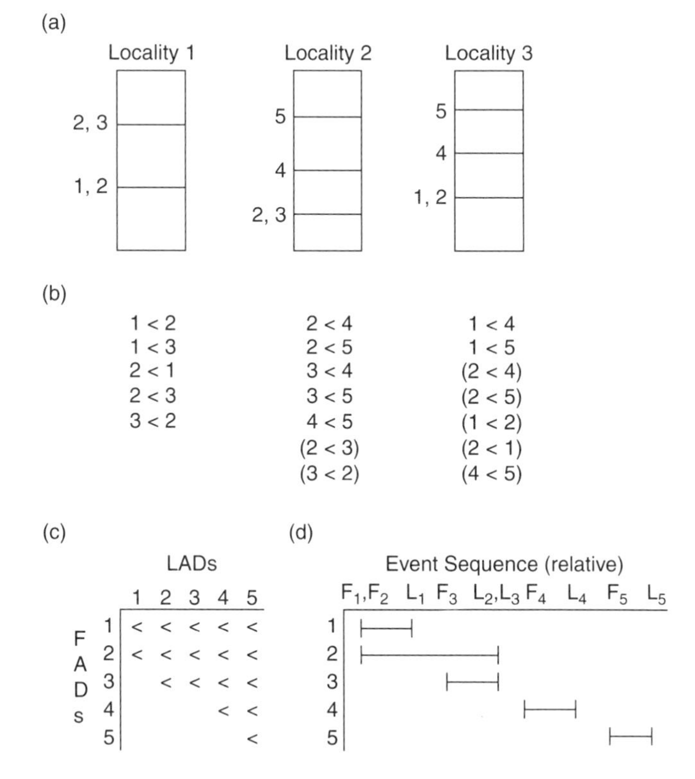


**FIGURE 6.7 Illustration of the principles of appearance event ordination (AEO).** (a) Occurrence data for five hypothetical taxa, numbered 1 through 5, at each of three hypothetical localities. Horizontal lines indicate the stratigraphic positions of collections at each locality, and the numbers to the left designate the taxa found in these collections. (b) A list of the F/L statements that can be made for each outcrop. Statements in parentheses repeat those already made at a previous locality. (c) A matrix that summarizes the information provided in part (b). (d) Composite event sequence for these taxa ($F_y$ = $FAD_y$; $L_y$ = $LAD_y$). *(Based on an example presented by Alroy, 1994a)*

*continued on next page*

*Box 6.2 (continued)*

with FADs that predate the first LAD; in this instance, we recognize in Figure 6.7c that FADs 1 and 2 predate LAD 1, so we depict these events sequentially in Figure 6.7d (i.e., FAD 1 and FAD 2 together, followed by LAD 1). We then repeat this process, determining which FADs precede the subsequent LADs, until our entire event sequence is determined.

LAD 1 predates FAD 3 because there is no evidence of stratigraphic overlap between taxa 1 and 3, and the stratigraphic range of taxon 1 resides entirely below that of taxon 3. By contrast, there is a stratigraphic overlap between taxa 2 and 3, and, therefore, FAD 3 predates LAD 2 (i.e., $3 < 2$). Note that LAD 2 and LAD 3 are grouped in tandem in the sequence because, based on the available evidence, we have no way of discriminating the relative timing of these two events (the same is true of FAD 1 and FAD 2). Their grouping should therefore *not* be taken as a definitive indication that the two events took place simultaneously.

In perusing the CS, it might not seem clear, at first, how we can be certain of the sequence depicted. For example, how do we "know" that FAD 4 postdates LAD 3? The answer is that we can never be completely certain of this because it is always possible that, at some later date, we will find evidence, say, of stratigraphic overlap in the distributions of taxa 3 and 4. However, for the moment, all available events point to the stratigraphic range of taxon 4 entirely postdating that of taxon 3, so FAD 4 is depicted as occurring after LAD 3. While, as we noted earlier, individual F/L statements are incontrovertible, the sequence of FADs and LADs exhibited by a set of taxa may continue to evolve with the addition of new data, and therefore we may well add F/L statements in the future.

In developing an *optimized* solution, a CS is established for the events preserved in aggregate among all localities. When one works with a large number of taxa and localities, it is almost unavoidable that individual localities will preserve some events in an order different from that of the CS because, as we noted earlier, there will likely be some contradictions in the relative orders of FADs and LADs for taxa shared among localities. An additional objective, therefore, is to develop an optimal (re)ordering of events at each locality that maintains the order established for the CS in the first step. A schematic description of how this is accomplished is presented in Box 6.3.

In delineating the optimal ordering of events, it is useful to determine the abundance of a taxon throughout its observed range. All else being equal, the stratigraphic range of a taxon will more closely approximate its true range when it is ubiquitous throughout its observed range [SEE SECTION 6.5]. In such instances, a researcher can have confidence that additional specimens discovered at some later date are less likely to fall outside the observed range of the taxon. This is perhaps best understood by considering the extreme opposite case, in which just a single specimen of a taxon is known from a locality, and, therefore, the FAD and LAD of the taxon coincide. In such a case, a researcher cannot be very confident that an additional specimen found at a later date will come from the same horizon as the first specimen. In fact, there is a strong possibility that the second specimen would come from a different horizon, thereby changing the FAD or LAD, depending on whether the second specimen is found stratigraphically below or above the first one. Thus, we would seek to weight the ranges of common taxa more heavily than those of rare taxa in deriving optimized solutions.

Continuing to build on these themes, a procedure known as **ranking and scaling (RASC)** can also be applied to stratigraphic data. The objective of RASC is to estimate the most *probable* ranges of taxa, under the assumption that inconsistencies among localities in the order of FADs and LADs result from some random combination of processes, such as those highlighted earlier, that potentially affect the stratigraphic ranges of taxa observed at any individual locality. In the simplest case, the relative orders of any given pair of events,

*Box 6.3*

## CONSTRAINED OPTIMIZATION (CONOP)

Using CONOP, determination of an optimal ordering of events at a locality, and of the CS itself, involves the use of weighting coefficients, based on quantification of the likelihood that some observed FADs and LADs of taxa are more dependable than others. In particular, it is important to take into account the truncations of stratigraphic ranges that are virtually inevitable in association with unconformities and changes in lithology.

Conversely, there are also predictable tendencies for the preserved ranges of taxa to be extended beyond their initial (true) preserved ranges. For example, fossil specimens in drill cores often tend to be smeared downward by physical processes associated with drilling and extrusion of the core. This should cause a downward smearing of the FAD of any taxon whose stratigraphically lowest occurrences are affected in this way. By contrast, LADs of common taxa are less likely to be affected by this process because, while some specimens will be transported downward in the core, other specimens will remain at their original stratigraphic positions in the core, thereby preserving the LAD. Of course, if the taxon is rare, this increases the likelihood that all specimens will be smeared downward, thereby altering the LAD as well. In any case, when unconformities and dramatic lithologic transitions characterize the data from a locality, or when the data come from a core, these factors will result in the downweighting, or even the elimination, of certain classes of data.

A and B, are compared at all localities, and the order that occurs most commonly—either A above B or B above A—is deemed to be the most probable sequence. However, it is sometimes the case that contradictions will arise in determining pairwise orderings among three or more events; in these cases, a determination is made with RASC of the most probable order of events, as discussed in Box 6.4.

Despite their differences, there is no reason to expect that any of the methods just highlighted should

*Box 6.4*

## RANKING AND SCALING (RASC)

As an example of the use of RASC in a case in which contradictions might arise in the apparent order of three or more events, consider the ordering of events A, B, and C at several localities. It might be observed that event A occurs most frequently below event B, and B most frequently below C. Therefore, it should follow that event A occurs below event C. Nevertheless, it is possible that, at some localities, event C will be observed to occur below event A. When these kinds of contradictions occur, the probability of pairwise event sequences are ranked, with greater weight given to event pairs that occur most commonly among the localities under investigation. In our three-event pair example, therefore, if observations of events A above B and B above C occur more frequently among the sampling localities than the observation of event C above A, then C above A is discounted in establishing the most probable sequence.

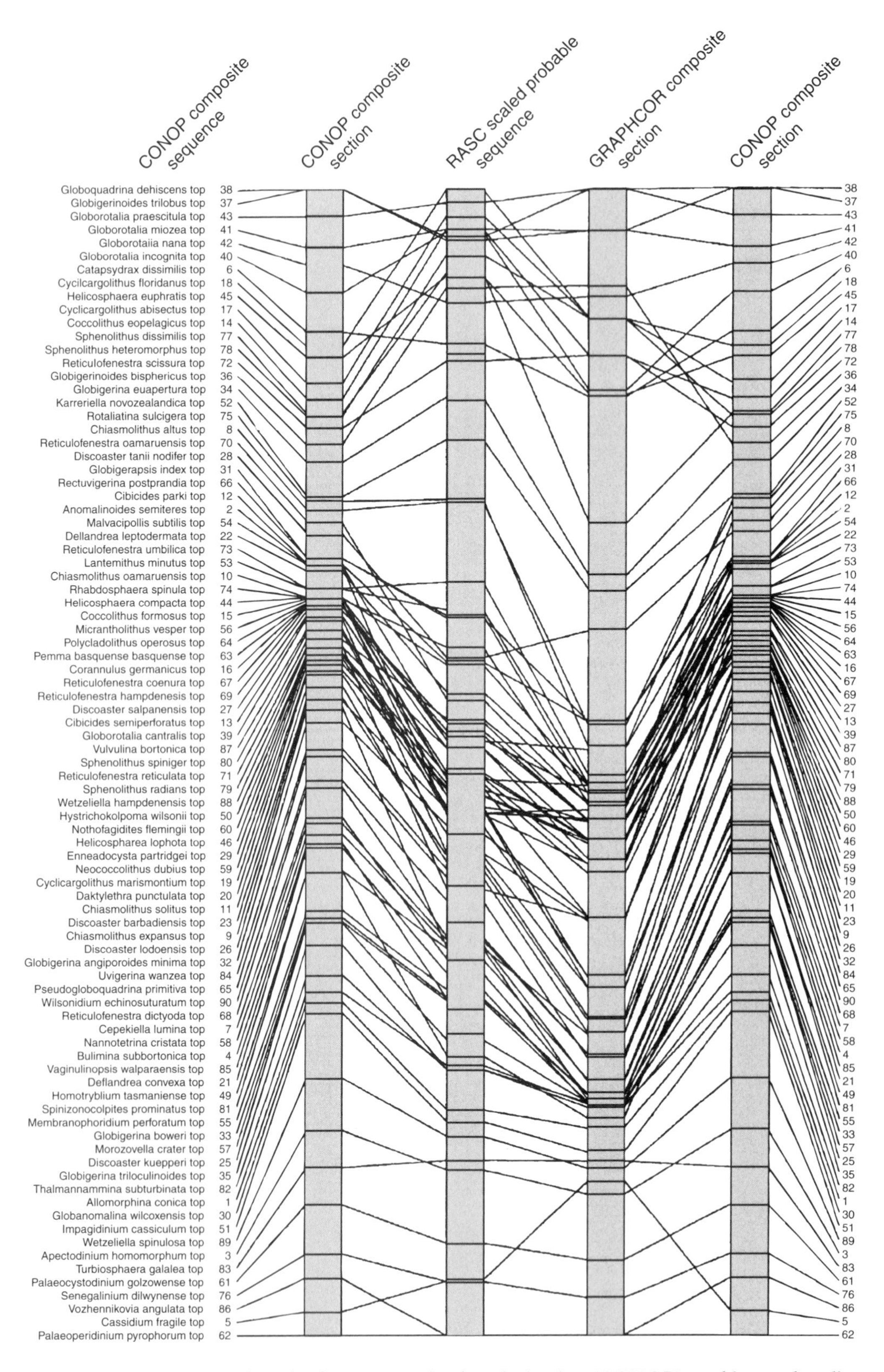


**FIGURE 6.8 Comparison of results from constrained optimization (CONOP), ranking and scaling (RASC), and graphic correlation (GRAPHCOR), applied to a set of eight cores from the Taranaki Basin of New Zealand.** A composite sequence of events (the sequence of species LADs, or "tops," at the left) and their spacing (the adjacent column, labeled "CONOP composite section") were determined for a culled data set that eliminated "questionable" events, including all FADs, because of concern that these were subject to downward smearing in the sample cores. The crossing of lines indicates instances in which there were contradictions from one technique to the next; note that these are concentrated most heavily in the central portion of the interval, where events are spaced most closely. *(From Cooper et al., 2001)*

produce wildly contradictory results when applied to the same data. A comparison presented by Roger Cooper and colleagues (2001) of GRAPHCOR (graphic correlation), CONOP, and RASC is illustrated in Figure 6.8, based on fossil data collected from eight wells in the Taranaki Basin of New Zealand, encompassing a complex set of Paleocene to Miocene strata. Since the data came from cores, the authors decided to exclude all FADs from the analyses because of the possibility of downward smearing, for reasons that were discussed in Box 6.3. Results generated with each method exhibit strong similarities, and, not surprisingly, most differences are concentrated in a central portion of the interval where there are many, narrowly spaced events.

The CONOP sequence was easily correlated directly to an existing regional timescale (Figure 6.9) because most of the events depicted in Figure 6.8 are well known outside the Taranaki Basin, and their broader stratigraphic significance throughout New Zealand had already been determined. Thus, based on these events, it was possible to correlate each of the localities individually to the regional timescale, as well as to one another (Figure 6.9).

## 6.3 REGIONAL CORRELATION WITH GRADIENT ANALYSIS

In Chapter 9, we discuss how fossil data are used in the spatial delineation of ancient ecological communities. Because the analytical approaches we will describe have also proven valuable for high-resolution correlation, we now briefly take up these methods before returning to them later in the context of paleoecology.

Paleontologists now recognize that, like biological communities in the present day, ancient communities did not consist of tightly interlocked sets of species separated by discrete boundaries [SEE SECTION 9.3]. Instead, just as environmental transitions along a marine transect or down the side of a mountain may be gradational, boundaries among the communities contained in these spaces are also gradational because species tend to be linked closely to the environments in which they are capable of living. The analysis of spatial variations in biotic composition has therefore come to be known as **gradient analysis,** which typically involves statistical comparisons of paleontological samples collected from different locations arrayed along a hypothesized gradient. While many different numerical techniques have been used in gradient analysis, they are similar in comparing samples on the basis of the occurrences—and, typically, the relative abundances—of the taxa they contain. Data reduction techniques like those discussed in Chapter 3 are then used to order the samples in a multidimensional space that reflects their compositional similarities. Samples with similar positions or *scores* in this space have similar biotic compositions.

The same analytical techniques can also be used to assess faunal variation through time rather than in space. Samples are collected from a series of horizons arrayed stratigraphically at a given locality, and the resulting sample scores from gradient analysis can be graphed in stratigraphic order. This paints a picture of temporal variations in biotic composition at a locality that may be related directly to paleoenvironmental transitions. An example of how these variations can then be used to accomplish high-resolution regional correlations is presented in Box 6.5 (on page 164), for a portion of the fossil-rich Upper Ordovician strata in the Cincinnati, Ohio, region.

Because changes in water depth relate in some (though certainly not all) instances to global fluctuations in sea level, one might expect that gradient analyses could be used for global-scale correlations. However, although local and regional stratigraphic patterns are partly influenced by global variations in sea level, this by no means suggests a one-to-one correlation of these variations with regional trends in water depth or other biologically significant environmental variables that are also expressed on a regional level. Individual regions are characterized by environmental factors (e.g., the availability of a source area for terrigenous sediments or the evolution of basin topography) that produce unique biotic transitions. Thus, while gradient analysis is a powerful tool for high-resolution correlation at the regional scale, it does not have much potential for global correlation.

## 6.4 SEQUENCE STRATIGRAPHY AND THE DISTRIBUTION OF FOSSILS

Earlier, we noted that post-mortem processes, or simply the tendencies of taxa to live in some environmental settings and not in others, make it highly unlikely that any two correlative intervals at different localities will contain

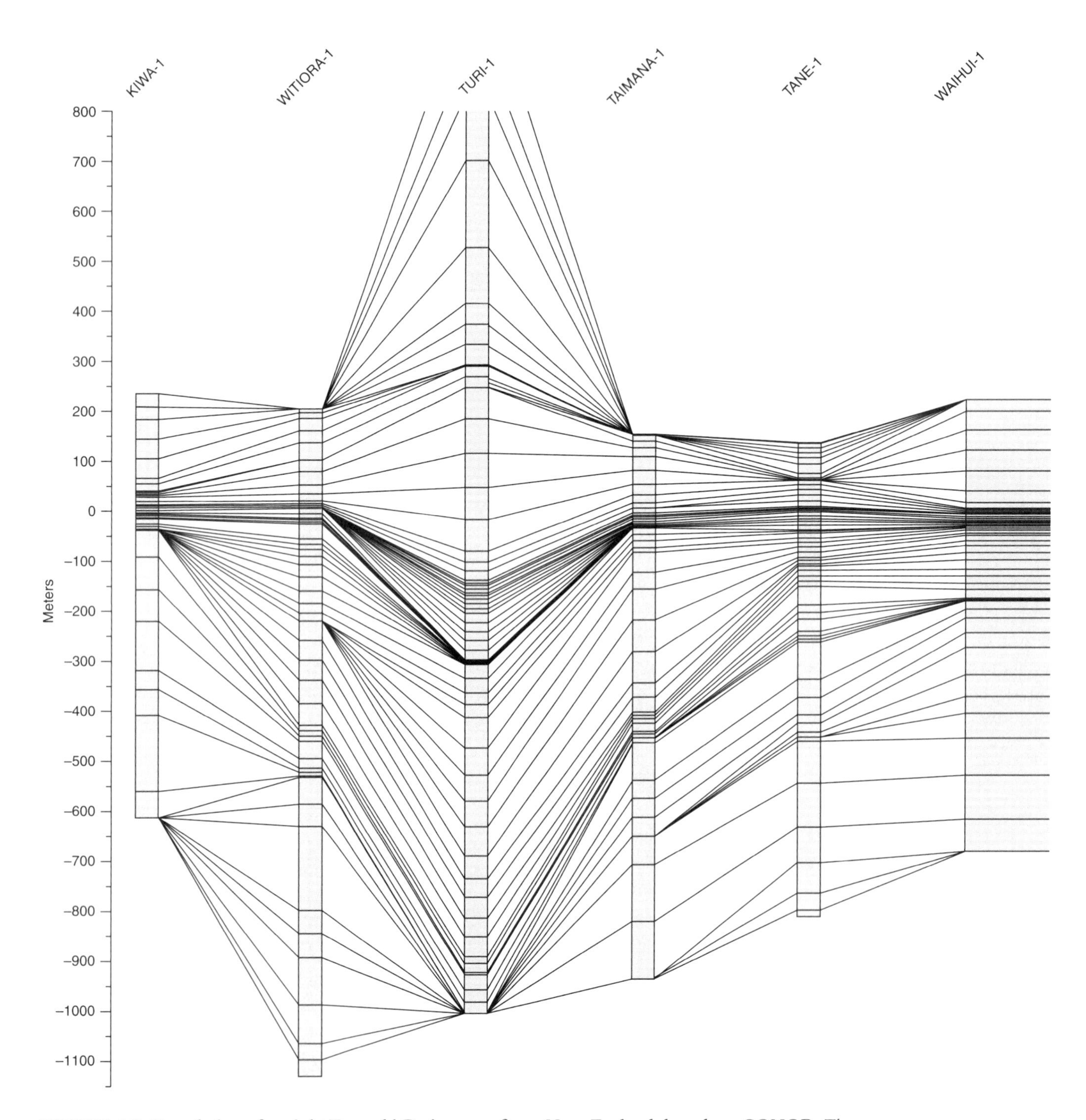


**FIGURE 6.9 Correlations for eight Taranaki Basin cores from New Zealand, based on CONOP.** The stratigraphic positions of events in each well were adjusted from their actual locations, based on the outcome of the procedure. *(From Cooper et al., 2001)*

exactly the same roster of taxa preserved in the same stratigraphic order. To some extent, these factors are expressed randomly in space and time: There is no telling precisely where a taxon will turn up in the fossil record throughout its paleogeographic, paleoenvironmental, or stratigraphic extent. However, there are also nonrandom factors that tend to produce a high concentration of taxonomic occurrences, particularly FADs and LADs, in specific kinds of horizons that are laterally extensive. In fact, stratigraphic patterns of first and last appearances, regionally and per-

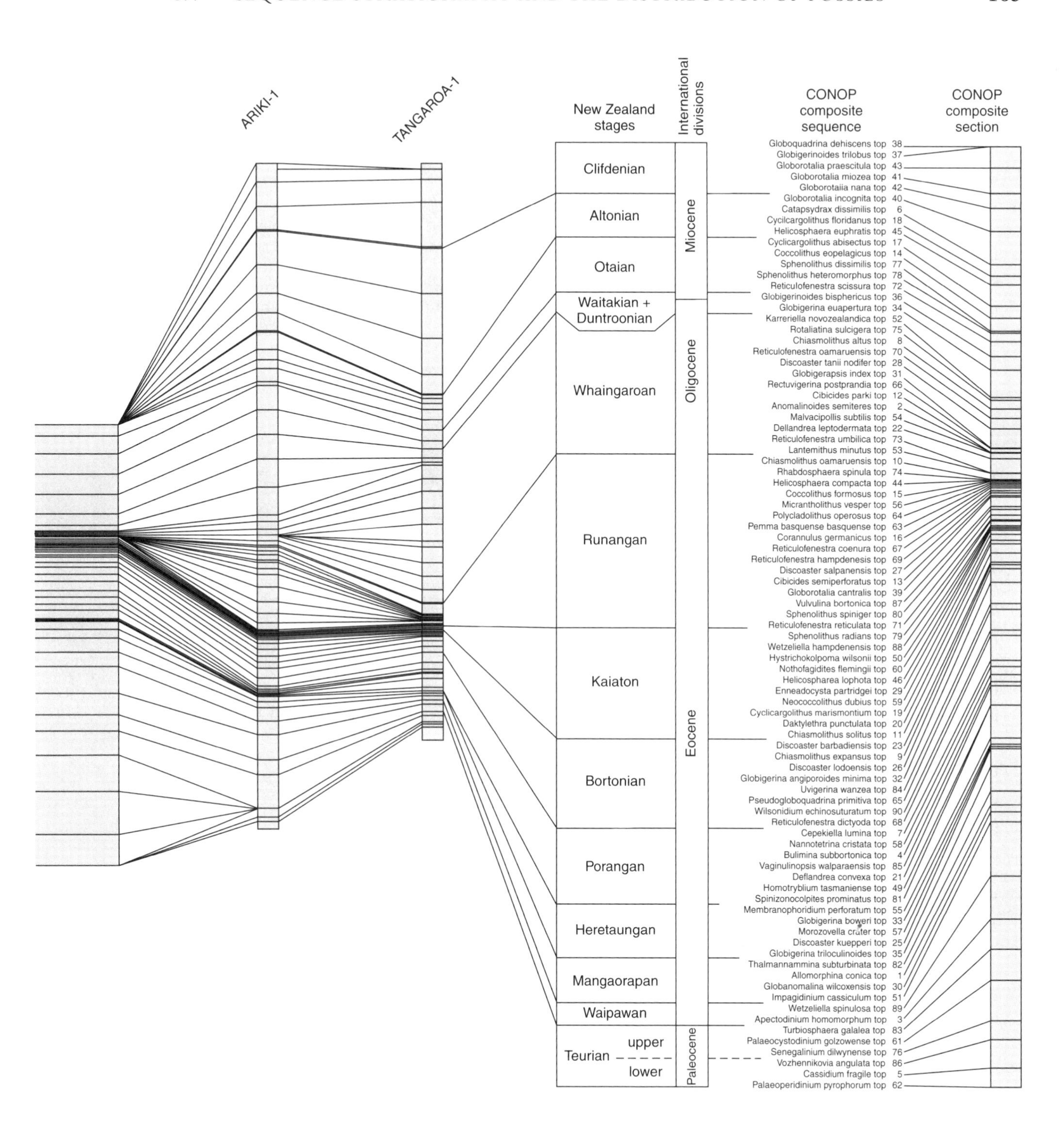


haps globally, may relate predictably to the physical stratigraphic record. The delineation of these relationships emanates directly from sequence stratigraphy.

A fundamental goal of sequence stratigraphy is to diagnose a hierarchy of stratigraphic units and the boundaries between them. Because the processes that produced these patterns were regional or, in some cases, global in scope, it follows that the units, or at least the boundaries between them, should be correlative over broad areas. To the extent that this is the

Box 6.5

## HIGH-RESOLUTION CORRELATION WITH GRADIENT ANALYSIS

The use of gradient analysis as a method of correlation is illustrated in Figures 6.10 and 6.11. In Figure 6.10, samples from the study area near Cincinnati are compared using a data-reduction technique called *detrended correspondence analysis* (DCA). Here, scores are graphed stratigraphically as a simple $x$–$y$ line plot in the middle column, for faunal census samples collected from every fossiliferous horizon at a locality in northern Kentucky. A smoothed version of these data, presented in the right column, is a moving average of the same data, constructed by averaging together for each horizon the scores of several of contiguous horizons (21, in this case) centered on the horizon in question.

In developing the moving average, there is no definitive guideline for deciding how many contiguous points to average together. As with the diagnosis of periodicity in harmonic analysis (see Box 2.2), the objective is to make it easier to see longer-term trajectories in the curve, while removing higher-frequency fluctuations that hinder the researcher's ability to recognize the broader pattern; typically, there is some trial and error involved in determining a suitable number of points to average. In the moving average curve, there are several broad "meanders" to the left (lower scores) and right (higher scores) that are associated with significant changes to the faunal compositions of samples.

Samples with higher scores are dominated by large, robust brachiopods and branching bryozoans.

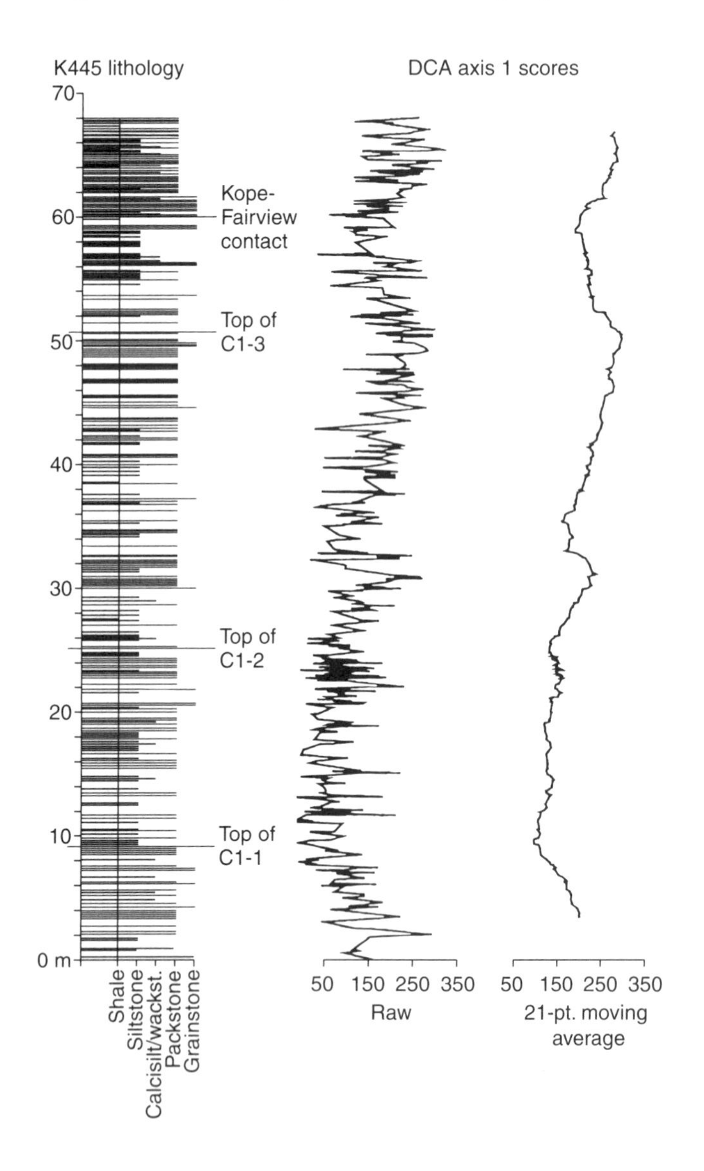


**FIGURE 6.10 Results of detrended correspondence analysis (DCA) of faunal census data collected bed by bed through a section of the Upper Ordovician Kope and Lower Fairview Formations, located in northern Kentucky (K445).** The column on the left summarizes the lithological variation through the section; the central column depicts the census scores from axis 1 of DCA; and the rightmost column is a smoothed version of the middle curve, based on a 21-point moving average. The value for each point on the smoothed curve is the average of the actual value at that horizon, plus those of the 10 sample horizons immediately below it and the 10 immediately above it. *(From Miller et al., 2001)*

case, it has motivated an intriguing pair of questions for paleontologists: (1) Are FADs and LADs concentrated at these boundaries, and, if so, (2) can these boundaries be taken as evidence for rapid evolutionary transitions? As we will see, an affirmative answer to the first question does not necessarily imply an affirmative answer to the second.

A depositional sequence can be defined as a cyclical unit bounded at its base and top by a **sequence boundary (SB),** which is marked by a marine or subaerially

Based on previous studies of their natural histories, these are known to have been associated in the study region with shallow, relatively turbulent waters, in which fine-grained sediments generally did not settle permanently on the sea floor. Samples with lower scores are dominated by smaller, more fragile brachiopods, as well as trilobites that were associated with quieter water and muddier (i.e., finer-grained) sea floors.

Therefore, inflections on the smoothed curve record transitions from intervals of deepening to intervals of shallowing, or vice versa, at the locality. Because these broad transitions were at least regional in scope, the inflection points can be correlated among localities throughout the region. This approach is illustrated for the Cincinnati-area localities in Figure 6.11. Here, the smoothed curve from the locality illustrated in Figure 6.10 was placed in the middle, and inflection horizons recognized in curves for several other localities were correlated directly to this curve. The correlations are supported by close correspondence to several distinctive horizons and surfaces that were already known on the basis of independent evidence to be correlative throughout the study area.

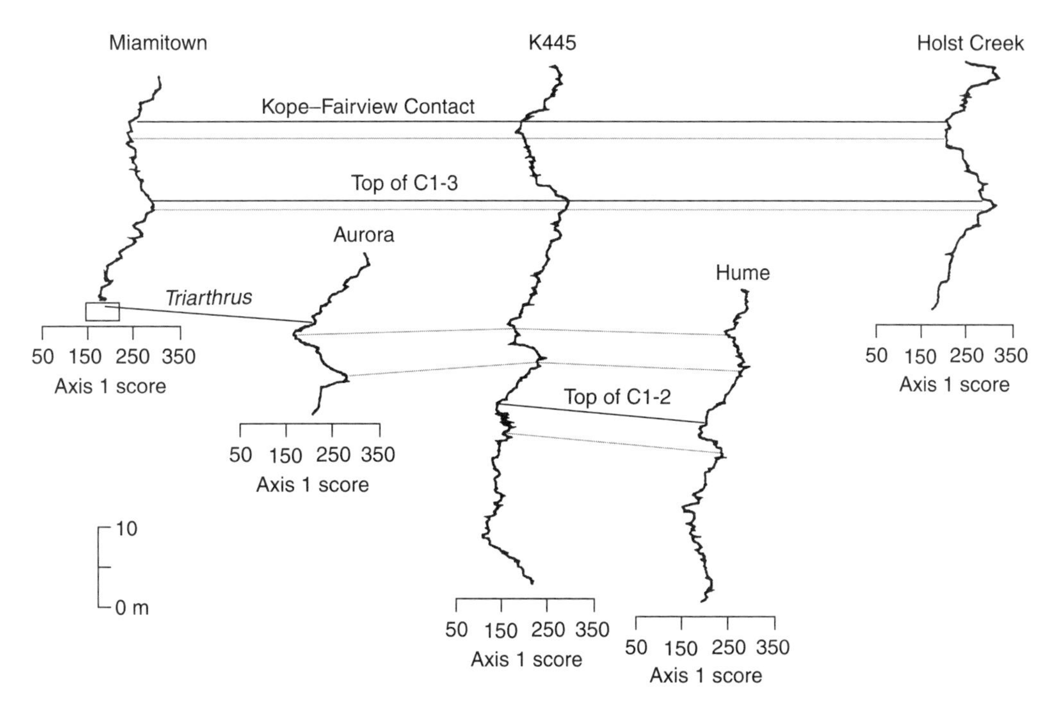


**FIGURE 6.11 High-resolution correlation based on DCA scores on axis 1 (smoothed curves) for Upper Ordovician localities in the Cincinnati, Ohio, area.** Gray lines indicate correlations based on major inflections at multiple localities (K445 is in the middle). Darker lines are independent correlations, for comparison, based on the tops of recognized stratigraphic cycles (labeled as "Tops of C1-2, C1-3"), the Kope–Fairview formational contact, and a unique stratigraphic occurrence of the trilobite *Triarthrus*. *(From Miller et al., 2001)*

exposed erosional surface (i.e., a hiatus), or a noneroded interval that correlates to a hiatus recognized at other localities (Figure 6.12a). Sequence boundaries are caused by relative falls in sea level. The internal architecture of sequences, in turn, may be controlled by several factors, including global fluctuations in sea level, the rate at which these fluctuations take place, local changes in water depth, tectonic controls on the geometry of the depositional basin, and the availability of sediment. Of all these factors, global sea level may have an overriding

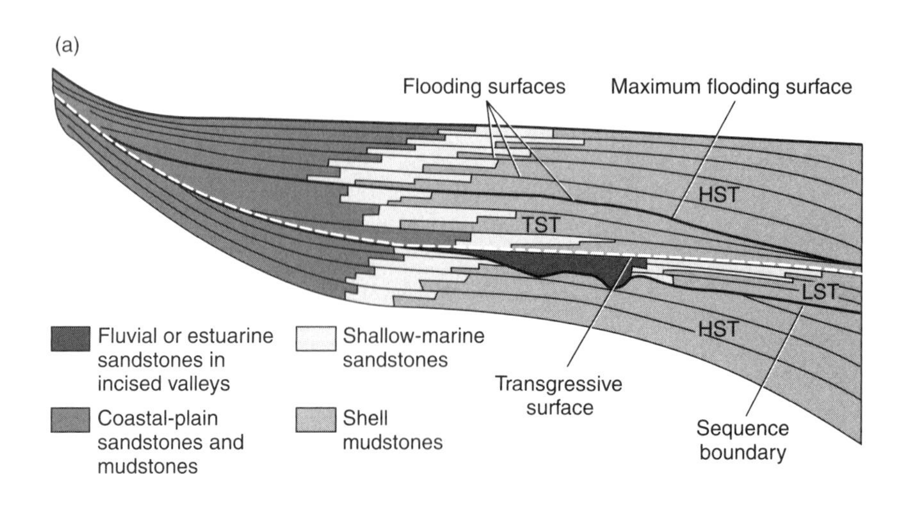


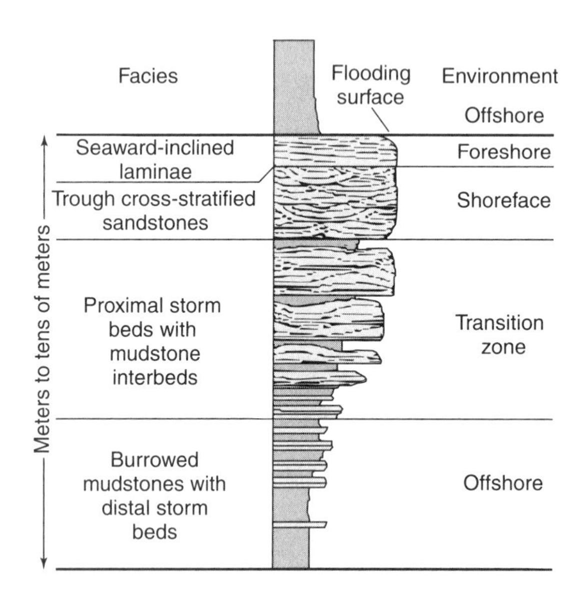


**FIGURE 6.12 Schematic illustrations of the major features of stratigraphic sequences and parasequences.** (a) Sequence architecture, including a lowstand systems tract (LST), a transgressive systems tract (TST), and a highstand systems tract (HST). (b) An idealized parasequence in a terrigenous setting, consisting of a shallowing-upward succession of strata, bounded at the top by a marine flooding surface. *(Modified from Van Wagoner et al., 1990, from an online guide to sequence stratigraphy, maintained by Steven Holland, www.uga.edu/~strata/sequence/seqStrat.html)*

effect, at least among large-scale sequences that may be several tens to thousands of meters thick; the factors controlling smaller-scale sequences have been more open to question. The anatomy of an idealized sequence is described in Box 6.6.

For our purposes here, the essential message of sequence stratigraphy is that the record preserves a hierarchical set of paleoenvironmental variations, and, within this hierarchy, it is possible to recognize particular horizons that mark significant gaps in the deposition of sediments and fossils. Any taxonomic first or last occurrence that took place during an unpreserved interval at a particular location cannot possibly be recorded in the fossil record at that location. Therefore, it stands to reason that the LAD of any taxon whose last occurrence falls during such an interval should be observed somewhere beneath the gap, and the FAD of any taxon whose first occurrence took place during such an interval should be observed somewhere above the interval. Moreover, FADs and LADs may be concentrated in a given section at times of rapid environmental change, when taxa that already exist follow their preferred environments and migrate into or out of the area where the section accumulated.

With these points in mind, Steven Holland has been investigating the relationship between the deposition of sequences and the stratigraphic distributions of fossils. Holland (1995) developed a set of models that simulate important depositional and taphonomic processes, and how they interact with the paleoecological properties of individual taxa to produce patterns that can be observed in the fossil record. Holland's models have two main facets. First, an environmental gradient [SEE SECTION 9.3] is simulated in which each modeled taxon is randomly assigned

*Box 6.6*

## THE ANATOMY OF A SEQUENCE

Because of the interplay of factors that control sedimentation, the nature of a sequence varies throughout its stratigraphic extent, and stratigraphers have designated three major sequence components. In stratigraphic order, working upward, these are the **lowstand systems tract (LST), transgressive systems tract (TST),** and **highstand systems tract (HST).** All three of these units, in turn, are built of sets of **parasequences** (Figure 6.12b). A parasequence is a shallowing-upward cycle that is bounded at its top by an abrupt deepening event, preserved as a **flooding surface,** which also defines the base of the next parasequence.

The LST consists of a progradational (i.e., net seaward stacking) set of parasequences or sequences that lie above the sequence boundary. Because the relative rate of sea-level rise during this initial transition is slow, the land-derived sediments that are deposited in shallow conditions—and the successive set of parasequences or higher-order sequences that are produced by these sediments—tend to build seaward, away from the shoreline (Figure 6.12a).

By contrast, during the subsequent deposition of the TST, an increase in the rate of sea-level rise causes successive parasequences, or higher-order sequences, to migrate landward—a pattern known as retrogradation (Figure 6.12a). The sea-level rise continues, but at a slower rate, in the early part of the HST. Because of the slowing rate, the stacking pattern reverses in the HST back to progradation. The late part of the HST is characterized by a relative fall in sea level.

Aptly, the boundary between the TST and the HST is referred to as the **maximum flooding surface,** which records the greatest water depth in the sequence and marks the transition from retrogradational to progradational stacking. As indicated by our discussion of the HST, the maximum flooding surface does not mark a cessation of sea-level rise; it forms at some point following a decrease in the rate of sea-level rise, when this rate is balanced by the rate of sedimentation. Thereafter, the rate of sedimentation exceeds the rate of sea-level rise, and progradation ensues.

To understand how this can be so, it is important to keep in mind that water depth is not synonymous with sea level. Although it may seem paradoxical, water depth can actually decrease in a situation where the sea level is rising if a significant amount of sediment is flowing into an area, which is what occurs during progradation. And, in fact, much of the sedimentary record is deposited under regional conditions of rising sea level. As progradation continues and sea level ultimately does begin to fall in the later stages of the HST, the area may become exposed, producing a hiatus that marks the next sequence boundary.

It should be kept in mind that the patterns illustrated in Figure 6.12 describe an idealized situation. The exact nature of any sequence may vary considerably from this model, contingent on such factors as the timescale over which the variations are observed; the nature of global sea-level changes; regional variations in water depth and the extent, if any, of subaerial exposure; the shape of the depositional basin; the rate of tectonic subsidence; and the availability of various kinds of sediments. For example, a rather different pattern than that illustrated here might be expected in the absence of terrigenous sediments.

---

attributes that define its abundance and environmental preferences in the landscape. In all, three numerical characteristics are used that govern the abundance of each taxon along a simulated water-depth gradient, with a distribution of individuals for the taxon that approximates a normal (i.e., bell-shaped) curve (Figure 6.13): a **preferred depth,** the center of the distribution where the taxon will exhibit its greatest probability of collection; a **peak abundance,** the maximum abundance achieved at the preferred depth; and a **depth tolerance,** the degree of spread in the distribution of the taxon. The latter variable is in keeping with the recognition that some taxa are highly tolerant of environmental variation and can subsist over a broad environmental region, whereas others cannot.

The second facet of the model simulates the occurrences of modeled taxa in a sequence stratigraphic

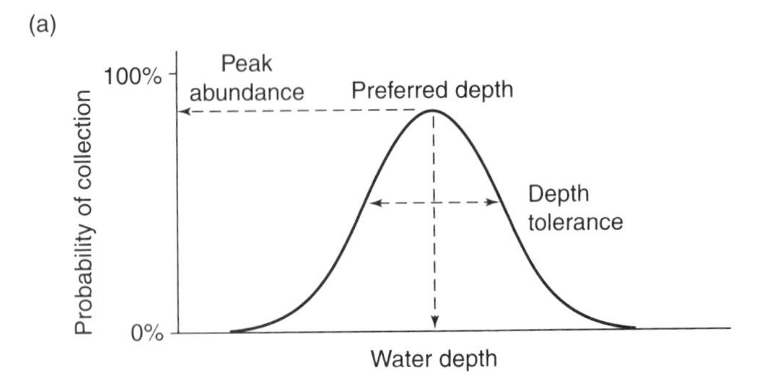


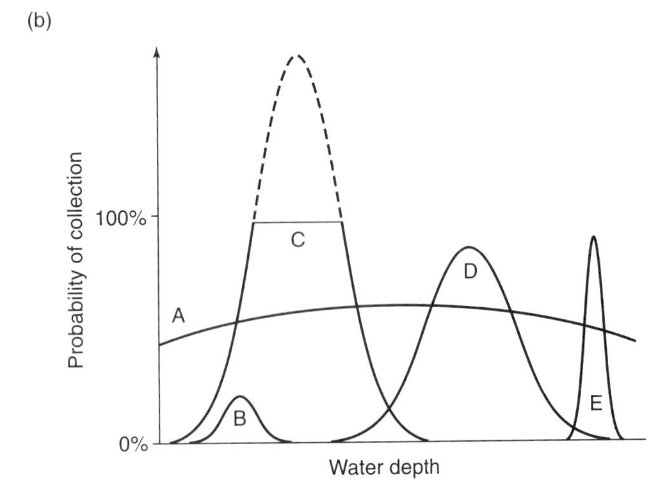


**FIGURE 6.13 Holland's protocol for determining the positions and collection probabilities of taxa included in his simulation of the preserved record of FADs and LADs in a sequence stratigraphic context.** (a) Each taxon is modeled to have a bell-shaped abundance distribution along a simulated water-depth gradient. Three characteristics are defined for each taxon: preferred depth, peak abundance, and depth tolerance. (b) Illustrations of the potential variability among different taxa, based on differences in these parameters. *(From Holland, 1995)*

context, based on their attributes as defined in the previous step. In Holland's model, two shallow-water sequences are simulated: Each has a duration of 3.5 million years, the scale of a typical regional sequence, and is broken into time steps that last 50,000 years. The time steps conform approximately to stratigraphic cycles recognized by field research; these are about 1 meter thick and compose the larger-scale sequences. Additional details of the model are described in Box 6.7.

In thinking intuitively about how the model works, it is useful to consider the depth (i.e., environmental) preferences of any taxon in the context of the water depth preserved in the strata. For a given taxon to be preserved in a stratum, one prerequisite is that the stratum represent a paleodepth in which the taxon could have occurred, based on its model parameters. However, this by no means guarantees that the taxon will be found in the stratum. If the taxon is comparatively rare, it will exhibit a probability of collection that is substantially less than 100 percent even at its preferred depth. If the taxon is common, it will nevertheless be encountered less frequently in a stratum representing a paleodepth near the fringes of the depth range that it can tolerate.

As Holland demonstrated, even fairly common, widespread taxa will not be encountered in every stratum that could potentially house them; rarer taxa will inevitably be distributed spottily throughout their stratigraphic ranges. And, as implied elsewhere, one outcome of this spotty distribution is that the preserved FADs and LADs of taxa will almost certainly be truncated, relative to the true intervals in which they first and last occurred. While this truncation will, on average, be especially acute for rare taxa, there is likely to be some degree of range truncation for virtually all taxa.

In this respect, the overprint of sequence architecture becomes particularly relevant. If the lower- or uppermost portion of the temporal range of a taxon coincides with a depositional hiatus, an inevitable outcome, noted earlier, is that the LAD or FAD will be truncated directly by the hiatus. Importantly, what Holland's model demonstrates is that we would expect LADs and FADs to coincide with the horizons immediately below (for a LAD) or immediately above (for a FAD) the hiatus. Thus, even in an interval where diversity remains stable and the true temporal ranges of taxa are not concentrated significantly at any horizon, we would expect to see a concentration of FADs and LADs at or near sequence boundaries. This is precisely what Holland observed as an outcome of his model.

In Figure 6.14, note the significant concentrations of observed first appearances through each TST at each of the first three flooding surfaces (horizons of rapid deepening). These mark the flooding surfaces at the bases of parasequences (Figure 6.12). The taxa that form these concentrations were of two basic types: (1) shallow-water taxa that originated during the preceding lowstand but were first observed in the section when the appropriate deposits are found for the first time during the transgression; or (2) deep-water taxa that also originated during the preceding lowstand *or earlier* in the shallow portions of the upper HST of the previous sequence, where they would not have been preserved because deep-water lithologies were not preserved at this location.

*Box 6.7*

## ADDITIONAL ASPECTS OF HOLLAND'S MODEL

Because of its relative proximity to shoreline, a shallow-water sequence often lacks the LST. Shallower areas will be subaerially exposed during lowstands, provided that tectonic subsidence rates are low. If subsidence rates are high, subaerial exposure will not occur and the LST will be preserved.

Holland's model excluded deposition during the LST (Figure 6.14) because much of the fossil record comes from slowly subsiding areas. In each of the two modeled sequences, the TST consisted of two parasequences and the HST consisted of six. In the case of the TST, its transgressive nature can be seen by the deepening (leftward), sawtooth pattern exhibited by the two parasequences; this conforms to the retrogradational pattern illustrated for the TST in Figure 6.12. In the case of the HST, a shallowing (rightward) pattern was exhibited, owing to the slowing and reversal of sea-level rise and the infilling of the basin with sediment; this conforms to the progradational pattern exhibited for the HST in Figure 6.12.

The entire depth range represented in the preserved portion of each modeled sequence was 65 m. Against this backdrop, standing taxonomic diversity was held constant at 1000 taxa, with a random extinction probability of 0.0125 in each 50,000-year time step. This corresponds to an empirically calibrated mean species duration of 4 million years [SEE SECTION 7.2]. To hold diversity fixed, each extinction was matched by the origination of a new taxon during the subsequent time step. Upon origination, each taxon was randomly assigned a preferred depth ranging from 0 to 65 m, a peak abundance ranging from a 25 to 100 percent probability of being collected, and a depth tolerance ranging from 1 to 21 m. This combination of model parameters produced the pattern of first and last occurrences illustrated in Figure 6.14.

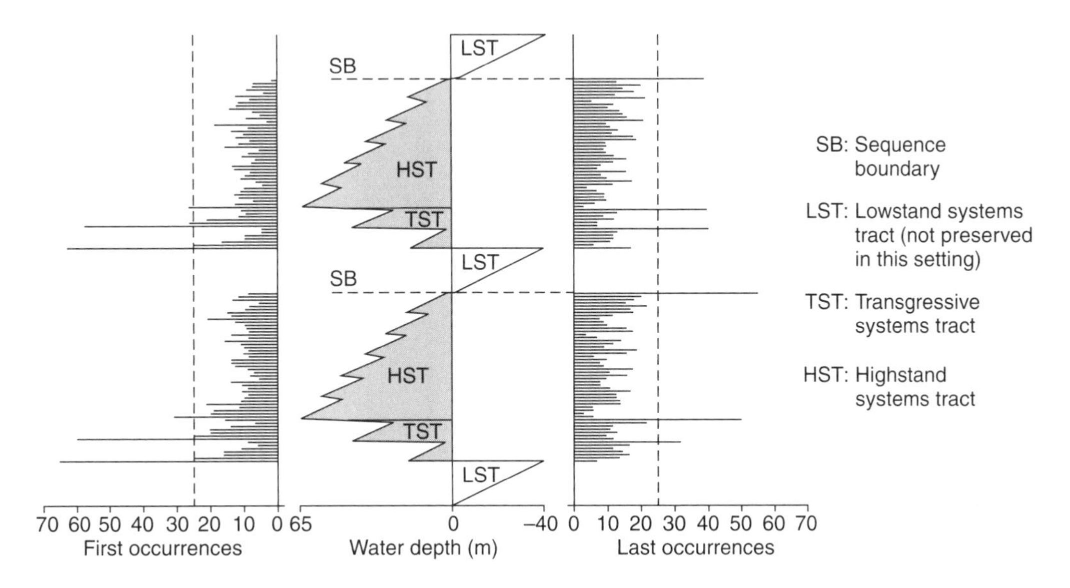


**FIGURE 6.14 Results from Holland's (1995) simulation of the preserved record of FADs and LADs in a sequence stratigraphic context.** Each TST consists of two parasequences, and each HST consists of six parasequences. The vertical dashed lines designate the threshold for statistical significance of peaks in first or last occurrences (see Figure 6.16 for more on statistical significance). *(Modified from Holland, 1995)*

Similarly, there is a significant concentration of observed LADs at the top of each sequence (the top of each HST). This is because taxa whose true time of extinction coincided with the sequence boundary or with the subsequent, unpreserved lowstand had their observed ranges truncated at the boundary. In addition, LADs are concentrated immediately beneath flooding surfaces in the TST. These represent shallow-water taxa with narrow depth tolerances that must have become extinct at some point after the deepening but prior to the recurrence of environments sufficiently shallow to house them.

Given the pervasiveness of sequence architecture in the stratigraphic record at several scales, it therefore becomes imperative to assess the possibility that observed local, regional, or global peaks in taxonomic origination or extinction may be inflated artificially by their association with sequence boundaries or by the loss of key paleoenvironments. For example, Daniel Goldman and colleagues (1999) reevaluated the stratigraphic zonation of graptolites in the Middle Ordovician strata of New York. Their analysis involved more extensive sampling than that of earlier work on these rocks, and they were able to place their biotic data in a stratigraphic framework developed independently using a network of volcanically derived K-bentonites (potassium-rich bentonites).

Comparison with the K-bentonites reveals that the boundaries between graptolite zones are fairly isochronous. At the same time, many common graptolite species were restricted to particular paleoenvironments, with some species occurring preferentially in shallow water and others in deep water. With respect to potential sequence stratigraphic overprints, this finding is particularly important because, as we saw when considering Holland's model, the restriction of species to certain environments enhances the likelihood that their ranges will be concentrated artificially at or near sequence or systems tract boundaries.

Indeed, this is precisely what Goldman and colleagues found, as illustrated in Figure 6.15. There is a clear association in the study interval of FADs with systems tract boundaries. Moreover, stratigraphic gaps in the observed ranges of some species are recognized in association with the loss, followed by the return, of appropriate paleoenvironments. In fact, in some instances, because of their extensive sampling, Goldman and his colleagues found several occurrences of key taxa in beds significantly below and above their previous known stratigraphic ranges. Thus, although the traditional zonation is upheld, there is

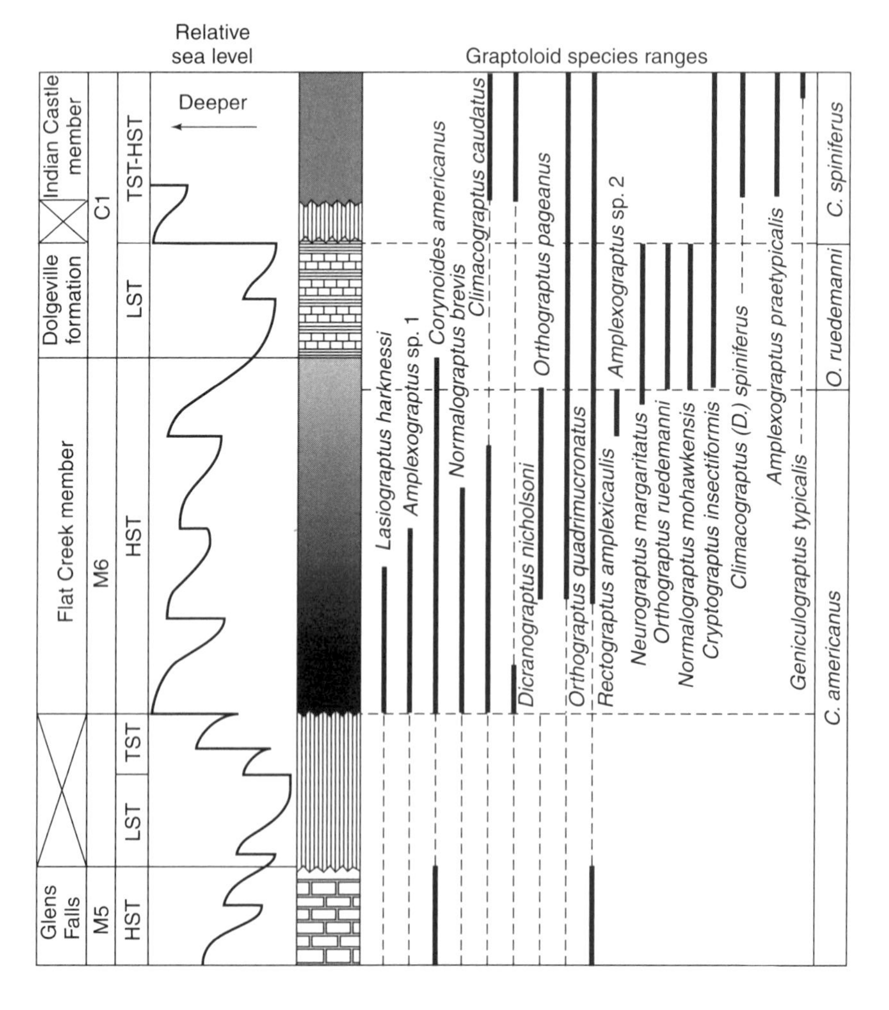


**FIGURE 6.15 Comparison of the stratigraphic ranges of graptoloid species to depictions of relative sea level and sequence-stratigraphic architecture in the Middle Ordovician Utica Shale of New York.** (Abbreviations for portions of sequences are as in Figure 6.14.) Solid lines indicate observed ranges, and dotted lines show inferred occurrences. Note the concentration of FADs at systems tract boundaries, and the gaps in the distributions of some species that relate to the availability of appropriate lithologies and the paleoenvironments they represent. *(From Goldman et al., 1999)*

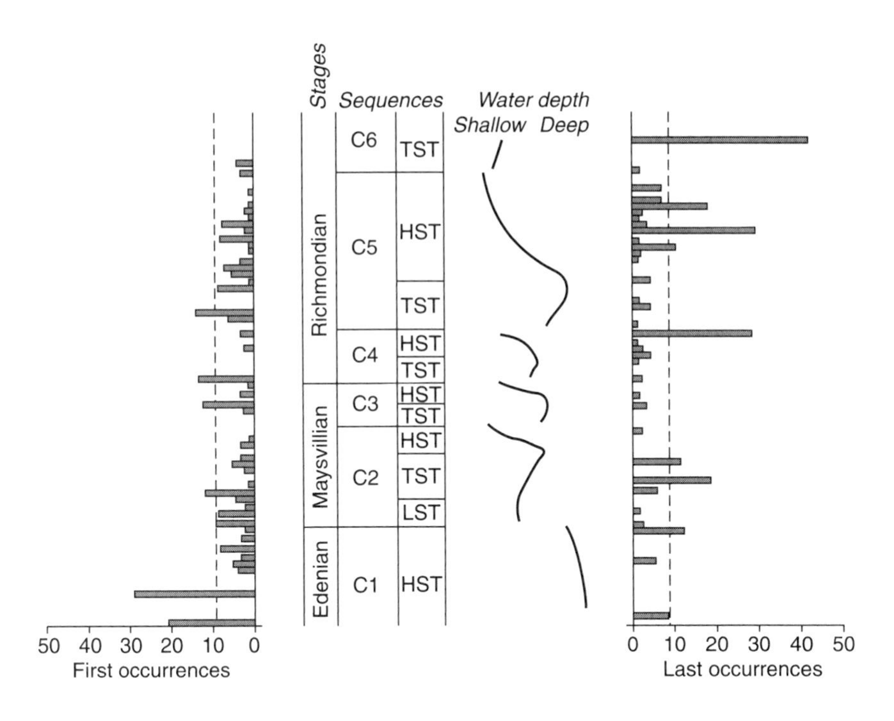


**FIGURE 6.16 Sequence stratigraphy and first and last appearances of species within a section from the Upper Ordovician of Indiana.** C1 through C6 denote sequences of the Cincinnatian Series. The vertical dashed lines indicate the statistical limits of how strongly clustered the FADs and LADs could appear to be if they were not in fact clustered; the probability is less than 0.001 that a spike would exceed the dashed lines if the first and last appearances were distributed randomly throughout the section. Some clusters of events occur where predicted by the sequence stratigraphic model (e.g., LADs at the ends of HSTs), but others cannot be easily explained by this model and are therefore more likely to reflect biologic events (e.g., LADs within the HST of sequence C5). *(From Holland, 1995)*

much more facies dependence and sequence-stratigraphic overprint than had previously been appreciated.

Sequence stratigraphy can aid paleontology much more positively than might be suggested by these notes of caution. When FADs and LADs cluster at levels where they would not be expected as artifacts based on position in the sequence, then they represent strong candidates for true biological turnover events. Figure 6.16 shows examples from an Upper Ordovician (Cincinnatian) section in Indiana. C1 through C6 in this figure denote sequences. Several clusters of events appear where they would be expected based on Holland's models—for example, the spike of LADs at the end of the HST of sequences C1 and C4. Others, however, appear where they are not predicted by the model and are therefore more likely to mark biological turnover—for example, the spike of LADs within the highstand of sequence C5.

Beyond the clustering of FADs and LADs, other paleobiological patterns are associated intimately with sequence stratigraphic architecture. For example, the nature of fossil preservation can vary significantly through a stratigraphic sequence, and even within a parasequence, because of variations in sedimentation rates, turbulence, geochemical conditions, and other parameters. With respect to correlation, these variations can be quite valuable: Beds marked by unique styles of preservation, or even a unique biotic composition produced by unusual depositional conditions, may be traceable throughout a region, providing an additional means of high-resolution, regional correlation.

## 6.5 CONFIDENCE LIMITS ON STRATIGRAPHIC RANGES

The recognition that depositional environments and their associated facies vary, and that the probability of collecting fossils of a given species varies with facies, can also be turned to advantage to estimate the probable size of the gap between the observed LAD and the true time of extinction, or between the observed FAD and the true time of origination. To see how this is possible, first consider the overly simplified situation in which the probability of sampling is assumed to be constant over time. Figure 6.17 depicts the stratigraphic range of a hypothetical species. There are 50 time increments between its first and last appearance, and it is actually represented by fossils in five of these increments. Thus, the estimated sampling probability of the species is 5/50 or 0.1 per time increment. Let us step 1 time increment beyond its observed last appearance and ask: How likely is it that the species was still alive at this time but simply was not preserved? The probability is $(1 - 0.1)$, or 0.9—a high enough value that it is quite reasonable to suppose that the species was still alive.

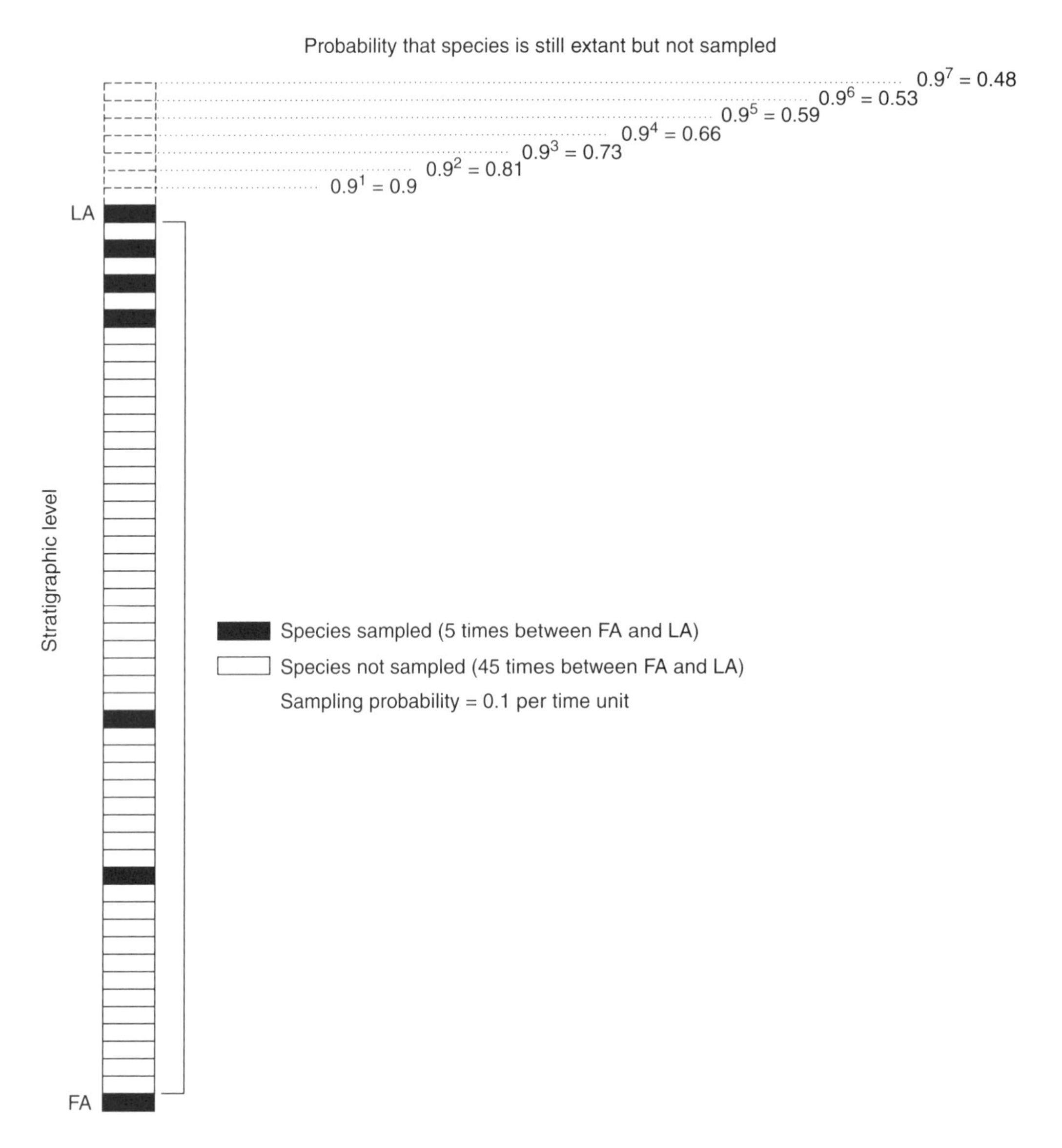


**FIGURE 6.17 Hypothetical example of confidence limits on the time of a species' extinction, assuming constant sampling probability.** FA and LA denote the first and last appearance. Filled boxes show time intervals in which the species is sampled; empty boxes show times when it is not sampled. The farther above the last observed appearance, the smaller is the probability that the species is still alive but simply has not been sampled.

If we go 2 time increments past the last appearance, then the probability that it was still alive but simply not preserved is equal to $0.9^2$, or 0.81. In other words, there is an 81 percent chance that the species is still alive and only a 19 percent chance that it has already become extinct. In this particular example, the probability that a species will fail to be preserved in seven successive time increments is $0.9^7$, or 0.48. At this point, it is about equally likely that the species is still alive as it is that it has already become extinct. The point in time at which the odds are even in this way is referred to as the 50 percent confidence limit; it is essentially a best guess for the true time of extinction. The size of the confidence interval varies from case to case, being smaller when the sampling probability is higher. Wider confidence limits, say 95 percent, can also be constructed if one wants to be even more certain that the true extinction falls somewhere between the last appearance and this limit. Confidence limits can be developed in the same way for first appearances.

The approach of confidence limits sketched out in Figure 6.17 is valid only if we can assume that the average sampling probability does not change significantly, as it might if the last appearance of the species coincided with a change in facies. If we knew how the probability of sampling varied with facies, we could easily substitute these variable probabilities into Figure 6.17. Box 6.8 illustrates one way that this can be done.

## 6.6 CONCLUDING REMARKS

Rather than accepting the stratigraphic record of fossil occurrences at face value, paleontologists are coming to understand why a particular fossil is likely to occur at a particular place in the record. And, just as important, they are developing methods to help reconcile differences in the stratigraphic distributions of taxa from locality to locality. Not only is this permitting more sophisticated approaches to high-resolution correlation; it is also a prerequisite to understanding the causes of many of the paleontological patterns described throughout this book.

For example, it is possible that the physical transitions responsible for sequence architecture and sequence boundaries were also capable of inducing real extinction and significant biotic turnover. However, as Holland's model illustrated for us, the *rate* at which these transitions are observed to have taken place in the stratigraphic record can exaggerate significantly the true rate of biological turnover. Recognizing these relationships, researchers are now developing methods that incorporate preservation criteria directly into the calibration of taxonomic origination and extinction rates.

*Box 6.8*

### CONFIDENCE LIMITS ON STRATIGRAPHIC RANGES WITH VARIABLE SAMPLING PROBABILITY

The curve on the left in Figure 6.18 shows estimated water depth through the Upper Ordovician Kope and Lower Fairview Formations; it is the same as the curve in Figure 6.10, with the *x* axis reversed so that deeper facies are to the right.

The bell-shaped curves in the upper part of Figure 6.18 depict estimated sampling probabilities for the trilobite *Cryptolithus* and the brachiopod *Sowerbyella*, relative to inferred water depth. These are analogous to the hypothetical sampling curves in Figure 6.13, but they are based on the actual frequency with which each genus is found in each environment. *Cryptolithus* preferred slightly deeper water than *Sowerbyella*, and overall its probability of sampling is only about half that of *Sowerbyella*. Knowing how water depth varies and how sampling probability varies with water depth, it is straightforward to determine how sampling probability varies through the section; this is indicated by the black bars. For example, *Sowerbyella* has its highest sampling probability, just under 50 percent, in relatively deep waters corresponding to a detrended correspondence analysis (DCA) score of about 100. Thus, whenever the left-hand curve hits 100, the sampling curve for *Sowerbyella* hits its maximum.

The dots in Figure 6.18 show stratigraphic levels at which *Cryptolithus* and *Sowerbyella* were actually sampled. The last appearance of *Cryptolithus* is at about 59 m in the section, and that of *Sowerbyella* is at about 32 m. The disappearance of *Cryptolithus* comes at a time of low sampling probability corresponding to water depths at which it did not prefer to live. It is therefore quite likely that its disappearance is not a true extinction. If the variable sampling probabilities are applied as in Figure 6.17, the 95 percent confidence limit is well beyond the top of the section. *Sowerbyella*, by contrast, disappears from the record despite the persistence of facies in which it would have a reasonably high probability of being sampled if it were still alive. It is therefore likely that the true extinction is close to the last appearance. Accordingly, the 95 percent confidence limit is just a short distance above the last appearance.

*continued on next page*

*Box 6.8 (continued)*

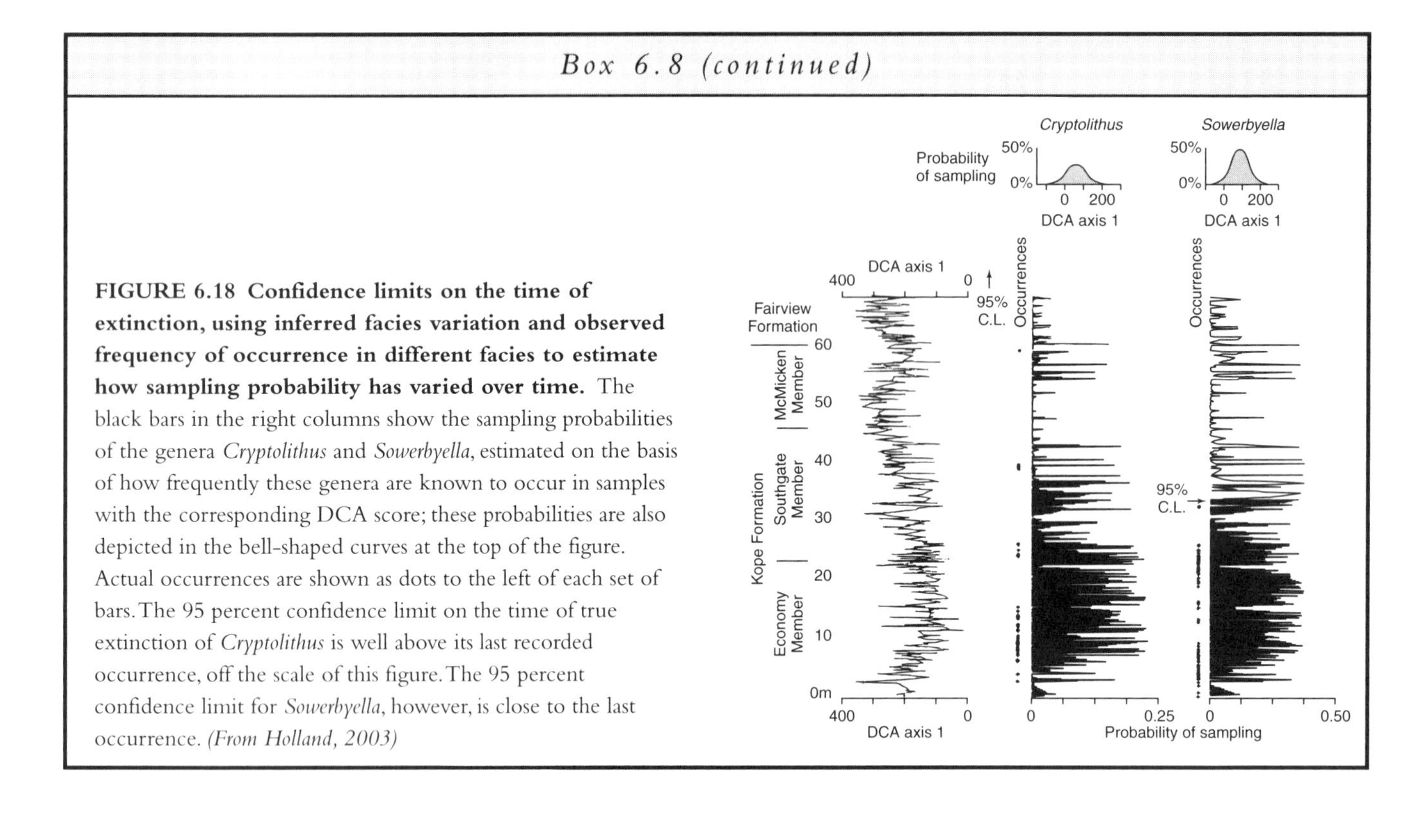


**FIGURE 6.18 Confidence limits on the time of extinction, using inferred facies variation and observed frequency of occurrence in different facies to estimate how sampling probability has varied over time.** The black bars in the right columns show the sampling probabilities of the genera *Cryptolithus* and *Sowerbyella*, estimated on the basis of how frequently these genera are known to occur in samples with the corresponding DCA score; these probabilities are also depicted in the bell-shaped curves at the top of the figure. Actual occurrences are shown as dots to the left of each set of bars. The 95 percent confidence limit on the time of true extinction of *Cryptolithus* is well above its last recorded occurrence, off the scale of this figure. The 95 percent confidence limit for *Sowerbyella*, however, is close to the last occurrence. *(From Holland, 2003)*

# SUPPLEMENTARY READING

Brett, C. E. (1995) Sequence stratigraphy, biostratigraphy, and taphonomy in shallow marine environments. *Palaios* **10:**597–616. [An overview of the relationship of sequence stratigraphy to biostratigraphy and taphonomy.]

Brett, C. E. (1998) Sequence stratigraphy, paleoecology, and evolution: Biotic clues and responses to sea-level fluctuations. *Palaios* **13:**241–262. [A follow-up to Brett (1995) that considers the paleoecological and evolutionary ramifications of sequence stratigraphy.]

Gradstein, F. M., Ogg, J. G., and Smith, A. G. (eds.) (2004) *A Geologic Time Scale 2004.* New York, Cambridge University Press, 589 pp. [A comprehensive overview and presentation of the latest global geologic timescale.]

Holland, S. M. (2000) The quality of the fossil record: A sequence stratigraphic perspective. *Paleobiology* **26** (Supplement to No. 4): 148–168. [An extensive discussion of the relationship between sequence architecture and the stratigraphic distributions of fossils.]

Mann, K. O., and Lane, H. R. (eds.) (1995) *Graphic Correlation. SEPM Special Publication* **53,** 263 pp. [An overview of graphic correlation and related techniques; includes the papers by Kemple et al., and Carney & Pierce, cited in the bibliography.]

# SOFTWARE

Holland, S. M. (1999) *Biostrat 1.7.* [A program for modeling the stratigraphic distributions of fossils within depositional sequences [SEE SECTION 6.4]. The program and documentation are available at www.uga.edu/~strata/software/Software.html.]

Sadler, P. M. (2003) *CONOP9, version 6.5.* [This program performs Constrained Optimization [SEE SECTION 6.2] on a user-defined data set. The program and documentation are available at www.usask.ca/geology/classes/ geol246/CONOP9.htm. A web-hosted version is currently under development at http://portal.chronos.org/gridsphere/gridsphere;jsessionid=A90316B3AF52C245C61C98C8B497ABAE?cid=tools_conop9.]

# Chapter 7

# EVOLUTIONARY RATES AND TRENDS

If natural selection is acting constantly, "daily and hourly scrutinising," to use Darwin's phrase, why do many species accumulate comparatively little evolutionary change over geologic time?

Why do some kinds of species live for greater spans of geologic time than others? For example, why do species of snails whose larvae feed on plankton often endure longer than species with nonfeeding larvae?

Why do many higher taxa exhibit striking evolutionary trends over time? For example, why are many groups of mammals larger in body size today than they were at their time of origin tens of millions of years ago?

These and many other fundamental questions in paleontology are informed by the study of evolutionary rates and trends. Put another way, we would like to know where evolution is headed, how fast it is getting there, and why. In this chapter, we focus on methods used to study rates and trends, illustrating them with case studies that provide some preliminary answers. The questions remain open for the most part, but the tools discussed here will be essential to students of paleontology who hope to address them.

## 7.1 MORPHOLOGICAL RATES

By the **morphological rate** of evolution we mean the rate of change of one or more anatomical traits, typically represented by quantitative measures. We commonly distinguish morphological rates and trends as **phyletic** versus **phylogenetic** (Figure 7.1). Phyletic change, also known as **anagenetic** change, occurs within a single species-level lineage. Phylogenetic change pertains to the more inclusive clade: Without reference to the evolutionary relationships within the clade, how rapidly and in what direction is the average form changing? Phyletic or anagenetic change also stands in contrast with **cladogenesis,** the splitting of lineages [SEE SECTION 3.3]. Phyletic evolution and cladogenesis are the two elements that give rise to phylogenetic patterns [SEE SECTIONS 7.3 AND 7.4].

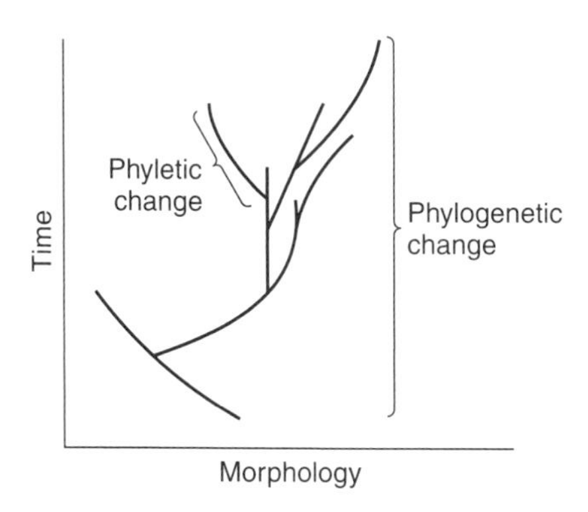


**FIGURE 7.1 Phyletic versus phylogenetic change.** Each line represents a single species-level lineage within a larger clade.

### Nature and Measurement of Morphological Rates

In a mathematical sense, the rate of change of a trait is equal to the slope of the curve that tracks the trait as a function of time; in other words, it is the first derivative of the curve with respect to time (Figure 7.2a). Such a rate is comparable to the speed we estimate by glancing for a moment at an automobile's speedometer. Paleontological

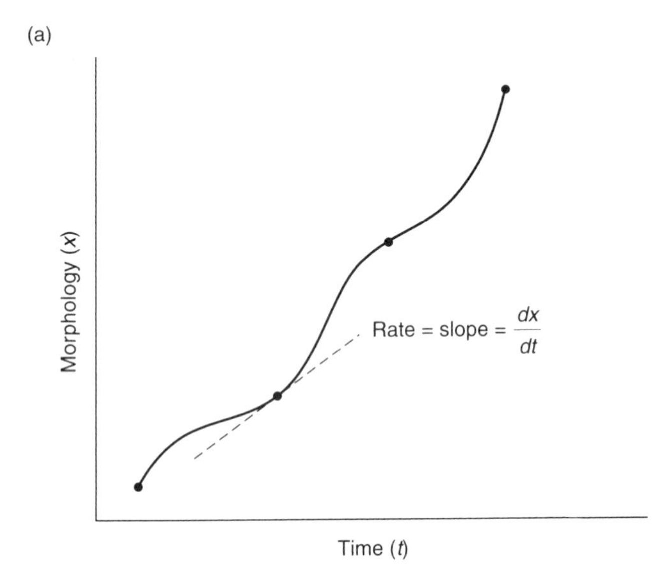


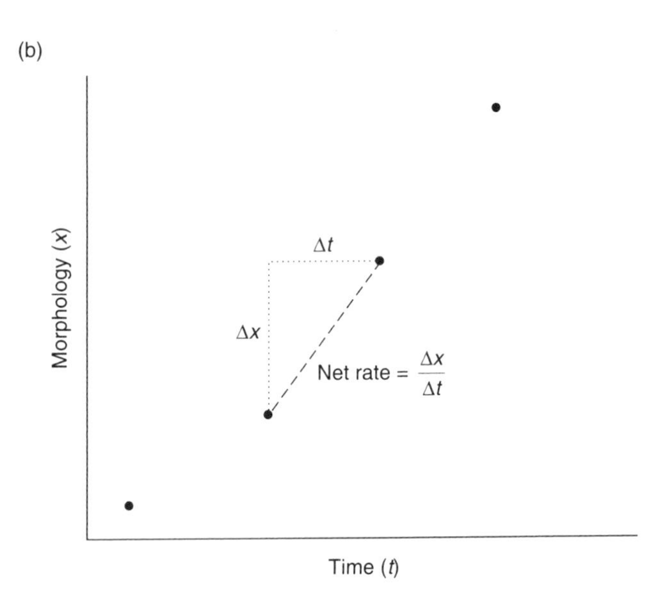


**FIGURE 7.2 Calculation of evolutionary rates.** (a) If change could be sampled continuously, the rate would be the slope of the curve at any point, in other words, the first derivative of the trait with respect to time, $\frac{dx}{dt}$. (b) In reality, the rate must be computed from the net change $(\Delta x)$ over some elapsed time $(\Delta t)$.

sampling is discontinuous in time, however. This reflects both gaps in the record and the discrete nature of sampling units, from beds up through stratigraphic zones and stages. Thus, it is generally not possible to observe evolutionary change at an instant in time. Instead, for a trait $x$, we observe a net change, $\Delta x$, over some interval of elapsed time, $\Delta t$ (Figure 7.2b). Measuring rates of evolution, therefore, is not like checking a speedometer. Rather, it is akin to estimating an automobile's speed from the time and distance between a few mileposts, knowing nothing about acceleration, coffee breaks, the details of the route, or time spent driving slowly through construction zones.

Given the net change and elapsed time, how we estimate the rate of change depends on how evolution is assumed to proceed between the sampled points. Consider two possible models of the evolution of a trait such as body size. In the first, we assume that the **absolute rate** of evolution in some lineage is constant—for example, body mass at $g$ grams per million years. In this model, the lineage will evolve in $\Delta t$ million years from size $x_1$ to size $x_2$, such that $x_2 = x_1 + g\Delta t$ (Figure 7.3a). The pattern of change will be linear over time. Assuming this model, we would estimate the rate from observed data as $g = (x_2 - x_1)/\Delta t$, or $g = \Delta x/\Delta t$.

In the second model, we assume that the **relative rate** or **proportional rate** of evolution is constant at a value $r$, such as 10 percent per million years. The process is a multiplicative one, in which the evolutionary increments in the trait are larger when the trait is larger. This leads to exponential change in which the lineage evolves from size $x_1$ to size $x_2$, where $x_2 = x_1 \times e^{r\Delta t}$ and $e$ is the base of natural logarithms (Figure 7.3b). Under this model, we would estimate the rate of evolution as $r = \ln(x_2/x_1)/\Delta t$, where ln denotes the natural logarithm. (Unless otherwise noted, logarithms referred to herein are natural logarithms.)

To emphasize proportional rather than absolute differences, traits are measured on a logarithmic scale [SEE SECTIONS 2.3 AND 8.3]. On this scale, an increase in size of 50 percent in a million years would represent the same rate of evolution whether the lineage were evolving from 10 to 15 g or from 100 to 150 kg. Because $\ln(x_2/x_1)$ is equal to $\ln(x_2) - \ln(x_1)$, the rate $r$ can also be expressed as $[\ln(x_2) - \ln(x_1)]/\Delta t$. If we compare this with the estimate of the absolute rate of change, $g = (x_2 - x_1)/\Delta t$, we see that measuring a proportional rate of change is the same as measuring the absolute rate if the trait values have first been transformed to logarithms. In other words, if the trait is measured on a logarithmic scale, a constant proportional rate yields linear change between sampled points (Figure 7.3c). For many biological problems, it makes sense to think of logarithms as a natural scale of measure rather than as a transformation of arithmetic measures.

The validity of the two models of change depicted in Figure 7.3 could not be tested if, as in that figure, the evolving trait were observed at only two points in time. To evaluate the assumed model of evolution, it is necessary to turn to sequences in which the trait is observed in many successive time intervals. For example, Figure 7.4a shows

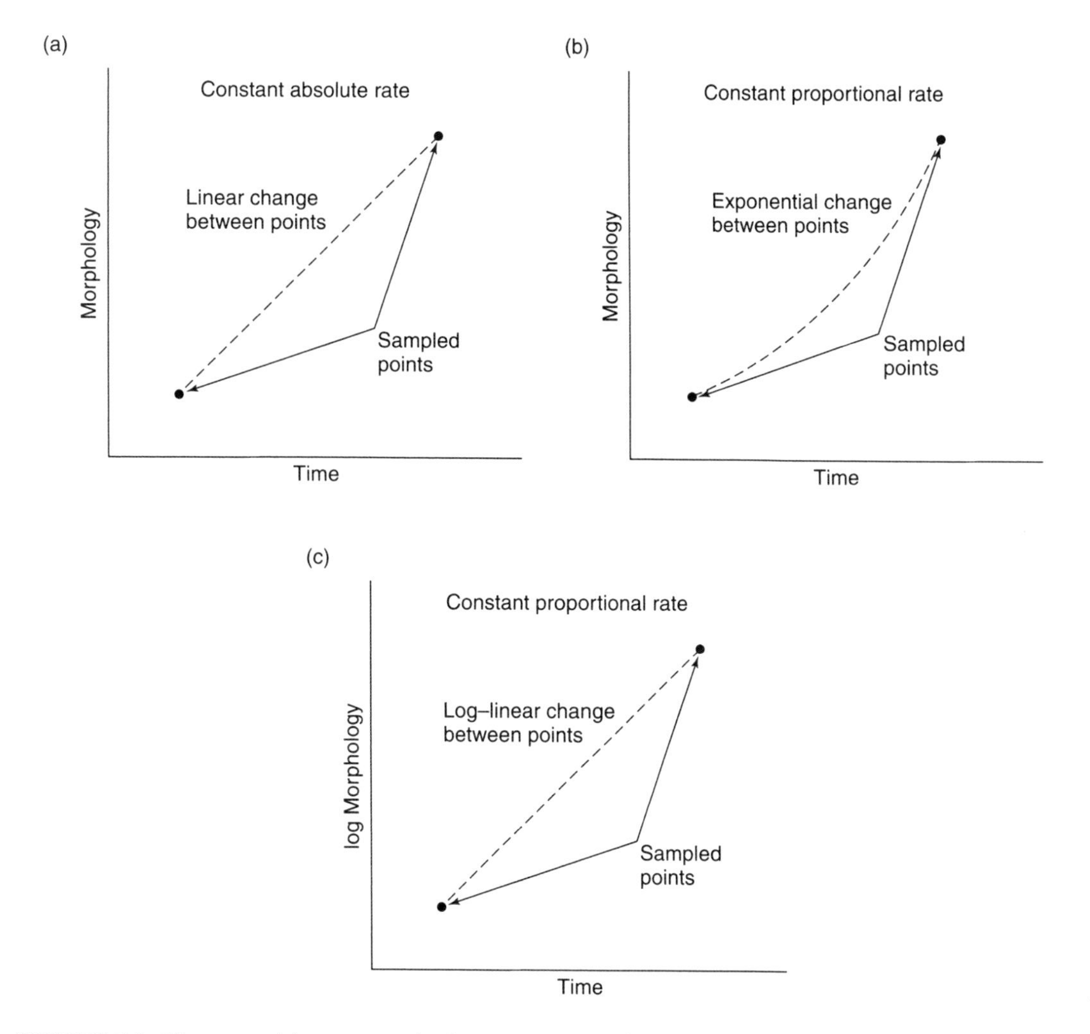


**FIGURE 7.3 Change at (a) constant absolute rate versus (b and c) constant proportional rate.** Change at a constant proportional rate is linear if the trait is measured on a logarithmic scale (c). The dashed line in each graph shows the assumed pattern of change between sampled points.

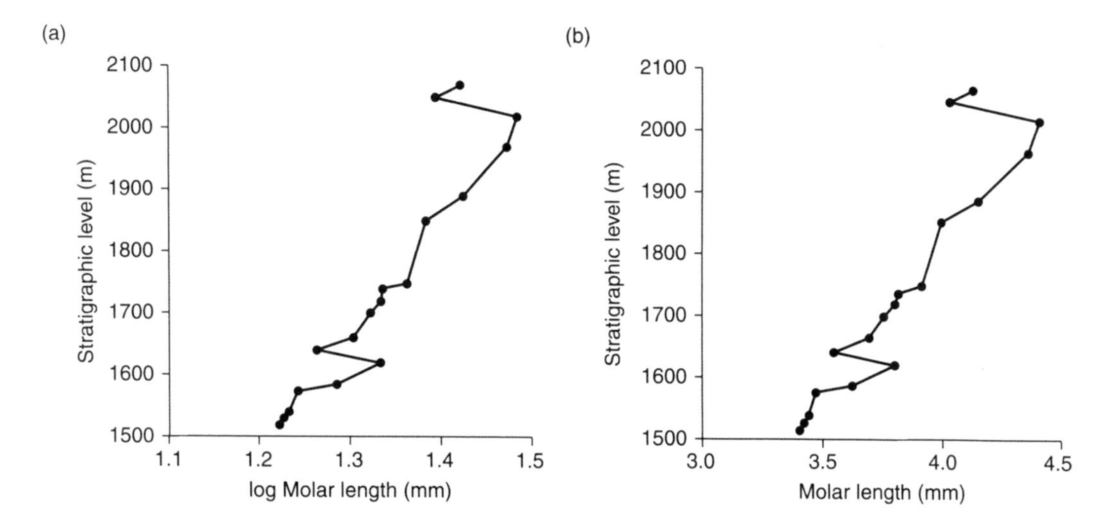


**FIGURE 7.4 Evolution of size, based on the length of the upper right first molar, in a succession of chronospecies of the Eocene primate *Cantius* from Wyoming.** Stratigraphic position is in meters above the base of the section. The elapsed time is about 1.5 m.y. The points show sample means. Part (a) depicts the logarithm of molar length, and part (b) depicts molar length on a linear scale. Because the net change is small, there is little difference between the patterns on logarithmic versus linear scales. *(Data from Clyde & Gingerich, 1994)*

an evolutionary sequence of molar tooth size in the Eocene primate *Cantius* over a span of about 1.5 million years. The sequence consists of a succession of five chronospecies [SEE SECTION 3.3] that make up a single evolutionary lineage. Molar length is measured on a logarithmic scale, and a line could reasonably be fit through the sequence of points. This would seem to justify the model of constant proportional change as opposed to the alternative of constant absolute change. In this instance, however, we cannot say with confidence which model is better supported. This is because the accumulated change is relatively small, with the result that the graph would not look very different if molar size were measured on an arithmetic scale, as in Figure 7.4b.

This example is rather typical. When the proportional rate of change is small and the elapsed time is relatively short, the multiplicative nature of proportional change is not very striking. The difference between arithmetic and logarithmic measures generally becomes important when we study evolution on longer timescales, however, and when we wish to compare the evolution of organisms that vary greatly in size.

The biologist J. B. S. Haldane (1949) suggested a standarized unit of measure for proportional rates of change, the *darwin* (denoted *d*), where one *d* is equal to one natural logarithmic unit per million years (m.y.). For example, a doubling of size in 2 m.y. would correspond to a rate of $\ln(2)/2$, or 0.347*d*. Figure 7.5 shows the variation in rates of size change within a group of related species of Eocene mammals, with a "darwinometer" to allow easy visual assessment of the rates. Even in this rather homogeneous group of species, evolutionary change varies substantially in rate and direction. Such variation is indeed one of the most general results that has come from the study of evolution in the fossil record.

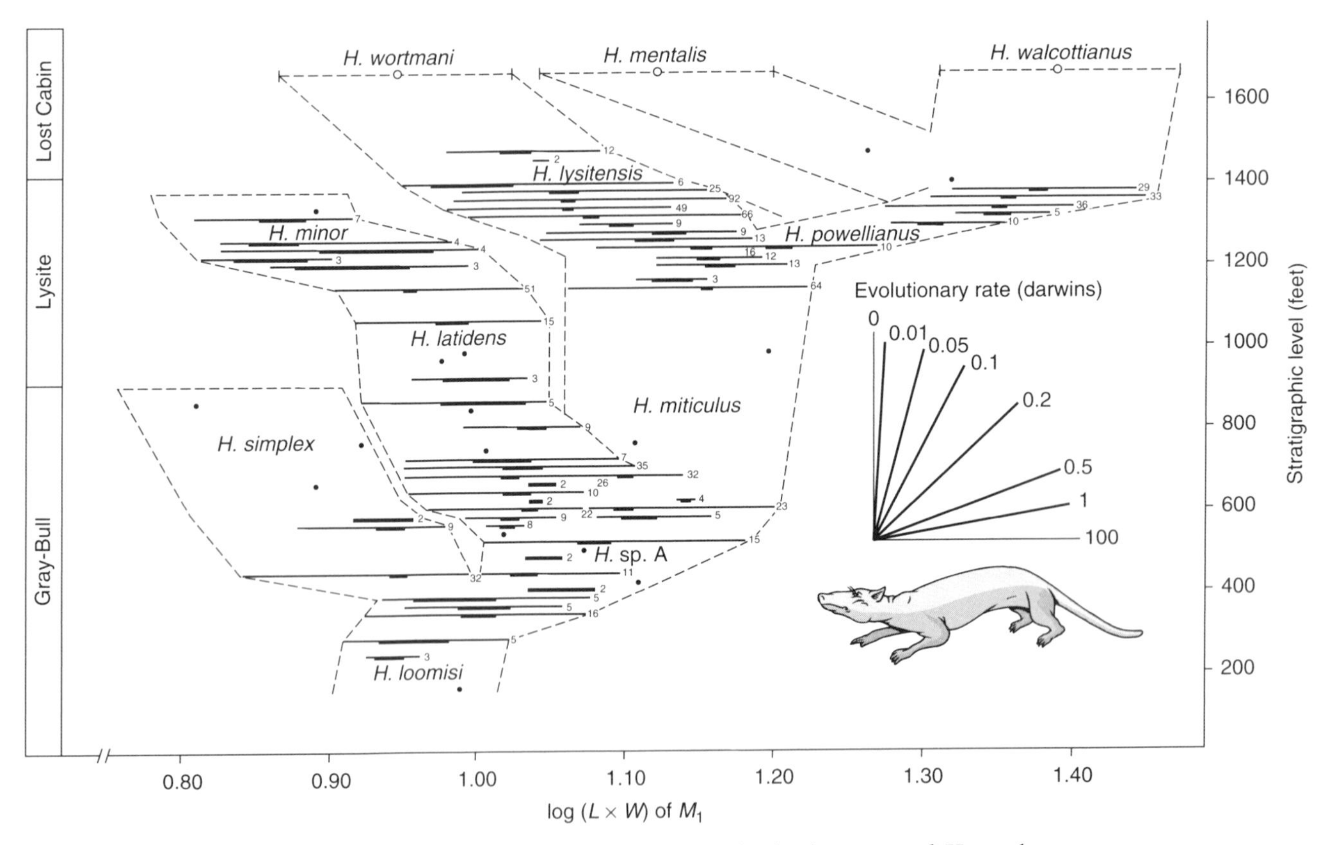


**FIGURE 7.5 Evolutionary change in size of the lower right first molar in the mammal *Hyopsodus* from the Eocene of Wyoming, represented as the base-10 logarithm of the product of length and width (in millimeters).** Stratigraphic position is shown as height in the section. The total elapsed time is about 5 m.y. For each sample, the light bar shows the range of values and the heavy bar the standard error of the mean. Number of specimens is given to the right of the light bar. Solid dots indicate single specimens. Open circles and dashed lines at the top are estimated means and ranges for species poorly sampled in this section but known from elsewhere. Net evolutionary rates can be assessed by comparing the slope of the line connecting two sample means with the reference slopes of the "darwinometer." *(From Gingerich, 1974)*

Two other considerations suggest an alternative measure of evolutionary rate. First, the raw material for evolution is heritable variation within populations [SEE SECTION 3.1]. The greater the variation, the greater is the potential for evolution. Think of two populations that undergo the same change in average size, $\Delta x$, via natural selection. If one of the populations has twice the variation of the other, then it requires only half the selection pressure to move its average by $\Delta x$. Second, the generation is a natural measure of time for evolving populations because new genetic variants and genetic combinations stem largely from gamete formation and reproduction. Paleontologist P. D. Gingerich (1993) has therefore proposed that rates be measured in terms of standard deviations per generation. Thus, the rate of evolution would be expressed as $(\Delta x/s)/(\Delta t/t_g)$, where $s$ is the standard deviation [SEE SECTION 3.2] of the trait value within a fossil population, $\Delta t$ is the elapsed time in years, and $t_g$ is the generation time, also in years.

Rates of evolution on the order of 0.1 standard deviation per generation are typical of modern populations. In practice, generation time is approximated in a number of ways for living organisms, such as the average time span between the birth of an organism and the production of its offspring. For extinct species, generation time is unknown, although it is often possible to approximate it using close living relatives.

We have discussed some assumptions of the three principal ways of measuring morphological rates: as change in arithmetic measures over time, as change in logarithmic measures over time, and as change in standard deviations per generation. There is an additional, tacit assumption underlying all of these measures. By their very nature, they compare net change to elapsed time (Figure 7.2). This approach is ideal only if the rate and direction of change are persistent over the time represented by the net change. Consider a hypothetical evolutionary sequence of body mass: 10 g, 15 g, 10 g, 15 g, 20 g. This sequence shows reversals in the direction of evolution. Thus, the net doubling of size from start to finish does not give a good idea of how rapidly evolutionary change can occur on shorter timescales.

## Temporal Scaling of Morphological Rates

We just saw with a hypothetical example that reversals in the direction of evolution may lead to lower net rates as evolutionary change is studied over longer spans of time. There are abundant empirical data confirming that longer time spans do in fact yield lower net rates on average. Figure 7.6a depicts several hundred net rates measured on timescales ranging from days to millions of years. Evidently, the longer the time span over which rates are measured, the lower is the net rate of change on average. This should not be the case if rate and direction are constant. If the rate and interval data of Figure 7.6a are plotted on logarithmic axes, a linear relationship results, with a slope near $-1$ (Figure 7.6b). These figures depict rates in darwins, but other rate measures yield a similar relationship between net rate and interval length.

The highest rates that have been measured—on the order of 1000 darwins—are for artificial selection experiments in the laboratory and rapid changes in modern populations. These are so large that, if sustained, they could in principle change something the size of a mouse into something the size of an elephant in much less than one million years. Yet sustained changes at such high rates are not in fact observed; net change does not increase in direct proportion to the amount of elapsed time.

Figure 7.6 also shows that very low rates are generally not recorded over short intervals of time. This is because the amount of change produced would be so small as to be essentially undetectable.

There are two principal reasons for the inverse relationship between net rates and the length of time over which they are measured. The first is biological. As already discussed, longer intervals of time tend to incorporate frequent reversals of direction (Figure 7.4). The second is a mathematical necessity. Rate measures have $\Delta t$ in the denominator; that is to say, they are proportional to $1/(\Delta t)$. Thus, unless the elapsed change increases in proportion to $\Delta t$, a plot such as Figure 7.6 will tend to have the form of $\Delta t$ versus $1/(\Delta t)$, which contributes to an inverse correlation. The relative importance of these two factors has not yet been determined; to do so would represent a most significant advance for paleontology.

The dependence of net rates on the length of time over which they are measured implies that meaningful comparisons—such as whether some species evolve more rapidly than others, or whether different morphological traits evolve at different rates—must be based on rates measured over the same amount of elapsed time. If we compare the rate for a fossil species over a million years with that for a living species from one year to the next, we will almost certainly find a lower rate for the fossil species, simply because the net rate is calculated

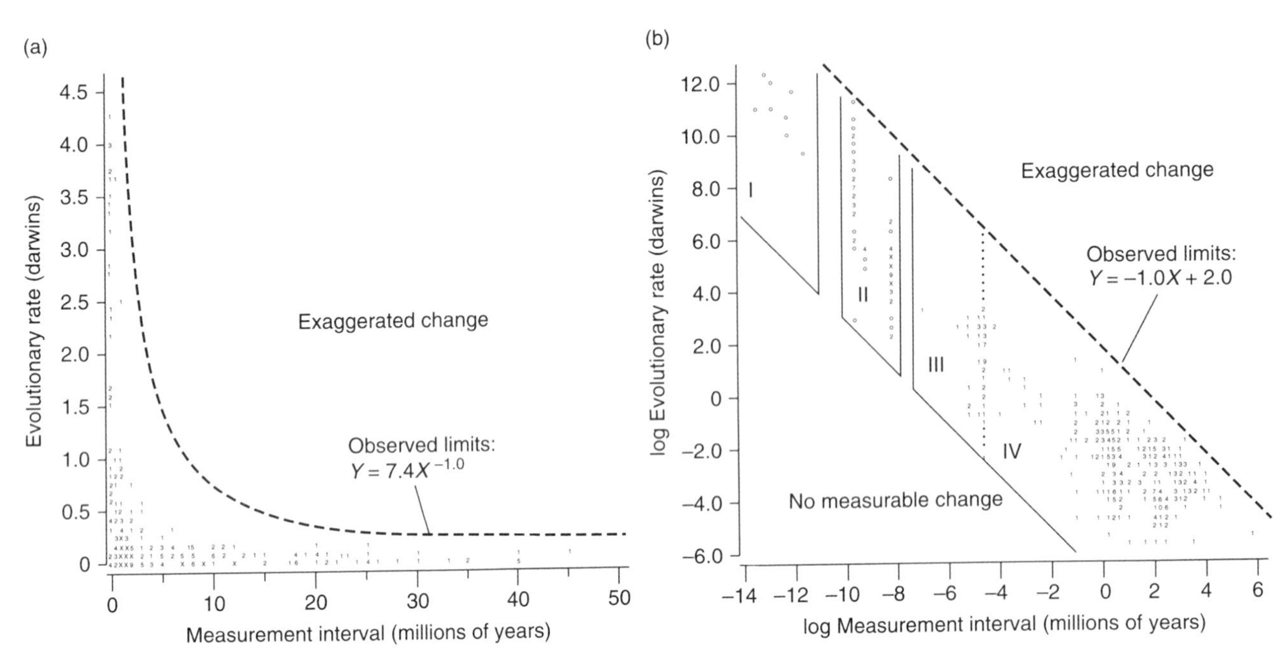


**FIGURE 7.6 The relationship between evolutionary rate in darwins and the interval of time over which the change is measured.** The graphs summarize over 500 calculated rates. Digits indicate the number of observations falling at that location on the graph; $X$ denotes 10 or more observations. (a) The middle part of the whole distribution shown on linear axes. (b) The entire distribution shown on logarithmic axes. The field marked I shows rates for laboratory selection experiments; II is for colonization events in historical time; III is for evolutionary change following the Pleistocene glaciation; and IV is for paleontological observations. *(From Gingerich, 1983)*

over a longer time span. Of course, if we compare coeval lineages from the same sedimentary beds, or different traits within the same lineage, the problem of temporal scaling disappears.

## 7.2 TAXONOMIC RATES

By **taxonomic rates** of evolution, we mean the rates at which new lineages originate and existing lineages become extinct. Consistent with our use of the term *speciation* in Chapter 2, origination refers to the splitting of lineages and extinction to their termination. We will generally speak of taxonomic rates for species, but they can also be studied for more inclusive taxa.

The rate of extinction, typically symbolized $q$, is generally thought of on a **per-capita** basis, meaning that the number of extinctions is scaled to the number of lineages at risk and to the amount of time they are at risk. A useful way to express this rate is as the number of extinction events per lineage-million-years (Lmy). This quantity is analogous to person-hours; five species that each live for 2 m.y. collectively span 10 Lmy, as does a single species that lives for 10 m.y.

Figure 7.7 illustrates this concept with a set of hypothetical lineages over a span of 10 m.y. The solid vertical lines show their durations within this interval, and the broken line indicates a partial duration outside the interval. The sum of Lmy for this interval is 43, and there are 14 extinctions within the interval. The rate of extinction is therefore 14/43, or 0.33 per Lmy. The rate of origination, denoted $p$, is defined in an analogous way as the number of origination events per Lmy. In this example, there are 13 originations, so the origination rate is 13/43, or 0.30 per Lmy.

It is also evident from Figure 7.7 that the mean duration, that is, the sum of durations divided by the number of lineages, is equal to 43/14, or 3.1 m.y. Clearly, this is simply the inverse of the extinction rate. In general, for a group of lineages that are all extinct, the mean duration is equal to $1/q$, and the extinction rate can therefore be estimated as the inverse of the tabulated mean duration.

Species durations of paleontologically important groups are typically on the order of 1 to 10 m.y., but these may vary from less than 100,000 years up to 20 m.y. or more. The typical species extinction rate—the inverse of mean duration—therefore generally varies

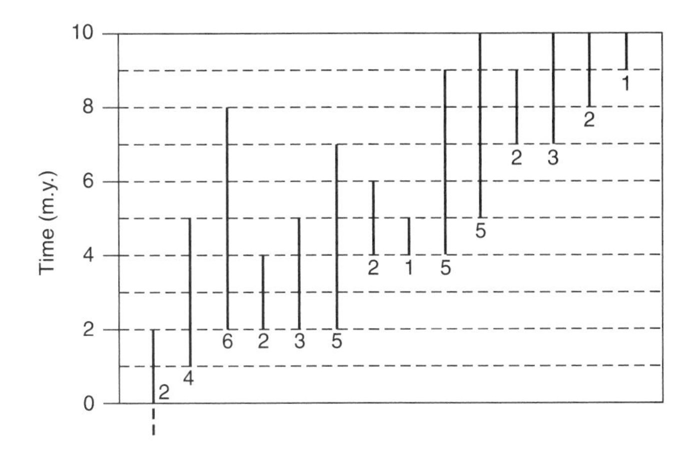


**FIGURE 7.7 Hypothetical species through a 10-m.y. interval of time.** The duration within the interval is shown near the base of each lineage. The number of lineage-million-years (Lmy) in this interval is the sum of durations, in this case, 43 Lmy. The dotted line at the base of the leftmost lineage indicates that it existed prior to this interval of time; this part of the duration is not counted in the tabulation. There are 13 originations and 14 extinctions in this interval. Thus, the origination and extinction rates would be 13/43 and 14/43, or 0.30 per Lmy and 0.33 per Lmy.

from about 0.1 per Lmy to 1.0 per Lmy. Genus durations typically vary from about 5 to 50 m.y., but there are long-lived genera that span well over 100 m.y.

If the per-capita extinction rate is the same for all species within a group and is also constant over time, then a constant *proportion* of the standing crop of species, rather than a particular absolute number of them, will become extinct in every unit increment of time. The proportion of species that survive *at least* to a duration of $T$ will be equal to $e^{-qT}$. This fundamental relationship of **exponential survivorship** is depicted schematically in Figure 7.8a, which shows the duration on the $x$ axis and the expected proportion of species with at least the given duration on the $y$ axis. If the $y$ axis is scaled logarithmically (Figure 7.8b), then a straight-line relationship results, and the magnitude of the slope of this line is equal to the extinction rate. This is analogous to the process of radioactive decay: If the amount of remaining parent material is plotted semilogarithmically against time, a linear relationship results, and its slope gives the decay constant.

Because distributions of durations tend to be skewed, with many short-lived taxa and a few that are long-lived, it is good practice to tabulate the median duration in addition to the mean [SEE SECTION 3.2]. Exponential survivorship implies that the median duration, or **half-life,** is equal to $\ln(0.5)/q$. Thus, another way of estimating the extinction rate is as $\ln(0.5)/T_{1/2}$, where $T_{1/2}$ is the tabulated median duration. This approach to rate estimation and most of the methods that we present below are underlain by the model of exponential survivorship.

## Long-Term Characteristic Rates for a Biologic Group

Methods for estimating taxonomic rates of evolution all depend to some extent on **taxonomic survivorship analysis,** the statistical study of the durations of taxa. Many approaches to survivorship analysis have been developed. We present just a few simple and general methods as examples.

In contrast to the view in Figure 7.7, we often know only the broad intervals of first and last appearance of each

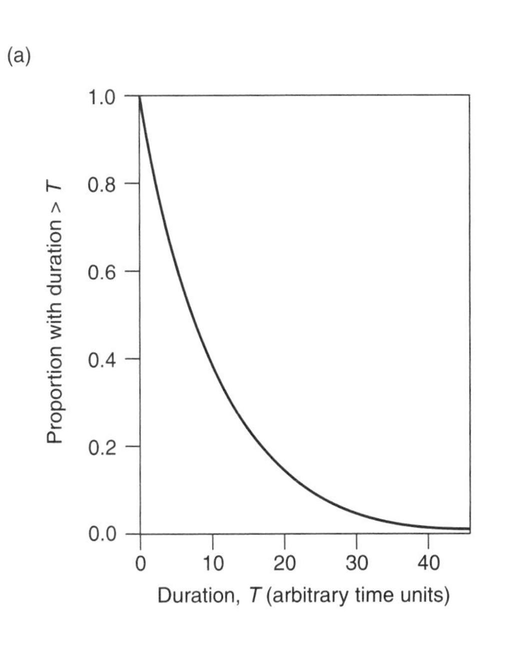


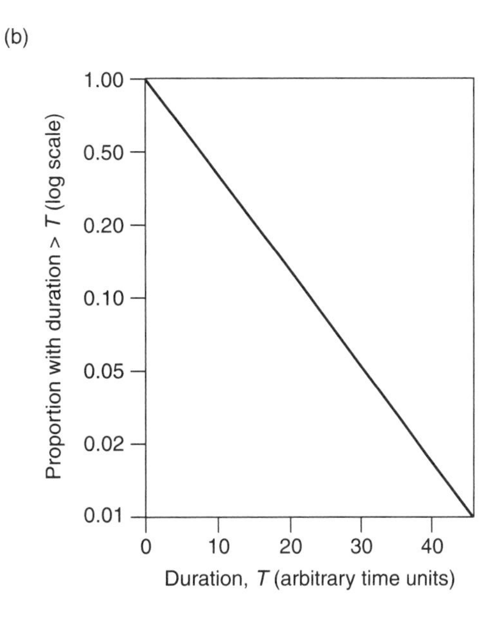


**FIGURE 7.8 Exponential survivorship.** The proportion of species with duration at least equal to $T$ declines exponentially with $T$. This gives a hollow curve in (a) linear coordinates, or a straight-line relationship when proportion of species is expressed (b) logarithmically. The magnitude of the slope of the line in part (b) is the extinction rate.

lineage, not precisely when during each interval these events took place. This allows us to specify only a range within which the actual duration falls. Fortunately, several survivorship methods take advantage of the fact that minimum duration can often be specified accurately.

***Dynamic Survivorship Analysis*** As we have already seen, the probability that a species will live at least to a duration $T$ is equal to $e^{-qT}$ (Figure 7.8). This reflects what is known as the cumulative probability distribution: the proportion of species with a duration of *T or longer.*

To use this relationship to estimate rates, we first tabulate species durations and construct from them the observed frequency distribution in the form of a **survivorship table.** Table 7.1 depicts survivorship data for species of North American Cenozoic mammals. This represents an unusual example, because the durations are specified to the nearest million years; temporal resolution is typically coarser than this. The data are depicted in two ways: as the number of species with a given duration and as the cumulative number with at least that duration. We express the frequencies on a logarithmic scale, then fit a line to the cumulative distribution (Figure 7.9a). Here the slope is $-0.458$; that is, the extinction rate is 0.458 per Lmy. The corresponding mean duration is 2.18 m.y. Such a figure would generally be reported in rough terms as 2.2 m.y. or even simply 2 m.y.

Since the expected proportion of species with duration $T$ or longer is equal to $e^{-qT}$, the expected proportion with duration between $T_1$ and $T_2$ is equal to $e^{-qT_1} - e^{-qT_2}$. This is referred to as the *differential proportion, raw proportion,* or simply the *proportion* to distinguish it from the cumulative proportion. An important property of exponential survivorship is that both the differential and cumulative proportions are expected to yield log–linear relationships with slope equal to the extinction rate. Fitting a line to the raw proportions in Table 7.1 yields a slope of $-0.476$, which is not very different from the value of $-0.458$ for the cumulative proportions.

Figure 7.10 shows another example of dynamic survivorship analysis, here for Paleozoic crinoid genera. The genera are divided into two groups: subclass Camerata, all of which possess fine extensions of the arms called *pinnules*; and nonpinnulate members of the remaining Paleozoic subclasses. This division was chosen to reflect an important functional difference. Crinoids are filter-feeders [SEE SECTION 5.2]; pinnulate crinoids, because they form finer filtration fans, depend on higher current speed to feed effectively. Nonpinnulate crinoids, on the other hand, can feed in either fast or slow currents. In effect,

TABLE 7.1

Survivorship Table for North American Cenozoic Mammal Species

| Duration (m.y.) | Number | Proportion | Cumulative Number | Cumulative Proportion |
|---|---|---|---|---|
| 0–1 | 1718 | 0.584 | 2941 | 1.000 |
| 1–2 | 429 | 0.146 | 1223 | 0.416 |
| 2–3 | 331 | 0.113 | 794 | 0.270 |
| 3–4 | 200 | 0.068 | 463 | 0.157 |
| 4–5 | 91 | 0.031 | 263 | 0.089 |
| 5–6 | 64 | 0.022 | 172 | 0.058 |
| 6–7 | 38 | 0.013 | 108 | 0.037 |
| 7–8 | 31 | 0.011 | 70 | 0.024 |
| 8–9 | 14 | 0.0048 | 39 | 0.013 |
| 9–10 | 6 | 0.0020 | 25 | 0.0085 |
| 10–11 | 7 | 0.0024 | 19 | 0.0065 |
| 11–12 | 2 | 0.0007 | 12 | 0.0041 |
| 12–13 | 5 | 0.0017 | 10 | 0.0034 |
| 13–14 | 3 | 0.0010 | 5 | 0.0017 |
| 14–15 | 1 | 0.0003 | 2 | 0.0007 |
| 16–17 | 1 | 0.0003 | 1 | 0.0003 |

*SOURCE:* Alroy (1994b)

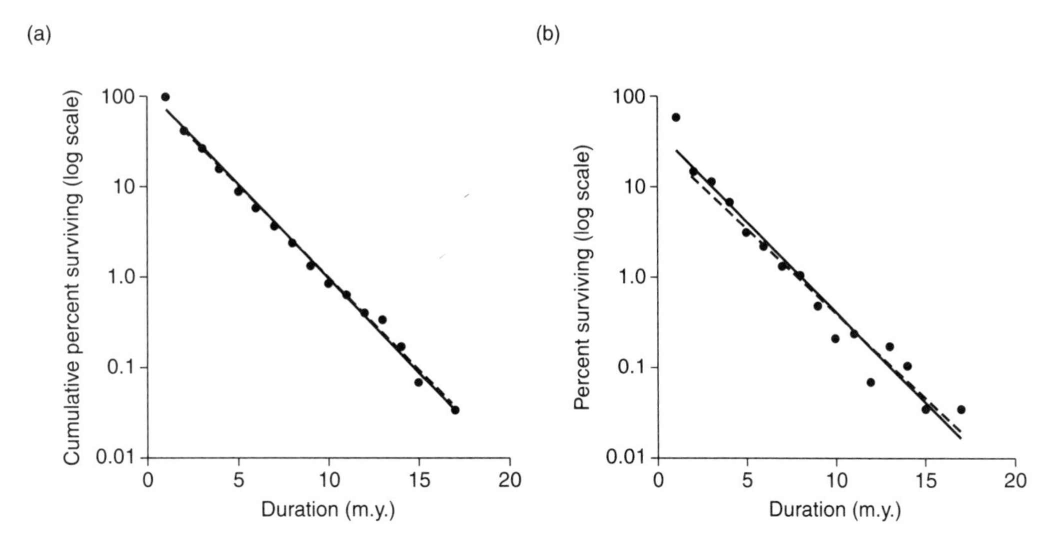


**FIGURE 7.9 Dynamic survivorship analysis of Cenozoic mammal species from North America.** Data are proportions from Table 7.1, converted to percentages. (a) Points show the percent of species with duration at least as long as the given value. (b) The percent with duration equal to the given value. The solid lines are fitted to the survivorship data; their slopes give estimates of the extinction rate. The dashed lines are fitted to all points except the left-most; these are explained in Box 7.2. *(Data from Alroy, 1994b)*

then, the pinnulate crinoids are ecologically more specialized. The survivorship curves show that nonpinnulate crinoid genera have an average duration nearly twice that of camerates—a difference that has been interpreted to reflect the degree of ecological specialization. The camerates, with stricter environmental requirements than nonpinnulates, may be more susceptible to extinction due to the loss of their preferred habitats.

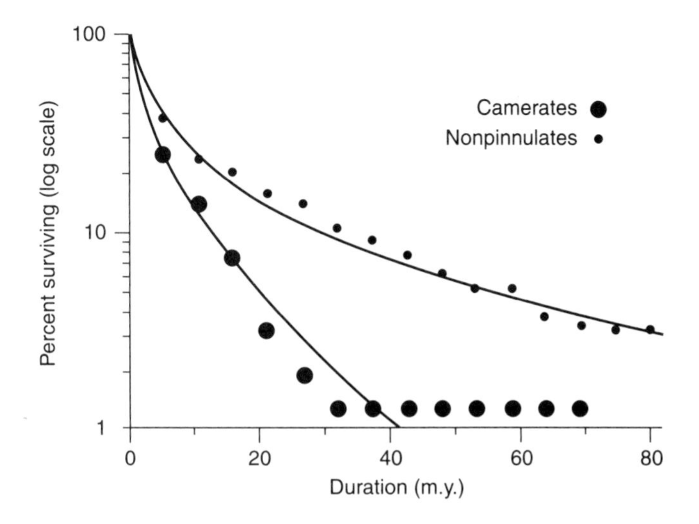


**FIGURE 7.10 Dynamic survivorship analysis of Paleozoic crinoid genera.** There are 245 genera of camerates and 324 genera of nonpinnulate crinoids. The points give percent of genera with duration at least as long as the time indicated, and curves are fitted to these points. Curves are nonlinear, indicating that the probability of genus extinction is not constant through the lifetime of a genus. *(From Baumiller, 1993)*

The curves in Figure 7.10 exhibit a feature often seen in the survivorship of genera and other taxa above the species level. These survivorship curves are apparently not log–linear. They become shallower with increasing duration, which means that the probability of extinction per unit time interval decreases as a genus becomes older. In other words, the longer a genus lives, the greater its chance of surviving yet another increment of time (Box 7.1). This pattern of age-dependent survivorship largely reflects the fact that older genera tend on average to contain more species and therefore to be less susceptible to extinction.

The method of dynamic survivorship analysis assumes that the full distribution of durations can be sampled. One of the main violations of this assumption occurs when the taxonomic group is young. A group that originated 10 m.y. ago, for example, can only have species with durations between 0 and 10 m.y. Also, as we discuss in Chapter 8, certain intervals of geologic time are marked by unusually high rates of extinction. Mass extinction events, as well as the periods of rapid origination that follow them, can keep the duration distribution from reaching a stable point.

***Cohort Survivorship Analysis*** We just saw that dynamic survivorship ideally requires a distribution of taxonomic ages that is not changing over geologic time: For the sake of analysis, all taxa with a given duration are classed together regardless of when they lived. If the age distribution changed over time, this could present problems for the dynamic method. We can avoid these

*Box 7.1*

## GENUS SURVIVORSHIP

Genus-level survivorship curves such as those of Figures 7.10 and 7.12 are typically nonlinear even on a semilogarithmic graph. The expected form of this nonlinearity can be determined with some simple mathematical modeling (see Raup, 1985). Assume that each genus originates with a single species at time $t = 0$, and assume that the species-level extinction rate ($q$) and the rate of origination of new species within the genus ($p$) are both constant. This last rate ignores those speciation events that give rise to new genera. Let $P_{s,T}$ be the probability that a genus survives *at least* until time $t = T$. Then

$$P_{s,T} \begin{cases} \dfrac{1}{1 + pT} & \text{if } p = q \\ \dfrac{(p - q)e^{(p-q)T}}{pe^{(p-q)T} - q} & \text{if } p \neq q \end{cases}$$

In practice, genus survivorship curves such as those of Figures 7.10 and 7.12 are produced with specialized curve-fitting techniques that yield the best-fit values of $p$ and $q$ (see Foote, 1988). Thus, by assuming a model of evolution—namely, that genera are monophyletic or paraphyletic and that species rates are constant—we are able to infer species-level rates even though the observed data are resolved only to the genus level.

problems by not combining taxa that lived at different times; this is the rationale behind **cohort survivorship analysis.** A cohort is a group of species or other taxa, all of which originated during a given time interval. A cohort is followed forward over geologic time, and the number of taxa still remaining is monitored as a function of elapsed time. Under the model of exponential survivorship, the expected proportion of taxa still extant at a time $T$ after the origin of the cohort is equal to $e^{-qT}$. Therefore, the slope of a straight-line fit to a cohort curve on a semilog graph gives an estimate of the extinction rate. A different form of survivorship curve is expected for taxa above the species level (Box 7.1).

Table 7.2 presents just the first five lines of a cohort survivorship table for Cenozoic mammals; the data here are from the same source as Table 7.1. For example, of the

**TABLE 7.2**

Partial Cohort Survivorship Table for North American Cenozoic Mammal Species

| Interval of Cohort Origin | Cohort Size | Number Extinct during Interval | | | | | | | | | | Number Surviving at End of Interval | | | | | | | | | |
|---|---|---|---|---|---|---|---|---|---|---|---|---|---|---|---|---|---|---|---|---|---|
| | | 65 | 64 | 63 | 62 | 61 | 60 | 59 | 58 | 57 | 56 | 65 | 64 | 63 | 62 | 61 | 60 | 59 | 58 | 57 | 56 |
| 65 | 54 | 24 | 26 | 4 | | | | | | | | 30 | 4 | 0 | | | | | | | |
| 64 | 31 | | 19 | 11 | 1 | | | | | | | | 12 | 1 | 0 | | | | | | |
| 63 | 41 | | | 19 | 3 | 4 | 11 | 2 | 2 | | | | | 22 | 19 | 15 | 4 | 2 | 0 | | |
| 62 | 49 | | | | 20 | 13 | 7 | 7 | 1 | 1 | | | | | 29 | 16 | 9 | 2 | 1 | 0 | |
| 61 | 71 | | | | | 38 | 12 | 6 | 13 | 1 | 1 | | | | | 33 | 21 | 15 | 2 | 1 | 0 |

*SOURCE:* Alroy (1994b)

*NOTE:* Interval 65 spans from 65 to 64 million years ago, and so on.

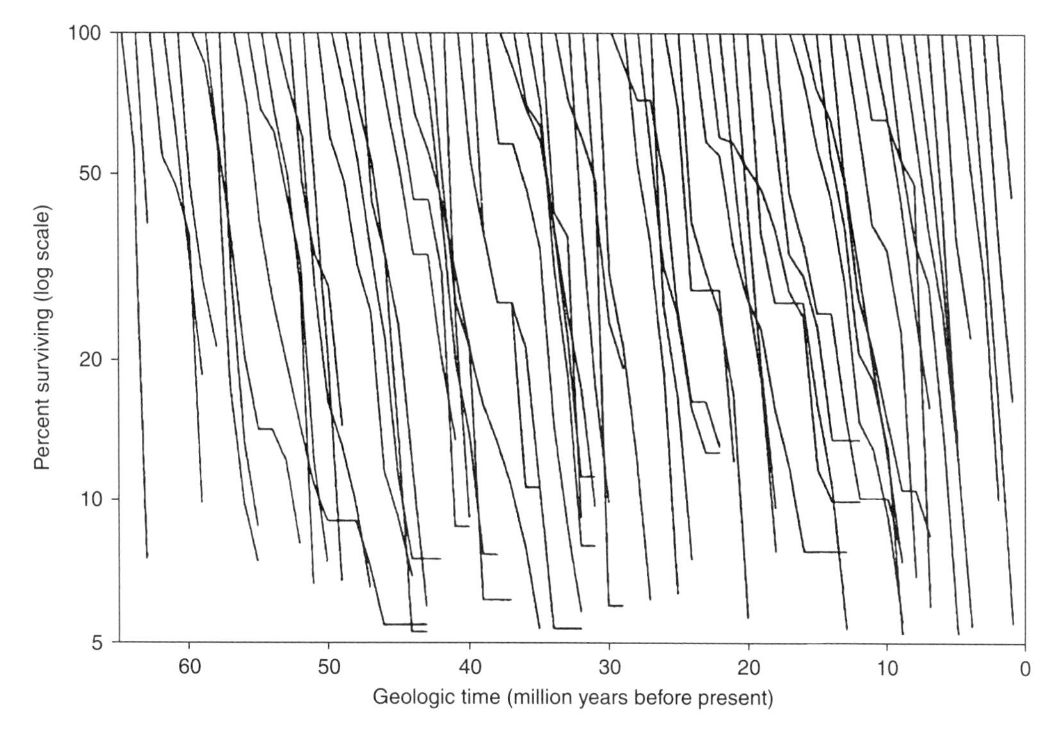


**FIGURE 7.11 Cohort survivorship curves for the species of Cenozoic mammals analyzed in Figure 7.9.** Each cohort consists of the species first appearing within a 1-m.y. increment of time. The proportion of species in each cohort that still survive is tracked forward through time. If this proportion is plotted on a log scale, the slopes of the cohort curves give the extinction rate. Curves are arbitrarily truncated at 5 percent survival, but many continue further than this. *(Data from Alroy, 1994b)*

41 species that first appear between 62 and 63 m.y. ago, 19 became extinct before 62 m.y. ago, 3 became extinct between 61 and 62 m.y. ago, 4 became extinct between 60 and 61 m.y. ago, and so on. In terms of cumulative survivorship, 22 species (54 percent) endured at least 1 m.y., 19 (46 percent) endured at least 2 m.y., 15 (37 percent) endured at least 3 m.y., and so on. Figure 7.11 shows the cohort survivorship curves corresponding to the cumulative data. If lines are fitted to these curves, they yield slopes ranging from $-0.2$ to $-1.3$, with a median of $-0.5$. This agrees well with the extinction rate of 0.46 to 0.48 per Lmy obtained earlier with the dynamic method (Figure 7.9).

Examples of genus-level cohort curves are shown in Figure 7.12. As with the dynamic curves of Figure 7.10, most of the curves become shallower as time goes on. Using the approach outlined in Box 7.1, it is possible to estimate species-level origination and extinction rates from the survivorship of genera. The speciation rate was about 0.40 per Lmy and the species extinction rate about 0.46 in trilobite genera that originated in the Cambrian; these rates were about 0.13 and 0.15 for genera that originated in the Ordovician. Based on the survivorship of genera, it is inferred that Ordovician species were about three times as long lived as Cambrian species.

Figure 7.12 shows another feature that is readily detected with cohort analysis. The sudden downturn of many cohort curves indicates a profound extinction event in the latest Ordovician [SEE SECTION 8.6], marked by the arrow. A temporal pattern such as this could not have been seen if all durations had been combined into a single distribution for dynamic survivorship analysis.

***Lyellian Proportions*** Charles Lyell, in his classic *Principles of Geology* (1833), tabulated the number of molluscan species known from various epochs of the Cenozoic and how many of these are still extant today. Of the species sampled at some census point in the past, the proportion still extant today is the **Lyellian proportion** (commonly expressed as a percentage). Exponential survivorship implies that the expected Lyellian proportion at an age $T$ before the Recent, $L_T$, is equal to $e^{-qT}$. Thus, the average extinction rate over the span of time is estimated as $q = -\ln(L_T)/T$.

The dynamic and cohort methods consider the entire frequency distribution of durations. Any one of the points

(a)

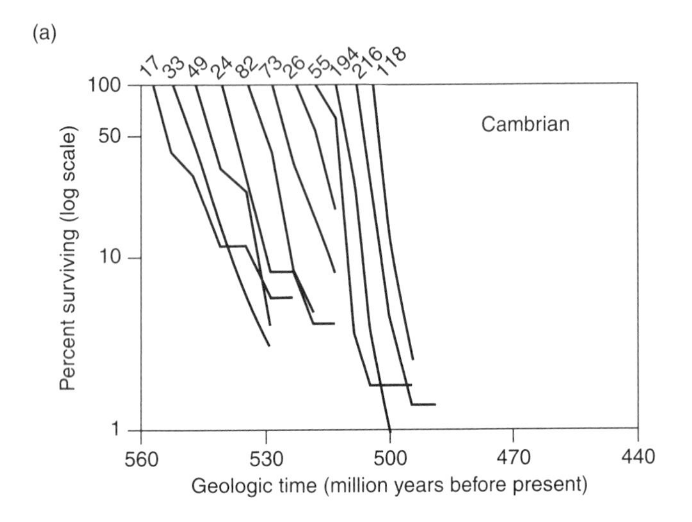


(b)

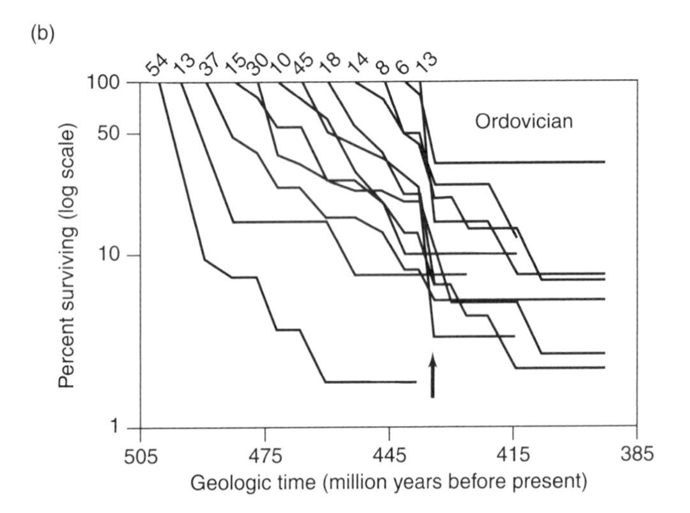


**FIGURE 7.12 Cohort survivorship curves for trilobite genera, divided into those first appearing during the Cambrian and those first appearing during the Ordovician.** Cambrian curves are steeper, indicating shorter genus durations. The arrow in the bottom figure marks the mass extinction at the end of the Ordovician. *(From Foote, 1988)*

in the distribution may be inaccurate because of sampling error, but these errors are averaged out by estimating a single rate for the whole distribution. By contrast, each Lyellian proportion is a single observation, rather than an entire distribution, and it may therefore yield an extinction rate that is not representative. For this reason, a number of Lyellian proportions will typically be computed. These may represent, for example, geographically distinct faunas of a given age or a series of faunas of different ages.

Figure 7.13 shows Lyellian proportions and ages for a number of Cenozoic molluscan faunas from Japan and California. The proportions for selected faunas are shown in Table 7.3, divided into bivalves and gastropods from the same fauna. There is some variation in the extinction rates calculated from these proportions, but it is clear that gastropods on average have had higher extinction rates than bivalves over the past 20 m.y. This is also evident from the fact that bivalves generally lie above gastropods in the plot of Lyellian proportions versus time (Figure 7.13). One postulated explanation for this difference is that bivalve species have broader geographic ranges and are therefore less susceptible to extinction by environmental perturbations (Stanley et al., 1980).

***Incompleteness of the Fossil Record*** Our treatment of survivorship analysis has implicitly assumed that the stratigraphic range of a taxon is a good proxy for its true duration. In reality, ranges are truncated by incomplete sampling [SEE SECTIONS 1.3 AND 6.1]. This makes extinction rates appear artificially high, and the effect is stronger as sampling becomes less complete. Lyellian proportions circumvent this problem by using the present day, where sampling is nearly complete, as the second census point for taxa known to be present at an earlier time. Another approach to the problem is given in Box 7.2.

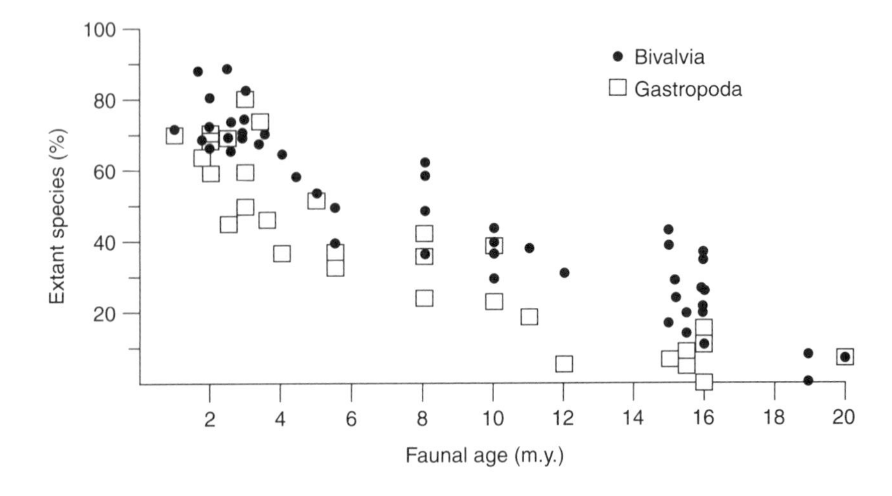


**FIGURE 7.13 Lyellian percentages for Neogene molluscan faunas from California and Japan, separated into bivalves and gastropods.** Each point shows the percent of species in the given fauna that are still extant today. Corresponding extinction rates for selected pairs of points are given in Table 7.3. *(From Stanley et al., 1980)*

TABLE 7.3

Lyellian Proportions for Species within Selected Neogene Molluscan Faunas

| | Lyellian Proportion | | Extinction Rate (per Lmy) | |
|---|---|---|---|---|
| Age of Fauna (m.y.) | Bivalves | Gastropods | Bivalves | Gastropods |
| 2 | 0.67 | 0.60 | 0.20 | 0.26 |
| 4 | 0.65 | 0.35 | 0.11 | 0.26 |
| 8 | 0.38 | 0.42 | 0.12 | 0.11 |
| 10 | 0.46 | 0.25 | 0.08 | 0.14 |
| 12 | 0.30 | 0.05 | 0.10 | 0.25 |
| 16 | 0.38 | 0.15 | 0.06 | 0.12 |
| 20 | 0.08 | 0.08 | 0.13 | 0.13 |

*SOURCE:* Stanley et al. (1980)

*Box 7.2*

## ESTIMATING TAXONOMIC RATES WITH INCOMPLETE SAMPLING

Figure 7.14 shows an ideal frequency distribution of stratigraphic ranges corresponding to the same simple model of sampling presented in Chapter 1 (Box 1.3, Figure 1.19). Extinction rate is assumed constant—in other words, survivorship is exponential. The probability of sampling a lineage in any given time interval is assumed to be less than 100 percent and to be constant throughout the history of the group in question. The frequency distribution of Figure 7.14 closely resembles the exponential case we have been considering all along—corresponding to 100 percent sampling—but there is one important difference. Here the distribution is log–linear, except that there is an excess of taxa with a stratigraphic range of one interval of time.

The proportion of such **singleton** taxa is a measure of the incompleteness of sampling; groups with less complete records will tend to have a higher proportion of singletons [SEE SECTION 1.3]. According to our simple model, if the singletons are disregarded, the remainder of the distribution is log–linear, and its slope gives the extinction rate. Therefore, we should ideally be able to fit a line to a distribution of stratigraphic ranges—*disregarding singletons*—to obtain an estimate of extinction rate that is not biased by incomplete sampling.

Dynamic survivorship data for mammal species were presented in Table 7.1 and Figure 7.9. The dashed lines in that figure are fitted to the survivorship data, omitting single-interval species. The lines through the raw and cumulative proportions yield extinction rates of 0.434 and 0.437 per Lmy. These are lower than the rates obtained earlier—namely, 0.458 and 0.476 per Lmy. This difference is in the direction we would expect; apparent rates are higher if the effect of incomplete sampling is not taken into account. In this case, however, paleontological completeness is comparatively high, thus the bias in average duration is small.

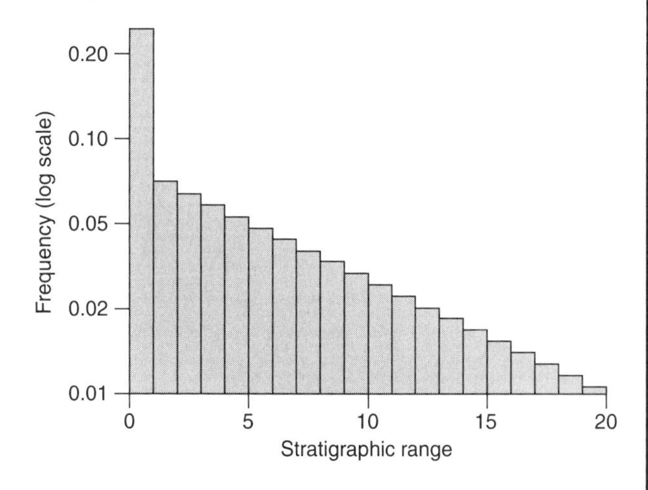


**FIGURE 7.14 Distribution of stratigraphic ranges of species, assuming incomplete but uniform sampling.** If single-interval species are disregarded, the rest of the distribution is log–linear with slope equal to the extinction rate. *(After Foote & Raup, 1996)*

***Long-Term Rates of Origination*** How far forward a species endures from some specified time, such as its origin or any other census point, depends on the rate of extinction. Similarly, how far backward in time a species extends depends on the rate of origination. Therefore, the principle behind cohort survivorship analysis can also be applied in reverse. The taxa that became extinct in a given interval of time are followed backward to determine what proportion of them had already originated by a given time. Under the assumption of constant origination rate, these "reverse survivorship" curves are log–linear with a slope equal to the origination rate.

We have focused in this section on long-term rates of extinction that characterize different groups of organisms. We have largely bypassed the measurement of orig-

*Box 7.3*

## TAXONOMIC RATE MEASURES WITHIN A TIME INTERVAL

All approaches begin with tabulations of taxa within an interval of time. Figure 7.15 shows a hypothetical time interval, with a number of taxa that are extant during all or part of the interval. These taxa fall into four categories: (1) those that make their first appearance sometime before the time interval and their last appearance during the interval; (2) those that make their first appearance during the interval and their last appearance afterwards; (3) those that make both their first and last appearance during the interval; and (4) those that make their first appearance before the interval and their last appearance afterwards.

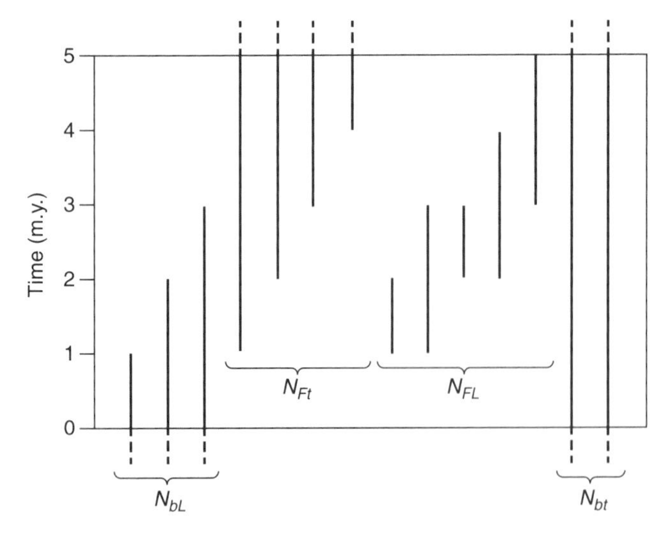


**FIGURE 7.15 Stratigraphic ranges of hypothetical species relative to a time interval of interest.** Dashed lines indicate ranges that extend beyond this interval. Species fall into four categories, depending on whether they cross the bottom ($b$) or top ($t$) interval boundary, and whether they have a first ($F$) or last ($L$) appearance within the interval.

Using the subscripts $b$ and $t$ to refer to taxa that cross the bottom and top boundaries of the time interval, and using $F$ and $L$ to refer to first and last appearance, we can denote the numbers of taxa in the four categories as $N_{bL}$, $N_{Ft}$, $N_{FL}$, and $N_{bt}$. The number of taxa making a first appearance in the interval is equal to $N_F = N_{Ft} + N_{FL}$ and the number making a last appearance is equal to $N_L = N_{bL} + N_{FL}$. The number alive at the beginning of the interval is equal to $N_b = N_{bL} + N_{bt}$, and the number alive at the end of the interval is equal to $N_t = N_{Ft} + N_{bt}$. Finally, the total diversity for an interval, including taxa sampled from the interval as well as those whose presence is inferred only because they are known both before and after the interval, is equal to $N_{\text{tot}} = N_{bL} + N_{Ft} + N_{FL} + N_{bt}$.

As discussed in Chapter 6, times of first and last appearance do not generally coincide with times of true origination and extinction, but let us consider how we would treat the data at hand with the assumption that first and last appearances represent a reasonable proxy for originations and extinctions.

The intensity of extinction could be measured simply by $N_L$, the number of last appearances (Table 7.4). This is problematic when we compare extinction intensity among different intervals of geologic time; we may well expect the number of extinctions to be greater when standing diversity is higher, simply because there are more taxa at risk. The simple number of extinctions is therefore seldom used, and it is much more common to measure the proportional extinction, $P_E$, defined as the number of extinctions divided by the total number of taxa in the time

ination rates because, on long timescales, rates of origination and extinction within a group are nearly equal to one another. Any group that is entirely extinct has had, by definition, equal origination and extinction rates on average. Moreover, the diversification process, like compound interest, yields multiplicative growth [SEE SECTION 8.4]. Thus, groups with many species have generally attained high diversity via a small excess of origination over extinction compounded over a long period of time.

### Interval Rates

In contrast to average rates over the entire history of a taxonomic group, origination and extinction rates within a single interval of time often differ profoundly. Box 7.3 discusses the estimation of rates for a discrete time interval such as a stratigraphic stage.

**TABLE 7.4**

Taxonomic Counts and Taxonomic Rate Measures (refer to Figure 7.15)

| Quantity | Symbol | Equivalence |
|---|---|---|
| Taxa at beginning of interval | $N_b$ | $N_{bL} + N_{bt}$ |
| Taxa at end of interval | $N_t$ | $N_{Ft} + N_{bt}$ |
| Total diversity in interval | $N_{tot}$ | $N_{bL} + N_{Ft} + N_{FL} + N_{bt}$ |
| Number of first appearances | $N_F$ | $N_{Ft} + N_{FL}$ |
| Number of last appearances | $N_L$ | $N_{bL} + N_{FL}$ |
| Proportional origination | $P_O$ | $N_F/N_{tot}$ |
| Proportional extinction | $P_E$ | $N_L/N_{tot}$ |
| Proportional origination per m.y. | $P_{Om.y.}$ | $P_O/\Delta t$ |
| Proportional extinction per m.y. | $P_{Em.y.}$ | $P_E/\Delta t$ |
| Per-capita origination rate per Lmy | $p$ | $-\ln(N_{bt}/N_t)/\Delta t$ |
| Per-capita extinction rate per Lmy | $q$ | $-\ln(N_{bt}/N_b)/\Delta t$ |

interval, $N_L/N_{tot}$. This is commonly expressed as percent extinction, $100 \times N_L/N_{tot}$. In the hypothetical case of Figure 7.15, the number of extinctions is 8, and the percent extinction is $100 \times 8/14$, or 57 percent. The corresponding numbers for origination are 9 and 64 percent.

Proportional extinction has the advantage of explicitly accounting for the number of taxa at risk. However, if extinction occurs fairly continuously throughout a time interval, then we expect more extinctions and a higher proportional extinction in longer time intervals, even if the extinction rate is constant. This expectation leads to the measurement of extinction as proportional extinction per million years, $N_L/N_{tot}/\Delta t$. This measure is also known as the *per-taxon rate of extinction,* which in Figure 7.15 is equal to 57 percent/5 m.y., or 11.4 percent per m.y. The corresponding origination rate is 12.9 percent per m.y.

Although percent extinction and the per-taxon rate are widely used, they have some undesirable properties that stem from the way they account for diversity. Because not all taxa present during a time interval exist throughout the entire interval, the total diversity of an interval of time overestimates the number at risk at any given time. This problem has been addressed by our last rate measure, which takes a census of taxa at precise moments in time—namely, the beginning and end of the interval—and which is based explicitly on the model of exponential survivorship.

*continued on next page*

*Box 7.3 (continued)*

Consider the lineages that are alive at the start of the interval; there are $N_b$ of these. They are at risk of extinction during the interval, and the higher the extinction rate, the smaller the number of lineages that will survive to the end of the interval. The proportion of lineages that survive through the entire interval will be equal to $e^{-q\Delta t}$. Note that this proportion is equal to $N_{bt}/N_b$. Thus, under our ideal model of survivorship, $N_{bt}/N_b = e^{-q\Delta t}$. We can rearrange this equation to solve for the **per-capita extinction rate** as $q = -\ln(N_{bt}/N_b)/\Delta t$. This is just like estimating an extinction rate with the Lyellian method. Here the ratio $N_{bt}/N_b$, which is the proportion of taxa found at a census point that are still extant at a later census point, is analogous to the Lyellian proportion.

To estimate origination rate, we use the reverse survivorship of lineages alive at the end of the interval. The proportion of them that were already extant at the start of the interval is equal to $N_{bt}/N_t$, which, under our ideal model, is equal to $e^{-p\Delta t}$. Thus, we solve for the **per-capita origination rate** as $p = -\ln(N_{bt}/N_t)/\Delta t$.

In Figure 7.15, $N_{bt}/N_b$ is equal to 2/5, or 0.4. Therefore, the per-capita extinction rate is equal to $-\ln(0.4)/(5 \text{ m.y.})$, or 0.18 per Lmy. The origination rate is equal to $-\ln(2/6)/(5 \text{ m.y.})$, or 0.22 per Lmy.

Most measures of taxonomic rate rest on the idea that the number of first and last appearances should increase with interval length and that interval length must therefore be accounted for. If there is substantial uncertainty in estimates of numerical ages, however, then normalizing by a poorly known interval length may introduce more problems than it solves. For this reason, rates will often be expressed on a per-interval rather than per-m.y. basis, and workers will attempt to compare intervals of time that are roughly equal in duration. Moreover, normalizing by interval length tacitly assumes that the originations and extinctions are spread throughout the interval. The extent to which turnover is continuous, as opposed to being clustered at a small number of time horizons such as the boundaries between time intervals, is still an unsettled question.

## Determinants of Taxonomic Rates

In the example of crinoid survivorship, taxa that are inferred to be ecologically specialized, and therefore more susceptible to environmental fluctuations, are shorter-lived. This illustrates one of the factors that is thought to be important in producing differences in taxonomic rates among groups of organisms.

Another factor that influences taxonomic rates is geographic range. Species with narrow geographic ranges will tend to be more vulnerable to environmental perturbations. This effect was invoked to explain the difference in survivorship between bivalves and gastropods (Figure 7.13). One factor that contributes to geographic range is dispersal ability. Species that do not disperse far during some stage of ontogeny will tend to have narrow geographic ranges. In such species, rare dispersal events may be more likely to produce new, geographically isolated populations. Poor dispersers may therefore also tend to have higher rates of speciation. We will see an example of this effect later in this chapter. In general, factors that lead to fragmented geographic range and population structure are likely to contribute both to speciation and to extinction. This may be one reason that groups with higher rates of origination also have higher rates of extinction.

Despite these successes in understanding variation in rates, there is still much to learn, particularly why different groups of organisms have characteristically different rates. Why, for example, do cephalopod molluscs have such high rates of taxonomic evolution relative to bivalves and gastropods, or trilobites relative to crustaceans? The answers may well lie in ecological differences such as specialization, dispersal, and population structure, but this is still not known with certainty in the majority of cases.

## 7.3 RELATIONSHIPS BETWEEN MORPHOLOGICAL AND TAXONOMIC EVOLUTION

One of the outstanding features of morphological evolution is that it can vary greatly in rate and direction. The search for regular patterns in this variation has been an

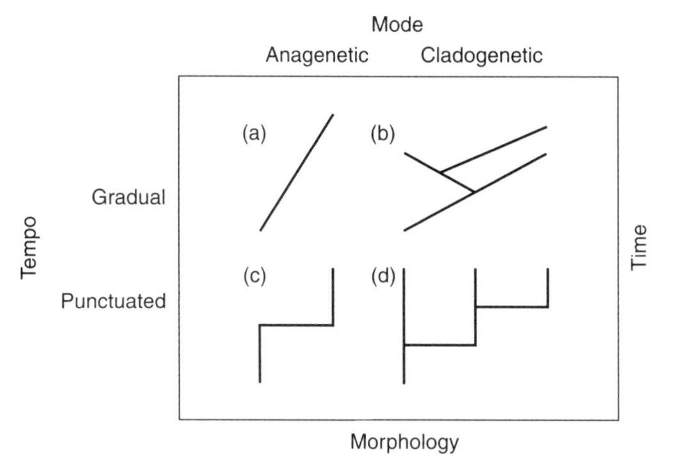


**FIGURE 7.16 Hypothetical combinations of evolutionary tempo and mode.** Morphology is depicted on the $x$ axis and time on the $y$ axis. These figures portray idealized patterns continuously with respect to time; in practice, these would be sampled discontinuously. *(Courtesy of David Jablonski)*

important endeavor in paleontology: Is change steady and continuous on average, with little variation in rate? Or is it highly abrupt, with long intervals of little or no change punctuated by brief intervals of substantial change? These two alternatives of evolutionary tempo represent extreme end-members. It is also useful to contrast two alternative evolutionary modes: anagenetic change, which involves evolution within lineages, and cladogenetic change, which involves the splitting of lineages.

Focusing on these end-members, there are four conceivable combinations of tempo and mode (Figure 7.16). Evolution can be smooth and gradual within a lineage, with no cladogenesis (Figure 7.16a); this has been referred to as **phyletic gradualism.** This same pattern of evolution within lineages can be superimposed on a cladogenetic pattern (Figure 7.16b). Highly abrupt change in the absence of cladogenesis has been termed **punctuated anagenesis** (Figure 7.16c), while the pattern of stasis interrupted by short-lived intervals of abrupt change that coincide with cladogenesis is referred to as **punctuated equilibrium** (Figure 7.16d).

## Macroevolution and the Importance of Tempo and Mode in Paleontology

Almost since the start of paleontology as a science, pronounced stasis within lineages and abrupt morphological change have been observed repeatedly in the fossil record. These patterns have generally been attributed to imperfections in the geological record. In fact, Charles Darwin (1859, pp. 280–293) argued that the temporal span of a species in the fossil record is generally a small fraction of its true duration, and that there are extensive gaps in the recorded history of species. Darwin's arguments together imply a model for the generation of stasis and punctuation as artifacts of the stratigraphic record (Figure 7.17). Little change would accrue during the preserved parts of a species' duration, and much change would accrue in the intervening gaps, only to be observed at the discontinuities between sedimentary beds.

In the century after Darwin published *On the Origin of Species*, the idea of punctuated change was often advocated. The idea did not take hold, however, because the idea that punctuations are artifacts held sway, and because the proposed mechanisms for punctuation were considered inadequate. Stasis was often observed in the fossil record, even for hundreds of thousands to millions of years—essentially the full duration of a typical species. However, because natural selection had been recognized as a major factor in evolution, it came to be expected that substantial change within species would generally

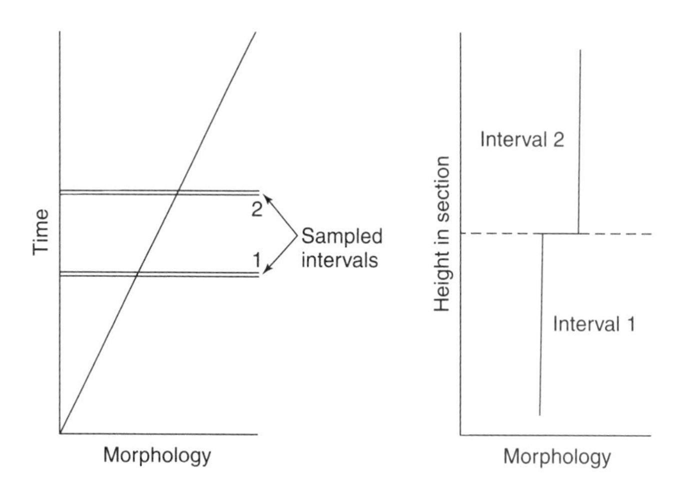


**FIGURE 7.17 A model to explain stasis and abrupt change as artifacts of discontinuous sedimentary accumulation.** The left side of this figure depicts the true evolutionary pattern; the right side depicts that pattern as seen in the stratigraphic record. The lineage evolves gradually over a long span of time, but only two short segments, indicated by the arrows, are actually sampled. These will be superimposed in a stratigraphic section, as on the right. The amount of change during the sampled intervals is the same on the left and the right. The change is so slight, however, that the lineage appears to be in approximate stasis during the sampled intervals. Change that accumulates during the span of time between the two sampled intervals appears as a punctuation.

be found if fossil lineages were followed in detail. The observation of stasis was therefore seen as something of a disappointment, even an embarrassment.

In the late 1960s and early 1970s, a number of paleontologists began to suspect that the fossil record supports the punctuated equilibrium model (Eldredge & Gould, 1972). This view was based on a few lines of reasoning. First, stasis had been seen over spans of time long enough to rule out the explanation of Figure 7.17. Second, some of the classic cases of gradual evolution did not stand up to closer scrutiny. Third and most important, it was reasoned that punctuated change associated with cladogenesis is exactly what one should expect if speciation were to be viewed through the lens of the fossil record. Speciation may be a slow process on ecological timescales, but even many thousands of years would seem like an instant when compared with species durations of millions of years. Provided that the morphological differences between species tend to evolve during the period of time when they are becoming reproductively isolated from one another [SEE SECTION 3.3], or shortly thereafter, speciation would be seen in the fossil record as a relatively abrupt change associated with branching. Later in this chapter we will return to the question of why morphological evolution may be associated with cladogenesis.

It is difficult to overstate the significance of the punctuated equilibrium hypothesis in the development of paleontology. This is partly because it forced a theoretical and empirical reevaluation of the combinations of evolutionary tempo and mode that best characterize the history of life. Moreover, it stimulated thought on the mechanisms of long-term evolutionary change and of the relationship between long-term change and short-term evolution within species. Evolutionary biologists and paleontologists have long distinguished between **microevolution** (change within populations and species) and **macroevolution** (change at or above the species level). Macroevolution includes, among other things, phylogenetic rates and trends (Figure 7.1) and changes in the relative diversity of different clades over geologic time. Paleontology has always been essential for the documentation of macroevolutionary *patterns*. The discussion that followed the proposal of punctuated equilibrium focused on *mechanisms* that entail paleontological research and cannot be fully understood solely from the study of evolution within living populations.

In the section on trends, for example, we discuss the possibility that large-scale evolutionary patterns may be attributable to differences in speciation and extinction rates, whereby some branches of the evolutionary tree accumulate diversity more rapidly than others. Such patterns have been collectively referred to as **species sorting** (Gould, 2002). This is a descriptive term that refers to the differential net production of species. **Species selection** (Stanley, 1975) is a more interpretative term intended to draw a parallel with natural selection among organisms, as outlined in Table 7.5. Just as natural selection describes a cause-and-effect relationship between organismal properties and their reproductive success, species selection describes a cause-and-effect relationship between properties of species and their tendency to

**TABLE 7.5**

Parallels between Natural Selection and Species Selection

| Feature | Organism Level | Species Level |
|---|---|---|
| What is the individual? | Organism | Species |
| What is a collection of individuals? | Population, species | Clade |
| Production of individuals | Birth | Speciation |
| Elimination of individuals | Death | Extinction |
| Biased direction of production | Mutation pressure (some mutations occur more often than others) | Directed speciation (daughter species differ from parent species in nonrandom ways) |
| Differential replication | Natural selection | Species selection |
| Higher production | Higher fecundity | Higher speciation rate |
| Lower elimination | Preferential survival | Lower extinction rate |

*SOURCE:* Based on Gould (2002)

produce daughter species or to become extinct. It is therefore important to consider what is meant by a species-level property. The issue is more complex than suggested by the following discussion; certain sources listed at the end of this chapter (Stanley, 1979; Jablonski, 2000; and Gould, 2002) should be consulted for a more thorough treatment.

If one kind of species is relatively resistant to extinction simply because individual organisms are very well adapted to their environment, then this is usually seen as organism-level selection; it can lead to species sorting, but it is not species selection. On the other hand, if the difference in taxonomic rates between two kinds of species results from a difference in a trait that logically cannot be expressed by individual organisms but only by species, this is generally seen as species selection. An example, which is discussed below, concerns higher speciation rates in gastropods with poorly dispersed as opposed to widely dispersed larvae. Although the dispersal of larvae is an organism-level trait, poor dispersal leads to species with narrower geographic distributions and a more subdivided population structure. Geographic range and population structure are properties of species rather than individuals. Thus, if differences in these properties contribute to differences in speciation rate, this is an example of species selection. Earlier in this chapter, we saw that differences in geographic range and population structure are likely to contribute to differences in extinction rate as well.

Punctuated equilibrium is not necessary for macroevolutionary patterns to be caused by species sorting; trends within clades that consist of gradually evolving lineages can be caused by the differential diversification of various branches within the clade. But if punctuated equilibrium predominates, then phylogenetic trends are quite likely to arise from species sorting rather than evolutionary trends within species. Because the punctuated equilibrium hypothesis is of such great importance to the subject of macroevolution, we must first discuss some of the problems involved in testing this hypothesis.

## Testing for Punctuated Equilibrium

A case study that tests the hypothesis of punctuated equilibrium should ideally satisfy several criteria (Jackson & Cheetham, 1999):

1. *A relatively continuous morphological record.* Of course, the record is always incomplete to some extent, but the strata should not contain major gaps that would lead to the spurious appearance of punctuations. The pattern of stasis is less sensitive to this problem, provided that the strata span a large part of the species' durations.
2. *Sufficient stratigraphic resolution.* The shorter the intervals of time in which morphological change is concentrated, the finer the stratigraphic resolution must be to record this change. It may not be necessary, however, to capture these intermediate steps in order to have some confidence that change was rapid. If the stratigraphic record is fairly complete and if stasis greatly predominates, then these facts together suggest that change, when it occurred, must have taken place over a short period of time. If change were more protracted, then more intermediate stages should be found.
3. *Well-resolved estimate of phylogeny.* Under the punctuated equilibrium hypothesis, punctuations correspond to lineage branching. An understanding of genealogy is especially important to determine whether both ancestor and descendant coexist after the observed morphological change, in order to have confidence that the change was cladogenetic rather than anagenetic.
4. *Adequate geographic control.* It is insufficient to study evolution in a single stratigraphic section, because what appears in one section as a punctuated morphological change may represent an immigration event of a species that had been diverging slowly elsewhere. For this reason, it is important to study the fossil record broadly. The possibility of prolonged evolution outside the area of study can never be ruled out conclusively, but this possibility becomes ever less likely as the study area increases.
5. *Assessment of ecophenotypic variation.* A sudden environmental change could cause punctuations in morphology that are not evolutionary and that do not represent transitions between parent and daughter species. It is therefore important to rule out ecophenotypic change [SEE SECTION 3.3] as a plausible explanation for observed punctuations.

In addition to these basic criteria, there are other features that could help to test for punctuated equilibrium in particular cases. For example, if some lineages are undergoing continuous, gradual change in the same stratigraphic sections and at the same levels at which others exhibit punctuated change, it is implausible that the punctuations are artifacts of gaps in the record. This is because such gaps would produce jumps in the gradually evolving lineages as well (Fortey, 1985b).

## A Case Study: Neogene Caribbean Bryozoans

An exemplary study of the genus *Metrarabdotos* by paleontologist Alan Cheetham (1986b) illustrates the steps needed to test for punctuated equilibrium. One noteworthy aspect of this work is that the punctuated pattern was found despite a number of operational protocols that bias against detecting it. The evidence for punctuated equilibrium is therefore especially compelling.

***Morphology and Phylogeny*** Morphological species were defined on the basis of a number of measured traits, using multivariate methods described in Chapter 3, and their phylogenetic relationships were inferred using the method of **stratophenetics.** This method, which can be useful when stratigraphic sampling is dense and fairly continuous, begins by forming an array of samples arranged by stratigraphic intervals. From one interval to the next, samples that are most similar morphologically are linked together. This is shown schematically in Figure 7.18. Like all phylogenetic methods, stratophenetics has both strengths and weaknesses. For purposes of this discussion, what is important is that stratophenetics tends to bias against the punctuated pattern. This is because it minimizes the change between ancestor and descendant and thus also minimizes the implied evolutionary rate associated with the transition.

The reconstructed evolutionary tree of *Metrarabdotos* is shown in Figure 7.19, in which the $x$ axis represents a multivariate measure of morphological dissimilarity

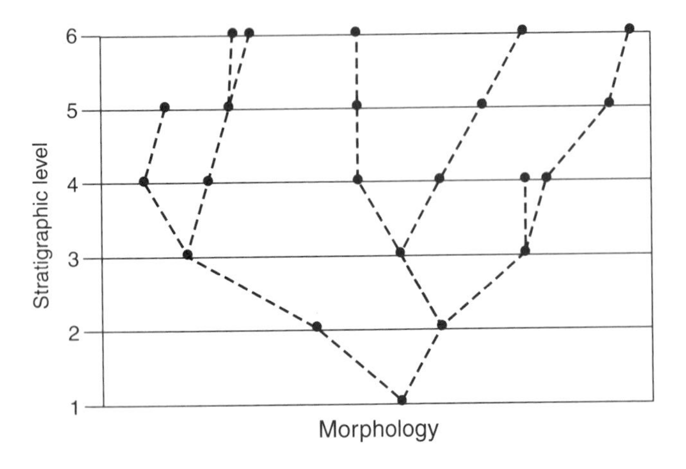


**FIGURE 7.18 Stratophenetic reconstruction of phylogeny illustrated with hypothetical data.** Each point represents a population mean. At each stratigraphic level, populations are connected with the most similar populations in adjoining levels.

[SEE SECTION 3.2]. Clearly, there is a multiplication of species and a persistence of pairs of related species following morphological changes. Thus, unless the new species have migrated in from elsewhere, the mode of evolution must be cladogenetic. Visually, the pattern of evolution in Figure 7.19 suggests that the punctuated equilibrium model holds. A statistical test, described in Box 7.4, supports this inference.

***Completeness and Resolution*** The study focuses on the more densely sampled part of the record—after about 10 m.y. ago (Figure 7.19). The temporal spacing between samples varies from about 20,000 years to about 1 m.y., with an average of about 160,000 years.

In Chapter 1, we discussed methods for estimating the proportion of species that have left a fossil record. There are analogous approaches for estimating the proportion of elapsed time that is represented by preserved sedimentary beds. In this case, such methods have been used to determine that the record is greater than 60 percent complete at the level of temporal resolution of the samples. What this means is that, of all the 160,000-year time intervals that elapsed during the time span covered by this study, more than 60 percent of them are represented by some preserved sediment. Sampling is uneven over time, however. In many cases, the sampling levels are closely spaced, so that it is possible to rule out protracted divergence between ancestor and descendant over many hundreds of thousands of years. In other cases, sampling is sparser, and the duration of the ancestor–descendant transition is difficult to infer directly—for example, the transition from *M. lacrymosum* to *M. unguiculatum*. The statistical test described in Box 7.4 takes this fact into consideration.

***Geographic Control*** Although the majority of data for this study are from the Dominican Republic (D.R.), samples were collected broadly from within the Caribbean region, including Haiti, Jamaica, Trinidad, and the Gulf and Atlantic coasts of the United States. A few long-lived species are found throughout the Caribbean, but most are restricted to the D.R. There is no indication that the derived species evolved over a long period of time outside the D.R. and migrated in, and studies of the fossil record outside the Caribbean region do not reveal species that appear to be closely related to those found in the Caribbean.

Geographic sampling always leads to an asymmetrical test. It is possible to demonstrate that what appears to be

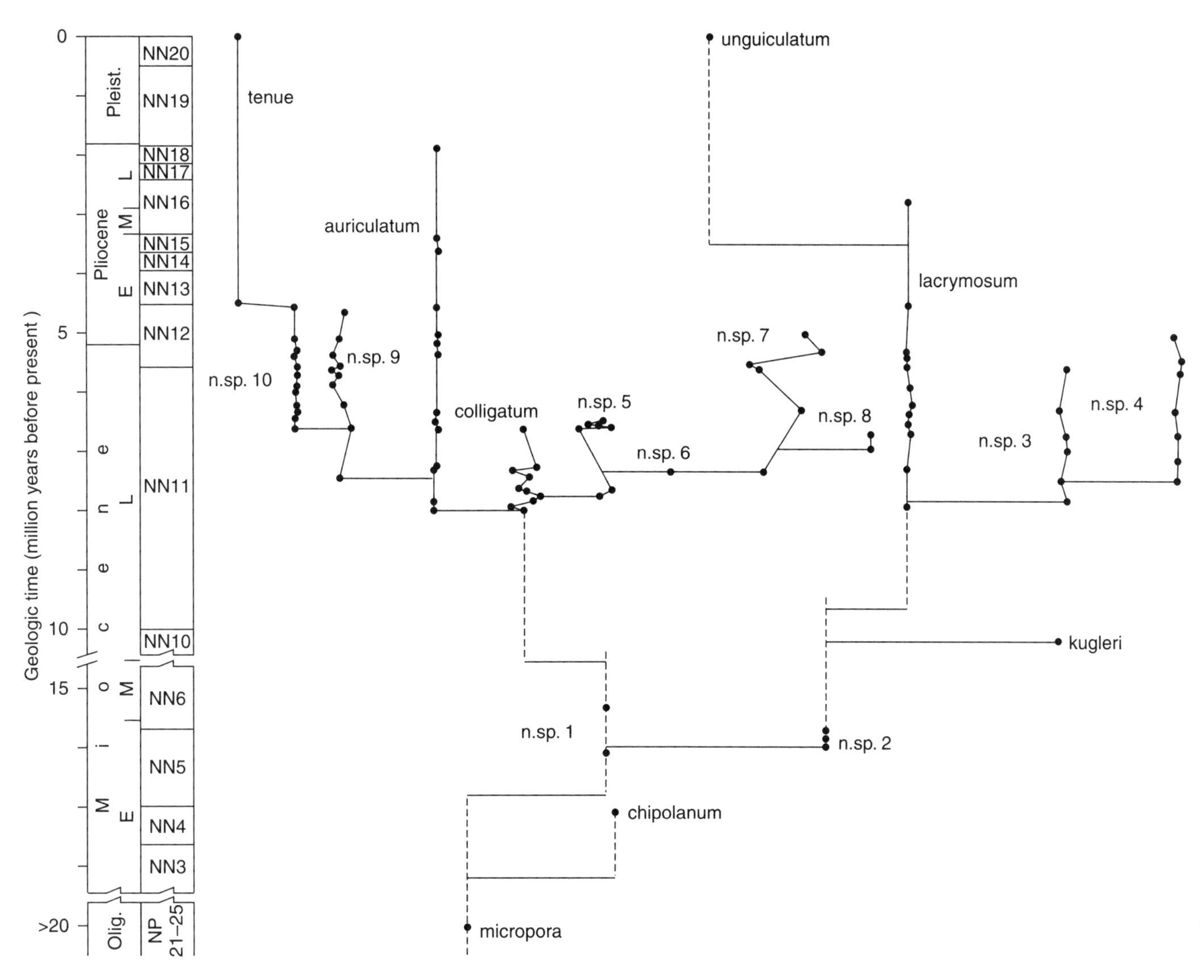


**FIGURE 7.19 Stratophenetic tree of species of the Neogene bryozoan *Metrarabdotos* from the Caribbean region.** The $x$ axis summarizes multivariate information on morphology in such a way that the morphological distances between closely related species are represented with little distortion. Each point shows the average of a sample. Intervals on the timescale marked NP and NN are biostratigraphic zones based on microfossils [SEE SECTION 6.1]. *(From Cheetham, 1986b)*

a punctuation in fact represents a migration event, but it is never possible to rule this out conclusively. Just how extensively one must sample remains a point of contention. Nevertheless, in this case, the breadth of geographic sampling leaves us with considerable confidence, if not complete certainty, that the rapid morphological changes observed in samples from the D.R. represent evolutionary rather than migrational events.

***Ecophenotypic Variation*** Two lines of evidence suggest that the observed morphological punctuations are genetic rather than ecophenotypic. First, if these changes were induced by an environmental shift, we should expect to see many or most species changing in form simultaneously in response to the shift. This is not the case (Figure 7.19). In general, we see many lineages in stasis while others are changing. In particular, ancestors remain in stasis while their descendants change rapidly. Second, there is a strong correlation between morphological and genetic variation, with little ecophenotypy, in cheilostome bryozoans related to those studied here [SEE SECTION 3.3].

In general, morphologically defined species are thought to be genetically distinct. Cryptic species may be

Box 7.4

## OPERATIONAL TEST FOR PUNCTUATED EQUILIBRIUM

An upper limit on the duration of a speciation event is defined by assuming that the event extends from the first appearance of the ancestor to the first appearance of the descendant (Figure 7.20). By assuming that the ancestor and descendant originated at the same time, the implied rate of evolution between ancestor and descendant is minimized. This protocol therefore biases against the punctuated pattern.

For each ancestral species, the mean rate of within-species evolution is calculated from the long-term net change over the entire duration of the species, as shown in Figure 7.20. Next, the rate of change between each pair of successive samples within a species is tabulated to determine the variance in within-species rate. The ancestor–descendant rate is then compared to the average within-species rate, scaled to the variance in within-species rate.

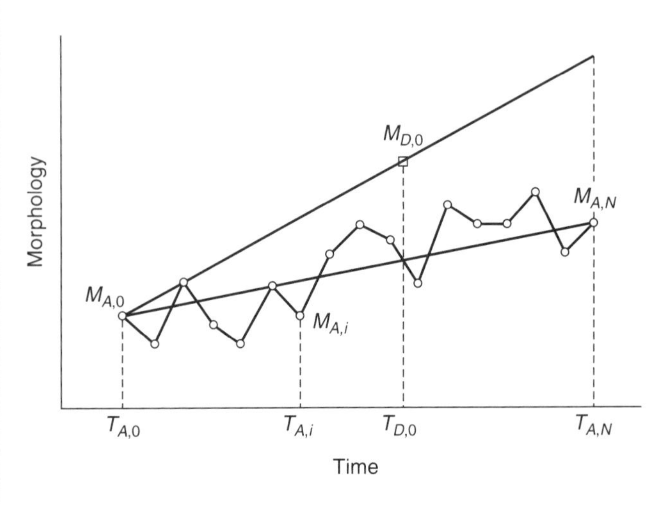


**FIGURE 7.20 Calculation of within-species and between-species rates of evolution.** Time and morphology are in arbitrary units. Times $T_{A,i}$ (from $T_{A,0}$ to $T_{A,N}$) are all the times at which the ancestor is sampled, its first appearance being at $T_{A,0}$. The first appearance of the descendant is at time $T_{D,0}$. Circles represent the ancestor; the one square is the descendant at its time of first appearance in the fossil record. The average within-species change in the ancestor is the slope of the line from point $(T_{A,0}, M_{A,0})$ to point $(T_{A,N}, M_{A,N})$, in other words, $(M_{A,N} - M_{A,0})/(T_{A,N} - T_{A,0})$. The variance in within-species rate is based on the slopes of all lines between adjacent points. The between-species change is the slope of the line from point $(T_{A,0}, M_{A,0})$ to point $(T_{D,0}, M_{D,0})$, or $(M_{D,0} - M_{A,0})/(T_{D,0} - T_{A,0})$. This assumes that the descendant has been evolving gradually since the first appearance of the ancestor and therefore yields a minimum estimate of the rate of between-species change. *(From Cheetham, 1986b)*

common, however; the absence of morphological distinction need not imply the lack of genetic difference between populations [SEE SECTION 3.3]. For the assessment of punctuated equilibrium, cryptic species are generally thought not to present a problem. This is because the punctuated equilibrium hypothesis states that change, *when it occurs*, tends to be associated with cladogenesis. Cladogenesis without morphological change is compatible with punctuated equilibrium, but significant morphological change without cladogenesis is not.

Considering how much is involved in the *Metrarabdotos* study, it is not surprising that there have been few entirely conclusive demonstrations of either the presence or absence of punctuated equilibrium. Stasis is easier to document, and it has been very widely observed in the fossil record. Although the overall predominance of punctuated equilibrium remains largely unknown, the case of *Metrarabdotos* is not unique. Next, we briefly consider some of the mechanisms that may contribute to punctuated equilibrium when it occurs.

## Mechanisms of Stasis

Several plausible explanations for stasis have been proposed by paleontologists and biologists. All are likely to play a role, but their relative importance still has not been determined.

According to the mechanism of **habitat tracking,** populations do not remain in stasis in a single locality. Rather, as environmental conditions change over time,

This statistical comparison starts with the assumption that the between-species rate is drawn from the same distribution as the within-species rates. This assumption, or null hypothesis, must be explicitly rejected before the alternative of punctuated equilibrium can be accepted. Gradualism is therefore treated as the preferred hypothesis, and the test biases against the punctuated model.

The result of this analysis is that the rates of change between ancestors and descendants are overwhelmingly too large to be part of the same statistical distribution that characterizes rates of change within species (Table 7.6). The gradualistic hypothesis is rejected in favor of the alternative of punctuated equilibrium.

TABLE 7.6

Evolutionary Rates for Ancestor–Descendant Pairs of Species in the Bryozoan *Metrarabdotos*

| Ancestor | Descendant | Within-Species Rate | Between-Species Rate |
|---|---|---|---|
| *M. auriculatum* | *M.* n. sp. 9 | 0.002 | 30.20 |
| *M.* n. sp. 9 | *M.* n. sp. 10 | 0.083 | 9.34 |
| *M.* n. sp. 10 | *M. tenue* | −0.031 | 5.38 |
| *M. colligatum* | *M.* n. sp. 5 | 0.169 | 50.37 |
| *M.* n. sp. 5 | *M.* n. sp. 6 | −0.158 | 27.96 |
| *M.* n. sp. 7 | *M.* n. sp. 8 | 1.065 | 51.74 |
| *M. lacrymosum* | *M.* n. sp. 3 | 0.014 | 118.9 |
| *M.* n. sp. 3 | *M.* n. sp. 4 | −0.009 | 69.13 |
| *M. lacrymosum* | *M. unguiculatum* | −0.021 | 7.56 |

*SOURCE:* Cheetham (1986b)

*NOTE:* Calculation of rates is explained in Figure 7.20. Rates are expressed in terms of synthetic morphological units (Figure 7.19) per m.y.

the geographic distributions of populations and species shift to track local conditions to which they are adapted. Habitat tracking underlies the facies dependence of species [SEE SECTION 6.4]. Detailed evidence comes from the late Cenozoic fossil record, where temporal resolution is fine enough to document changes in distribution that take place in as little as 1000 years. Figure 7.21 shows an example for a number of insect species. Although this process clearly occurs, there are cases of stasis within a single geographic area, and therefore other mechanisms are needed. One that is commonly invoked is **stabilizing selection.** Observations on many living species have shown that, in stable environments, there tends to be a preferred, intermediate phenotype that is better adapted than others. If environments were stable over long periods of time, this modal form would be maintained. Although there is no reason to doubt the operation of stabilizing selection in principle, it is difficult in the fossil record to identify the particular selective forces that favor certain phenotypes over others. More important, many cases have been documented in which species are in stasis despite major changes in climate and other aspects of environment. Additional mechanisms are therefore required.

A number of potential causes of stasis recognize abundant evolutionary change within species but postulate that, for various reasons, this change does not accumulate to yield substantial net evolution in the long run. First, if relevant environmental conditions fluctuate around a stable long-term average, and the species

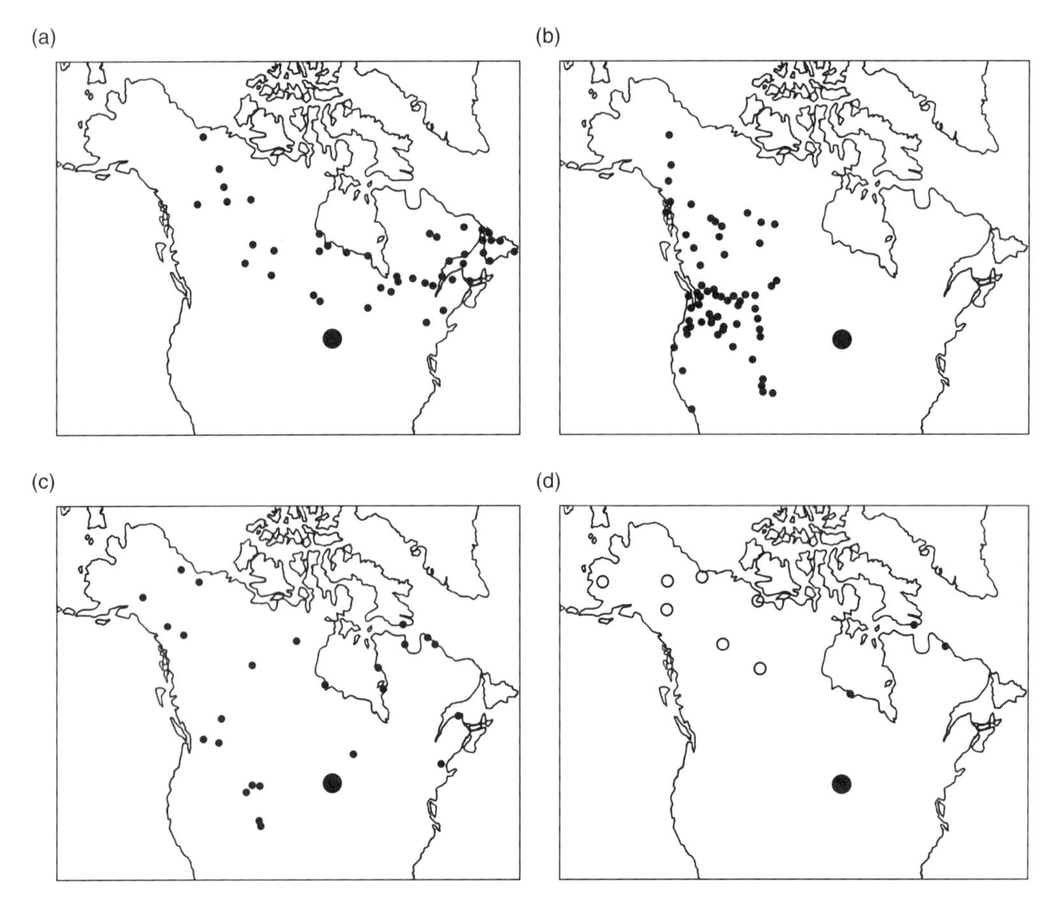


**FIGURE 7.21 Shifting distributions of insect species in northern North America.** Present-day occurrences of five species are given. Parts (a), (b), and (c) each show a single species (small filled circles); part (d) shows two species (small filled circles and open circles). The large filled circle in each part shows the same sample from lake sediments at a locality in Minnesota, about 12,000 years old. All five species are found together in that sample. Today, however, only one of the distributions includes this locality, and none of the species is found anywhere today with any of the others. *(From Bennett, 1997)*

evolves in response to these short-term environmental changes, the species may show little net evolutionary change. In a similar vein, abrupt environmental changes caused by variation in climate and other factors may alter which traits are favored by natural selection, effectively erasing any accumulated evolutionary change.

The subdivision of species into geographic populations provides another way for stasis to occur despite substantial evolution within species. Gene flow between geographically subdivided populations may contribute to stasis by hindering the accumulation of change in any one local population. Environmental fluctuations, by fostering geographic range shifts and therefore greater gene flow, would enhance the stability of species.

Finally, there is a class of related mechanisms that can be grouped together as developmental constraint [SEE SECTION 5.1]. Constraints are thought to bring about stasis when a species expresses extremely limited phenotypic variation, due to either a paucity of genetic variation or a tendency for developmental processes to produce a limited array of phenotypes. Although the explanation of constraint makes sense in principle and can be documented in living species, it is for most practical purposes untestable in fossil populations.

## Mechanisms of Punctuated Change

There are two components to abrupt change in the punctuated equilibrium model: It takes place over a time span that is short relative to the duration of a species, and it is associated with cladogenesis. Conventional biological views posit that the process of speciation is generally fast enough that it would appear to be instantaneous in the fossil record. The essential question, therefore, is why morphological change should be concentrated in events of cladogenesis. Rather than review the long and complex history of this question, we will present a cogent explanation offered by evolutionary biologist Douglas Futuyma (1987). This explanation agrees with what is known of evolution within populations and the population structure of species, and it is also closely tied to one of the mechanisms for stasis described previously.

A species consists of numerous geographic populations, with varying amounts of gene flow between them [SEE SECTION 3.1]. Because these populations are in different places, they may be adapted to local conditions. Therefore, evolutionary change may be possible anywhere and at any time in the history of a species. However, local populations do not generally become distinct; they either suffer extinction or maintain gene flow with other populations. Any change that accumulates within local populations will therefore tend to be short-lived. Only in those rare cases in which local populations do persist and attain reproductive isolation will the evolutionary change that accumulated through local adaptation be maintained. There may be abundant evolutionary change within the history of a species, but that change will be permanent only if it occurs in a population that founds a new species. Therefore, the net evolutionary change seen in the fossil record will be associated with cladogenesis.

It is important to keep in mind that much of the importance of stasis and punctuated change is independent of what causes these phenomena. The very pattern of punctuated equilibrium shapes our understanding of trends and other aspects of macroevolutionary change.

## 7.4 EVOLUTIONARY TRENDS

The punctuated equilibrium hypothesis has led to a reevaluation of the mechanisms of macroevolution. Nonetheless, many paleontologists had long been aware that large-scale trends in the history of life may be caused by mechanisms other than the tendency for individual species to evolve in the same direction. Appreciation for the complexity of macroevolutionary patterns is captured well in the following passages from George Gaylord Simpson's *The Major Features of Evolution* (1953), on the subject of horse phylogeny (Figure 7.22) and trends in body size, limb and foot form, and skeletal and dental anatomy:

> Even in some of the most recent works . . . this phylogeny is [incorrectly] presented as a single line of gradual transformation of *Hyracotherium* into *Equus* [the only living genus in the family]. It has been well known to the better informed for more than two generations that the phylogeny includes considerable branching, and for the last ten or fifteen years it has been increasingly evident that the really striking and characteristic part of the pattern is precisely its repeated and intricately radiating splitting. Its botanical analogue would be more like a bush than like a tree, and even if the tree figure of speech were used, *Equus* would not correctly represent the tip of the trunk but one of the last bundles of twigs on a side branch from a main branch sharply divergent from the trunk. (p. 260)

> The Equidae had no trends that: (1) continued throughout the history of the family in any line, (2) affected all lines at any one time, (3) occurred in all lines at some time in their history, or (4) were even approximately constant in direction and rate in any line for periods longer than on the order of 15 to 20 million years at most (usually much less). . . . The whole picture is more complex, but also more instructive, than the orthogenetic progression that is still being taught to students as the history of the Equidae. It is a picture of a great group of real animals living their history in nature, not of robots on a one-way road to a predestined end. (p. 264)

Like the history of horses (Figure 7.22), the broader history of life is replete with evolutionary change. In some cases, evolution is persistent in direction, toward larger size or more efficient food capture, for example. We single out these cases as **evolutionary trends.** An understanding of a specific trend often starts with a statistical consideration of just how persistent it is.

### Tests for Persistent Directionality

A number of statistical tests for persistence have been developed, and here we present just two of the simplest ones.

To understand the rationale behind these tests, consider a highly simplified model of evolutionary change known as a **random walk.** According to this model, a trait value has a certain probability of increasing or decreasing at every time step, and which way it goes is a matter of chance, independent of previous changes.

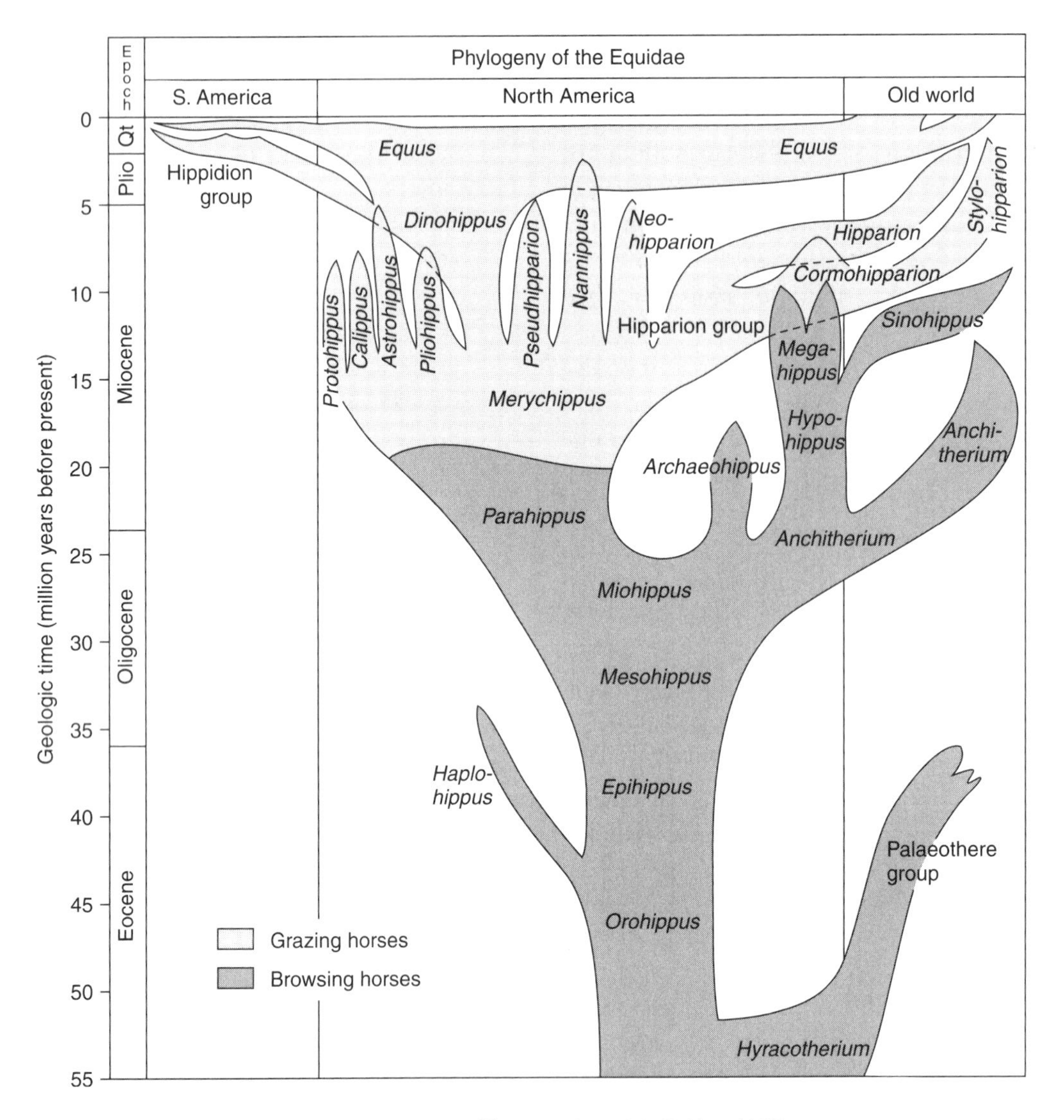


**FIGURE 7.22 Inferred evolutionary tree of horses.** *(From MacFadden, 1985)*

Random walks with a probability of increase substantially less than or greater than one-half would tend to produce trends downward or upward. In a **symmetric random walk,** the probability of increase is exactly one-half, so there is no inherent directionality.

Figure 7.23 shows a few examples of evolutionary patterns simulated by a computer programmed to generate symmetric random walks. In the first case, the random walk appears to have a tendency to stay near its starting point ("stasis"). In the second, it is relatively stable, then shifts over a brief interval of time ("punctuation"), then continues to be stable at a new point. In the third, the overall appearance is that there is a strong directionality. This exercise reveals that sequences that appear to have a striking pattern may potentially be produced by a process with no inherent directionality or stability. Thus, the impression created by the pattern should not by itself be taken as strong evidence for prevailing evolutionary tendencies.

Figure 7.23, especially the third panel, also illustrates why it is generally inappropriate to interpret correlation coefficients [SEE SECTION 3.2] in which one of the variables is time. Temporal series of data generally consist of successive values that are not independent of one another, each one being the previous value plus some increment. The standard statistical analysis of correlation coefficients, in contrast, assumes independent data points.

One test for directionality considers the mean size of evolutionary steps between successive time horizons. If there is a tendency for evolutionary change to proceed in a particular direction, then the mean evolutionary step

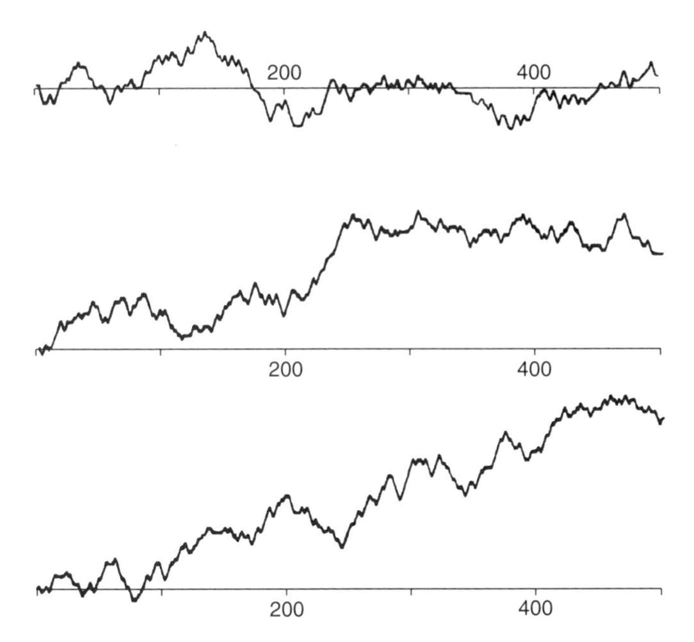


FIGURE 7.23 **Examples of computer-generated, symmetric random walks of 500 steps.** In each example, time is on the *x* axis and morphology is on the *y* axis. At each time step, morphology has an equal chance of moving up or down one step, independent of prior history. Striking patterns may emerge even though the process has no inherent directionality or stability. *(From Raup, 1977)*

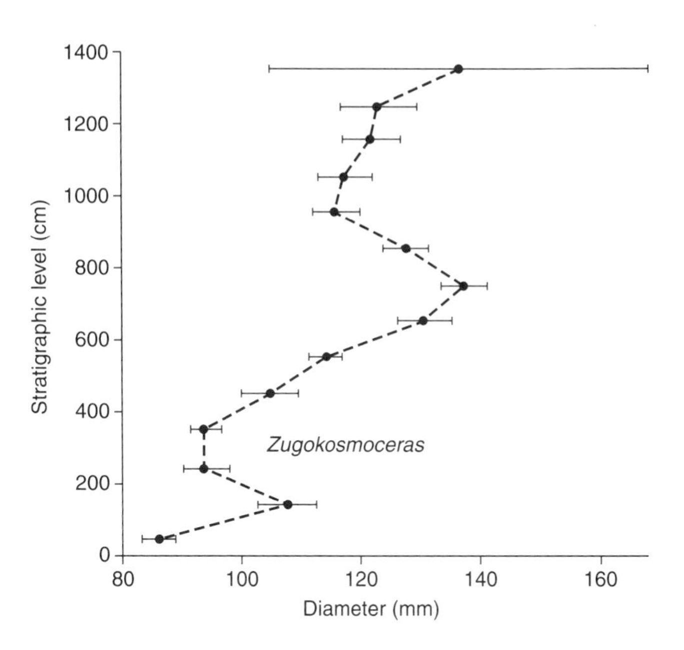


**FIGURE 7.24 Shell diameter in the Jurassic ammonite *Zugokosmoceras* through 14 m of section.** Each point shows the mean ±2 standard errors of the mean, for a 100-cm interval. *(From Raup & Crick, 1981)*

should differ appreciably from zero. For example, consider shell diameter in the Jurassic ammonite *Zugokosmoceras*. Figure 7.24 shows a summary of this lineage through 14 meters of section. There is an overall increase in size, but there are several reversals as well; thus, it is not clear just how strong the tendency is toward size increase. Each point in Figure 7.24 depicts the mean of all specimens within a 100-cm interval. In fact, the section was sampled much more finely than this, with nearly 300 distinct, *Zugokosmoceras*-bearing horizons.

Figure 7.25 gives the frequency distribution of changes in shell diameter between adjacent horizons at the finer level of temporal resolution. There is a slight preference for positive steps, but there are many reversals, and the mean step size is not substantially different from zero. Thus, there does not appear to be a strong directionality.

A second test for directionality ignores the magnitudes of change and simply tabulates the number of positive and negative changes. This test determines the probability that a symmetric random walk would yield at least as many changes in the same direction as were actually observed in the data (Box 7.5). In the sequence of molar sizes in Figure 7.4, there are 17 steps, of which 15 are increases. The probability of obtaining 15 or more steps in the same direction, if the probabilities of increase and decrease are truly equal, is only 0.0023. We therefore reject the random-walk model in favor of the hypothesis of persistent change.

The tests for a persistent tendency within a single evolutionary sequence require that we observe multiple

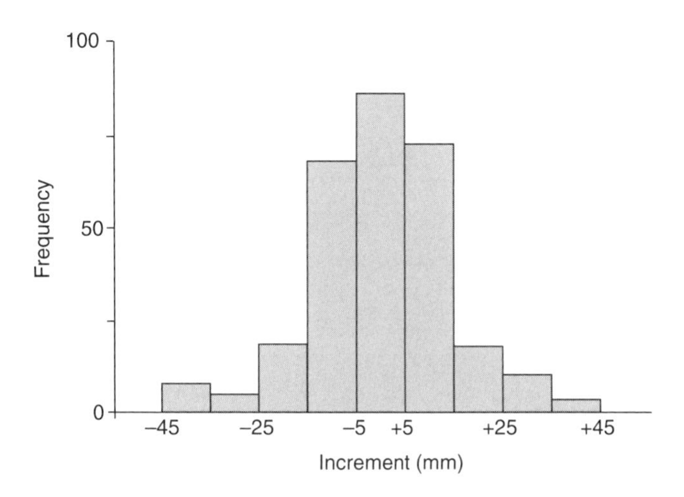


**FIGURE 7.25 Frequency distribution of evolutionary steps in shell size between adjacent, *Zugokosmoceras*-bearing horizons in the same section depicted in Figure 7.24.** The mean is near zero, suggesting that there is not a strong directionality. *(From Raup & Crick, 1981)*

Box 7.5

## TESTING FOR DIRECTIONALITY OF EVOLUTIONARY SEQUENCES

Here we give details on statistical assessment of the number of changes in a given direction and make some general points about testing for trends.

Suppose there are $n$ evolutionary steps, of which $m_i$ are increases and $m_d$ are decreases. Under the symmetric random-walk model, which serves as the null hypothesis of the test, the probabilities of increase and decrease are each equal to 0.5. The probability of exactly $m_i$ increases, if increases and decreases are in fact equally likely, is given by:

$$P(m_i, n) = \frac{n!}{m_i!(n - m_i)!} 0.5^n$$

where $n!$ ("$n$ factorial") is the product $n \times (n-1) \times (n-2) \times \cdots \times 1$, and $0! = 1$ by convention. The derivation of this binomial equation can be found in any elementary textbook on probability and statistics. Referring to the example of molar tooth size in the mammal *Cantius* (Figure 7.4), the probability of obtaining exactly 15 increases out of 17 steps is equal to

$$P(15, 17) = \frac{17!}{15!(17 - 15)!} 0.5^{17}$$

which is approximately equal to 0.00103.

For a sequence of $n$ steps, the number of increases can be anything from 0 to $n$, and the number of ways the steps can be arranged is enormous. The exact probability of any particular sequence of steps is minuscule. In statistical analysis, we are generally not interested in such isolated probabilities; these have no special significance other than as part of a whole family of outcomes with something important in common. In this case, a given number of increases may seem higher than expected under the random-walk model; if so, then any number of increases greater than the observed value would also seem high. We are therefore interested in the sum of probabilities of all outcomes at least as extreme as the one observed. By "extreme" we mean deviating from the expectation of an equal number of increases and decreases.

We thus take the larger of $m_i$ and $m_d$, which we denote simply $m$, and calculate the sum of a number of changes *greater than or equal to* the number observed:

$$P_{\geq m(n)} = \sum_{k=m}^{n} P(k, n)$$

If this is a small number, conventionally less than 0.05, we reject the null hypothesis in favor of the alternative that change in one direction is more likely than in the other. Again taking the mammal example, the probability of 15 or more increases out of 17 steps is equal to:

$$P_{\geq 15(17)} = \frac{17!}{15!(17 - 15)!} 0.5^{17} + \frac{17!}{16!(17 - 16)!} 0.5^{17} + \frac{17!}{17!(17 - 17)!} 0.5^{17}$$

steps. Sometimes we simply have before-and-after comparisons, but we have them for numerous lineages. In such a case, we can apply the same tests for persistence.

It has been proposed, for example, that there is a general tendency for body size to increase in evolution—a generalization sometimes referred to as *Cope's Rule*. Figure 7.26 shows estimated body size for species of fossil horses, a group which, as a whole, has increased in size over time. To determine whether there is such a tendency at the lineage level, we study ancestor–descendant comparisons. In Figure 7.27, the rate of evolution between inferred ancestors and descendants is shown, with increases and decreases indicated separately as solid and open circles. Of 24 observed changes in body size, 19 are increases. The probability of this many increases, if increases and decreases are equally likely, is only 0.0033. Thus, it is reasonable to reject the hypothesis of a symmetric random walk in favor of the alternative that increases in body size are more likely.

which is equal to 0.00103 + 0.00013 + 0.000008, or about 0.00116.

As with many statistical tests, it is important to determine whether the test should be one-sided or two-sided. If we had no prior reason to test for a preponderance of positive or negative steps, we would perform a two-sided test. Suppose, for example, that there were more positive than negative steps. We first set $m$ equal to $m_i$ and calculate $P_{\geq m(n)}$, which tells us the probability of at least $m$ positive steps. However, a result at least this extreme in the opposite direction might also have occurred, and, since we had no prior reason to test for positive rather than negative deviations, the probability of both alternatives must be taken into consideration. What we wish to evaluate, therefore, is $2P_{\geq m(n)}$. Only if this is sufficiently small can we reject the null hypothesis.

We used a two-sided test to evaluate the sequence of steps in the case of mammal tooth size (Figure 7.4) and the mean step size in the case of ammonite shell size (Figure 7.24), because we were simply testing for change in a preferred direction, *either direction*. In the mammal case, for instance, the probability of 15 or more increases out of 17 steps is equal to 0.00116, as was just shown. Therefore, the probability of 15 or more changes in one direction is twice this number, or approximately 0.0023. This is a rather small probability, so we would reject the idea that there is no directionality in favor of the alternative that there is a preferred direction of change.

On the other hand, if we are testing a specific evolutionary hypothesis that predicts change in a particular direction, we perform a one-sided test. We tabulate the number of changes in this direction and calculate the probability of at least this many changes *in that direction*, regardless of whether the observed number is greater or less than expected under the null hypothesis. We used a one-sided test to check for an increase in the body size of horses (Figure 7.27) because there is a specific hypothesis that body size tends to increase in evolution. This hypothesis was supported by the test. If fewer than half of the observed changes had been increases, we would simply have considered the hypothesis of body-size increase not to be supported by the data.

There is an important asymmetry in these and other statistical tests. If we can reject the null hypothesis, we can be confident that there is some directionality in the evolutionary sequence. If we cannot reject it, however, that does not imply that the sequence was in fact the result of a symmetric random walk. It is possible that there is a preferred direction of change but that it has not been detected statistically because the tests lack sufficient power. This can happen if the observed sequence is too short or if within-species variance makes the estimation of successive average trait values too uncertain.

## Mechanisms of Phylogenetic Trends

The average morphology of a clade can change directionally over time if many of the lineages in that clade are evolving directionally. That is, a phylogenetic trend could in principle be reducible to a series of phyletic trends. When evolutionary stasis overwhelmingly predominates, however, this possibility is logically ruled out. It is therefore important to consider some of the ways that evolutionary trends within a clade may result if lineages within the clade are in stasis (Figure 7.28). For the most part, these mechanisms are simply logical possibilities that have not yet been well documented, and determining the causes of trends in particular cases remains an important task for paleontology. The mechanisms we will discuss do not presuppose that punctuated equilibrium predominates; they can produce trends even if most evolutionary change is gradual. But because we are discussing clade-level trends, some cladogenesis is assumed.

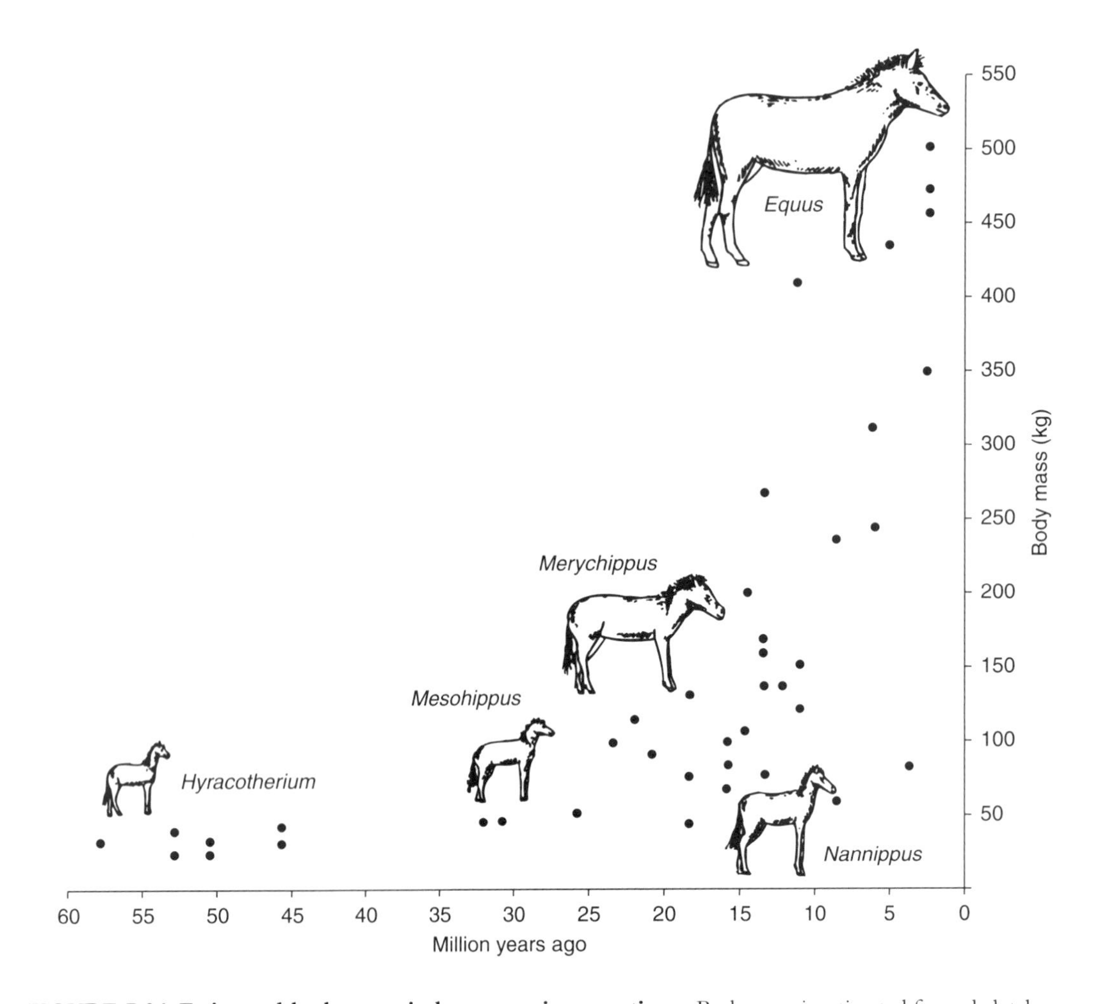


**FIGURE 7.26 Estimated body mass in horse species over time.** Body mass is estimated from skeletal measurements, using methods described in Section 3.2. Sketches show examples of genera. Compare with the evolutionary tree of Figure 7.22. *(From MacFadden, 1986)*

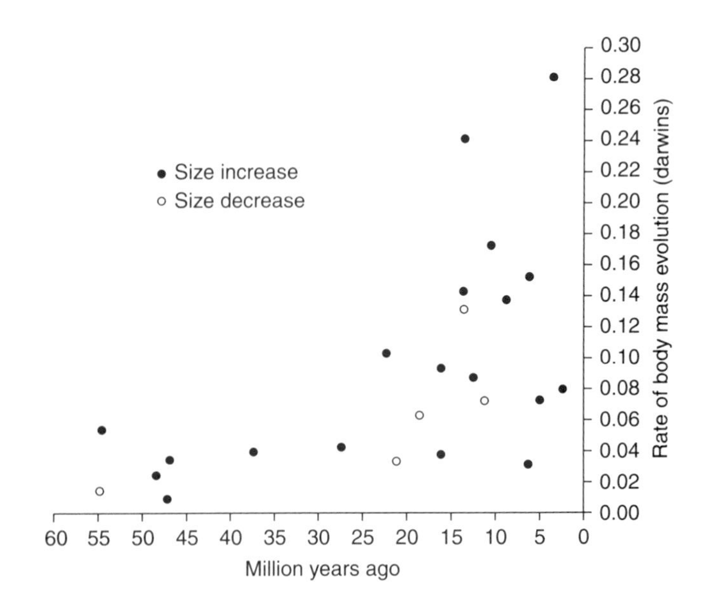


**FIGURE 7.27 Evolutionary rates (in darwins) for ancestor–descendant pairs of horse species in Figure 7.26.** Closed circles show increases in body mass, and open circles show decreases. There are many more increases than decreases, suggesting directionality in the evolution of body size. *(From MacFadden, 1986)*

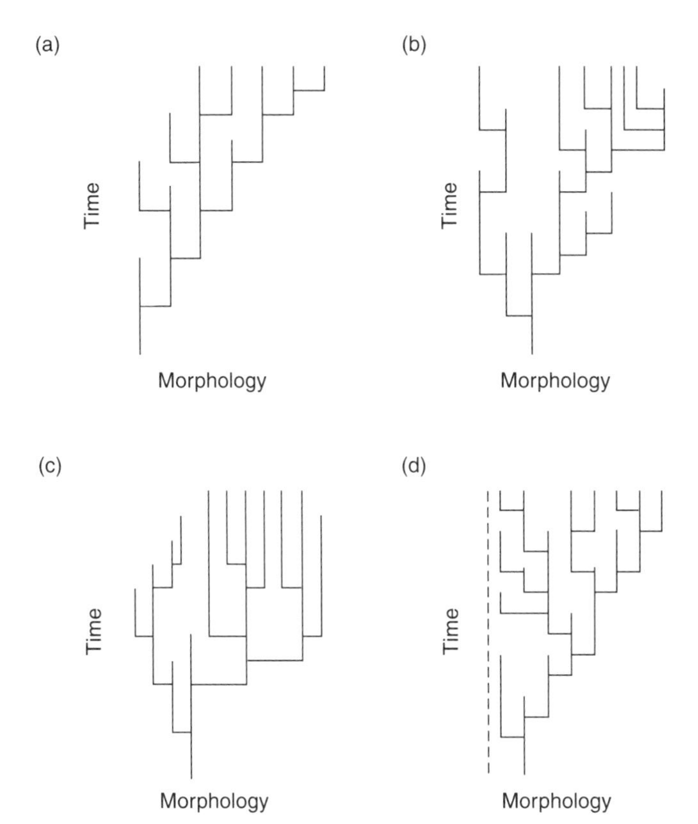


**FIGURE 7.28 Hypothetical evolutionary trees showing different ways that phylogenetic trends can develop even if species are in stasis.** These can also apply if species are evolving gradually. (a) *Directed speciation.* Cladogenesis produces daughter species that are preferentially to the right of their ancestors. (b) *Speciation-rate bias.* The rate of production of new species is higher on the right, leading to an accumulation of diversity and a trend in the average morphology of the clade. (c) *Extinction-rate bias.* The rate of extinction is lower on the right, also leading to an accumulation of diversity. (d) *Asymmetric increase in variance.* The clade originates near a lower limit, indicated by the dashed line. The maximum and mean morphology increase because there is more room for evolutionary change to the right of the starting point. *(a–c: After Gould, 1982; d: After Stanley, 1973)*

***Directed Speciation*** This mechanism posits that change is concentrated at speciation events and that descendant species tend to shift in a preferred direction relative to their ancestors (Figure 7.28a).

A possible example concerns larval mode in certain gastropod taxa. The larvae of marine gastropods can be roughly divided into planktotrophic and nonplanktotrophic forms. Planktotrophs swim freely and feed on plankton. Some nonplanktotrophs do swim, but many are entirely benthic and all depend on yolk or other food supplied with the egg. The planktotrophic mode generally involves complex swimming and feeding organs that tend to be absent in nonplanktotrophs. Fortunately for paleontologists, observations on living species show that some characteristics of the larval shell, including size and number of whorls, allow planktotrophic and nonplanktotrophic species to be distinguished fairly accurately from preserved shell features (Figure 7.29) [SEE SECTION 8.6].

A number of gastropod families show an increase in the proportion of nonplanktotrophic species during the Early Tertiary (Figure 7.30)—in other words, a phylogenetic trend toward nonplanktotrophy. Planktotrophic mode is thought to be primitive within these families. Although the nonplanktotrophic larval mode was presumably selected for in those particular lineages in which it evolved, there is no compelling reason to believe that either mode is *generally* advantageous relative to the other.

The complex swimming and feeding organs used by planktotrophs tend to be lost in the transition to the nonplanktotrophic mode. Because complex organs, once lost, may be difficult to reevolve, it is generally thought that the transition from planktotrophic to nonplanktotrophic mode is more likely than the reversal. In fact, some phylogenetic analyses bear this out. Thus, there is

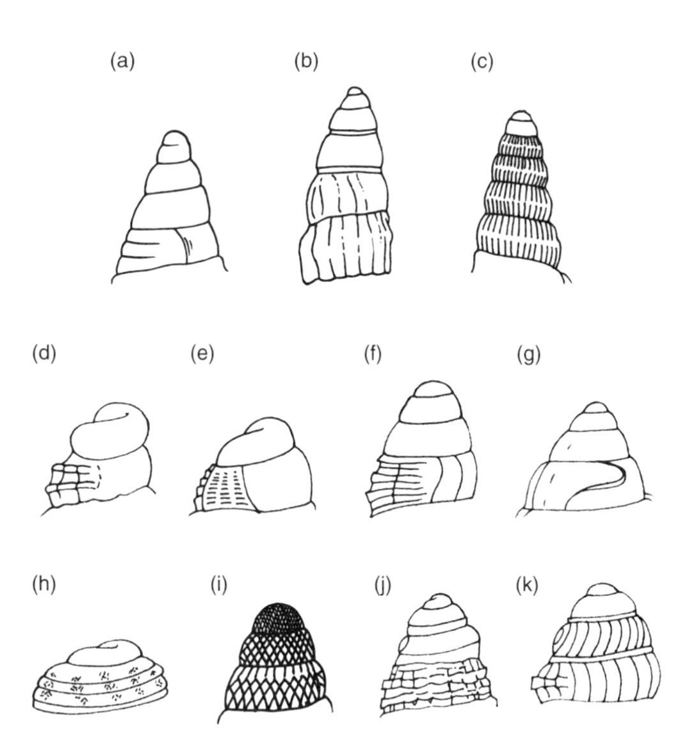


**FIGURE 7.29 Larval shells of some gastropods.** Parts (a) through (c), (f), (g), and (i) through (k) show planktotrophic shells. Parts (d), (e), and (h) show nonplanktotrophic shells. In contrast to nonplanktotrophic forms, planktotrophic larvae tend to have more whorls and a high, conical shape. *(From Shuto, 1974)*

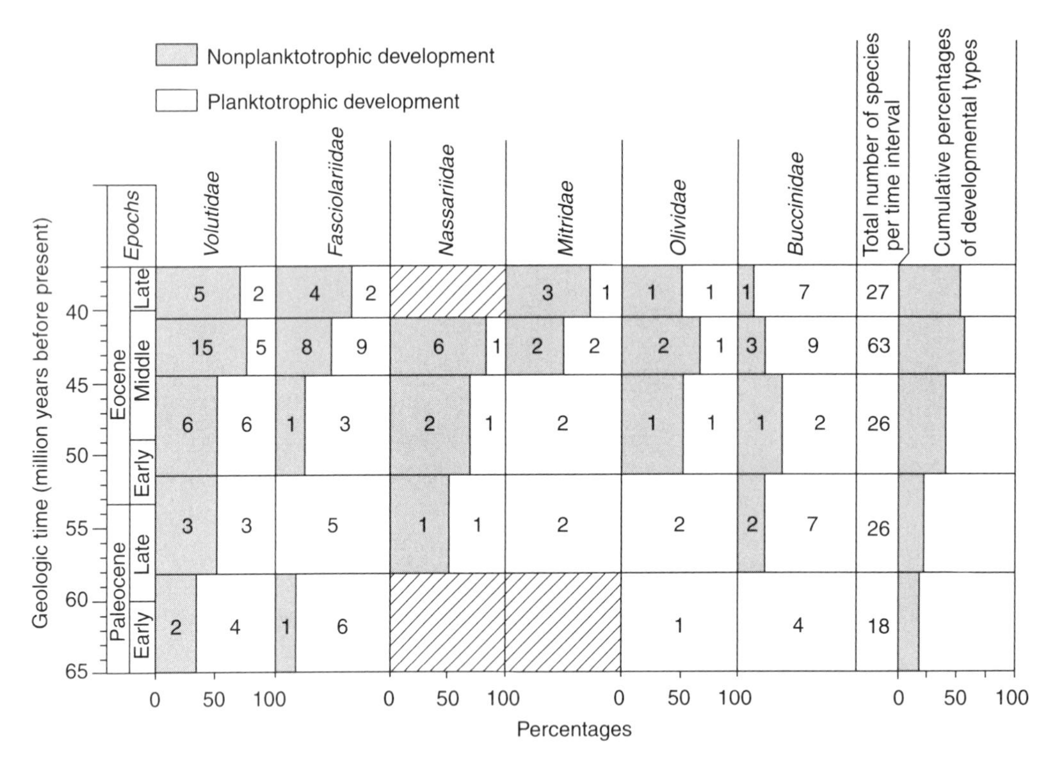


**FIGURE 7.30 Number of planktotrophic and nonplanktotrophic species within several gastropod families during the Early Tertiary of the Gulf and Atlantic coasts of the United States.** Numbers are counts of species for which larval mode could be assessed. Hatched boxes indicate cases in which species are known but there are no preserved protoconchs that would allow larval mode to be determined. The percentage of nonplanktotrophic species increases over time. *(From Hansen, 1982)*

a biased direction of change, and the phylogenetic trend toward increasing nonplanktotrophy appears to be influenced by directed speciation.

***Speciation-Rate Bias*** If species of a certain form have a higher rate of production of daughter species than do species of an alternative form, the number of the first kind will tend to increase over time, shifting the clade average as it does so (Figure 7.28b).

This mechanism may also play a role in the increase in nonplanktotrophy in gastropods. Given their lower dispersal abilities, we would expect nonplanktotrophs to be more likely to form isolated populations and thus to have a higher speciation rate. This difference in speciation rate can be demonstrated in a rather simple way. Figures 7.31a and 7.31b show the durations of planktotrophic and nonplanktotrophic species in the family Volutidae. Nonplanktotrophic species have about half the average duration of planktotrophic species, implying that the extinction rate of nonplanktotrophs is about twice as high.

The higher extinction rate of nonplanktotrophs may be a consequence of their smaller geographic ranges (Figures 7.31c and 7.31d), which would make them more susceptible to chance environmental fluctuations. Regardless of the reason why nonplanktotrophs are more extinction-prone, they increase in diversity at a higher rate despite their higher extinction rate. It is therefore a logical necessity that they also have a higher speciation rate. If the results for volutids are typical, this may help to account for the increasing phylogenetic trend in the proportion of nonplanktotrophic species.

There are two potential complications with this example. First, to explain a trend in terms of differences in speciation rate, we must tacitly assume that like species give rise to like: planktotrophs to planktotrophs and nonplanktotrophs to nonplanktotrophs. Although some switching between larval modes is known, as we saw earlier, a number of phylogenetic studies suggest that switching is limited. However, much more work is needed before we can generalize.

Second, the inference of higher speciation rate in nonplanktotrophs depends on the accurate estimation of extinction rates. Because nonplanktotrophic species have smaller geographic ranges, it is conceivable that they have a less complete fossil record than planktotrophs and therefore exhibit artificially short durations and an

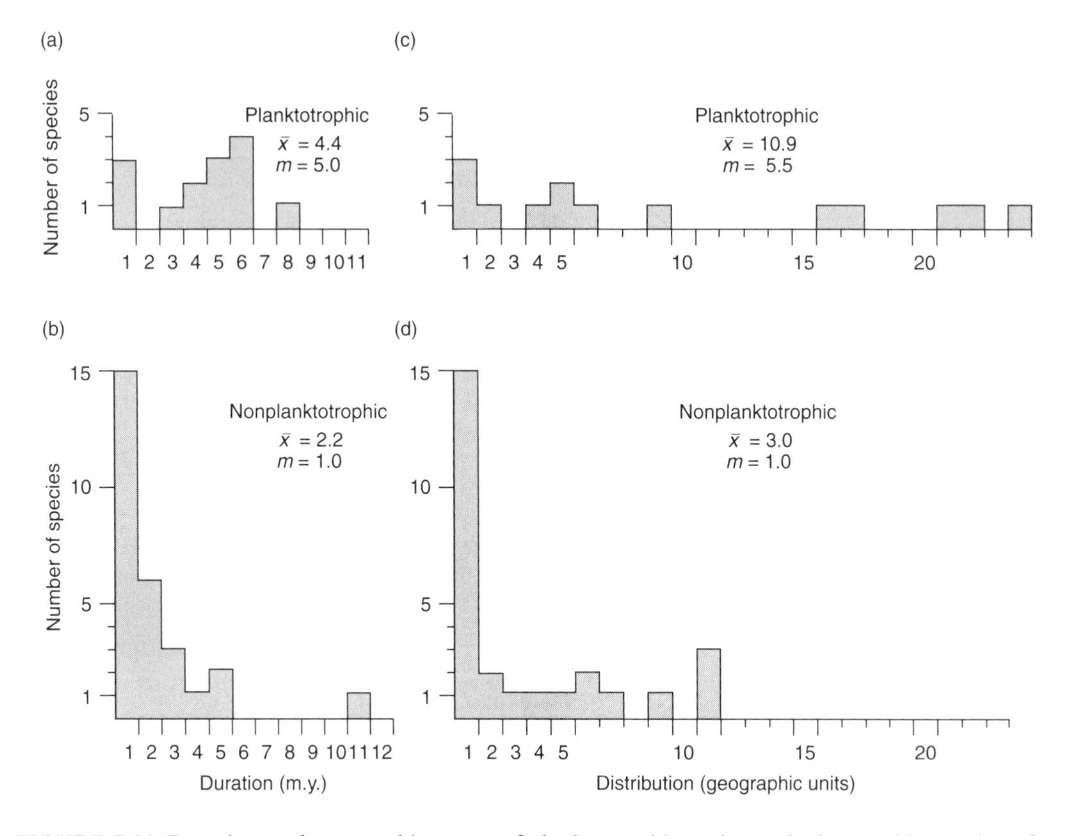


**FIGURE 7.31 Duration and geographic range of planktotrophic and nonplanktotrophic species of the gastropod family Volutidae from the Early Tertiary of the United States.** The outcrop belt of the Gulf and Atlantic coasts was divided into units roughly 75 km wide. Geographic range is measured as the number of these units spanned by the species. The mean and median longevity and geographic range are indicated by $\bar{x}$ and $m$. Nonplanktotrophic species have smaller geographic ranges and shorter durations on average.
*(From Hansen, 1980)*

artificially high extinction rate relative to planktotrophs. This is a classic conundrum in paleontology: We observe a striking pattern—in this case, a correlation between geographic range and longevity—which can seemingly be explained as either a true biological effect or an artifact. Fortunately, we can resolve this problem.

Earlier we showed that the effects of incompleteness on extinction rates can be mitigated by disregarding species known from a single stratigraphic interval (see Box 7.2). If we omit the leftmost bars in the histograms of Figures 7.31a and 7.31b, we find that the planktotrophic species have a mean duration of 5.3 m.y., or an extinction rate of 0.19 per Lmy, while nonplanktotrophs have a mean duration of 3.5 m.y, or an extinction rate of 0.28 per Lmy. Median durations are similar. Thus, the higher extinction rate of nonplanktotrophs is not an artifact of a less complete fossil record. Moreover, the magnitude of the difference—roughly 50 percent higher extinction rate in nonplanktotrophs—is biologically significant. It is comparable, for instance, to differences between groups of crinoids, and between gastropods and bivalves, that were presented earlier in this chapter.

***Extinction-Rate Bias*** We just saw that a phylogenetic trend can be produced by a speciation differential that increases the relative numbers of one form versus another. Likewise, a trend can also result from a difference in extinction rates (Figure 7.28c). A likely example of such a trend is seen in the planktonic Foraminiferida (Norris, 1991). Several times during the Late Cretaceous and Cenozoic, planktonic forams went through cycles of increasing and decreasing diversity. During the increasing phases in the Paleogene and Neogene, forms lacking a pronounced keel became predominant relative to keeled forms (Figure 7.32). As shown in Figure 7.33, the durations of unkeeled species are longer on average than those of keeled species. In other words, unkeeled forms have a substantially lower extinction rate. Although

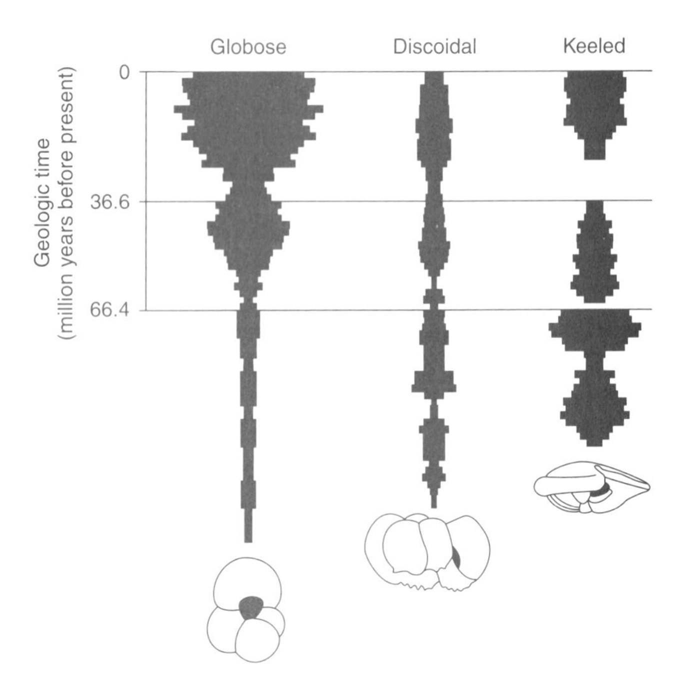


**FIGURE 7.32 Species diversity of keeled and unkeeled (globose and discoidal) planktonic foraminifera during the Late Cretaceous and Cenozoic.** The width of each bar is proportional to the number of species. The proportion of unkeeled forms increases over the course of the Cenozoic. *(From Norris, 1991)*

the reasons for this are still unclear, the lower extinction rate of unkeeled species evidently contributes to their preferential diversification.

***Asymmetric Increase in Variance*** It is possible that trends within clades are explained not so much by what happens during the evolution of the clade but rather by the form possessed by its founding member—in other words, not by where the clade is headed but by where it began (Figure 7.28d). As an explanation for Cope's Rule, Steven Stanley (1973) noted that orders of mammals tend to originate at small body size and to reach larger sizes over time as the number of species increases (Figure 7.34). There is a minimum size that a mammal must attain in order to function properly—for example, a size below which it is impossible for a tiny mammal to garner sufficient food to support its high metabolism. Other biologic groups also face similar limits. Starting near the lower limit, there is more opportunity for a clade to increase in body size (or some other trait) than to decrease. Both the maximum and mean trait value of such a clade will tend to increase over time.

The reason for origination of mammal orders at small size may be that large body size often brings with it

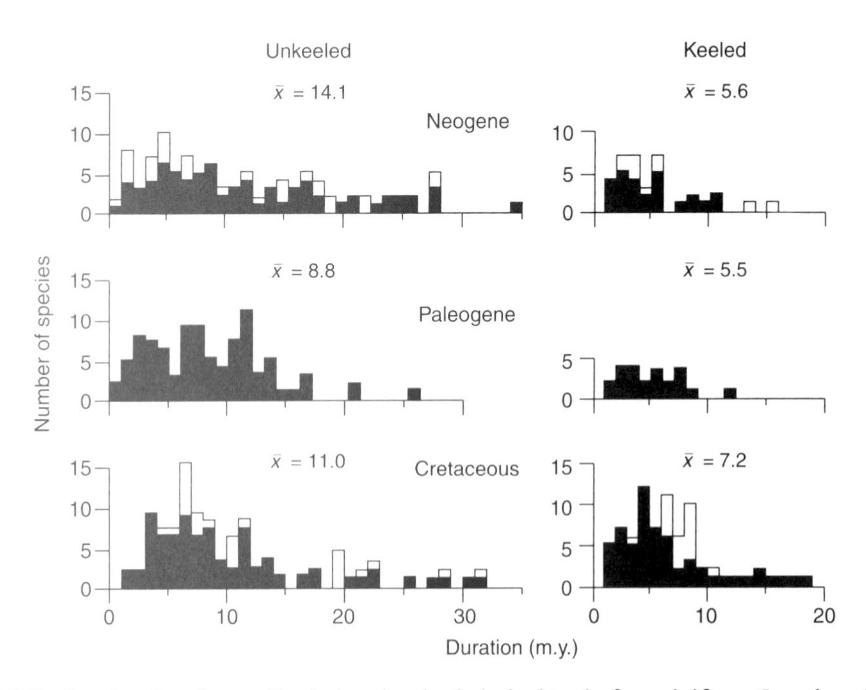


**FIGURE 7.33 Species durations of keeled and unkeeled planktonic foraminifera.** Open bars show durations that are truncated either by the end-Cretaceous extinction event or by the present day. Mean durations are indicated by $\bar{x}$. Average duration of unkeeled forms is longer than that of keeled forms. *(From Norris, 1991)*

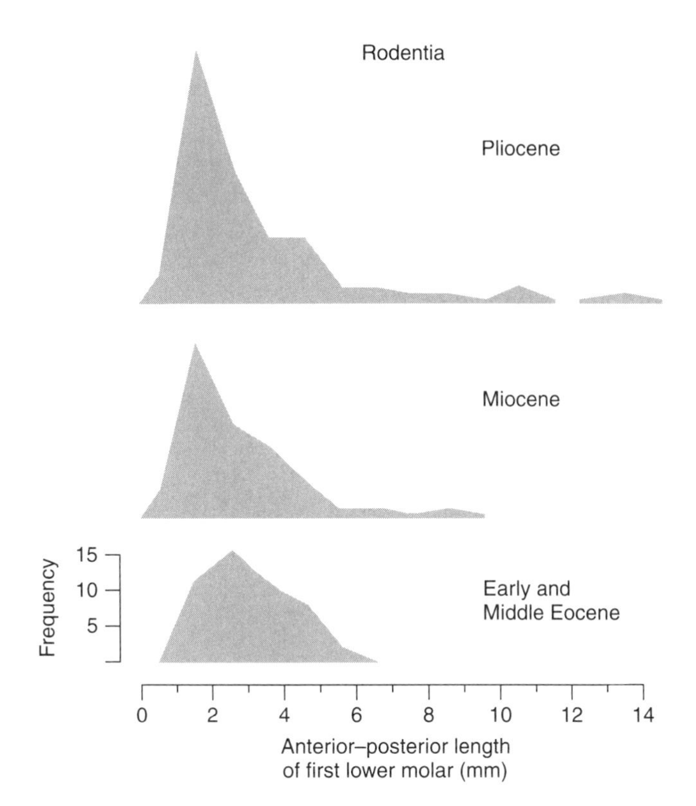


**FIGURE 7.34 Body size (estimated by molar length) in Tertiary rodents.** Each species is represented by the largest recorded size. Frequency distributions for species in different time intervals show that the mean and maximum size increase while the modal and minimum sizes stay roughly constant. *(From Stanley, 1973)*

many structural and ecological specializations. A transition to a major new way of life, which would characterize the origin of a new order, is generally thought to be unlikely in a very specialized lineage.

A number of important trends in the history of life may be at least partly attributable to the origin of major biologic groups near some lower limit. In addition to the increase in body size just mentioned, there is the case of structural complexity. Although complexity can be difficult to define, a number of reasonable operational measures have been developed, including number of cell types, number of kinds of organelles within cells and organs within organisms, amount of DNA, and differentiation among serial structures such as vertebrae. The first organisms of which we have a fossil record were single-celled and morphologically simple. Even though bacteria still dominate the earth's ecosystems today, the maximal complexity attained by organisms seems to have increased throughout earth history.

The fact that a clade may have originated near some structural limit does not necessarily mean that this is the *cause* of a phylogenetic trend. Returning to the horse example, this family originates at small body size and seems to spread out asymmetrically (Figure 7.26). Yet we have already seen that there is a strong tendency for body size to increase from ancestors to descendants (Figure 7.27). Thus, the origin at small size may contribute to the trend in horses, but it cannot be the sole explanation.

## 7.5 CONCLUDING REMARKS

This last point underscores the complexity that underlies evolutionary rates and trends. Out of convenience and necessity, we generally contrast alternative, simple models. But in actual cases, it is almost certain that a mixture of mechanisms was at work. A single lineage sometimes evolves directionally and sometimes is static. A taxonomic group whose average extinction rate we wish to estimate actually consists of some lineages that are highly susceptible to extinction and others that are nearly immortal. Several causes of trends may be acting simultaneously—for example, directed speciation and biased speciation rate in the case of larval mode in gastropods—and they may in principle act in opposition. The simplicity of models should not be seen as a flaw, however. Without them, we would have abundant data on morphology and stratigraphic occurrence, but we would be hobbled in our ability to interpret these data.

We have conspicuously ignored the role of adaptation in phylogenetic trends. Yet in our discussion of theoretical morphology in Chapter 5, we assumed that adaptation is a major cause of the relative numbers of species with different morphological features. In some cases, it is clear that phylogenetic trends are likely to be adaptive in nature—most confidently when they are underlain by phyletic trends in the same direction. In other cases, this is not so obvious. We might posit that superior adaptation leads to a decrease in extinction rate, and perhaps even to an increase in origination rate through the reduced extinction probability of isolated populations that have the potential to become new species [SEE SECTION 3.3]. One difficulty with this line of reasoning is that it implies a negative correlation between origination and extinction rate, when the two are in fact positively correlated in general. Moreover, many documented differences in origination and extinction rates appear to be underlain by differences in geographic range and other aspects of population structure, which are not organism-level adaptations.

In summary, it is likely that adaptation often plays a role in phylogenetic trends. Given the shortage of detailed case studies, however, we do not know just how strong its role is in general. The broader issue here is that of bridging the gap between the microevolutionary timescale, where adaptation may be most important, and the timescale of macroevolution.

We have taken a brief look at some of the fundamental ways in which paleontology has provided unique insights into evolution, especially on geologic timescales. Our examples illustrate the richness of data and approaches brought to bear on macroevolutionary questions: morphology, geologic duration, geographic range, and phylogeny, to name but a few. Our understanding of large-scale evolution has been significantly enhanced by considering mechanisms such as species selection that involve processes other than accumulated, generation-by-generation evolution within populations. Insights from the fossil record have helped both paleontologists and biologists to understand better the relationships between micro- and macroevolution. Macroevolution must, of course, be consistent with microevolution. But, practically speaking, it is often impossible to predict particular outcomes of macroevolution from microevolutionary observations alone.

Other reasons why this is so are discussed in the next chapter, when we consider the possible role of mass extinction episodes in eliminating accumulated evolutionary change within lineages, removing ecologically established taxa, and generally changing the rules of evolutionary success and failure that operate at other times.

## SUPPLEMENTARY READING

Erwin, D. H., and Anstey, R. L. (eds.) (1995) *New Approaches to Speciation in the Fossil Record.* New York, Columbia University Press, 342 pp. [A series of papers illustrating a variety of ways that speciation can be studied in the fossil record, with particular emphasis on the question of punctuated equilibrium.]

Gould, S. J. (2002) *The Structure of Evolutionary Theory.* Cambridge, Mass., Harvard University Press, 1433 pp. [Comprehensive overview of punctuated equilibrium and related topics in macroevolution.]

Jablonski, D. (2000) Micro- and macroevolution: Scale and hierarchy in evolutionary biology and paleobiology. *Paleobiology* **26** (Suppl. to No. 4):15–52. [Comprehensive overview of major themes in macroevolution, including species selection and determinants of taxonomic longevity.]

Levinton, J. (2001) *Genetics, Paleontology, and Macroevolution*, 2nd ed. New York, Cambridge University Press, 617 pp. [Treatment of macroevolution in the fossil record from the standpoint of biology.]

McNamara, K. J. (ed.) (1990) *Evolutionary Trends.* Tucson, Ariz., University of Arizona Press, 368 pp. [Series of papers on theoretical issues in the analysis of trends and case studies from the fossil record.]

McShea, D. W. (1994) Mechanisms of large-scale evolutionary trends. *Evolution* **48:**1747–1763. [Critical assessment of models for explaining phylogenetic trends.]

Raup, D. M. (1985) Mathematical models of cladogenesis. *Paleobiology* **11:**42–52. [A useful summary of models of orignation and extinction, with a detailed appendix giving equations for survivorship and other important properties of species and clades.]

Roopnarine, P. D. (2003) Analysis of rates of morphological evolution. *Annual Review of Ecology, Evolution and Systematics* **34:**605–632. [Overview of technical problems in the measurement and interpretation of rates of evolution.]

Simpson, G. G. (1953) *The Major Features of Evolution.* New York, Columbia University Press, 434 pp. [Landmark in the quantitative study of rates and trends.]

Stanley, S. M. (1979) *Macroevolution: Pattern and Process.* San Francisco, W. H. Freeman and Company, 332 pp. [Important treatment of the factors that give rise to macroevolutionary patterns.]

Van Valen, L. M. (1973) A new evolutionary law. *Evolutionary Theory* **1:**1–30. [Pioneering application of the exponential survivorship model to paleontological data.]

*Chapter 8*

# GLOBAL DIVERSIFICATION AND EXTINCTION

The investigation of global diversity has flourished in paleontology during the past quarter century. In a sense, this can be viewed as the culmination of efforts spanning several centuries to catalogue the contents of the fossil record. But there is little doubt that diversity studies have also been spurred on by additional factors, including (1) the advent of computers, which permit the assembly and analysis of large databases; (2) a growing interest in the history of global diversity through geologic time, which relates to environmental change coupled with evolutionary patterns and processes highlighted throughout this book; and (3) present-day concerns about the ongoing crisis in diversity (or, as it is commonly known, **biodiversity**), for which the fossil record provides historical perspective at timescales far exceeding the human life span.

In this chapter, we consider the major features of global diversification and extinction, including large-scale transitions in taxonomic composition; time intervals characterized by significantly elevated extinction rates (**mass extinctions**); regional variations in global diversity trends; and the analysis of **morphological diversity** as a complement to the more traditional reliance on taxonomic diversity.

## 8.1 THE NATURE OF BIOLOGICAL DIVERSITY

The term **diversity** has taken on a variety of meanings in biological and paleontological research. For example, ecologists often characterize the diversity of a given set of taxa with metrics that account jointly for the number of unique taxa in a sample (**taxonomic richness**) and the abundances of each taxon. Recently, paleontologists have also formalized the concept of morphological diversity, which provides an alternative to the strict assessment of taxonomic composition as a measure of macroevolutionary dynamics and trends [SEE SECTION 8.10].

With respect to global diversity, the main focus has been on the calibration and explanation of trends through time in global taxonomic richness. As we will see, these efforts for the Phanerozoic Eon, the interval of earth history characterized by an abundant record of multicellular organisms, are complemented by consideration of diversity trends at local and regional scales and how these combine to produce the patterns that we observe at the global level. But much of the motivation for local and regional studies came from the recognition of a set of intriguing patterns first observed at the global level. Thus, the initial focus in this chapter is on the tools that permit global-scale analyses, including the development of global taxonomic databases and the use of these data to construct global diversity curves. We have already considered, in Chapter 7, methods for measuring origination and extinction; these measurements will also be important to our discussions in this chapter.

## 8.2 GLOBAL TAXONOMIC DATABASES

Before constructing a graph that depicts the global history of diversity through successive intervals of geologic time, fossil taxa must first be catalogued in a database that lists, for each taxon, the interval of its first and last known global appearances. Although there are various methods for actually tabulating a diversity curve (see below), the basic objective is to determine from these data the number of taxa that were extant from interval to interval.

Substantial efforts therefore have focused on the development of databases that accurately capture information on fossil occurrences collected worldwide. This has been aided for more than a century by the development of encyclopedic compilations of known fossil occurrences over broad regions, or, in the case of publications like the *Treatise on Invertebrate Paleontology*, the entire world. Historically, the development of regional or global diversity curves has followed closely on the assembly of these compilations. Among the earliest examples was John Phillips's pair of graphs (Figure 8.1), produced in 1860, for fossil marine biotas of Great Britain, based on John Morris's publication *A Catalogue of British Fossils* (1854). Despite the geographic limitation of the data that Phillips depicted, the clear implication was that he was capturing patterns of broader global significance, and, indeed, the major features of these graphs are shared by global compilations produced more than a century later.

The best-known databases used to compile Phanerozoic global marine diversity curves in recent times were developed by Jack Sepkoski, at the family and genus levels. Other efforts have been equally ambitious, resulting in diversity curves that are quite similar to those derived from Sepkoski's databases. A portion of Sepkoski's genus-level compendium is shown in Figure 8.2, which illustrates two important features of the database. First, the degree of stratigraphic resolution in first and last known global appearances is highly variable, but, in general, Sepkoski sought to resolve these appearances to fine-scale subdivisions of the geologic timescale, such as subepochs. Second, while Sepkoski initiated his data compilations by extracting information from the *Treatise on Invertebrate Paleontology*, his efforts moved far beyond the *Treatise*. The collection of new data from the fossil record and the refinement of taxonomic designations is an ever-evolving process for even well-known taxa, and any single volume of the *Treatise* inevitably becomes outdated shortly after, or even during, its publication. Thus, Sepkoski

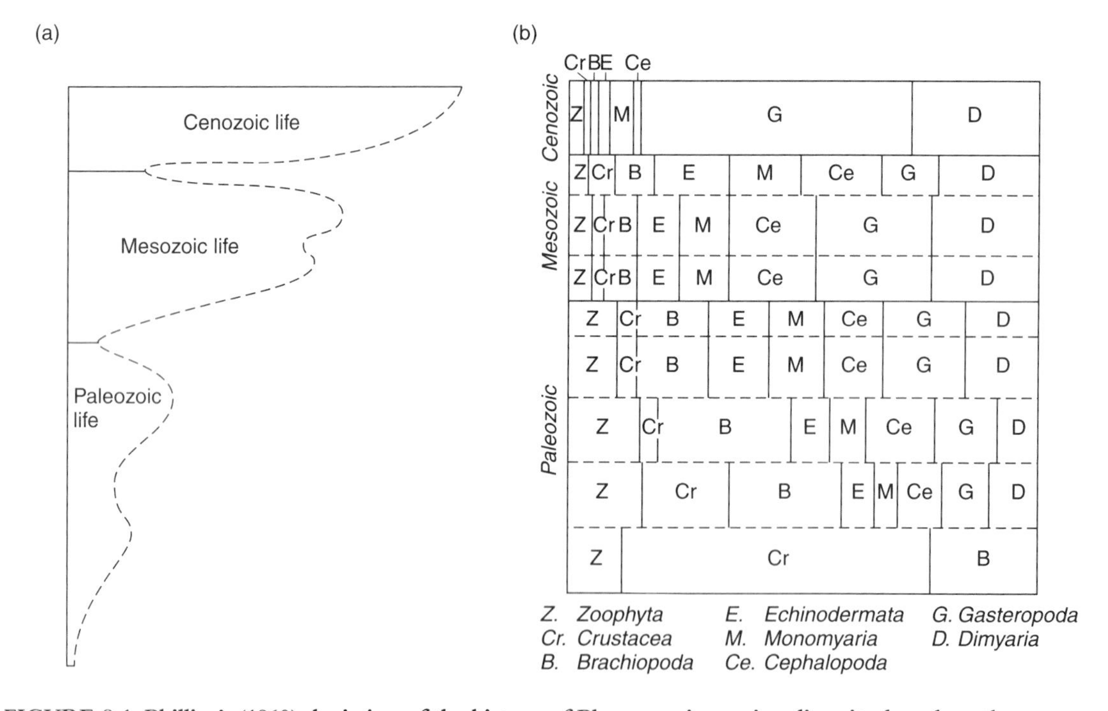


**FIGURE 8.1 Phillips's (1860) depiction of the history of Phanerozoic marine diversity, based on the occurrences of marine fossils in Great Britain.** (a) An illustration of changes through time in the number of species. (b) A summary of changes through time in the relative contributions of major taxa to the overall composition of the marine biota. Although some of the names are outdated (e.g., "Crustacea" refers primarily to trilobites and "Dimyaria" are bivalve molluscs), the transitions that Phillips depicted closely approximate those illustrated more than a century later by other researchers. *(From Phillips, 1860)*

| | |
|---|---|
| *Rugosowerbyella* | O (Ashg-l) - O (Ashg-u) |
| *Rurambonites* | O (Cara-u)?- O (Ashg-u) |
| *Rutrumella* | O (Llvi) |
| *Sampo* | O (Cara-u) - O (Ashg-u) |
| *Sanjuanella* | O |
| *Schedophyla* | O (Aren-l) - O (Llvi)? |
| *Sentolunia* | O (Ashg-l) |
| *Sericoidea* | O (Llde-u) - O (Ashg-u) |
| *Shlyginia* | O (m) |
| *Sowerbyella* | O (Llvi-u)?- S (Ldov-l) |
| *Sowerbyites* | O (Llde-u) - O (Cara-u) |
| *Spanodonta* | O (Trem-u) - O (m)? |
| *Strophomena* | O (Llvi-l) - S (Ldov-l) |
| *Syndielasma* | O (Llvi-l) |
| *Taffia* | O (Aren-l) - O (Llvi) |
| *Taphrodonta* | O (Llvi-u) |
| *Teratelasma* | O (Cara-l) |
| *Tetraodontella* | O (Llvi-l)?- O (Cara-m) |
| *Tetraphalerella* | O (Llde) - O (Ashg-m) |

**FIGURE 8.2** A small portion of Sepkoski's (2002) genus-level global compendium, depicting the global stratigraphic ranges of several strophomenide brachiopods that were extant during the Ordovician Period, abbreviated in the compendium by the letter O. Intervals of first and last known global appearances are indicated, except when the genus is confined to a single interval, in which case only a single designation is provided (e.g., *Rutrumella*). Note that the stratigraphic resolution is variable. Designations in parentheses are abbreviated names for series or stages of the geologic timescale, some of which have changed in updated versions of the timescale. In many cases, finer subdivisions of these intervals are designated as lower (−l), middle (−m) and upper (−u).

relied on the primary literature to continually expand and refine his compendia.

This last point serves as a general reminder that, in the study of fossil biodiversity, the data are never complete because new discoveries and analyses that affect the data are almost continuously forthcoming in the literature. Nevertheless, as discussed in Chapter 1, there is substantial evidence that, for broad perspectives on Phanerozoic global diversity trends, databases like Sepkoski's compendia have matured sufficiently to capture accurately the main biotic transitions preserved in the fossil record [SEE SECTION 1.5].

## 8.3 CONSTRUCTION OF GLOBAL DIVERSITY CURVES

Methods for constructing global diversity curves are conceptually straightforward, but variations have been proposed over the past several years, as described and illustrated in Box 8.1. The objective is to characterize the number of taxa extant in consecutive stratigraphic intervals and then to depict these values on an $x$–$y$ plot. As explained in Box 8.1, there are many ways to count the number of taxa, all of which rest on the convention that each taxon *ranges through* the entire stratigraphic interval between its first and last appearances, as depicted in the database.

A potential problem with Phanerozoic-scale diversity curves is the possibility that values are inflated toward the present day, a pattern termed **Pull of the Recent.** Two separate aspects to this phenomenon concern us here. The first, as discussed in Chapter 1, is the increase, particularly in the Cenozoic, in the amount of sedimentary rock and hence the number of fossils available for sampling. The second is a direct consequence of the range-through assumption and the likelihood that the Recent (i.e., the present day) is far better sampled than even well-preserved parts of the fossil record (Figure 8.4). As the Recent is approached, it is probably not uncommon for taxa that occur in just one stratigraphic interval in the fossil record (stratigraphic singletons), or that are limited to a narrow stratigraphic window, to have a Recent representative. Whenever this is the case, the taxa in question are credited to the entire interval between the Recent and their first, and perhaps only, appearance in the fossil record. This inflates diversity above the level that would be achieved if no data from the Recent were available. This is of little or no consequence, say, for Paleozoic taxa, because they are unlikely to have Recent representatives. But it may happen frequently with late Mesozoic and, especially, Cenozoic taxa.

The actual inflation induced by the Pull of the Recent is difficult to quantify. One way of assessing this problem would be to develop a database that includes information not only on the global first appearance of a taxon that ranges to the Recent, but also on known occurrences that fall after the first appearance. With these data, it could be determined for a given extant taxon whether there remains a significant gap between its *last* known fossil appearance and the Recent.

In a pioneering analysis of bivalve molluscs that adopted this approach, David Jablonski and colleagues (2003) studied 958 fossil genera and subgenera that are still extant today. Of these, 906 (95 percent) have fossil representatives in Pliocene and/or Pleistocene strata, suggesting that, at least for this important Cenozoic group, the Pull of the Recent is minimal. Questions remain concerning the possibility that the Pliocene and Pleistocene records are, in themselves, so unusually outstanding that they produce a "Pull of the Plio-Pleistocene."

*Box 8.1*

## CONSTRUCTION OF DIVERSITY CURVES

This box illustrates the construction of global diversity curves using methods discussed in the main text. The data for these curves, presented in Table 8.1, consist of 40 hypothetical genera whose first and last appearances fall within the epochs of the Cenozoic; 21 of the taxa are extant (i.e., their "last appearance" is in the Recent). These data are used to construct a set of values, presented in Table 8.2, that provide the basis for the curves depicted in Figure 8.3.

An important assumption in the construction of all diversity curves, known aptly as the **range-through** assumption, is that a taxon remains extant for the entire interval between its first and last known occurrences in the fossil record. Importantly, if a taxon that is represented in the fossil record is also extant today, it is therefore assumed that the taxon ranged through the entire interval from its first appearance in the fossil record through the present day, even if its first appearance is its only known fossil occurrence (i.e., it is a fossil singleton).

To calculate diversity by the methods illustrated here, we must first determine the number of originations (N. $\text{orig}_t$) and the number of extinctions (N. $\text{ext}_t$) in each interval. For example, in perusing Table 8.1, we can see that there were nine first appearances and one last appearance in the Paleocene. Therefore, in Table 8.2, N. $\text{orig}_t$ for the Paleocene is 9, and N. $\text{ext}_t$ is 1. Once these values have been determined for all intervals, diversity can be calculated for the standard and boundary-crosser methods (Figures 8.3a and 8.3b)

**TABLE 8.1**

Global Stratigraphic Ranges for a Set of Hypothetical Genera

| Genus | First Appearance | Last Appearance | Genus | First Appearance | Last Appearance |
|---|---|---|---|---|---|
| a | Paleocene | Recent | u | Pliocene | Recent |
| b | Miocene | Miocene | v | Oligocene | Pleistocene |
| c | Paleocene | Eocene | w | Eocene | Eocene |
| d | Eocene | Eocene | x | Pleistocene | Recent |
| e | Oligocene | Recent | y | Paleocene | Paleocene |
| f | Pleistocene | Recent | z | Miocene | Pliocene |
| g | Oligocene | Miocene | aa | Paleocene | Miocene |
| h | Paleocene | Eocene | bb | Eocene | Recent |
| i | Pliocene | Recent | cc | Pliocene | Recent |
| j | Eocene | Eocene | dd | Pleistocene | Pleistocene |
| k | Oligocene | Pleistocene | ee | Pleistocene | Recent |
| l | Pleistocene | Recent | ff | Paleocene | Eocene |
| m | Pleistocene | Pleistocene | gg | Miocene | Recent |
| n | Pliocene | Recent | hh | Oligocene | Recent |
| o | Miocene | Recent | ii | Paleocene | Eocene |
| p | Paleocene | Oligocene | jj | Pleistocene | Recent |
| q | Eocene | Recent | kk | Pliocene | Recent |
| r | Miocene | Recent | ll | Miocene | Recent |
| s | Paleocene | Eocene | mm | Pleistocene | Recent |
| t | Pleistocene | Recent | nn | Pleistocene | Pleistocene |

**TABLE 8.2**

Values and Equations Used to Calculate Diversity Curves for the Hypothetical Data from Table 8.1

| Interval | Ma | N. $orig_t$ | N. $ext_t$ | $N_{st}$ | $d_t$ with All Data Included | N. $orig_t$ without Singletons | N. $ext_t$ without Singletons | $d_t$ without Singletons | N. $orig_t$ with Recent Occurrences Ignored | N. $ext_t$ with Recent Occurrences Ignored | $d_t$ with Recent Occurrences Ignored | Boundaries (for Boundary-Crosser Method) |
|---|---|---|---|---|---|---|---|---|---|---|---|---|
| Paleocene | 60 | 9 | 1 | 1 | 9 | 8 | 0 | 8 | 9 | 2 | 9 | Paleocene/Eocene |
| Eocene | 45 | 5 | 8 | 3 | 13 | 2 | 5 | 10 | 5 | 10 | 12 | Eocene/Oligocene |
| Oligocene | 28 | 5 | 1 | 0 | 10 | 5 | 1 | 10 | 5 | 3 | 7 | Oligocene/Miocene |
| Miocene | 14 | 6 | 3 | 1 | 15 | 5 | 2 | 14 | 6 | 7 | 10 | Miocene/Pliocene |
| Pliocene | 4 | 5 | 1 | 0 | 17 | 5 | 1 | 17 | 5 | 6 | 8 | Pliocene/Pleistocene |
| Pleistocene | 1 | 10 | 5 | 3 | 26 | 7 | 2 | 23 | 10 | 12 | 12 | Pleistocene/Recent |

**Key:** $d_t$ = diversity in interval $t$
$d_{t-1}$ = diversity in interval $t - 1$
$d_{t/t+1}$ = diversity at boundary between intervals $t$ and $t + 1$
N. $orig_t$ = number of originations in interval $t$
N. $ext_t$ = number of extinctions in interval $t$
N. $ext_{t-1}$ = number of extinctions in interval $t - 1$
$N_{st}$ = number of singletons in interval $t$

**Standard Method for Calculating Diversity:**
$d_t = d_{t-1} + \text{N. orig}_t - \text{N. ext}_{t-1}$

Example: When all data are included,
$d_{\text{Oligocene}} = d_{\text{Eocene}} + \text{N. orig}_{\text{Oligocene}} - \text{N. ext}_{\text{Eocene}}$;
thus, $d_{\text{Oligocene}} = 13 + 5 - 8 = 10$

**Boundary-Crosser Method for Calculating Diversity:**
$d_{t/t+1} = d_t - \text{N. ext}_t$
Example:
$d_{\text{Oligocene/Miocene}} = d_{\text{Oligocene}} - \text{N. ext}_{\text{Oligocene}}$;
thus, $d_{\text{Oligocene/Miocene}} = 10 - 1 = 9$

*continued on next page*

*Box 8.1 (continued)*

using the equations presented in the table. In addition, the table presents values based on variations of the standard method that exclude singletons (Figure 8.3c) and Recent occurrences (Figure 8.3d). The elimination of Recent occurrences, advocated by some researchers, automatically relegates to singleton status all genera that range to the Recent. In practice, a decision to ignore these occurrences should be coupled with an attempt to discover the last known fossil occurrences of these taxa to avoid artificial inflation of the number of singletons (see text for further discussion).

It should also be noted that many of the values described here and depicted in Table 8.2 are also relevant to our discussion of evolutionary rates in Chapter 7, albeit with a slightly different terminology (see Box 7.2 for a comparison of these terms).

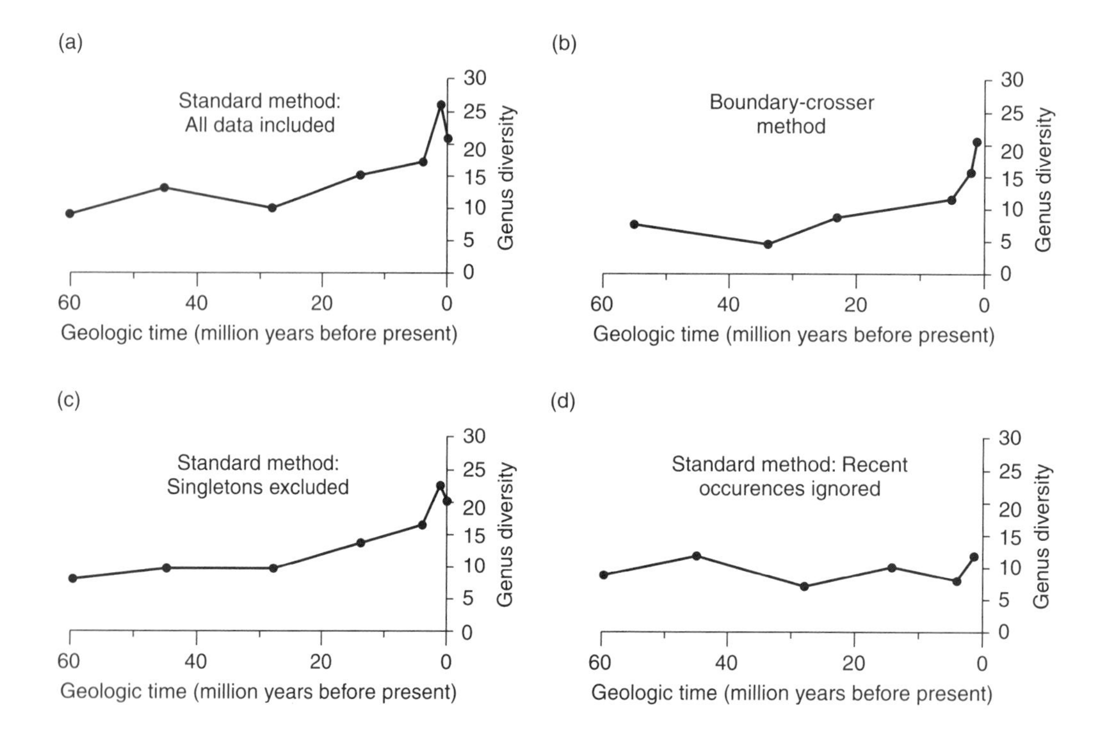


**FIGURE 8.3 Four depictions of taxonomic diversity through the Cenozoic Era, based on four alternative treatments of the hypothetical data presented in Table 8.1.** (a) Standard method, with all data included. (b) Boundary-crosser method. (c) Standard method with singletons excluded. (d) Standard method with Recent occurrences excluded.

This possibility has yet to be evaluated, and it also remains to be determined whether the pattern documented by Jablonski and colleagues will hold up for other taxa. Nevertheless, this analysis illustrates a promising, if time-consuming, procedure for addressing what has been a persistent question in paleontology.

The Pull of the Recent is but one feature that might cause a global diversity trend to depart from the true biological signal. Even in intervals far removed from the Recent, variations from interval to interval in the availability of fossil samples, the intensity of sampling, and interval duration could all cause further distortion. Sepkoski reasoned that one way to help overcome these problems would be to exclude singletons from consideration. In Sepkoski's view, variations in the number of singletons from interval to interval relate directly to variations in sampling. If a particular interval has significantly more singletons than the intervals that surround it, this could be taken to indicate that the size of the sample for that interval is inflated in some way. Thus, in

| Taxa | Paleocene | Eocene | Oligocene | Miocene | Pliocene | Pleistocene | Recent |
|---|---|---|---|---|---|---|---|
| A | F | | L | | | | X |
| B | F | L | | | | | X |
| C | | F | L | | | | X |
| D | | F | | L | | | X |
| E | | | F | | L | | X |
| F | | | | F | L | | X |
| G | | | | | F | L | X |
| H | | | | | | F | X |
| Standing diversity: Recent excluded | 2 | 4 | 4 | 3 | 3 | 2 | |
| Standing diversity: Recent included | 2 | 4 | 5 | 6 | 7 | 8 | |

**FIGURE 8.4 A schematic illustration of the Pull of the Recent.** The first (F) and last (L) known fossil occurrences are illustrated for eight hypothetical taxa that are all known to be extant. Using the "standard" method for calculating diversity (illustrated in Box 8.1), the inclusion of Recent occurrences significantly inflates standing diversity, beginning in the Oligocene, relative to what it would have been without the inclusion of Recent occurrences.

producing his genus-level depiction of Phanerozoic diversity (Figure 8.5b), Sepkoski did not include singletons. This procedure, which dampens interval-to-interval variations in diversity (see Box 8.1), has been adopted by several other authors, reflecting a growing consensus that this helps to mitigate variations in sampling intensity.

A diversity curve can be built at any taxonomic level. Family- and genus-level depictions of Phanerozoic diversity are illustrated in Figure 8.5 and are broadly similar to one another, but they differ in two respects that relate to the hierarchical nature of taxonomic classification. First, and not surprisingly, there were substantially more genera than families during most Phanerozoic intervals. Second, the genus-level curve is more volatile than the family-level depiction, punctuated by increases and decreases that are more exaggerated. This is because a net change in family diversity during a given time interval will *necessarily* be accompanied by a change in genus diversity that is at least

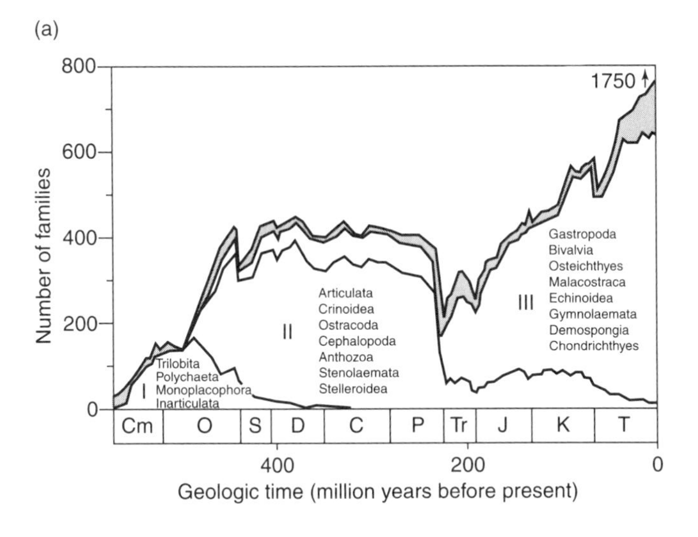


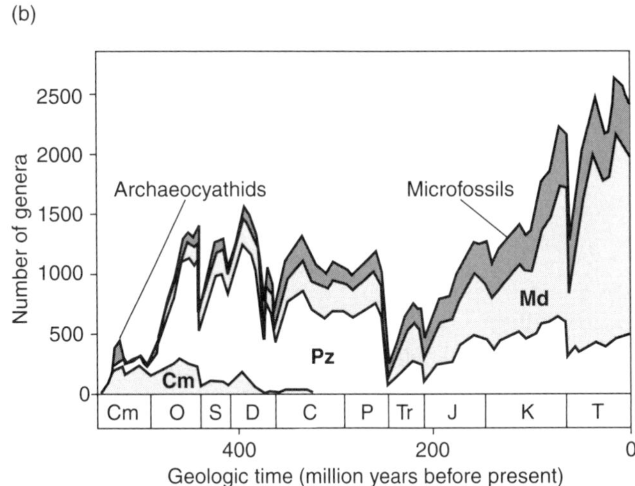


**FIGURE 8.5 Sepkoski's depictions of the Phanerozoic history of global marine diversity.** In this and subsequent figures that depict multiple groups, the curves are "stacked," meaning that they are cumulative. (a) Family diversity: The gray portion is for taxa classified by Sepkoski as poorly preserved. Subsets of the graph designated with roman numerals I through III delineate Sepkoski's three evolutionary faunas (see text and Figure 8.8). (b) Genus diversity: The evolutionary faunas are designated with abbreviations: Cm for Cambrian, Pz for Paleozoic, and Md for Modern. The dark gray portion depicts archaeocyathid diversity in the Cambrian, and microfossils thereafter. *(a: Sepkoski, 1981; b: Sepkoski, 1997)*

as large: Every family contains at least one genus, and, given that many families contain more than one genus (sometimes many more), the accompanying change at the genus level will likely be more pronounced.

In contrast, a net change in genus diversity *need not* be accompanied by a change at the family level. In the most extreme cases, an increase in genus diversity could take place entirely within families that are already extant, and a decrease in genus diversity could take place without the extinction of any families to which the genera belong. Major intervals of extinction are marked by significant declines at both the genus and family levels, but the declines are inevitably more pronounced, on a percentage basis, at the genus level.

To further dampen the effects of variations in sampling and interval length (see Box 7.3), Richard Bambach and other researchers have advocated the use of only **boundary-crossing** taxa in Phanerozoic diversity compilations (Figure 8.6), in contrast to the "standard method" (Box 8.1) used by Sepkoski and many other workers. Bambach tabulated standing diversity at boundaries between intervals by determining the total number of genera extant in the older interval and then subtracting from that value the number of genera whose ranges ended in the older bin. If done sequentially for a set of boundaries, this conveys for each interval the change in the number of genera entering it and leaving it (see Section 7.2 and Box 8.1 for an illustration of this method in comparison to others). Undoubtedly, the methods used to reconstruct the history of Phanerozoic global diversity will continue to evolve as researchers seek to extract meaningful biological signals from the raw material of the fossil record.

The main focus in this chapter thus far has been on the development of graphs depicting Phanerozoic diversity for marine animals and protists. Corresponding compilations and graphs have also been developed for terrestrial animals and for plants. Examples that illustrate Phanerozoic patterns among marine and terrestrial vertebrates and among terrestrial plants are illustrated in Figure 8.7. In these depictions, and in those of marine diversity that we looked at earlier, there are major transitions in taxonomic composition throughout the Phanerozoic, as well as significant increases in total diversity in the approach to the Recent. While the Pull of the Recent and other sampling issues might inflate the appearance of the Mesozoic–Cenozoic increase, there is little doubt that the underlying taxonomic transitions took place at the approximate times indicated by these graphs. A major challenge confronting paleontologists has been to explain these transitions, and to determine whether macroevolutionary themes common to all realms and taxa are responsible for them. We turn to these issues in the next section.

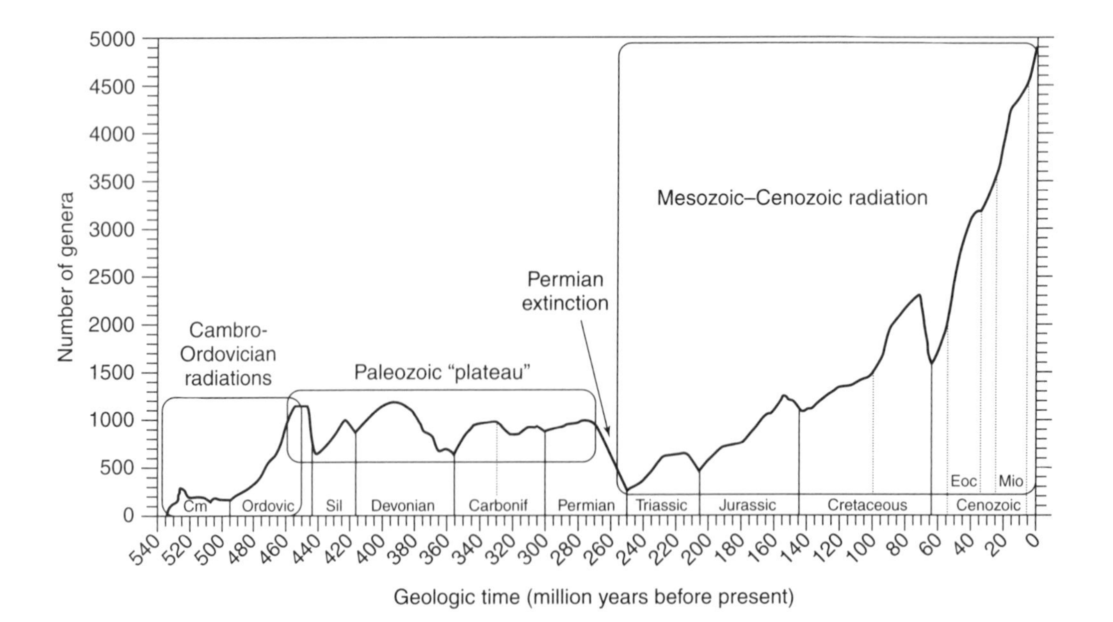


**FIGURE 8.6 Bambach's depiction of the Phanerozoic history of global marine genus diversity, using only boundary-crossing genera.** *(Bambach, 1999)*

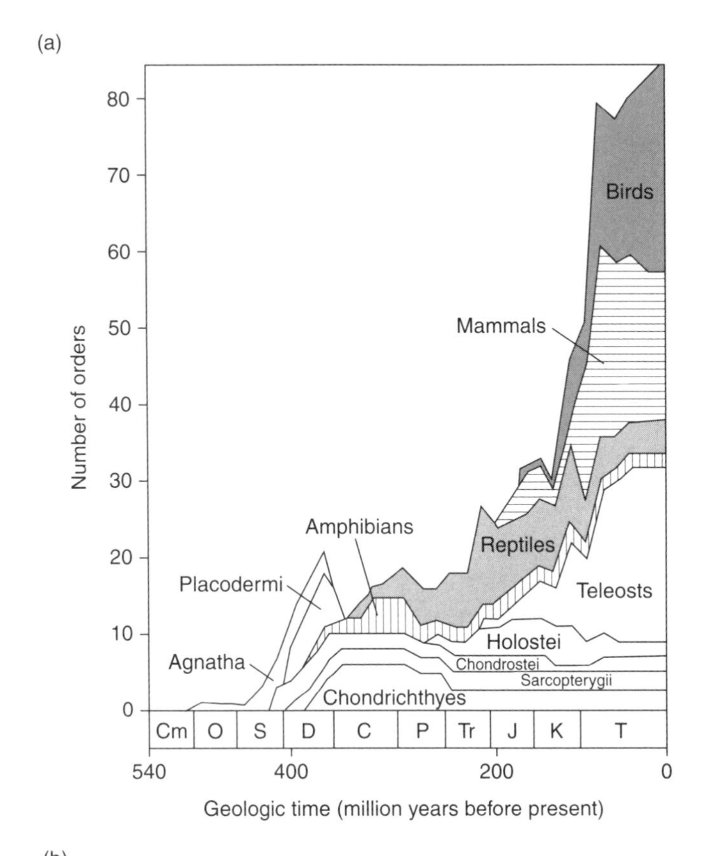


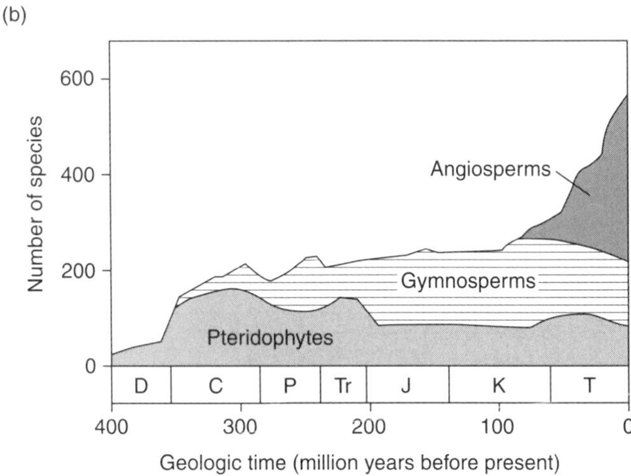


**FIGURE 8.7 Other examples of Phanerozoic global diversity curves.** (a) Vertebrate orders. (b) Terrestrial plant species. *(a: Padian & Clemens, 1985; b: Niklas, 1997)*

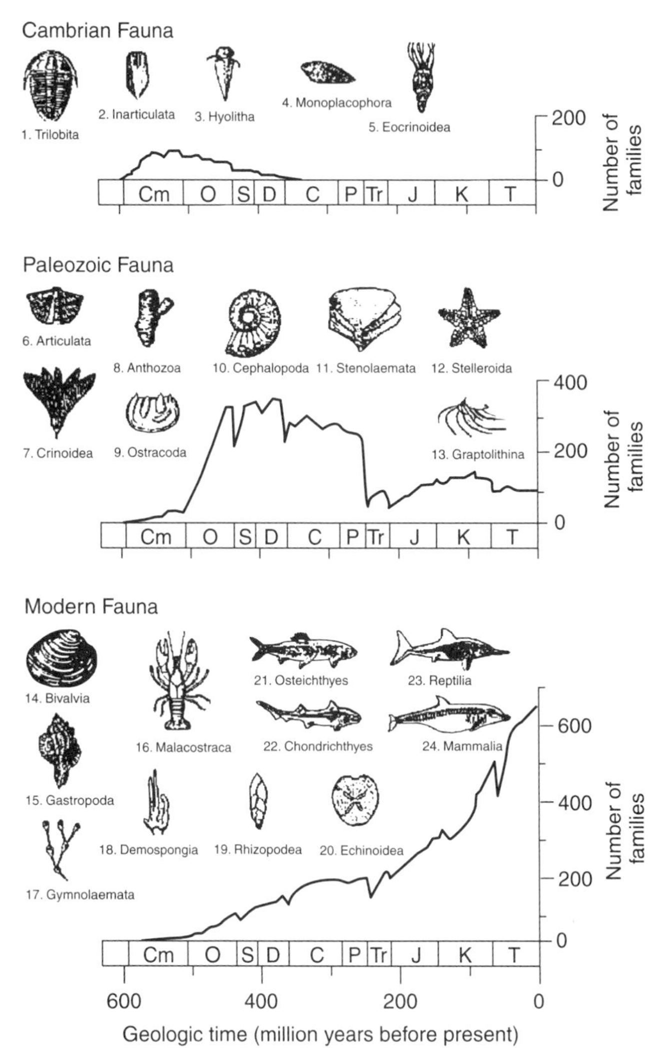


**FIGURE 8.8 Sepkoski's (1984) depictions of the major taxa in each of his three Phanerozoic marine evolutionary faunas.** *(Sepkoski, 1984)*

## 8.4 PHANEROZOIC TRANSITIONS IN TAXONOMIC COMPOSITION

### The Marine Realm

Building on the pioneering efforts of Karl Flessa and John Imbrie (1973), Jack Sepkoski (1981) presented a quantitative description of transitions among marine taxa through the Phanerozoic. Sepkoski coined the term **evolutionary fauna** to describe broad sets of taxa that were globally dominant through extended geologic intervals. He recognized three Phanerozoic evolutionary faunas (Figures 8.5 and 8.8). The **Cambrian Fauna** dominated marine settings throughout the Cambrian following the initial burst of diversification at the start of the Phanerozoic known as the **Cambrian Explosion** [SEE SECTION 10.2]. The **Paleozoic Fauna** diversified significantly during the **Ordovician Radiation,** when global diversity attained unprecedented levels that were maintained through much of the remaining Paleozoic. The **Modern Fauna** exhibited limited diversity throughout the Paleozoic Era but diversified appreciably in the post-Paleozoic to become the dominant biota of the Mesozoic and Cenozoic Eras.

Sepkoski argued that the three evolutionary faunas were more than just coincidental collections of taxa. He viewed them as functional units that interacted with one another, *causing* the major global biotic transitions observed through the Phanerozoic. This perspective has been challenged by many paleontologists, in part because it is now understood that evolutionary faunas were neither as internally cohesive nor as distinct from one another in space and time as Sepkoski once envisioned [SEE SECTION 9.4]. Nevertheless, it is important to consider Sepkoski's quantitative perspective of transitions among evolutionary faunas because of its central role in developing a large-scale, synthetic outlook in the investigation of biodiversity.

Sepkoski's goal was to develop a mathematical description of the major features of Phanerozoic marine

*Box 8.2*

## DEVELOPMENT OF THE COUPLED LOGISTIC MODEL

The use of the coupled logistic model by Sepkoski and others to simulate Phanerozoic diversity patterns can be explained in three steps. We begin by first considering the simple case of a diversity trajectory exhibited during an exponential diversification. To simulate an exponential diversification, we can use the following equation:

$$d_t = d_{t-1} + rd_{t-1}$$

where $d_t$ represents diversity (the number of taxa) in interval $t$, $d_{t-1}$ represents diversity in interval $t - 1$, and $r$ denotes a constant rate of increase (sometimes referred to as the *intrinsic* rate of increase). In simulations of Phanerozoic diversification, Sepkoski set the value of a simulated time unit to 1 million years.

An example of the application of this equation is presented in Figure 8.9. Here, the starting diversity at time interval 0 is set to unity, and the intrinsic rate of increase, $r$, is set to 2. For example, we can calculate the diversity at time 1:

$$d_1 = d_0 + rd_0$$
$$d_1 = 1 + (2 \times 1)$$
$$d_1 = 3$$

(a)

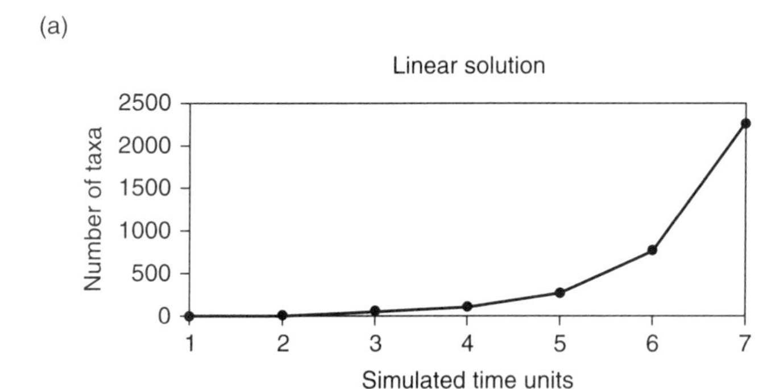


(b)

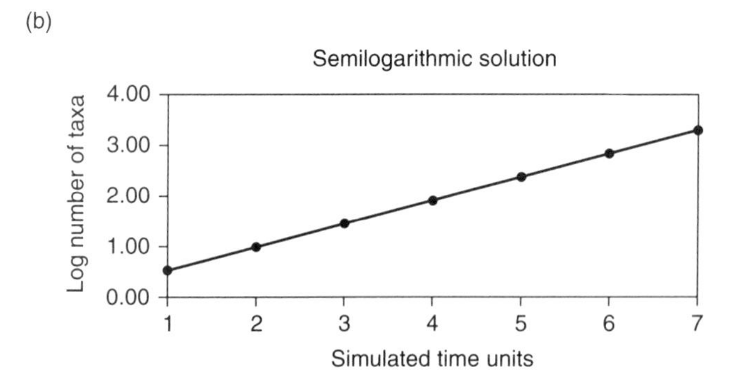


**FIGURE 8.9 Trajectory of an exponential diversification for the example discussed in Box 8.2.** (a) Linear axis for diversity. (b) Logarithmic axis for diversity. Note that when diversity is plotted logarithmically, the trajectory is a straight line.

In Figure 8.9, the results of the simulation are plotted for time units 1 through 7, both linearly and semilogarithmically. The semilog plot illustrates an important attribute of an exponential diversification, indicated here by the straight line—a constant per-taxon rate of diversification. By this we mean that the rate of diversification exhibited by a single taxon remains unchanged throughout the simulation. In this example, diversity always triples from one time unit to the next. For instance, as we move from time unit 6 to time unit 7, each of the 729 taxa extant in time unit 6 triples on average in number, thereby resulting in 2187 taxa in time unit 7.

Next, we consider simple logistic diversification (Sepkoski, 1978). The equation for a simple logistic diversification can be developed by first altering

diversification at the family level. At the heart of the model that he developed for this purpose is the **logistic equation,** which produces a sigmoidal (i.e., s-shaped) curve, describing an initial, nearly exponential growth in species richness, followed by a continuous decline in the rate of growth until an *equilibrium* level is approached (see Box 8.2 for a detailed explanation of the logistic model). Building on earlier research by R. H. MacArthur and E. O. Wilson (1967) on the colonization of newly emergent islands, Sepkoski extended the concept of equilibrium to global marine diversity by reasoning that the earth's oceans collectively constitute a finite space with limited resources in which marine diversity cannot continue to increase indefinitely.

Sepkoski first suggested that a *simple* logistic equation (see Box 8.2) adequately describes the Phanerozoic

slightly the structure of our equation for exponential diversification:

$$d_t = d_{t-1} + (k_s - k_e)d_{t-1}$$

where $k_s$ represents the rate of taxonomic origination, and $k_e$ the rate of taxonomic extinction. This is in recognition of the fact that diversification (characterized earlier as the constant $r$) is the product of a balance between the origination and extinction of taxa: $r = k_s - k_e$. If $k_s - k_e > 0$, then diversity will increase; if $k_s - k_e < 0$, then diversity will decrease.

The premise of the simple logistic equation is that rates of origination and extinction are both affected by the number of taxa already present, because there is only a finite amount of space available for the subsistence of taxa. Thus, the rate of origination is thought to decrease, and the rate of extinction is thought to increase, as diversity increases. In their simplest forms, the relationships between evolutionary rates and diversity can be modeled as linear functions:

$$k_s = k_{s0} - ad \quad \text{and} \quad k_e = k_{e0} + bd$$

where $k_{s0}$ and $k_{e0}$ are the initial rates of origination and extinction at the start of the simulation, and $a$ and $b$ are constants that describe the slope (rate) of the decrease in the origination rate and the increase in the extinction rate as functions of increasing diversity. The right-hand sides of these equations can be substituted directly for $k_s$ and $k_e$ in our earlier equation:

$$d_t = d_{t-1} + [(k_{s0} - ad_{t-1}) - (k_{e0} + bd_{t-1})]d_{t-1}$$

This equation describes simple logistic diversification. While it may appear a bit intimidating, it is actually fairly straightforward. The primary difference from the equation for exponential diversification is that instead of remaining fixed for the entire simulation, the rates of origination and extinction converge as diversity increases. Because of this convergence, the rate of diversification, which is nearly exponential at the start of the simulation, will begin to decrease until we reach equilibrium diversity—that is, the point at which the rate of origination is equal to the rate of extinction. From then on, diversity will remain unchanged.

An example of a curve generated with the simple logistic equation is illustrated in Figure 8.10a. This solution was produced using the parameters provided in the figure and by setting diversity at time $0(d_0)$ to 4. Note that the curve is sigmoidal, reflecting the decrease in the total rate of diversification associated with the approach to equilibrium diversity. All simple logistic curves are sigmoidal in shape. However, several aspects of this curve vary from solution to solution, including the rate at which diversity increases initially, the rate at which diversification begins to decline as equilibrium is approached, and the actual value of the equilibrium. These attributes of the curve are contingent on the parameter values, like those used to generate Figure 8.10a. In general, a greater initial rate of diversification will be associated with a greater difference in the initial rates of origination ($k_{s0}$) and extinction ($k_{e0}$). The rate at which diversification decays as equilibrium is approached will be greater in cases where slopes are greater in the decay of the origination rate $a$ and the increase in the extinction rate $b$.

Finally, we consider coupled logistic diversification (Sepkoski, 1979, 1984), which entails the simultaneous diversification of two or more groups. The premise of the model is that the rates of origination and extinction

*continued on next page*

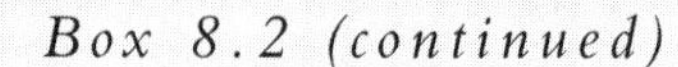


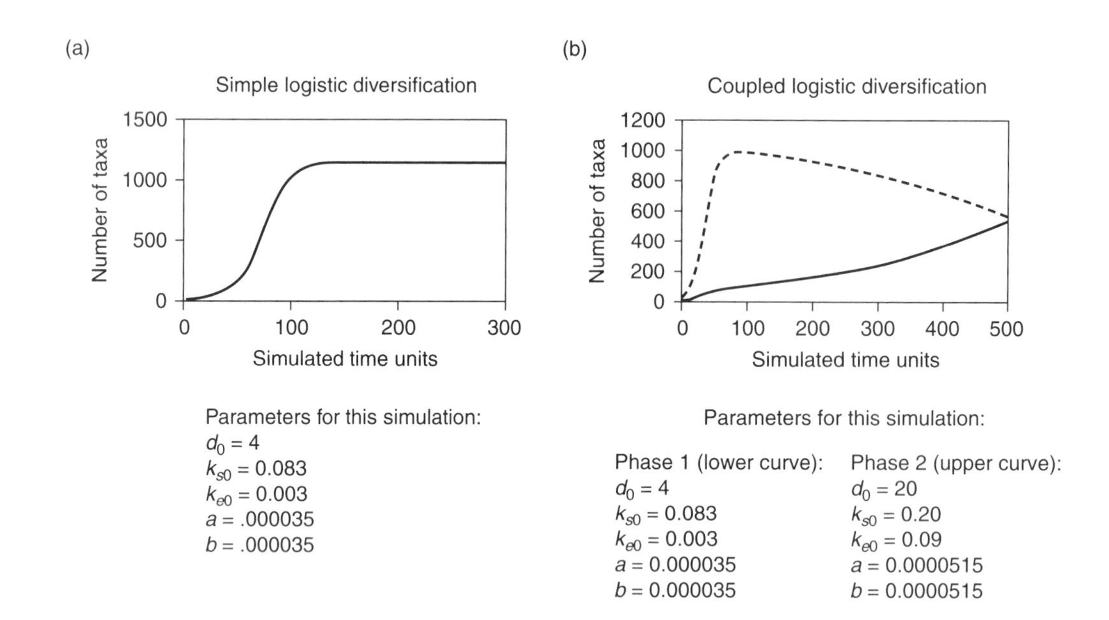


**FIGURE 8.10 Trajectories for logistic diversification for the simple and coupled models.** (a) Curve for a simple logistic model, based on parameters designated beneath the graph. (b) Curves for a two-phase (i.e., two-curve), coupled logistic model, based on parameters for each phase designated beneath the graph. Note that the parameters for the lower curve are identical to those used for the lone curve in the simple logistic solution.

for a given group are affected not only by the number of constituent taxa already present for the group, but also by the number of constituent taxa belonging to other groups diversifying at the same time. Thus, this model is viewed as interactive, or "coupled," because the diversification of one group affects that of the others.

In a coupled logistic model, each modeled group is referred to as a phase; the diversification of a phase is governed by parameters defined in the coupled logistic equation:

$$d_{x,t} = d_{x,t-1} + [(k_{s0} - a\mathrm{DTOT}_{t-1}) - (k_{e0} + b\mathrm{DTOT}_{t-1})]d_{x,t-1}$$

where $d_{x,t}$ represents the diversity of phase $x$ in interval $t$, $d_{x,t-1}$ represents the diversity of phase $x$ in interval $t - 1$, and $\mathrm{DTOT}_{t-1}$ is the total diversity of all phases in interval $t - 1$. All other parameters are identical to those of the simple logistic equation, but the actual values for the four constants—$k_{s0}$, $k_{e0}$, $a$, and $b$—usually vary among the phases.

history of marine diversity at the taxonomic level of orders (Figure 8.11). The diversity of orders increased dramatically through the Cambro-Ordovician, but then leveled off and maintained a fairly steady state thereafter. However, as we have already seen (Figure 8.5), this is a clear departure from the pattern of Phanerozoic diversification observed at the family and genus levels, both of which exhibit significant increases through the post-Paleozoic. This difference reflects the observation that the plurality of taxa at the order level and higher originated during the late Precambrian through Paleozoic, whereas an increasing number of families, genera, and species continued to originate through the whole of the Phanerozoic.

To accommodate the post-Paleozoic family-level increase, Sepkoski developed a series of three *coupled* logistic equations (see Box 8.2), which corresponded to the three Phanerozoic evolutionary faunas. The central premise of coupling is that the level of diversity achieved at any point in time by any one of the components depends not only on its standing diversity, but also on the summed diversity of all three components. Using the

Comparison with the equation for simple logistic growth indicates only two differences between the simple and coupled equations: (1) the inclusion in the coupled version of $DTOT_{t-1}$ in place of $d_{t-1}$ in the terms describing the decline in the rate of origination and the increase in the rate of extinction; and (2) the possibility that the rate of extinction can exceed the rate of origination as the simulation proceeds. This causes the diversification of a phase to be impeded by that of the other phases, in accord with the premise of the coupled model. Just as importantly, as illustrated by the following example, it opens up the possibility that a phase will experience a decline in total diversity in the event that the extinction rate does indeed exceed the origination rate.

Examples of two simultaneously diversifying curves generated with coupled logistic equations are presented in Figure 8.10b. The parameters for Phase 1 are identical to those used in the example of simple logistic diversification. Note, however, that the trajectory of Phase 1 is rather different from that in the simple logistic example. The diversification of Phase 1 is impeded early in the simulation, and its subsequent growth is much slower than in the simple case. This is a consequence of its numerical interaction with Phase 2, which initially diversifies much more rapidly than Phase 1 and thus adds significantly to the DTOT term. The added diversity included in the DTOT term accelerates the rate of decline in Phase 1's origination rate and the rate of increase in its extinction rate, relative to the simple model.

Despite the initial lag in Phase 1, its diversification slowly accelerates, quite literally at the expense of Phase 2. Ultimately, it would have overtaken Phase 2 had the simulation been continued beyond 500 time units. Why? It is a direct consequence of the relative values of the parameters for the two phases. Comparison of the initial origination and extinction rates for the two phases shows that the initial rate of diversification $(k_s - k_e)$ is greater for Phase 2 than for Phase 1 (0.11 versus 0.08). Thus, early in the simulation, Phase 2 diversifies at a rate that far exceeds that of Phase 1. However, the rate of decay in origination and growth in extinction ($a$ and $b$) for Phase 1 is also less than that for Phase 2 (0.000035 versus 0.0000515). Thus, as the summed diversity (DTOT) of both phases grows, the cost to Phase 2 is greater than that to Phase 1: It experiences a more rapid decline in origination and a more rapid growth in extinction, moving beyond the point where extinction exceeds origination, causing a steady decline in the diversity of Phase 2 (Miller & Sepkoski, 1988).

In folktale parlance, Phase 1 is the "tortoise" and Phase 2 the "hare." Much the same relationship exists among the three phases of Sepkoski's (1984) coupled logistic model that describes the Phanerozoic diversity trajectories of his three evolutionary faunas: the modeled Cambrian Fauna has a greater initial rate of diversification, but a greater rate of decay in diversification than the modeled Paleozoic Fauna. The modeled Modern Fauna, in turn, initially diversifies at an even slower rate than the modeled Paleozoic Fauna, but the rate of decay in its diversification rate is less than that of the modeled Paleozoic Fauna.

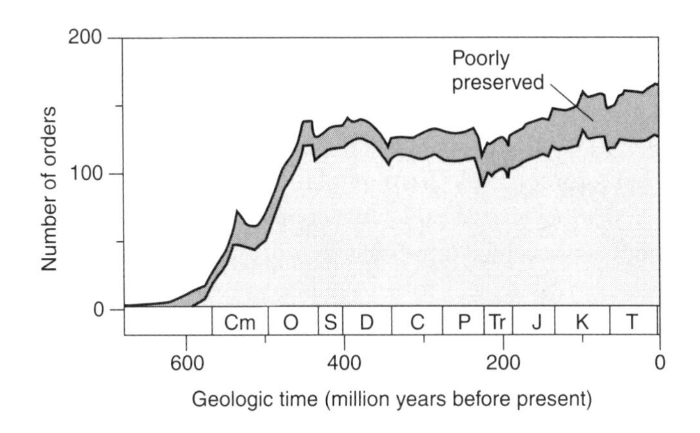


**FIGURE 8.11 Sepkoski's depiction of Phanerozoic marine diversity at the order level.** The darker portion is for orders that Sepkoski described as poorly preserved. *(Sepkoski, 1978)*

(a)

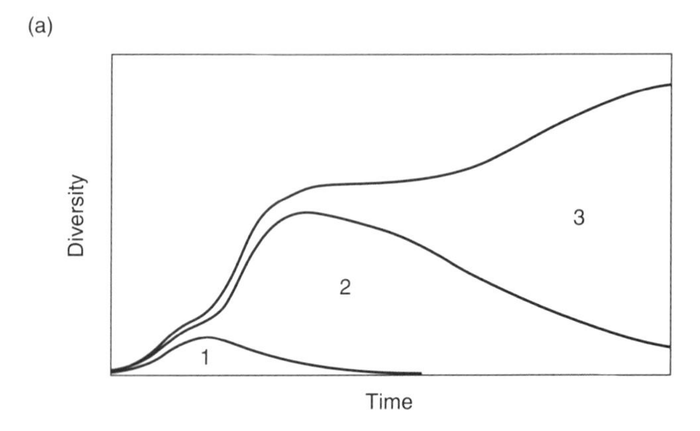


(b)

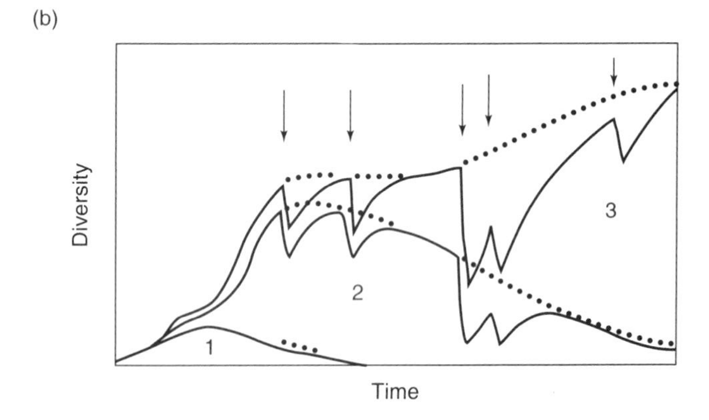


**FIGURE 8.12 Sepkoski's simulation of the Phanerozoic history of global marine family diversity, based on the three-phase coupled logistic model.** (a) Without mass extinctions. (b) With the "big five" Phanerozoic mass extinctions included in the simulation, as indicated by the arrows. Dotted curves show the trajectory without mass extinctions [i.e., the trajectory in part (a)] for comparison; solid curves show the trajectory with mass extinctions. *(Sepkoski, 1984)*

coupled approach, Sepkoski produced a simulated family-level trajectory (Figure 8.12) that closely approximated the actual pattern, particularly when major extinction events [SEE SECTION 8.6] were imposed in the simulation (compare Figure 8.12b with Figure 8.5a). The details of the simulated diversity trajectories in Figure 8.12 depend on whether mass extinctions are imposed, but note *that the ultimate fates of the three phases are the same whether or not there are mass extinctions.*

The ability of the model to depict diversity trajectories for the three faunas rests on the assumption that the Paleozoic Fauna could attain a higher level of equilibrium diversity than the Cambrian Fauna, and that the Modern Fauna could achieve a higher level than the Paleozoic Fauna. This was accomplished by selecting model parameters for each of the three curves that corresponded to observed, relative characteristics of the Cambrian, Paleozoic, and Modern Faunas: an average decline from one fauna to the next in initial diversification rates, a similar decline in average origination and extinction rates, and an increase in equilibrium diversity (see Box 8.2).

Although the volume of the earth's oceans has varied through the Phanerozoic, there is little reason to think that there has been a significant net increase in their total volume or in the area of sea floor available for colonization. Thus, the increase in equilibrium diversity for successive faunas, if real, could not have resulted from a simple increase in the amount of space available for colonization. Instead, an increase in ecological space, or the variety of ways that organisms make a living, may have been responsible for the differences among evolutionary faunas. We will return to this topic in Chapter 9.

While the success of the model in describing the observed pattern of marine biotic transitions opens up the possibility that biological interactions among members of the three evolutionary faunas caused the transitions (e.g., through competition for resources or for space), the close match certainly *does not demonstrate* that this was the case. To strengthen the argument of a key role for interactions as agents of long-term change, it is important to document both the precise nature of the interactions and the likelihood that these interactions persisted for extended intervals of geologic time. It is virtually impossible to do this at the broad level of evolutionary faunas, but we might do better if we restrict our analyses to small sets of taxa for which the nature of interactions can be understood.

With this in mind, Sepkoski and colleagues developed a coupled logistic model that described the global transition from cyclostome to cheilostome bryozoans through the Cretaceous Period and the Cenozoic Era (Figures 8.13a and 8.13b). In this case, not only did the model successfully describe the global diversity trajectories of the two groups, but there is also evidence from the fossil record that cheilostomes overgrew cyclostomes in a significant majority of observed cases (Figure 8.13c), opening up the possibility that over the long term, cheilostomes might have outcompeted cyclostomes.

As we suggested earlier, the coupled logistic model has not been accepted universally by paleontologists, and several alternative scenarios have been proposed in recent years. These range from the suggestion that the entire Phanerozoic trajectory of marine diversity is best described as an exponential diversification (Box 8.2; Benton,

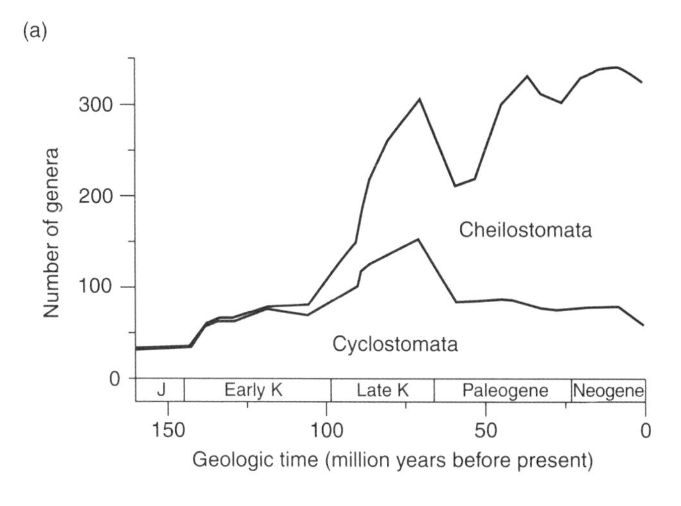


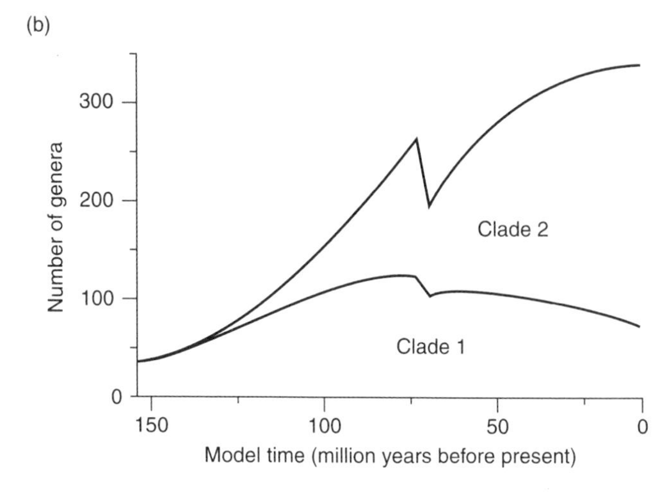


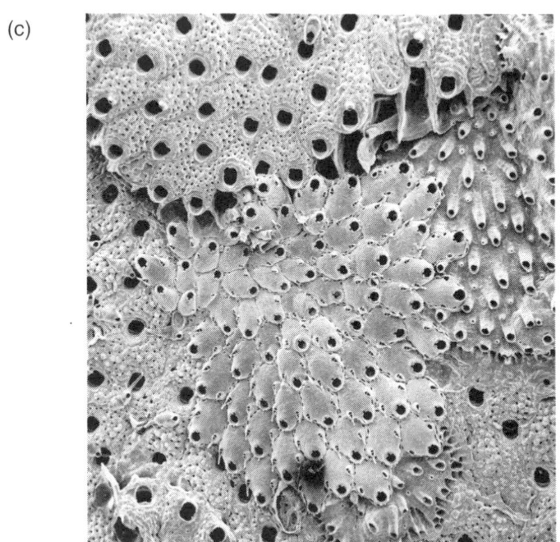


**FIGURE 8.13 Coupled logistic model, applied to the diversification histories of cheilostome and cyclostome bryozoans.** (a) Actual genus diversity graphs. (b) Outcome of the coupled model. (c) An illustration of a cheilostome (center, *Escharina vulgaris*) overgrowing cyclostome colonies (*Diplosolen obelia*) during the present day in the northern Adriatic Sea. The ability of cheilostomes to overgrow cyclostomes in the majority of the cases in which they interact has remained consistent throughout their histories and is thought to relate to the more rapid development of zooids along their colony margins. *(a, b: Sepkoski et al., 2000; c: Photo and interpretation from McKinney, 1992)*

1995), punctuated and sometimes impeded by major extinction events (Stanley, 1999), to the hypothesis that it is best represented as a sequence of simple logistic (Box 8.2) diversification intervals, demarcated and reset by major extinctions (Courtillot & Gaudemer, 1996).

In addition, it has been demonstrated that the turnover rates of taxa in the Modern Fauna increased significantly in the Mesozoic and Cenozoic, which is not accounted for in Sepkoski's set-up of the coupled logistic model (Alroy, 2004). However, even though in the end it may be demonstrated that Sepkoski's coupled model is not appropriate as an explanation of Phanerozoic marine diversification, the model endures as the intellectual starting point for virtually all alternative views. It also continues to provide students of paleontology with a unique opportunity to understand how biodiversity trends can be assessed numerically.

## The Terrestrial Realm

Figure 8.7 exhibits a series of transitions among vertebrates and plants that can be likened to the transitions among Sepkoski's evolutionary faunas. However, the sequential pattern of decreasing turnover rates from one fauna to the next that characterizes the marine realm, and which underlies Sepkoski's coupled logistic model for marine biotas, is not clearly exhibited by the major biotas of the terrestrial realm. For example, whereas all major plant groups may have experienced an initial burst of speciation followed by a rapid decline in speciation rates, there is no evidence that the "evolutionary floras" illustrated in Figure 8.7b exhibited successively lower turnover rates.

In fact, there are indications that, if anything, the opposite was the case, as demonstrated by James Valentine and colleagues (1991). While not ruling out the possibility that a coupled logistic or some alternative model could be developed to describe transitions among terrestrial plants, its numerical dynamics would obviously be different from those for marine animals. Whether these differences would reflect anything meaningful about differences in the evolutionary dynamics of marine and terrestrial systems remains to be seen, as paleontologists have only recently begun to attempt these kinds of comparisons.

## 8.5 PHANEROZOIC DECLINE IN ORIGINATION AND EXTINCTION RATES

Given the observation that marine diversity increased during the Phanerozoic, it follows that overall, the number of originations must have exceeded the number of extinctions, particularly during intervals marked by major increases in diversity (see Box 8.2). Likewise, during intervals of declining diversity, the number of extinctions exceeded the number of originations. But what of longer-term patterns of origination and extinction? Even if there was a long-term increase in diversity, this does not necessarily mean that origination rates must have increased in parallel with diversity. An increase in diversity could also have been caused by a long-term decline in extinction rates. In fact, origination rates might have even declined through the Phanerozoic, so long as extinction rates declined at a greater rate.

An initial representation of average extinction rates for marine families was presented by David Raup and Jack Sepkoski in 1982 (Figure 8.14). Although the main intent of the analysis was to provide a statistical basis for delineating mass extinctions (see the next section), another important aspect of extinction was also documented: a significant, long-term decline in extinction rates through the Phanerozoic. This decline has also been documented at the genus level, and a similar Phanerozoic long-term decline has been shown for origination rates (Figure 8.15) as well. There was a temporary rebound in origination rates at the beginning of the Mesozoic Era in the aftermath of the most extensive extinction event in the history of life [SEE SECTION 8.6], but origination rates then began to decline anew.

A long-term decline in origination rates can also be recognized in several groups of animals and plants in terrestrial settings (Figure 8.16). Therefore, any explanation

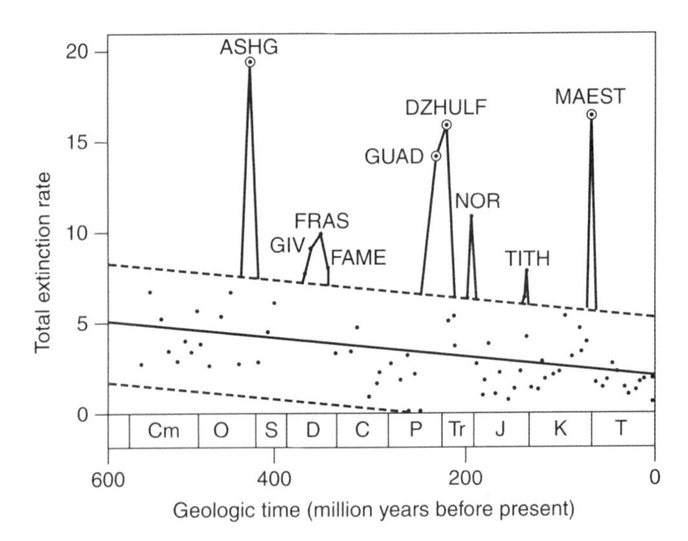


**FIGURE 8.14 Raup and Sepkoski's depiction of extinction rates of marine families, as the number of extinctions per million years, stage by stage through the Phanerozoic.** The solid line is a linear regression fit to the data, and the dotted lines define a 95 percent statistical confidence interval around the regression. Abbreviations for stage names are provided in cases where extinction rates for the stage fall outside the upper 95 percent confidence band (i.e., they are judged to be statistically significant by the standards of this analysis). The circled points (e.g., the Ashgillian interval of the Ordovician) fall outside the upper 99 percent confidence band (i.e., they are highly significant). Note the long-term decline in extinction rates. *(From Raup & Sepkoski, 1982)*

for a decline in origination or extinction rates should transcend the fundamental biological differences among individual higher taxa or the places where they live. Some researchers have proposed that the rate decline indicates a general change through time in the nature of interactions among taxa. For example, as we will see in Chapter 9, the history of life may have been marked by profound "arms races" in marine and terrestrial settings between predator and prey lineages. One potential outcome of these arms races over the long term could have been an increased resistance to extinction, but this does not explain why origination rates also declined.

In marine settings, the decline in origination and extinction rates appears to reflect the overall biotic transition, discussed earlier, from taxa exhibiting high turnover rates (members of the Cambrian and Paleozoic Faunas) to others exhibiting lower turnover rates (members of the Modern Fauna). This, of course, raises the question: Why is it that different higher taxa exhibit turnover rates

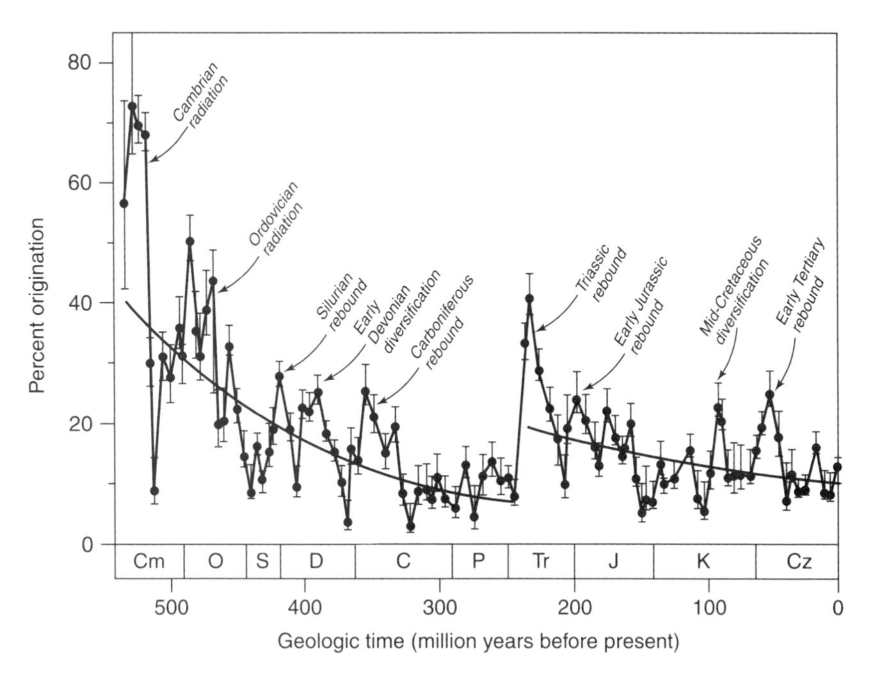


**FIGURE 8.15 Sepkoski's depiction of declining genus origination rates through the Phanerozoic.** Origination was measured in this case as the percentage of genera extant in a given substage that had their first appearance in that substage. Percent origination is simply 100 times proportional origination (see Table 7.4). Note the temporary rebound in rates following the Late Permian mass extinction. *(Sepkoski, 1998)*

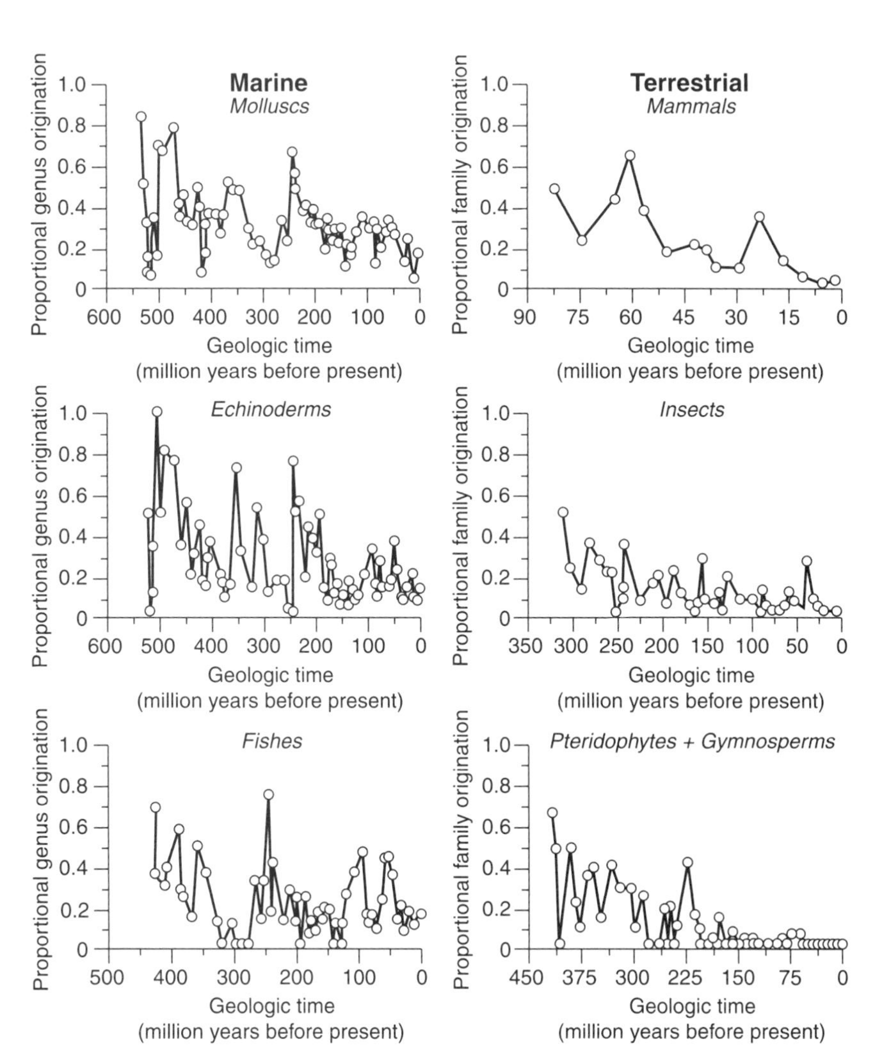


**FIGURE 8.16 Eble's depiction of declining genus origination rates in several marine and terrestrial groups during the Phanerozoic.** These are measured as proportional origination (see Table 7.4). *(Eble, 1999)*

that can be so strikingly different from one another but that can nevertheless be maintained fairly consistently among their constituent families, genera, and lineages? (See Chapter 7 for additional discussion.) This remains one of the unsolved mysteries of evolution, and it is certainly worthy of intensive paleontological research in the future.

## 8.6 MASS EXTINCTIONS

### The Diagnosis of Mass Extinctions

In the past quarter century, perhaps no subject has received more attention from paleontologists than **mass extinctions.** This interest and excitement was spurred on by a finding reported by Nobel prize–winning physicist Luis Alvarez and his son Walter, a geochemist. The upshot of the discovery, which we will discuss later, was that a major, global extinction event at the end of the Cretaceous Period may have been caused by the impact of a large asteroid or comet. The scientific debate that was triggered by the Alvarez investigation helped paleontologists recognize that catastrophic events, whatever their causes, have probably affected life profoundly. More broadly, it has motivated paleontologists and other geologists to investigate the complex relationships between physical and biological processes on earth throughout the history of life.

A mass extinction can be defined as *an unusually large extinction of the earth's biota that takes place in a relatively short interval of geologic time.* While this definition is easy to understand, it inevitably motivates one to ask *how large* an extinction and *how short* a time interval. Indeed, considerable effort has been aimed at providing an unambiguous definition of a mass extinction that would enable someone to label an interval of apparently elevated extinction definitively as a mass extinction.

Perhaps the most straightforward means of diagnosing a mass extinction is to assess the percentage of taxa that became extinct (or, more precisely, exhibited their last global appearances), interval by interval, throughout the Phanerozoic, and then determine whether there were intervals during which these percentages were elevated substantially. Although this might not seem like a very precise method, paleontologists have nevertheless been able to designate five intervals in which percentages stood well above the levels exhibited at most other times; these constitute the "big five" mass extinctions of the Phanerozoic. By far, the largest of these occurred during the Late Permian [SEE SECTION 10.3], when upwards of 40 percent of families and 60 percent of genera became extinct. Other major extinctions, with extinction percentages on the order of 20 percent of families and 50 percent of genera, took place at or near the ends of the Ordovician, Devonian, Triassic, and Cretaceous periods.

In an analysis discussed in Section 8.5 in relation to the long-term decline in origination and extinction rates, Raup and Sepkoski also suggested a statistical definition for a mass extinction, based on an assessment of the number of extinctions per million years for the 76 Phanerozoic stages included in their analysis. A mass extinction was defined as any interval during which the extinction rate exceeded the statistical confidence interval for a regression line that was fitted to the data (Figure 8.14). By this standard, of the big five, only the Late Devonian extinction did not stand significantly above "background" levels. However, as Raup and Sepkoski pointed out, the Late Devonian appears to be unique relative to the other extinctions in that extinction rates were elevated for three successive stages.

This was not the only statistical definition of a mass extinction that Raup and Sepkoski offered. In a series of subsequent analyses on the extinction record from the Late Permian to the Recent, they defined a mass extinction as any interval during which the percentage of families or genera becoming extinct stood significantly above the percentages for the intervals that immediately surrounded it (Figure 8.17). On this basis, several additional mass extinctions were delineated, particularly at the genus level, beyond the two post-Paleozoic members of the big five. Interestingly, nearly all of these were episodes that paleontologists had recognized previously as significant extinctions based on their experiences with regional patterns of faunal change.

### Causes of Mass Extinctions

In 1980, the Alvarez team published a paper in which they reported elevated levels of the element iridium (Ir) across the Cretaceous–Tertiary (K/T) boundary at a locality in Italy (Figure 8.18a). Because Ir is rare at the earth's surface but is more common in meteorites, the team suggested that the source of the Ir was a large comet or asteroid that impacted the earth at the end of the Cretaceous Period. Moreover, based on the amount of Ir found at the K/T boundary, the team estimated that the impacting body had a diameter of about 10 km. The team

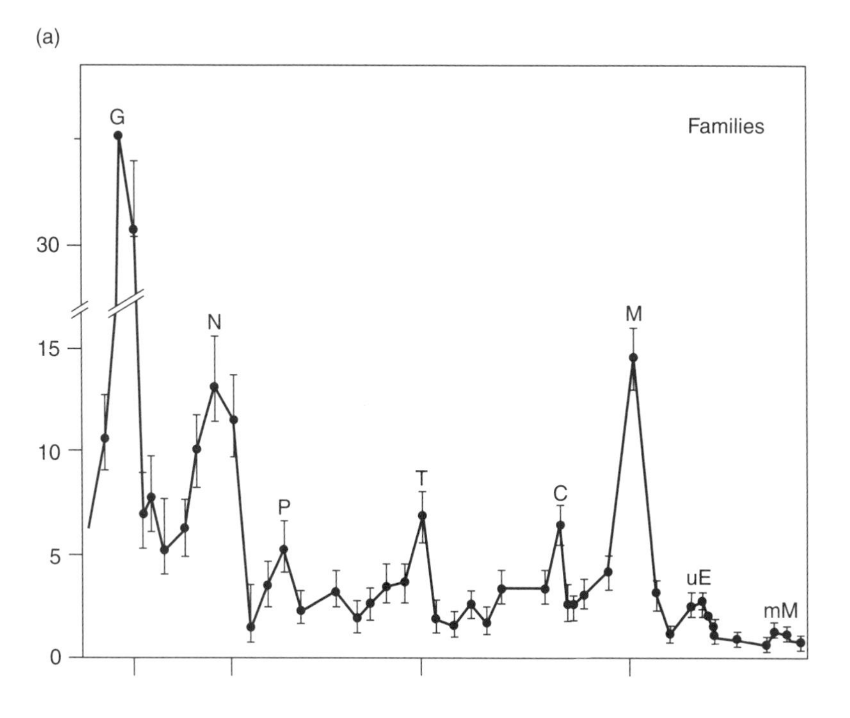


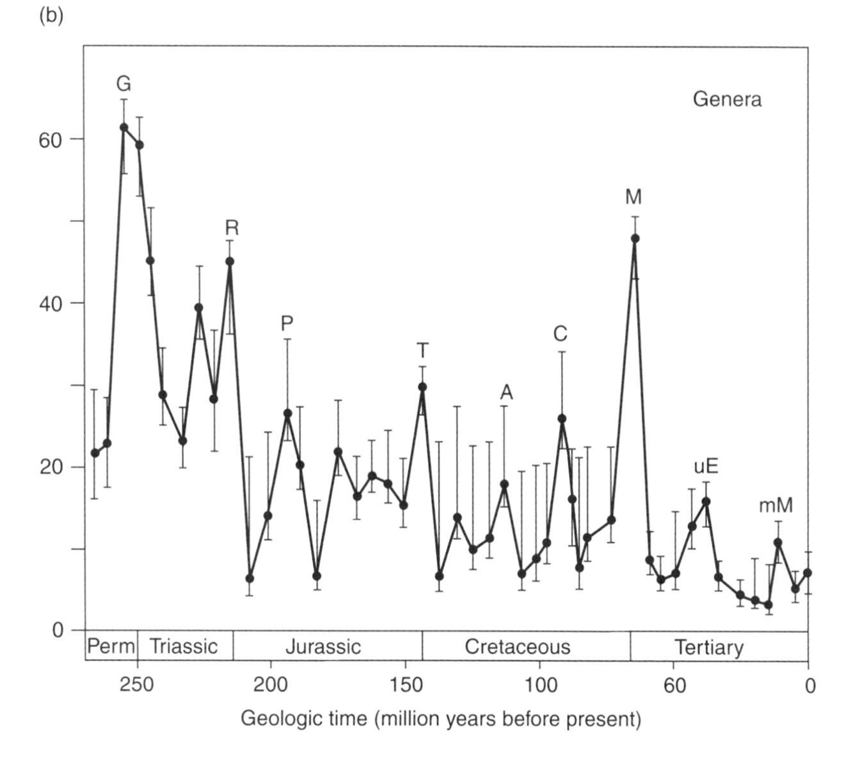


**FIGURE 8.17 Raup and Sepkoski's delineation of mass extinctions in the Late Permian and after.** (a) Families. (b) Genera. Error bars give 1 standard error on either side of the observed percentage extinction. Mass extinctions (labeled with lettered abbreviations for the stages in which they occurred) were defined as intervals ("peaks") during which extinction percentages stood significantly above those in immediately surrounding intervals. *(From Raup & Sepkoski, 1986)*

proposed a mechanism linking the impact to the K/T mass extinction that involved a collapse of the food chain globally, caused by a severe reduction in the amount of sunlight reaching the earth's surface. Among the ramifications of this and related mechanisms is that the mass extinction took place rapidly, an issue that has been debated and analyzed in detail by paleontologists.

Subsequent findings at the K/T boundary have firmly supported the hypothesis of a major impact at the end of the Cretaceous, including elevated iridium levels worldwide, microspherules (impact droplets; Figure 8.18b), and quartz marked by shock features (Figure 8.18c). Sedimentary deposits consisting of coarse, angular fragments, likely produced by impact-related

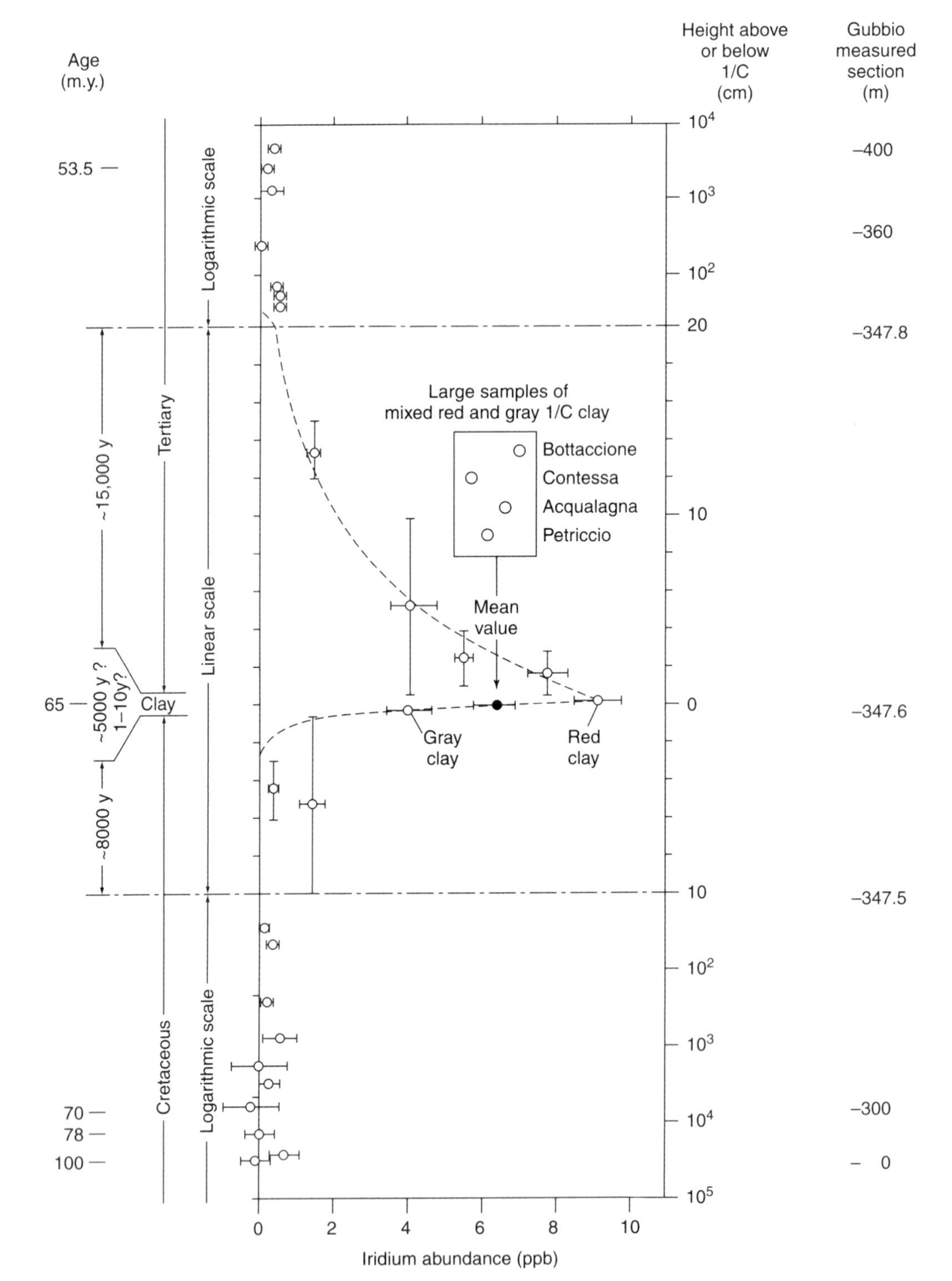


**FIGURE 8.18**
**Evidence for the impact of a large meteorite at the K/T boundary.**
(a) Dramatic increase in the abundance of the element iridium at the K/T boundary near Gubbio, Italy. Iridium is rare at the earth's surface but is more common in meteorites.
*(a: From Alvarez et al., 1980)*

tsunamis, have been found at Late Cretaceous localities near the present-day coast of the Gulf of Mexico. And deep sea cores extracted from the K/T boundary in the western Atlantic Ocean reveal evidence of massive submarine flows that were likely induced by the impact. But the real "smoking gun" in this case was the discovery of the likely site of impact, off the coast of the Yucatan Peninsula of Mexico. There, geophysical soundings of the sea floor revealed a concentric-ringed structure with a diameter of approximately 200 km (Figure 8.19). Detailed sampling of the site has recovered melt rocks that return a radiometric date of 65 million years and have compositional similarities to microspherules collected at the K/T boundary in Haiti. This similarity implies that

(b)

(c)

**FIGURE 8.18** ***(cont.)*** (b) Glassy microspherules and other glassy objects from the K/T boundary in northeastern Mexico (2 to 3 mm in diameter) that formed from molten droplets produced during the impact. (c) Photomicrograph of shocked quartz from a K/T boundary section in southern Colorado (field of view is approximately 0.2 mm); parallel striations imparted on quartz grains by the cataclysmic explosion are associated with the impact. *(b: Courtesy of David A. Kring, Lunar and Planetary Laboratory; c: From Kerr, 1987)*

the bedrock from the putative crater region was the source of the Haitian microspherules. Thus, while the hypothesis of a K/T impact was initially quite controversial, the evidence is now considered overwhelming that such an impact actually took place.

Given the geophysical, geochemical, and sedimentological underpinnings of much of the evidence, paleontologists have played only marginal roles in gathering data to test the hypothesis that an impact took place at the end of the Cretaceous. However, paleontologists have worked to assess the possible link between an impact and the extinction itself. Initially, many paleontologists resisted suggestions of such a linkage, in part because of indications that species became extinct gradually, rather than catastrophically, prior to the K/T boundary, or even that some nonavian dinosaurs survived into the Tertiary. These views have changed dramatically over the past two decades, motivated by reconsideration of the pattern that we should expect to see at a boundary if extinction were rapid rather than gradual.

Owing to issues of preservation and sampling, the last appearances of taxa in the fossil record are likely to predate their actual extinctions [SEE SECTION 6.1]. In other words, many taxa will disappear from the fossil record below the horizon in which they actually became extinct. Because the interval of disappearance will likely differ from taxon to taxon—contingent, in part, on variations in the abundance or rarity of taxa [SEE SECTION 6.5]—this should cause any major extinction event to appear more gradual than it actually was. Philip Signor and Jere Lipps (1982) were the first paleontologists to fully articulate this view, which predicts that the nature of preservation will cause an abrupt extinction to appear gradual (the so-called **Signor–Lipps effect**).

Some researchers have begun to quantify the likelihood of this kind of local and regional artificial-range truncation at the K/T boundary and in association with

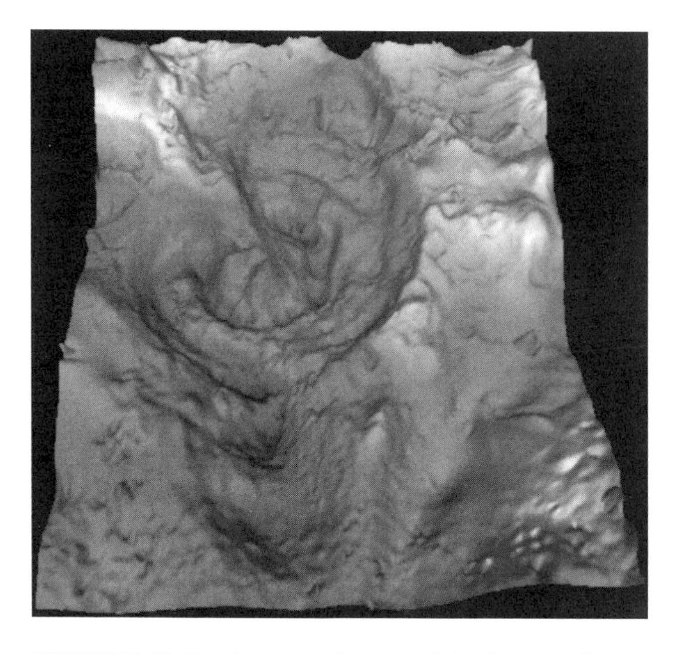

**FIGURE 8.19 Gravity anomaly map of the feature off the coast of the Yucatan Peninsula that is thought to be the site of a large impact at the end of the Cretaceous.** *(From http://solarsystem.nasa.gov/multimedia/display.cfm?IM_ID=791). (Courtesy of Lunar and Planetery Institute, Houston, TX)*

other mass extinctions, but it is still too soon to make definitive statements about the relative abruptness of any of these events on a global scale. Certainly, knowledge of whether a given mass extinction was abrupt or gradual would be helpful in determining what caused it. Whereas the K/T impact scenario appears to call for a fairly abrupt extinction, several earthbound mechanisms (e.g., sea-level changes or global cooling) might have been tied to extinctions that were more gradual, perhaps taking place over millions of years.

We will discuss this issue further in Chapter 10, when we consider the Late Permian extinction in more detail. In the same way that the K/T extinction has provided a focal point for consideration of possible links between impacts and extinctions, the Late Permian event has been a focus for the investigation of earthbound extinction mechanisms [SEE SECTION 10.3]. In the wake of the research presented by the Alvarez team, geoscientists began to look for evidence of major impacts in association with several extinction horizons. To date, these efforts have met with only limited success, and in cases such as the Late Permian, for which evidence of an impact is still being pursued actively, the quality and interpretation of the primary data have been controversial.

However, the possibility of a broad linkage between mass extinctions and impacts was given impetus during the mid-1980s, when David Raup and Jack Sepkoski statistically analyzed the Phanerozoic extinction record from Late Permian, using harmonic analysis (see Box 2.2) and other methods. Raup and Sepkoski suggested that post-Paleozoic peaks in extinction (Figure 8.17) have been spaced roughly at 26-million-year intervals (e.g., Raup & Sepkoski, 1984, 1986). This diagnosis immediately fueled a burst of research and speculation among scientists about its possible causes and ramifications. Following, as it did, on the initial research suggesting a K/T impact, it was not surprising that researchers began to propose possible astronomical mechanisms that would dramatically increase the probability of large bodies hitting the earth during narrow time intervals every 26 million years.

The most notorious of these proposed mechanisms was the *Nemesis theory*, named for a hypothetical, unseen solar companion. The orbit of this solar companion would take it near the Oort Cloud of comets once during its 26-million-year orbit around the sun, disturbing the orbits of some comets sufficiently to cause them to move into the inner solar system, thereby increasing the probability that at least one comet would strike the earth (Davis et al., 1984).

Although it is purely hypothetical, the Nemesis theory predicts, with some precision, the characteristics of the unseen companion. Armed with this information, researchers began to search star catalogues in the hopes of finding an object with the appropriate characteristics. To date, no such object has been found. Beyond that, the actual diagnosis of periodicity has been challenged on statistical grounds and because of possible problems with absolute dates of some of the extinction events.

Regardless of whether extinction periodicity is a reality, the entire exercise has raised the consciousness of paleontologists regarding the likelihood that large-body impacts have affected the history of life on earth. In this vein, Raup (1991, 1992) developed a numerical curve depicting the average waiting time between impacts of various sizes and showed that it was rather similar to a curve depicting the average expected waiting time between extinctions of varying magnitudes. While this certainly does not demonstrate that large-body impacts have been responsible for most extinctions during the Phanerozoic, it does remind us that impacts, particularly of objects that span up to a few kilometers in diameter, have been very common throughout earth history. Their role in mediating the history of life should therefore be explored further.

## Selectivity of Mass Extinctions

Various extinction mechanisms would be expected to preferentially affect taxa living in particular environments or climatic regimes, or exhibiting certain life habits. Therefore, knowing whether mass extinctions were selective would help researchers determine what caused them. For example, episodes of global cooling might be particularly severe among tropical taxa because it would not be possible for these taxa to migrate to warmer climates during a cooling episode. By contrast, a large-body impact above a certain size threshold, such as that implicated in the K/T event, would likely have a global reach, with no real expectation of a latitudinal gradient in extinction rates. In the case of an impact, however, researchers have suggested that an associated reduction in sunlight penetration for an extended period of time would cause a collapse in primary productivity, with an associated extinction of primary producers and taxa dependent on them as sources of food.

In much the same way that extinction rates can be measured for the earth as a whole, rates can also be measured and compared among different areas of the world by amassing and analyzing databases separately for the fossils preserved within each area. In the case of the K/T extinction, David Raup and David Jablonski (1993) evaluated previous suggestions that the extinction was concentrated in the tropics by dividing the world into bins, each with a dimension of 10 degrees latitude by 10 degrees longitude, and then analyzing variations in the extinction rates of bivalve genera among the bins (Figure 8.20).

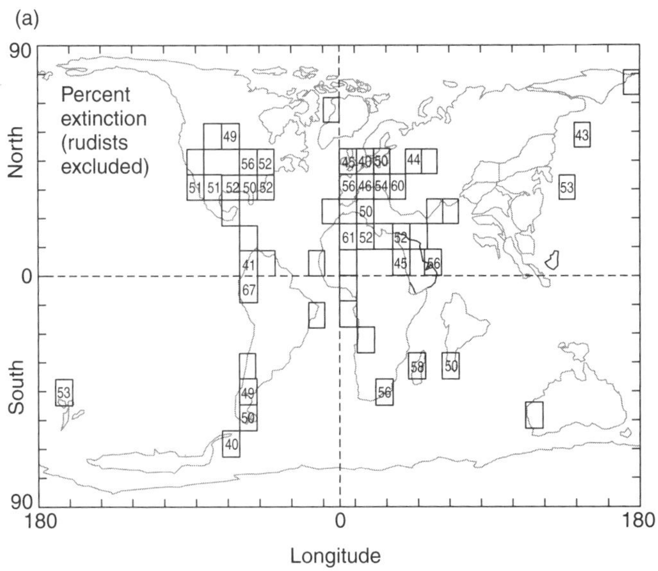


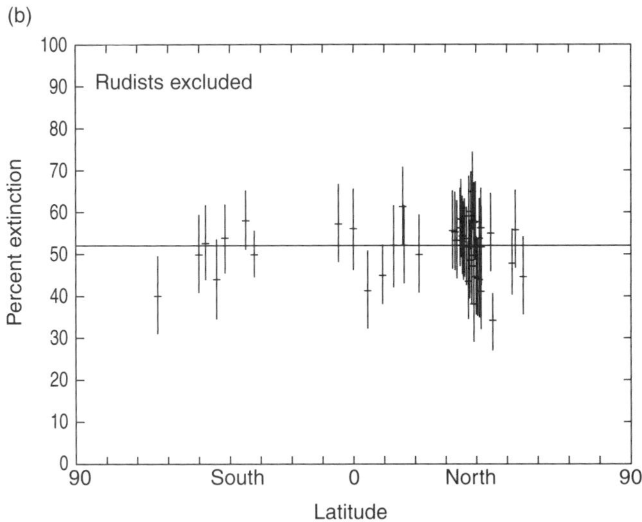


**FIGURE 8.20 Assessment of geographic patterns in bivalve extinction (excluding rudists) during the K/T extinction.** (a) Extinction rates (expressed as percentages) in 10-by-10-degree grids for which sufficient data were available. (b) Extinction rates depicted with respect to latitude. In either grouping, there is no discernable geographic selectivity in extinction rates. *(From Raup & Jablonski, 1993)*

When rudists, an extinct group of reef-building bivalves limited to the tropics, were excluded from the analysis, there was no evidence of latitudinal selectivity in extinction rates. In addition, Raup and Jablonski found no evidence of K/T extinction selectivity among bivalves with respect to body size, position along an onshore-to-offshore bathymetric gradient, or life position above or below the sediment–water interface. They did find that the extinction rate among deposit-feeders was significantly less than that among suspension-feeders, but there were indications that this was a taxonomic rather than an ecologic effect: Certain taxonomic groups of deposit-feeders exhibited extinction rates that were significantly higher than others and that were comparable to rates exhibited by suspension-feeders. Raup and Jablonski also found that genera with broader geographic distributions exhibited significantly lower extinction rates than did more localized genera, indicating that a wide geographic range provided something of a buffer against extinction, at least at the genus level.

Peter Sheehan and colleagues (1996; Sheehan, 2001) described a two-phase pattern of extinction among brachiopods during the Late Ordovician mass extinction that bears similarity to, as well as striking differences from, the K/T event. During the Late Ordovician, there was a major southern hemisphere glaciation and associated drop in sea level that partially drained epicontinental seas, the broad, shallow bodies of water that covered large portions of several continents at that time. This caused the extinction of significant percentages of the genera restricted to the areas that were drained. By contrast, genera that were geographically widespread preferentially survived the event, a pattern that was comparable to that observed among bivalves during the K/T event.

However, in the wake of the initial Ordovician event, a broadly distributed biota (the so-called Hirnantian Fauna) became established worldwide in association with the global climatic cooling. Major elements of this biota, in turn, became extinct during a distinct, second phase of extinction as the glaciation waned, sea level rose, and the climate warmed. In this case, therefore, broad geographic range did not provide an extinction buffer, and extinction was apparently triggered by the global climatic shift back to warm water conditions.

One of the most intriguing instances of selective extinction known to paleontologists has now been documented for the Late Pleistocene extinction, during which large mammals were particularly hard hit. This extinction is unique in the geologic record because it may represent the first instance in which *Homo sapiens* was implicated directly as a significant agent of extinction. Because of its particular relevance to the question of extinction in the present day, we provide a detailed overview of the Late Pleistocene extinction in Chapter 10.

***The Evolutionary Significance of Mass Extinctions*** While the study of selectivity is informative with respect to extinction mechanisms, there is another important reason to investigate selectivity. Mass extinctions have arguably played major roles in causing biotic transitions throughout the Phanerozoic, by abruptly causing the extinctions of some taxa while leaving others relatively unscathed. In fact, Stephen Jay Gould (1985) and others have suggested that mass extinctions were capable of undoing accumulated evolutionary change that took place during the intervals between them. By contrast, as we noted earlier [SEE SECTION 8.4], in constructing his three-phase coupled logistic model (Figure 8.12), Jack Sepkoski suggested that global biotic transitions among evolutionary faunas were caused by interactions among taxa, and that mass extinctions did not significantly alter diversification patterns among the main groups of higher taxa in the three evolutionary faunas.

There is a middle ground between these two end-member views that may best explain the evolutionary role of mass extinctions—that these events and smaller-scale, regional extinctions resulted in biotic transitions primarily because they remove *incumbent* taxa (*the taxa that are already present*), thereby freeing up **ecospace** for the diversification of other taxa. In contemplating this possibility, it is helpful to consider the role of incumbency in electoral politics. It is well understood that in a political election, incumbent candidates enjoy significant advantages over their challengers, even in cases where a dispassionate observer might determine that the challenger would do a better, more effective job if elected. Thus, even a "competitively superior" challenger stands little chance of dislodging an incumbent, unless some unexpected event, such as a political scandal, intervenes to severely weaken the fortunes of the incumbent or, better yet, removes the incumbent from office shortly before the election!

Similarly, in the evolutionary arena, a well-entrenched incumbent taxon, particularly one that is abundant and widespread, is thought to enjoy a major advantage over a less entrenched taxon, even in cases where the less entrenched taxon appears to possess features that, if all else were equal, would provide it with a clear competitive advantage over the incumbent. As an example, consider the case of evolutionary transitions among turtles. Michael Rosenzweig and Robert McCord (1991) investigated the replacement of turtles incapable of retracting their heads into their shells by turtles with flexible necks that permitted head retraction (Figure 8.21).

Rosenzweig and McCord argued that the advent of neck flexure constituted an important advance that was responsible for the radiation of taxa that possessed this feature. However, they also noted that the actual replacement took place at different times in different regions of the world, related in each case to a regional extinction event that first removed incumbent groups of turtles that could not flex their necks. For example, in the case of western North America, the percentage of neck-flexing turtles increased significantly in association with the K/T mass extinction, but replacements elsewhere took place at other times. That the same replacement occurred not just once, but several times in different regions, provides strong support for the argument that neck flexure constituted a competitive advantage, ultimately leading to the global demise of turtles that do not flex their necks.

The turtle example reminds us that there is much more to the investigation of extinction events than the quantitative assessment of extinction rates. In considering their possible roles as agents of evolutionary change, it is also important to ask whether mass extinctions differ

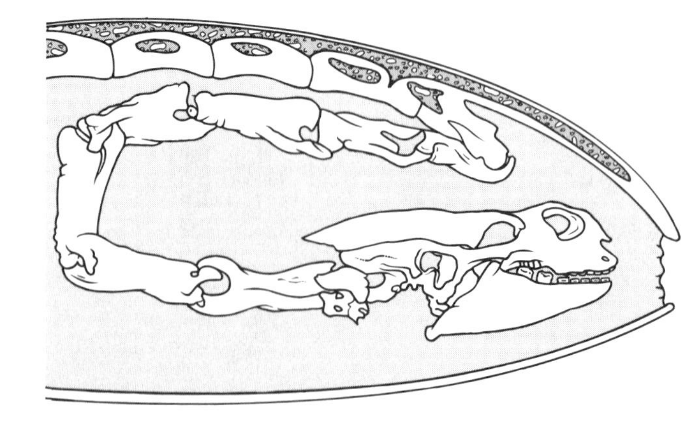

**FIGURE 8.21 A sketch of the turtle *Trionyx*, illustrating the flexibility of the neck, which is a prerequisite for head retraction.** Rosenzweig and McCord (1991) argued that straight-necked turtles (Amphichelydia) were replaced in several venues worldwide by turtles capable of head retraction (Pleurodira and Cryptodira), but the timing of replacement varied among different paleocontinents, contingent on events that first decimated incumbent, straight-necked species.

*qualitatively* from extinctions that take place at other times. For instance, we can ask whether a mass extinction eliminates advantages that some taxa enjoyed prior to its onset.

As an example, David Jablonski investigated gastropod extinction and survival before and during the K/T mass extinction. The early growth stages preserved in the gastropod shell permit a paleontologist to determine whether the larval stage of a species was planktotrophic or nonplanktotrophic [SEE SECTION 7.4] (Figure 8.22a). By examining the shells of numerous Cretaceous gastropods, Jablonski classified the species into these two groups. He also determined the geologic durations of each species and demonstrated that species with planktotrophic larvae tended to have longer durations than species with nonplanktotrophic larvae (Figure 8.22b). Thus, in general, a planktotrophic larval stage was associated with a decreased likelihood of extinction because the absence of a planktotrophic stage reduces the opportunity for widespread dispersal, thereby increasing the likelihood that genetic discontinuities will develop among populations of a species—factors that enhance both speciation and extinction [SEE SECTION 7.4]. Inversely, the presence of a planktotrophic stage lessens the likelihood of isolation.

During the K/T extinction, however, there was no significant difference in extinction rates with respect to larval type (Figure 8.22b). Therefore, an important factor affecting extinction rates during background times did not operate during the mass extinction. This lack of a difference may reflect the global reach of the K/T event. Even widespread species were susceptible to

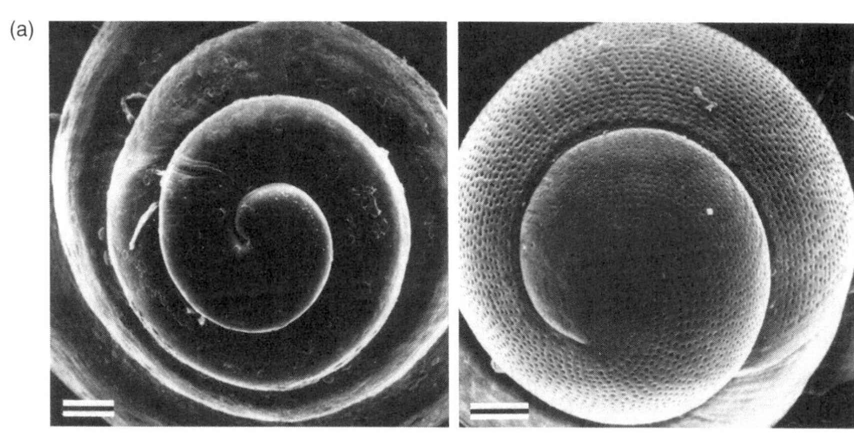


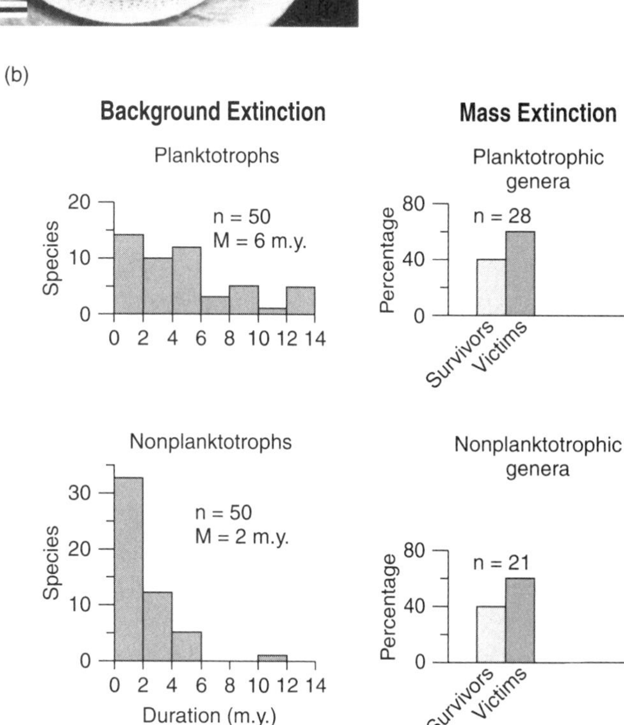


**FIGURE 8.22 Larval mode and extinction in gastropods.** (a) Comparison of protoconch morphology in two gastropod species from the family Rissoidae, illustrating distinctions in larval stage. The scale bar in each figure is 100 microns. In the species at right, Protoconch I (the initial shell, comprising the first two whorls or less) is much more inflated than it is in the species at left, indicating that the species at right had a yolk-rich egg and a nonplanktotrophic larval stage, whereas the species at left had a yolk-poor egg and a planktotrophic larval stage. (b) Jablonski's comparison of species duration and extinction rates in Cretaceous gastropods exhibiting a planktotrophic larval stage versus Cretaceous gastropods exhibiting a nonplanktotrophic larval stage: $n$ = number of taxa, $M$ = median duration (in millions of years). On average, as indicated by the histograms in the first column, species with planktotrophic larvae had longer geologic durations, indicating lower rates of extinction. However, during the K/T extinction, the two groups exhibited no significant difference in extinction rates. *(a: Courtesy of Catherine Thiriot-Quiévreux, C.N.R.S/Jablonski & Lutz, 1980; b: Jablonski, 1986)*

extinction at that time, an inference supported by Jablonski's additional findings that neither the species richness of a genus nor the geographic ranges of its constituent species affected its chances of surviving the K/T event.

## 8.7 THE NEXT GENERATION OF PALEONTOLOGICAL DATABASES

The question of regional versus global transitions, such as that highlighted in the Rosenzweig and McCord example, is one of many issues motivating the development of new paleontological databases that include data not available in earlier global compendia. Although older databases (e.g., Figure 8.2) permitted the construction of global diversity curves, it is not possible to determine from them whether diversification patterns during any interval varied regionally with respect to paleogeographic, paleoenvironmental, or tectonic settings. To address these possibilities, newer paleontological databases have moved beyond earlier efforts in two significant respects:

1. Instead of including information on just the first and last known appearances of fossil taxa, newer databases catalogue multiple occurrences of taxa, wherever they occur globally or within the regions to which the database may be limited.
2. When possible, a variety of additional information is collected for each catalogued occurrence. Typically, these data include geographic location, stratigraphic interval, characteristics of the enclosing sediments, and the inferred paleoenvironmental and tectonic settings.

Two examples of major database initiatives in paleobiology are illustrated in Figures 8.23 and 8.24. The objective of *The Paleobiology Database* (PBDB, found at http://paleodb.org; Figure 8.23) is to produce an exhaustive compilation of Phanerozoic marine and nonmarine paleofauna and flora for the entire world. As suggested by its title, the scope of the *Neogene Marine Biota of Tropical America* database (NMITA, found at http://porites.geology.uiowa.edu/; Figure 8.24) is more limited, given its focus on research questions related to the marine evolutionary history of tropical America during the past 25 million years. Both projects endeavor to collect similar classes of subsidiary information. While the PBDB emphasizes the collection of existing taxonomic and geologic information from the literature, NMITA has focused on the collection of new data directly from strata in the region of focus. As such, NMITA has significantly augmented the relatively scarce data available from the region previously.

## 8.8 DISSECTING DIVERSIFICATION AND RECOVERIES FROM MASS EXTINCTIONS

As the PBDB, NMITA, and similar large-scale databases continue to grow, they will inevitably include the information available in earlier compendia (Figure 8.2) on the first and last global appearances of taxa. However, the additional data that they contain on the geography, lithology, and environments of individual taxonomic occurrences permit researchers to assess in more detail the nature of diversification within and among regions during important transitions in the history of life. Several examples are illustrated in this section, and this is a persistent theme throughout the remainder of the book.

### The Ordovician Radiation

As we have already seen, there was a major global diversification of marine organisms during the Cambrian and Ordovician periods. Although it was once commonplace to view the entire interval as a single "event," it is now understood that the biological nature of the Cambrian Explosion was rather different from the subsequent Ordovician Radiation. The Cambrian Explosion represented the initial, major diversification of multicellular animal life, during which most present-day animal phyla observed in the fossil record first evolved their characteristic anatomical features and began to diversify, although they may have originated much earlier. Because it has been a focus of intensive, multidisciplinary research, we provide an extended discussion of the Cambrian Explosion in Chapter 10.

From a taxonomic perspective, the Ordovician Radiation was most pronounced at the order level and below (Figures 8.5 and 8.11), and it was characterized by a major transition in marine settings from the Cambrian evolutionary fauna to the Paleozoic evolutionary fauna (Figures 8.5 and 8.8). Moreover, by the end of the Ordovician, standing diversity, measured at the genus and family levels, had increased three- to fourfold, relative to Cambrian levels. Because of these profound changes, coupled with the excellent preservation of Ordovician rocks

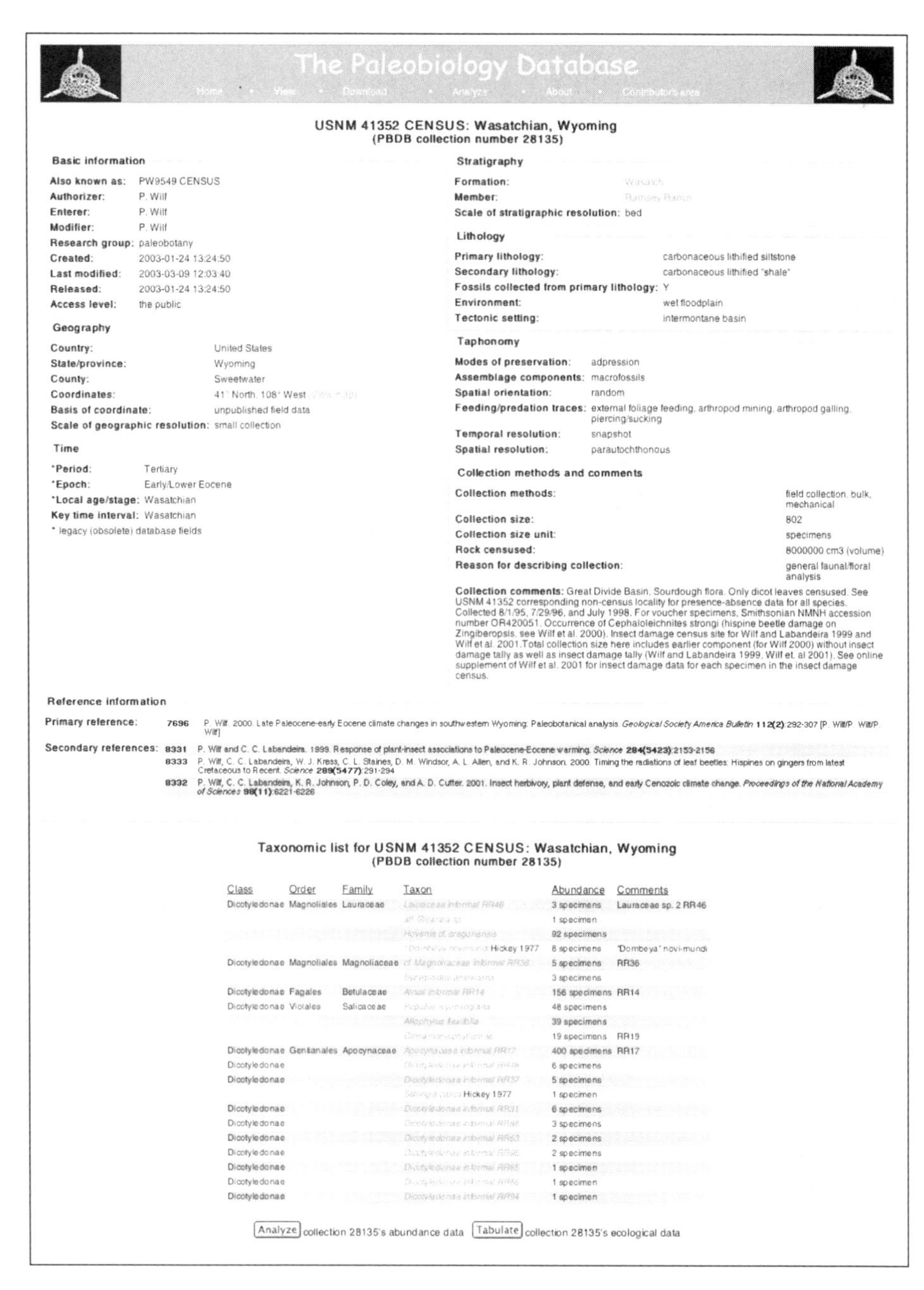


FIGURE 8.23 **An example of a collection contained in the Paleobiology Database (PBDB: http://paleodb.org).** This is a paleobotanical collection from Lower Eocene strata of Wyoming. Any of the information contained in a collection can be searched and downloaded.

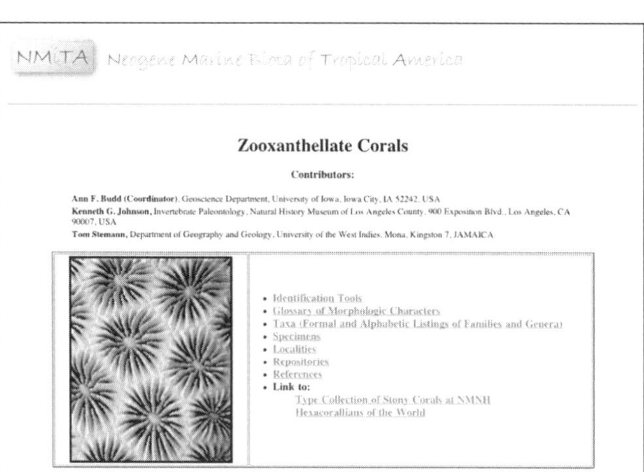


FIGURE 8.24 **A portion of a Web page for the NMITA database (Neogene Biota of Tropical America: http://porites.geology.uiowa.edu/), illustrating a partial range of the data available for searching.**

and fossils in several places around the world, the Ordovician has become a focal interval for the assessment of biotic patterns at local and regional levels in comparison to the global signal. In this research, paleontologists have sought answers to two important questions:

1. Did diversification patterns at regional levels simply mirror what we observe at the global level or, alternatively, were there significant differences between global and regional patterns, as well as from region to region?
2. If there were substantial differences in diversification patterns among regions, can we explain what caused these differences?

In a series of analyses, Arnie Miller compared the diversification of major higher taxa in different regions worldwide. The example in Figure 8.25, for bivalve molluscs, illustrates regional origination rates in comparison with the global signal. The methodology for determining regional evolutionary rates was the same as that used

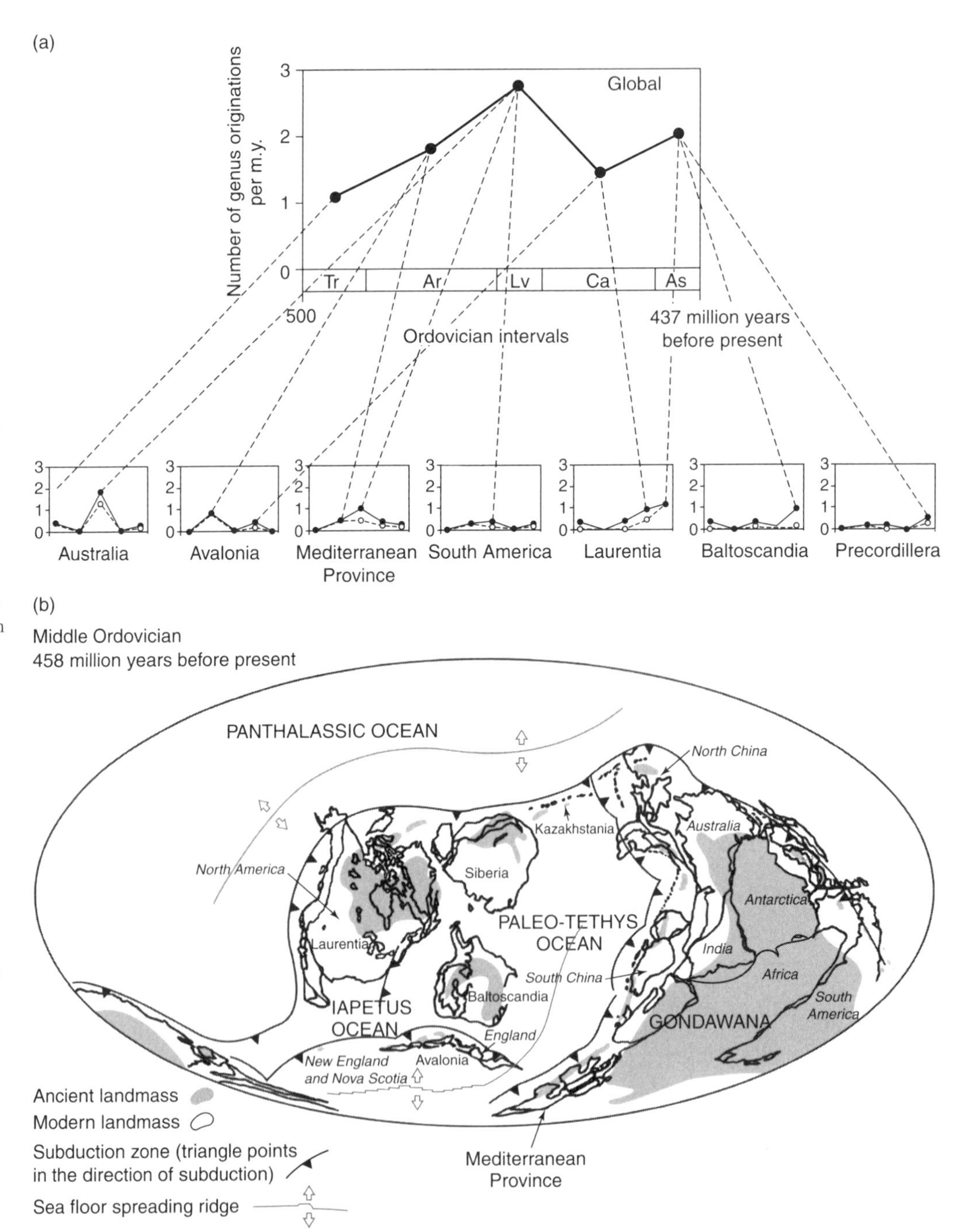


**FIGURE 8.25 Variations among Ordovician paleocontinents in bivalve origination rates throughout the Ordovician period.** (a) Graphs depicting origination rates for several Ordovician paleocontinents, in comparison to the aggregate, global pattern. The dashed lines connect portions of the global curve to peaks of origination in various paleocontinents that contribute most significantly to global origination during each Ordovician series. For example, significant bivalve diversification did not take place in Laurentia (much of present-day North America) until late in the period. In the graphs for each paleocontinent, the dashed curve is the number of true originations. The solid curve includes genera that originated elsewhere but were making their first appearances on the paleocontinent. (b) A paleogeographic reconstruction for the Middle Ordovician, illustrating the positions of paleocontinents, including those highlighted in Figure 8.25a. *(a: From Miller, 2001; b: From a map available from the Web site of Christopher Scotese: www.scotese.com/newpage1.htm)*

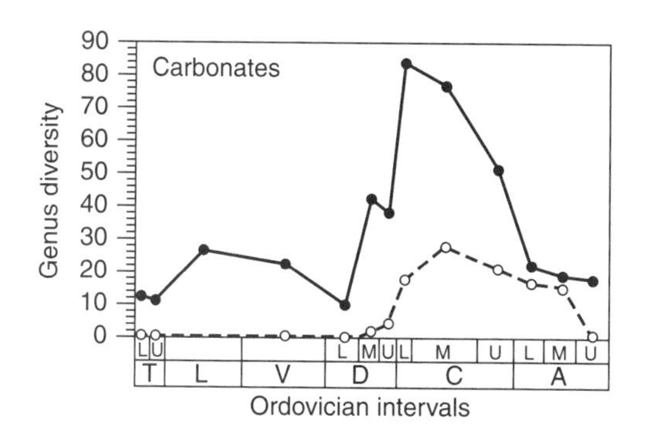


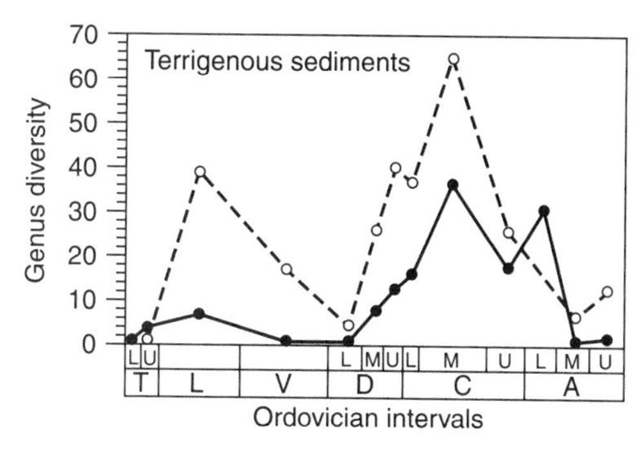


**FIGURE 8.26 Diversity trajectories of bivalves (dashed lines) and gastropods (solid lines) in terrigenous and carbonate settings worldwide through the Ordovician.** Note that bivalves were far more diverse in terrigenous sediments than in carbonates through most of the period. The opposite was the case for gastropods. *(From Novack-Gottshall & Miller, 2003)*

for global rate determination [SEE SECTION 7.2], except that the data on first and last appearances were confined to the regions in question. Based on this figure, it can be seen that bivalve diversification varied significantly throughout the Ordovician and that different regions contributed to global diversification at different times.

The regions contributing diversity early in the Ordovician were located mainly in high southern latitudes, whereas the low latitude setting of Laurentia (present-day North America) did not exhibit a bivalve diversification until later. Differences in the geological characteristics of these regions appear to have governed this pattern, as can be demonstrated by determining the lithologies in which bivalves occurred worldwide. Collectively, bivalves exhibited a significant preference throughout the Ordovician for substrates that were rich in terrigenous sediments eroded off of nearby landmasses (Figure 8.26). These substrates were readily available at high southern latitudes, such as in the Mediterranean Province, during the Early Ordovician but did not become widespread in Laurentian seas until the Middle and, especially, the Late Ordovician in association with the onset of a major episode of mountain building that provided a significant source of terrigenous sediments.

These regional variations raise a basic question: In general, were global Phanerozoic diversity trends (e.g., Figure 8.5) *caused* primarily by global-scale evolutionary processes, or did they represent the summation of environmentally mediated transitions operating mainly at regional scales? From the patterns that we just considered, one might get the impression that it is the latter. Nevertheless, we should bear in mind that, while these observations can explain why some higher taxa thrived when appropriate environmental conditions were available and, conversely, why others did not, they probably do not explain why total diversity increased so dramatically at that time. Like the Cambrian Explosion [SEE SECTION 10.2], it appears that diversity increased throughout the entire world during the Ordovician Radiation, so it seems probable that the ultimate explanation for this increase was also global in scope.

## Regional Marine Cenozoic Transitions in Tropical America

There is a particular urgency to understand the regional, evolutionary history of reef ecosystems throughout the Cenozoic Era, given their relevance to present-day concerns about the health of coral reefs in high-diversity tropical regions. The NMITA database, which we discussed earlier, is an outgrowth of these efforts. It has permitted paleontologists to understand the responses to environmental perturbations of fossil reef corals that are closely related to, and in some cases include, present-day species. After assembling these data, Nancy Budd (2000) examined the history of Early Eocene through Late Pleistocene Caribbean coral diversity in the light of region-wide environmental changes. The data for this analysis came from samples of 57 assemblages spanning the Eocene to Recent, depicting the presence of 294 coral species and 66 genera.

Compilations of stratigraphic ranges (Figure 8.27a illustrates genera) and representations of total diversity and evolutionary rates (Figures 8.27b and 8.27c) permitted the recognition of a series of diversity plateaus and peaks in the Middle to Late Eocene, Late Oligocene to Early Miocene, and Late Pliocene. Origination and extinction rates, particularly at the species level, appeared to be

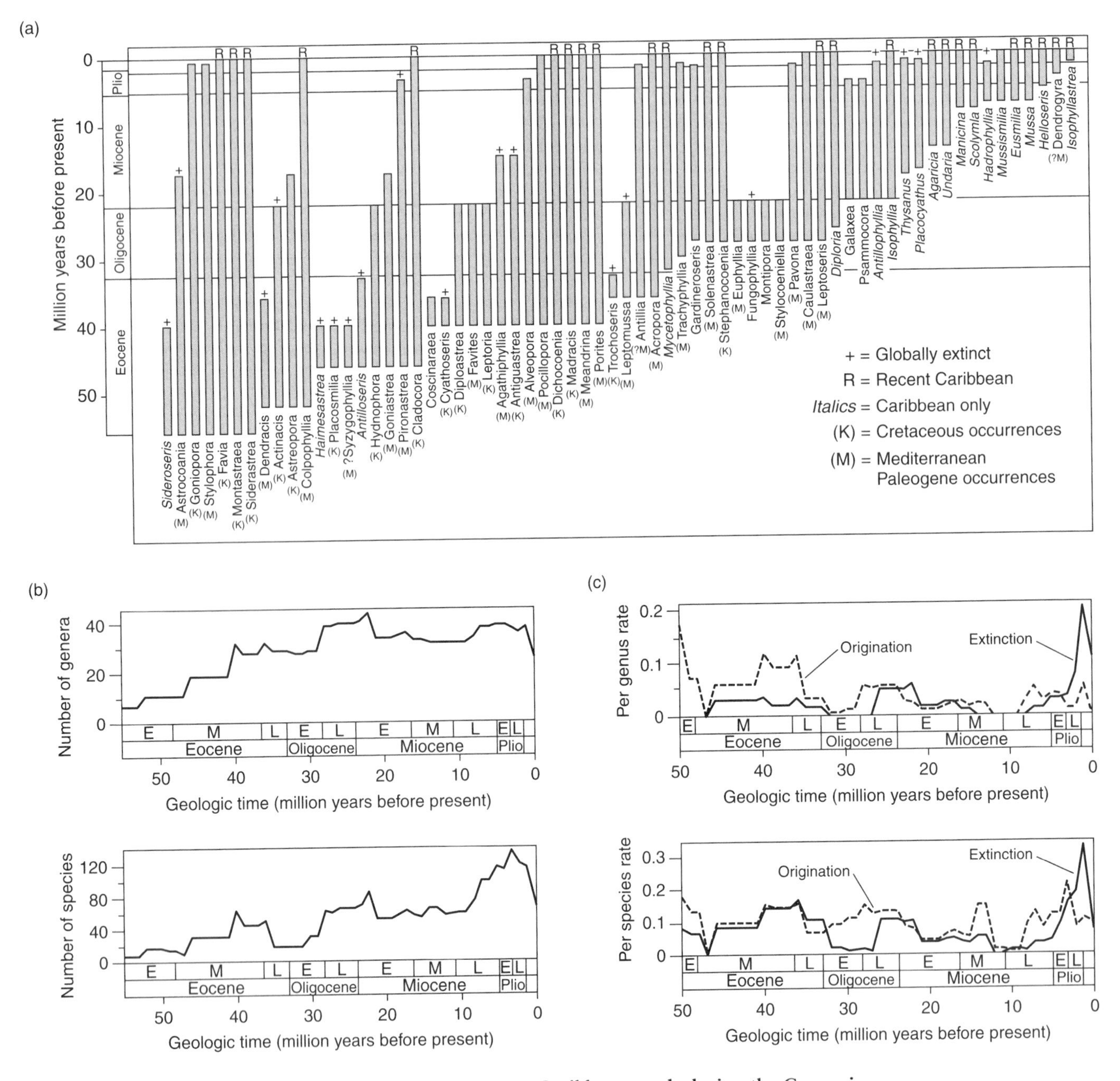


**FIGURE 8.27 Budd's analysis of evolutionary rates among Caribbean corals during the Cenozoic.** (a) Composite chart of stratigraphic ranges for Caribbean coral genera from the Eocene to the Recent. (b) Diversity of coral genera (top) and species (bottom) from the Eocene through the Pliocene. (c) Origination and extinction rates for coral genera (top) and species (bottom) from the Eocene through the Pliocene. Origination and extinction were measured as proportional rates per million years (see Table 7.4). *(Budd, 2000)*

independent of each other to a fair extent after the Eocene, with evidence of elevated extinction near the end of each diversity plateau.

Despite a broad increase in total species diversity throughout the study interval, Budd found that the maximum number of species contained within individual assemblages had leveled off by the Late Eocene. She suggested that stabilization in total numbers, if not composition (Figure 8.27a), may have been tied to the cessation of dispersal from the Mediterranean region during the Oligocene. Supporting this view is the observation that genera found in the Caribbean region from that time on were confined to the Caribbean, suggesting that they originated there rather than elsewhere.

In addition, Budd ascribed the intense genus-level extinction in the Plio-Pleistocene to climatic deterioration. At the species level, elevated extinction rates in the Middle–Late Eocene, Late Oligocene–Early Miocene, and Plio-Pleistocene coincided with episodes of cooling in the first and third cases, and the onset of regional upwelling and turbidity in the second case. Interestingly, there is evidence that Pleistocene species in a variety of terrestrial and aquatic settings were able to survive significant climatic variations through shifts in their geographic ranges [SEE SECTION 9.6], and data on present-day coral species off the coast of Florida suggests that they, too, have undergone range shifts in response to recent climatic changes [SEE SECTION 10.6]. Thus, there is still much to learn about why some species appear to have been quite resilient to climatic changes, whereas others apparently were not.

## Recoveries from Mass Extinctions

Global diversity rebounded rapidly in the aftermaths of several of the major extinctions of the Phanerozoic (Figure 8.5). Because the major mass extinctions were global in extent, it stands to reason that recoveries would also be observed on a global scale. However, there is no inherent reason to expect that the factors favoring the diversification of particular taxa should also be global. As we saw with the Ordovician Radiation, the physical characteristics of different regions or environments can yield different diversification pathways and, thus, the nature and trajectory of biotic recovery may vary markedly from region to region.

In an analysis of biotic recovery from the K/T mass extinction, David Jablonski evaluated previous suggestions that the initial recovery in the Gulf Coast of North America was dominated by species belonging to so-called bloom taxa—very widespread species that were considered capable of diversifying unusually rapidly after the extinction. To test this suggestion, Jablonski (1998) compared molluscan diversification for the Gulf Coast against three other major regions. He found that, in proportion to other taxa, the diversity of species recognized by previous authors as bloom taxa, did *not* increase in the other regions (Figure 8.28). Not only does this call into question the global significance of bloom taxa, but it also shows that the recovery from the K/T extinction was far from a globally uniform process.

Because of the magnitude of the global diversity increase in the aftermath of the Late Permian mass extinction (Figure 8.5), this diversification has received considerable attention from paleontologists. In terms of the number of new taxa, this Mesozoic Radiation appears to have rivaled the Cambrian Explosion and

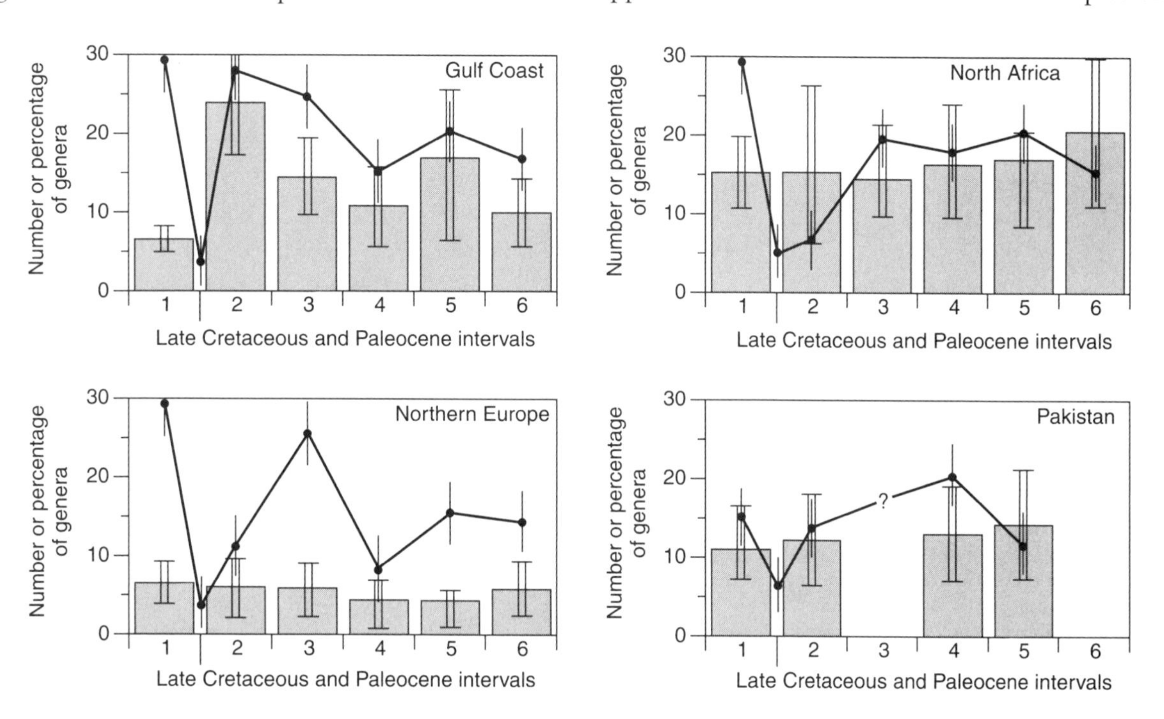


**FIGURE 8.28 Jablonski's depiction of regional differences in the diversity of "bloom taxa" across the K/T boundary and into the Paleocene.** Histograms depict percentages of total species diversity and points depict actual numbers of species. Among the four regions studied, the proportion of taxa categorized as bloom taxa increased following the mass extinction only along the North American Gulf Coast. *(Jablonski, 1998)*

Ordovician Radiation combined. However, the Mesozoic Radiation was arguably less profound from an evolutionary standpoint than its Paleozoic counterparts. During the Mesozoic, there was comparatively little origination of phyla and classes among marine taxa (Figure 8.29). Two principal explanations for this have been offered. First, by the end of the Paleozoic, genetic pathways may have become channeled to an extent that prevented the kind of "experimentation" that appears to have been rampant during the Late Precambrian and Early Paleozoic [SEE SECTION 10.2]. Second, despite the severe diversity decline of the Late Permian extinction, the actual ecospace occupied by the remaining taxa did not contract substantially, relative to pre-extinction levels.

Therefore, in marked contrast to the Early Paleozoic diversification—which was accompanied by the advent of organisms exhibiting a variety of novel modes of life above, at, and below the sea floor [SEE SECTION 9.4]—the amount of unoccupied ecospace available for the development of novel lifestyles in the early Mesozoic may have been more limited. Clearly, these two explanations are not mutually exclusive and may have acted together to inhibit the origination of phyla and classes after the most extensive mass extinction in the history of life.

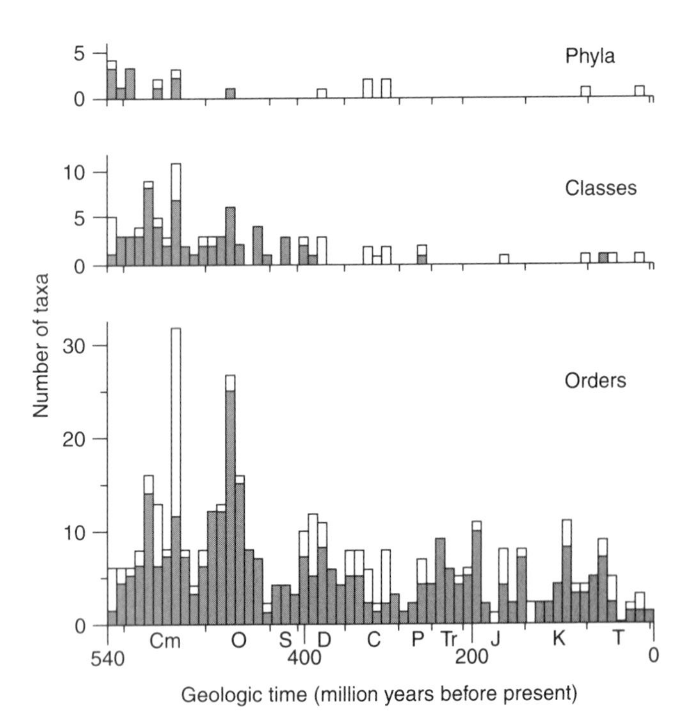


**FIGURE 8.29 Erwin et al.'s depiction of the time of origination of well-skeletonized (solid bars) and lightly skeletonized or unskeletonized (open bars) marine taxa.** Note that the vast majority of phyla and classes originated early in the Paleozoic. *(Erwin et al., 1987)*

## 8.9 A SCHEMATIC OVERVIEW OF BIOLOGICAL TRANSITIONS

As suggested throughout this chapter, paleontologists have proposed a variety of explanations for the major global transitions in taxonomic composition observed throughout the Phanerozoic. Typically, these have involved the classification of two or more groups with respect to biological attributes that give one group an advantage over the other through geologic time. In some cases, like that of cyclostome and cheilostome bryozoans [SEE SECTION 8.4], competition has been invoked to explain the long-term diversification of the "later" group at the apparent expense of the "earlier" group. As we have noted, however, this approach is typically not fruitful because it is difficult to demonstrate definitively that one higher taxon enjoyed a long-term competitive advantage that would be sufficient to cause biotic replacement through geologic time.

The topic of biotic transitions is explored further in Chapter 9, in our discussion of paleoecology. For the moment, to help summarize the variety of ideas proposed to explain the replacement of one group by another through geologic time, it is useful to consider the schematic diagram prepared by Michael Benton, illustrated in Figure 8.30. Explanations range from gradual replacement brought on by interactions in which one group outcompetes another over the long term (Type 1, left side of the figure) to rapid replacement fueled by fortuitous survival—or the lack thereof—in the wake of a mass extinction, accompanied by the rapid radiation of survivors into the ecospace vacated by organisms that became extinct (Type 5, right side of the figure). Intermediate scenarios involve the advent of morphological or functional innovation (a "key adaptation") in group B that affords it a competitive advantage over group A, but, to varying degrees in Types 2, 3, and 4, a mass extinction or perturbation plays a significant role in the transition by removing incumbents.

Among the examples that we have considered in this chapter, Sepkoski's coupled logistic model, and its use to explain the transition from cyclostome to cheilostome bryozoans, would be classified as a Type 1 replacement. Gould's (1985) view of mass extinctions as overarching arbiters of long-term transitions is a Type 5, and Rosenzweig and McCord's scenario for transitions from stiff- to flexible-necked turtles falls somewhere in between. Although Benton and others have sought to determine the extent to which each of these five models appropriately describes Phanerozoic biotic transitions, we are still a long way from knowing

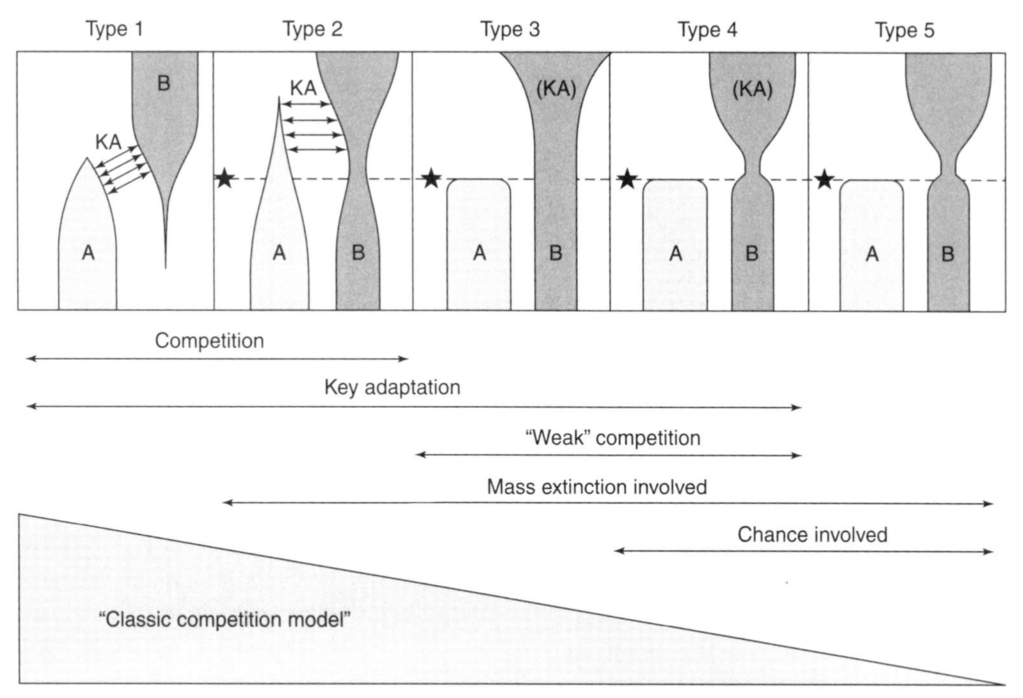


**FIGURE 8.30 A series of schematic models depicting possible modes of biotic replacement of one clade by another.** The role of competition diminishes, and that of physical perturbation increases, from left to right. The star in four of the figures denotes a mass extinction, and "KA" denotes a "key adaptation." *(From Benton, 1996)*

whether any of these models has dominated the history of life. However, there is reason to be optimistic that the approaches described throughout this book will ultimately help us to make these determinations.

## 8.10 MORPHOLOGICAL DIVERSITY

Our discussion of diversity has so far focused strictly on taxonomic diversity or richness. Yet many questions raised in this chapter involve morphological differences between species. These differences, which are referred to as **disparity,** represent another important aspect of biological diversity.

Because species in the fossil record are recognized on the basis of their form, it is natural to ask whether taxonomic and morphological diversity really are different for paleontological species. To illustrate the distinction between these two aspects of diversity, Figure 8.31 depicts morphological data for a number of specimens of blastoid echinoderms. The measurements are the $x$-, $y$-, and $z$-coordinates of selected homologous landmarks on the theca, as shown in Figure 2.8b. Because there are more measurements than can easily be visualized, the data have been converted to synthetic principal components [SEE SECTION 3.3]. In Figure 8.31, the specimens of

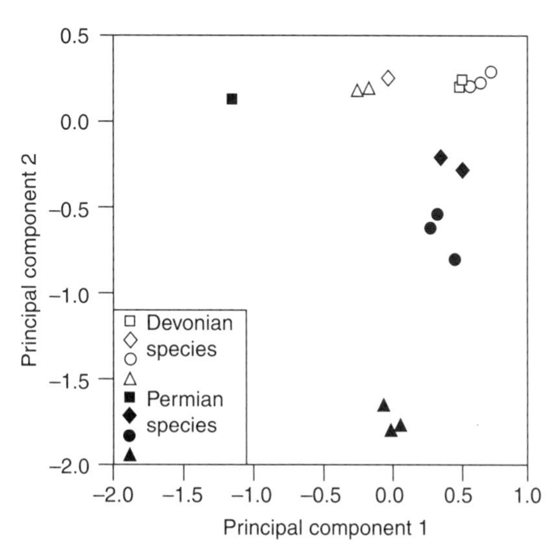


**FIGURE 8.31 Principal component scores for a sample of Devonian and Permian blastoids.** Each time period is represented by several specimens belonging to four species; thus, species richness is the same in these two periods. The Permian forms are more dispersed, however; they have greater disparity. *(Data from Foote, 1991)*

each species cluster together and tend to be separated from other species. This is not a circular exercise, since the species were not originally recognized on the basis of the measurement data shown here.

Figure 8.31 shows several species from the Devonian and Permian, with 1–3 specimens per species. Each period is represented by four species, but the Permian forms are more widely dispersed—they have greater disparity. As this simple example demonstrates, richness and disparity measure quite different aspects of diversity. How they relate to each other is one of the primary questions addressed by evolutionary studies of disparity. Box 8.3 describes some approaches to measuring disparity.

*Box 8.3*

## MEASURING DISPARITY

Studies of disparity typically start by measuring differences among related species within an interval of time. To illustrate this, Figure 8.32 portrays the mean of each species from Figure 8.31. One obvious way to measure the disparity of a sample of species would be to measure the area covered by the points in this figure—or the volume or "hypervolume," if there were three or more dimensions. This area is shown for the Permian sample by the hatched lines. This approach, while having a clear intuitive appeal, creates one problem, especially for paleontological studies: The area covered by the points depends on how many species are sampled. Thus, apparent disparity may reflect the completeness of sampling. If only half the species in Figure 8.32 had been sampled, the spread of points would have been considerably smaller.

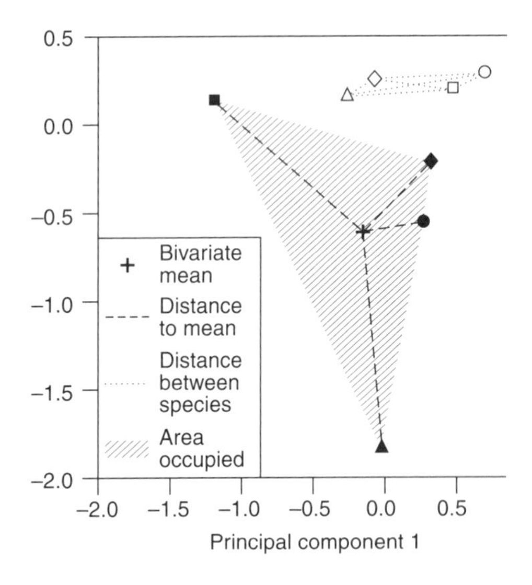


**FIGURE 8.32 Three ways to measure disparity.** The points represent the species means from Figure 8.31. (1) The hatched region shows the area occupied by the Permian species. (2) The dashed lines show the distances from these same species to the overall Permian average; the mean of the squared distances measures disparity. (3) The dotted lines show the distances between the Devonian species; the mean of these squared distances also measures disparity. *(Data from Foote, 1991)*

Another way to measure disparity is to extend the univariate measure of variance that we considered in Chapter 3. This is illustrated for the Permian species in Figure 8.32. Recall that, for a single trait, $x$, measured on $n$ species and having mean $\bar{x}$, the sample variance is defined as $s^2 = \Sigma(x - \bar{x})^2/(n - 1)$. With two or more traits, we first compute the mean of each variable and then plot the position of the resulting bivariate or multivariate mean, shown as the cross in Figure 8.32. The straight-line distance $d$ of each species from this mean is then calculated [SEE SECTION 3.2]. The bivariate or multivariate variance is the average of the squared distances from the mean: $\Sigma d^2/(n - 1)$. Like the univariate variance, this measure of disparity is generally unbiased by sample size.

Yet another way to measure disparity is to compute the squared distances between all pairs of species, as shown for the Devonian forms in Figure 8.32, and to take the average of these distances. (The average squared distance is in fact directly proportional to the variance.) This approach must be used when the traits are categorical or ordinal rather than continuous [SEE SECTION 3.2]. This is because the mean and variance make sense only for continuous variates. There are many ways to measure a distance between two species using discontinuous characters. One approach is simply to tabulate the total number of traits for which the two species have a different trait value.

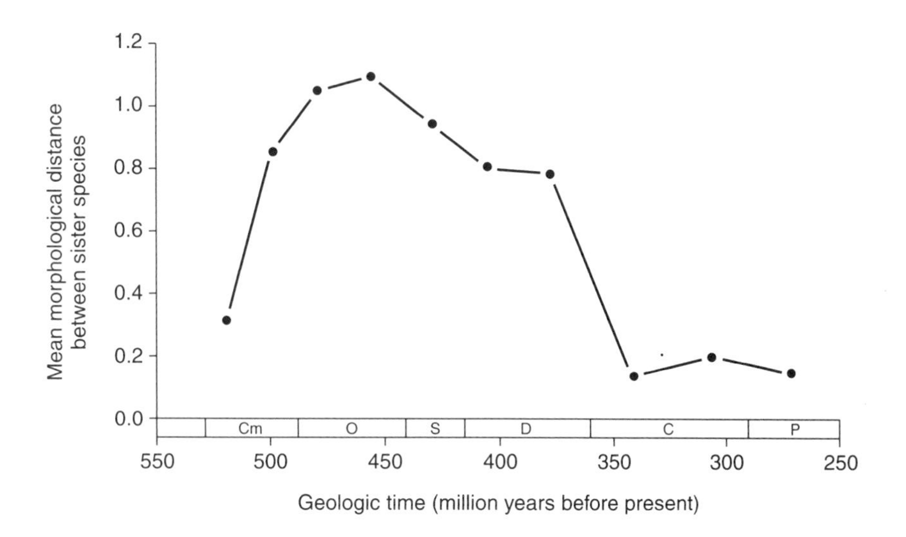


**FIGURE 8.33 Mean morphological distance between sister species of blastozoan echinoderms.** Larger distances imply larger evolutionary transitions. *(Data from Wagner, 1995)*

## Examples of Disparity Analysis

***Morphological Changes during Evolutionary Radiation*** Many paleontologists over the years have noted a general tendency for morphological changes to be relatively large during the early stages of diversification of a biologic group, and to become smaller as the evolutionary radiation proceeds. This pattern has generally been interpreted to reflect major morphological changes associated with the occupation of new ways of life. The tendency has been questioned, however, in part because it has sometimes been inferred from a subjective assessment of the morphological differences between taxa. Analysis of morphological data allows patterns of early radiation to be assessed more explicitly.

Echinoderms are one of the groups commonly thought to have undergone large evolutionary transitions in form early in their history. As an example, Figure 8.33 shows results of a study of the echinoderm subphylum Blastozoa, based on about 65 discontinuous morphological characters. Several phylogenetic analyses, all of them conducted independently of this study, were combined to produce a cladogram [SEE SECTION 4.2]. The size of morphological changes was estimated by computing the morphological distance between sister-species pairs on the cladogram. Of all the species sampled, sister species are most closely related, so the distances between them should reflect evolutionary changes. Consistent with the conventional view, the size of morphological transitions in blastozoan echinoderms is high in the Cambrian and Ordovician, and it declines over the rest of the Paleozoic.

In many cases, phylogenetic relationships within a group are not known; therefore, the direct measurement of evolutionary changes is not possible, as it was in the case of blastozoans. However, disparity among species, irrespective of their phylogeny, can be useful as an indirect guide to transition sizes. To infer the nature of transitions without phylogenetic information, it is necessary to rely on additional assumptions in the form of evolutionary models.

Figure 8.34 compares two highly idealized models of the diversification process. In the first (Figure 8.34a), the average size of morphological steps between ancestor and descendant does not change over time. As a result, the evolutionary tree continues to spread out as it branches and accumulates new taxa; thus, disparity and richness increase together. In the second model, morphological steps are substantially larger early in the radiation than later (Figure 8.34b). The evolutionary tree

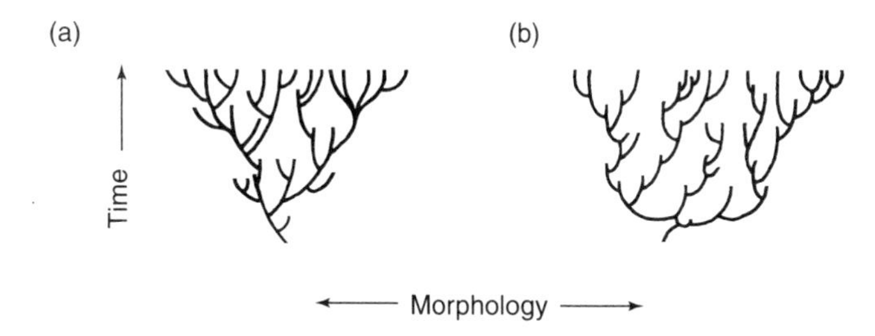


**FIGURE 8.34 Models of richness and disparity during an evolutionary radiation.** (a) The average size of evolutionary transitions between ancestor and descendant is constant over time. As a result, disparity and richness increase together. (b) The size of transitions is large early and small later. As a result, disparity initially increases more rapidly than richness. *(From Foote, 1993)*

spreads rapidly at first, then more slowly, and therefore disparity initially increases more rapidly than richness.

With these models in mind, let us return to the blastozoan data. These are depicted in another form in Figure 8.35a, as the average distance between all species extant within an interval of time rather than just between sister species. Like the distances between sister species, these distances increase through the Cambrian and Ordovician and then decline. Yet another way to display the data is as a series of scatterplots of the species, sorted by time interval (Figure 8.35b). Here each scatterplot represents synthetic variables similar to principal components. As suggested by Figure 8.35a, the dispersion of points increased through the Cambrian and Ordovician, declined through the Silurian and Devonian, and was largely stable through the Carboniferous and Permian.

The idealized models in Figure 8.34 suggest that it is helpful to interpret the evolution of disparity in relation to taxonomic richness. For comparison with blastozoan disparity, Figure 8.35c depicts the number of blastozoan genera. Evidently, disparity increased much more rapidly than richness during the Cambrian and Early Ordovician. In light of the model in Figure 8.34a, this suggests that morphological transitions were larger early in the history of blastozoans—which is exactly what was found by measuring transitions directly on the cladogram (Figure 8.33). Thus, the indirect and direct approaches largely agree in this example.

(a)

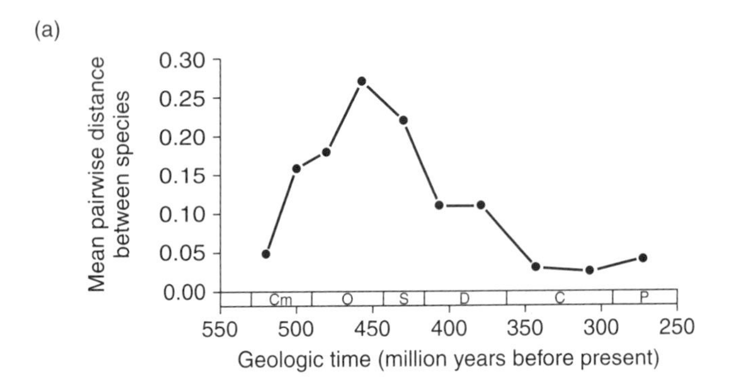


(b)

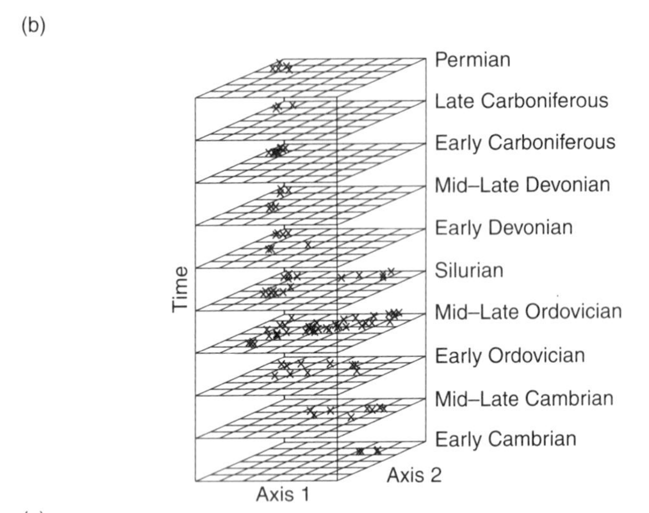


(c)

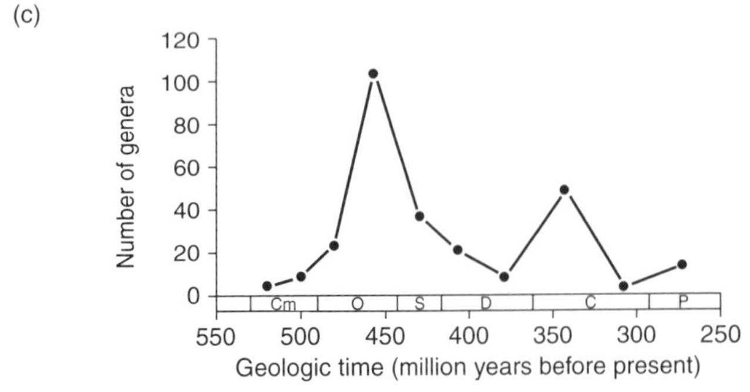


**FIGURE 8.35 Disparity and richness in blastozoan echinoderms.** (a) Disparity, measured as the average squared distance between species. (b) Scatterplots of species through time, showing a wide spread of points at times of high disparity. (c) Richness, measured as the number of genera. Note that disparity and richness do not change in the same way over time. *(Data from Foote, 1992)*

***Morphological Selectivity in Extinction*** The richness and disparity histories of blastozoans show a curious feature. There was a pronounced drop in richness from the Ordovician into the Silurian, yet disparity barely changed at this time. We can make some sense of this discordance with additional idealized models (Figure 8.36). During a decline in richness, if lineages are lost at random with respect to morphology, then the observed range of form, which is correlated with the number of species sampled, will decrease (Figure 8.36a). However, the branches of the tree will be thinned out rather than pruned back, and the average distance among them will largely be maintained. As a result, disparity, measured as variance, will not drop as severely as richness. If, on the other hand, extinction selectively removes certain main branches of the evolutionary tree, as in Figure 8.36b, then disparity may drop along with richness.

Returning to the blastozoan data of Figure 8.35 with these models in mind, the transition from the Ordovician to the Silurian suggests a loss of lineages that is largely nonselective. From the Silurian to the Devonian, on the other hand, the drop in richness is accompanied by a large decline in disparity, suggesting that certain branches of the evolutionary tree are selectively removed. In fact, by the Carboniferous, all classes of blastozoans except the blastoids had become extinct. Thus, disparity is lower in the Carboniferous than in the Devonian, even

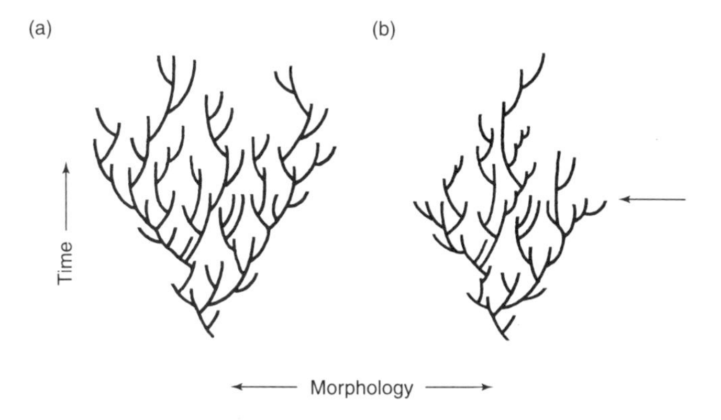


**FIGURE 8.36 Models of richness and disparity during a decline in richness, which begins at the point in time marked by the arrow.** (a) Lineages are lost at random with respect to morphology. As a result, disparity is largely maintained. (b) Morphologically extreme lineages are lost preferentially. As a result, disparity and richness drop together. *(From Foote, 1993)*

though there was in fact a major increase in genus richness in the Carboniferous (Figure 8.35c).

Therefore, whether or not disparity and richness decline together provides an indirect test for morphological selectivity in extinction. It is important to bear in mind that selectivity depends to some extent on the scale of analysis. In saying that extinction from the Ordovician to the Silurian appears to be nonselective, we do not mean that there is no good reason, related to particular details of their form, why some species survived and others did not. We mean instead that there is not a simple, overarching relationship between survival and form, such that one could have predicted the survivors based on where they lie in a scatterplot like Figure 8.35b.

## 8.11 CONCLUDING REMARKS

It should be evident from the topics covered in this chapter that the paleontological study of global biodiversity is still in its formative stages, even though it has been a focal point of extensive paleontological research for more than a quarter century. To be sure, several topics that we considered in this chapter, such as the causes of global taxonomic transitions and mass extinctions, remain controversial. At the same time, perhaps more than any other themes in paleontology, these issues have captured the interests and imaginations of a wide array of geoscientists, biologists, and even astronomers.

In this context, one of the more promising developments in the past few years has been the growth in cross-disciplinary collaborations among researchers interested in understanding the causes of biological transitions throughout earth history. Databases such as NMITA and the PBDB are outgrowths of this interest, as paleontologists have come to recognize the importance of understanding how local and regional diversity trends combine to produce the patterns observed at the global level. Beyond the assessment of diversity as an end in itself, however, highly resolved regional and taxonomic data can be coupled with information on morphology, lithology, and geochemistry to develop a more intricate picture of the relationship between biological and physical transitions throughout earth history. Clearly, this is one of the major growth areas in paleontology, several examples of which are highlighted in Chapters 9 and 10.

# SUPPLEMENTARY READING

Foote, M. (1997) The evolution of morphological diversity. *Annual Review of Ecology and Systematics* **28:**129–152. [A review of research on morphological diversity, emphasizing paleontological examples.]

Hallam, A., and Wignall, P. B. (1997) *Mass Extinctions and Their Aftermath.* Oxford, U.K., Oxford University Press, 320 pp. [Chapters summarizing each of the major mass extinctions in the history of life.]

Jablonski, D., Erwin, D. H., and Lipps, J. H. (eds.) (1996) *Evolutionary Paleobiology.* Chicago, University of Chicago Press, 484 pp. [A collection of papers emphasizing broad approaches to the study of macroevolutionary patterns.]

Rosenzweig, M. L. (1995) *Species Diversity in Space and Time.* Cambridge, U.K., Cambridge University Press, 436 pp. [An overview of large-scale patterns of species diversity and the factors that govern it, with an emphasis on the present day.]

Ryder, G., Fastovsky, D., and Gartner, S. (eds.) (1996) *The Cretaceous–Tertiary Event and Other Catastrophes in Earth History. Geological Society of America Special Paper 307*, 569 pp. [A comprehensive collection of papers on mass extinction; a successor to Silver & Schulz, 1982.]

Silver, L. T., and Schulz, P. H. (eds.) (1982) *Geological Implications of Impacts of Large Asteroids and Comets on the Earth. Geological Society of America Special Paper 190*, 528 pp. [A collection of papers focused on early research spurred on by the Alvarez et al. impact scenario for the K/T extinction.]

Valentine, J.W. (ed.) (1985) *Phanerozoic Diversity Patterns: Profiles in Macroevolution*. Princeton, N.J., Princeton University Press, 441 pp. [A collection of papers that consider a variety of features of Phanerozoic diversification in marine and terrestrial settings.]

Wills, M.A. (2004) Morphological disparity—A primer. In J. M. Adrain, G. D. Edgecombe, and B. S. Lieberman (eds.), *Fossils, Phylogeny and Form: An Analytical Approach*. New York, Kluwer Academic/Plenum Publishers, pp. 55–144. [A detailed overview of methods and applications of disparity analysis.]

*Chapter 9*

# PALEOECOLOGY AND PALEOBIOGEOGRAPHY

**Ecology** is the study of interrelationships between extant organisms and the environments in which they live, so it follows that **paleoecology** is the study of the interrelationships between ancient organisms and the paleoenvironments in which they lived. Despite this distinction between the present-day and the geological past, it should be understood that ecological patterns are the consequences not only of present-day agents, but also of evolutionary processes that unfolded over millions of years. In this respect, paleontologists now recognize that the investigation of evolutionary transitions (or, in some cases, stability) through protracted intervals of geologic time, when melded with ecological perspectives, provides a dimension not available to ecologists working in the present day. The investigation of these long-term patterns and processes falls in the realm of **evolutionary paleoecology.**

On a parallel front, new analytical techniques, such as the geochemical assessment of skeletal compositions, have significantly improved the precision of paleoenvironmental and paleoclimatological reconstruction based on fossil material. These techniques have also enabled researchers to evaluate changes in ambient conditions at high resolution through extended intervals of geologic time. Thus, from a practical standpoint, paleoecology is now yielding data of importance to investigations that span a wide range of the geosciences.

In this chapter, we focus on several aspects of the ever-changing subdiscipline of paleoecology, emphasizing (1) the growing importance of the geologic time dimension to evolutionary paleoecology, and (2) new approaches to paleoenvironmental and paleoclimatological reconstruction. Before doing so, however, it is important to review basic operational considerations and principles related to the development of **community paleoecology,** which was a significant, earlier focus of paleoecological research. The infusion of the community concept into paleoecology provided scientific and methodological foundations on which the development of evolutionary paleoecology was strongly dependent.

**Paleobiogeography,** the study of the ancient geographic distribution of life, also focuses on the spatial dimension. While paleontologists recognize fundamental differences between the environmental dimension that paleoecology highlights and geographic space, which is the focus of paleobiogeography, these differences are not always perfectly obvious. We will conclude this chapter by considering the relationship between environmental and geographic distributions and by discussing the role of paleobiogeography in assessing biotic distributions at regional and global scales.

## 9.1 THE NATURE OF PALEOECOLOGICAL DATA

In Chapter 1, we discussed general principles of sampling the fossil record, including the need to assess whether differences in the quality or volume of fossiliferous samples affect our ability to detect meaningful biological differences among them. Because many paleoecological analyses involve direct comparisons of the biotic compositions of a set of samples, paleontologists often seek to standardize the method of data collection used for any given study, to ensure that differences in sampling protocol are not, in themselves, responsible for measurable variations in sample composition. Given that the nature of preservation and accessibility of fossil material are highly variable throughout the fossil record, the methods used to extract, identify, and count fossils may differ appreciably from study to study.

The choice of sampling method is sometimes dictated by the nature of preservation, such as when a researcher encounters fossils in crystalline limestones or sandstones that are so hard that a jackhammer or dynamite blast is required to remove them (not that this necessarily stops an ambitious paleontologist!), or when fossils are located on public lands where removal of any material is illegal. These circumstances might call for an *in situ* census of fossil material. Moreover, even in cases where it is feasible to remove fossils, *in situ* methods may still be desirable when it is believed that the spatial arrangement of fossils at the outcrop reflects a true biological distribution rather than post-mortem transport (Figure 9.1a). Finally, when strata are not exposed at outcrops, samples may be obtained from a core collected during drilling (Figure 9.1b). This has proven particularly useful for sampling the strata preserved beneath oceans and lakes.

In cases where samples are removed for fossil identification and counting, it is common practice to remove approximately the same volume of material for each sample—referred to as a **bulk sample**—to lessen the likelihood that differences among samples in the abundances of fossil elements are direct consequences of volumetric differences in the amount of sampled material. Another way to standardize the sizes of collections when working with fossils that are plentiful is to count the *same number* of individuals in each sample.

Although an *in situ* census is most easily conducted on a bedding plane (Figure 9.1a), a census can also be carried out on the lateral face of a stratum in wall-like exposures when bedding planes are not available. In the former case, a rectangular or square template, also known as a **quadrat,** can be placed directly on the bedding plane, and the individual fossils contained inside the template are identified and counted. If there is reason to believe that the spatial arrangement of fossils within the quadrat preserves a biologically meaningful pattern, a

(a)

(b)

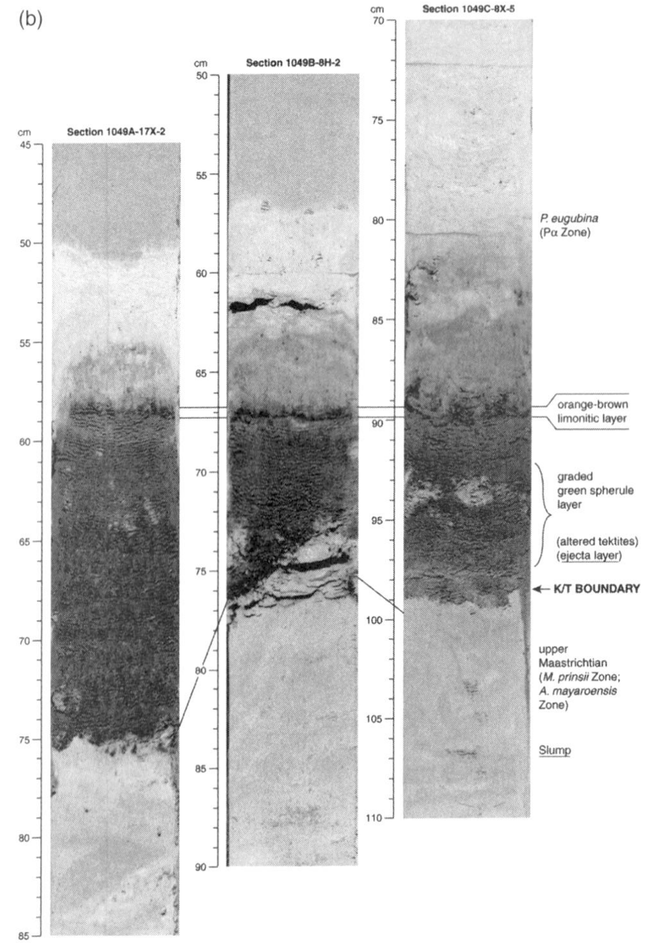


**FIGURE 9.1 Methods of collecting samples for paleoecological analysis.** (a) A researcher uses the quadrat method to assess the spatial distributions of fossils preserved directly on a bedding plane exposed at a Devonian locality in New York. (b) A correlated set of core samples collected under the auspices of the Ocean Drilling program from a region known as Blake Nose, in the western Atlantic Ocean, approximately 500 km east of northern Florida. The cores straddle the Cretaceous/Tertiary (K/T) boundary and illustrate the effects and remnants of a large-body impact. *(a: Courtesy of D. L. Meyer; b: From Norris et al., 1998)*

researcher might also develop a map of all the fossils contained in the quadrat.

An alternative to a quadrat, which can be used on a bedding plane or when conducting a census on the lateral face of a stratum, is a stretched line, also known as a **transect.** The abundances of taxa along the transect can be assessed by either counting up the number of individuals that the transect passes over or by direct measurement of the lengths of the transect covering each specimen. The latter approach yields estimates of how much of the transect is covered by each fossil taxon. It is particularly appropriate when working with clonal organisms such as corals, where the delineation and counting of "individuals" becomes problematic, so estimates of areal or volumetric coverage may instead be more appropriate.

Often paleontologists must also determine how many individuals in a sample contain disarticulated skeletal pieces—such as shell fragments, bones, or leaves—that may be derived from the *same* individual. Methods for making these determinations are considered in Box 9.1.

*Box 9.1*

## DETERMINING THE NUMBER OF INDIVIDUALS IN A SAMPLE OF FRAGMENTED OR CLONAL SKELETAL ELEMENTS

When working with a sample that contains disarticulated skeletons, it is useful to consider whether there is reason to believe that multiple elements were derived from the same individual, and then to reduce the estimated abundances of these individuals accordingly to the minimum number of individuals that could have produced the elements in question. For example, when considering the preserved valves of bivalve molluscs, it is appropriate to determine the number of left and right valves present for a particular species. It can then be assumed, *at a minimum*, that the number of biological individuals of the species in the sample is equal to whichever of the abundances of the two valves is *greater*.

However, some researchers have viewed this kind of minimum approximation as too conservative, and instead advocate the opposite approach by assuming that every identifiable skeletal element is derived from a different individual. The number of individuals of a given taxon, then, is equal to the total number of identifiable elements of that taxon present in the sample. In our bivalve example, the number of individuals—a *maximum* estimate, really—would therefore be the number of left valves *plus* the number of right valves. While this would obviously be an inappropriate approximation in some cases—such as when the left and right valves of a bivalve are found as articulated pairs or when the bones that obviously comprise a single dinosaur skeleton are found adjacent to one another—there are many instances when the use of this approximation may actually be quite reasonable.

In fact, it has been shown statistically that, in cases where the elements that make up the skeleton of a taxon do not remain articulated to one another, there is a good possibility that all or most of the elements in a confined sample are derived from different individuals. While this may seem surprising, it is actually intuitive if one considers, for example, that the bivalve elements present in a typical bulk sample would likely have been derived from a living assemblage that covered a far larger volume of the sea floor. Given the taphonomic processes that we discussed in Chapter 1, once the two valves of a bivalve become disarticulated, it is highly *unlikely* that they will remain in close enough proximity to become part of the same two- or three-liter bulk sample, even if there has not been a significant amount of post-mortem transport.

When working with clonal organisms or highly disarticulated nonclonal organisms in a bulk sample composed of lithified material, an alternative to approximating the number of individuals is to estimate each taxon's areal or volumetric coverage, through a technique known as **point counting.** One way to conduct a point count is to draw a rectangular or square Cartesian grid with a fixed number of points on a transparent sheet that can be overlain directly on the surface of the bulk sample. The number of grid points that overlie each fossil element can then be counted, providing estimates of the relative densities of the taxa that occur beneath at least one grid point. Because the same grid can be used for all samples in a given investigation, point counting can help standardize the collection of information from one sample to the next.

Our discussion of paleoecological data has focused on field-based assessments of the occurrences and abundances of taxa, which remain at the heart of many paleoecological studies. As we have already noted, however, paleoecology in recent years has grown to include investigations that require other kinds of data and analyses, such as determination of the geochemical compositions of skeletal elements and the interpretation of the life habits and food preferences of fossil taxa that are encountered in a sample. We consider these additional sources of data later in the chapter.

## 9.2 COMMUNITIES

The development of paleoecology as a biologically based subdiscipline was dependent on the earlier research of biologists, working in isolation from paleontology. Thus, it is important to consider the scientific themes that emanated from this work and their eventual impact on paleontology.

While many researchers preceding him had been interested in the ways that environmental attributes affect the distributions of living organisms, the Danish biologist C. G. J. Petersen was among those who first applied the concept of a **community** to marine settings. Petersen's views emanated from his empirical observations and data collected while working on marine bottom faunas in Scandinavian waters, where he established a systematic program of faunal sampling and measurement of physical attributes of the sea floor along transects extending from the shoreline into waters several hundreds of meters deep. Petersen used bulk samples collected from the sea floor with a mechanical device dropped from a ship. His biological data consisted of lists of species identified from each sample, with abundances reported as both the number of individuals and the collective weight of these individuals.

Petersen recognized that, within his study area, species did not occur randomly in all possible combinations with one another. Rather, he observed that they could be found in a finite number of recurring assemblages and that similar sequences of such assemblages could be recognized from onshore to offshore in transects established at several locations (Figure 9.2). Based on his observations, Petersen viewed a community as a regularly recurring combination of numerically common species. He delineated communities in his study area based mainly on the species that were characteristic of a particular suite of samples. In this regard, Petersen distinguished between *abundant* species, which were truly characteristic of an assemblage, and *attendant* species, which were also abundant but too widespread in their distributions to be diagnostic of a particular community.

In Petersen's view, the chief factor governing the distributions of bottom communities along depth gradients was not depth itself, but rather temperature changes associated with the transition to deep water. Petersen argued that this explained why the depth limitations of certain communities varied from transect to transect: At similar depths along different transects, water temperatures were not the same. He noted that other physical factors were also likely to be important in governing the distributions of species in his study area, and he acknowledged a possible role for biological interactions as well. Clearly, however, Petersen favored physical factors over biological interactions as explanations for the community patterns he delineated.

Although Petersen may have viewed the recognition of communities as straightforward from an operational standpoint, it is clear that among many of his contemporaries, the characterization of communities was controversial—and remains so even today. In particular, biologists have long debated whether the web of interactions that link species together (e.g., Figure 9.3) cause a community to function in some ways as a cohesive unit. Two end-member views, based on the study of terrestrial plants, illustrate the debate.

One was the **individualistic concept** championed by H. A. Gleason (e.g., 1926), who argued that plant species distributions were governed by two primary agents: (1) the migration of seeds and other propagules; and (2) local environmental conditions. Gleason noted that both factors were subject to substantial fluctuation and suggested, therefore, that chance plays a significant role in the assembly of associations (communities). To Gleason, it came as no surprise that species compositions in any two locations are unlikely to be the same, even in cases when the species in question might be considered representative of the same association, because no two locations are identical physically. Gleason downplayed the role of biotic interactions in effecting the distributions of species, and he questioned the proliferation of community classification schemes that, in his view, tended to mask differences among assemblages from place to place.

A decidedly different view was espoused by F. E. Clements (e.g., 1916), who wrote on the subject of **succession,** a general term for changes through time in the

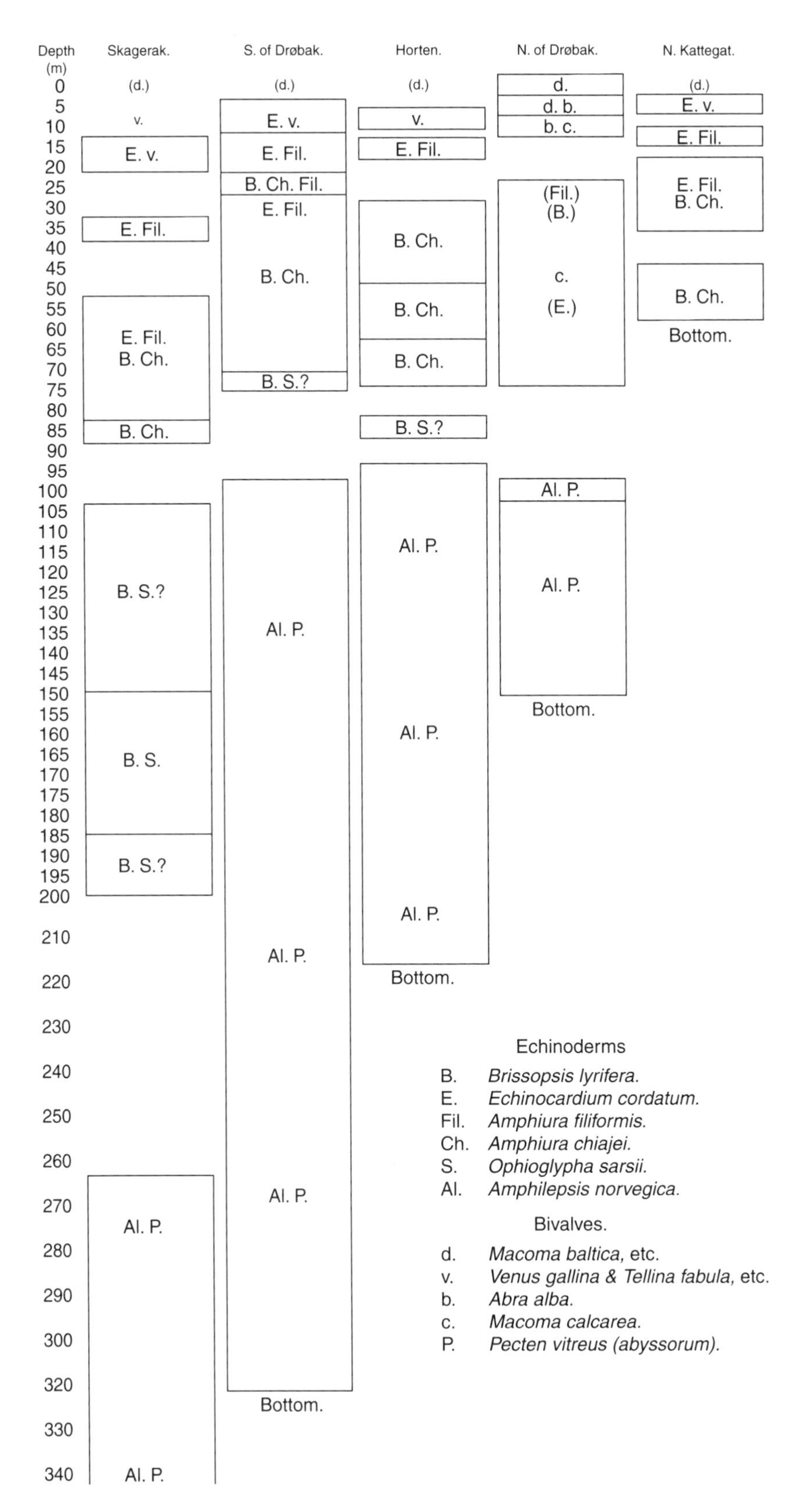


**FIGURE 9.2 Distributions of marine communities along five underwater transects located in waters off the coast of Scandinavia.** Communities are named for dominant bivalve and echinoderm species; a key to abbreviations is provided in the lower right-hand corner. Note that several of the transects exhibit similar, but not identical, patterns. *(From Petersen, 1915)*

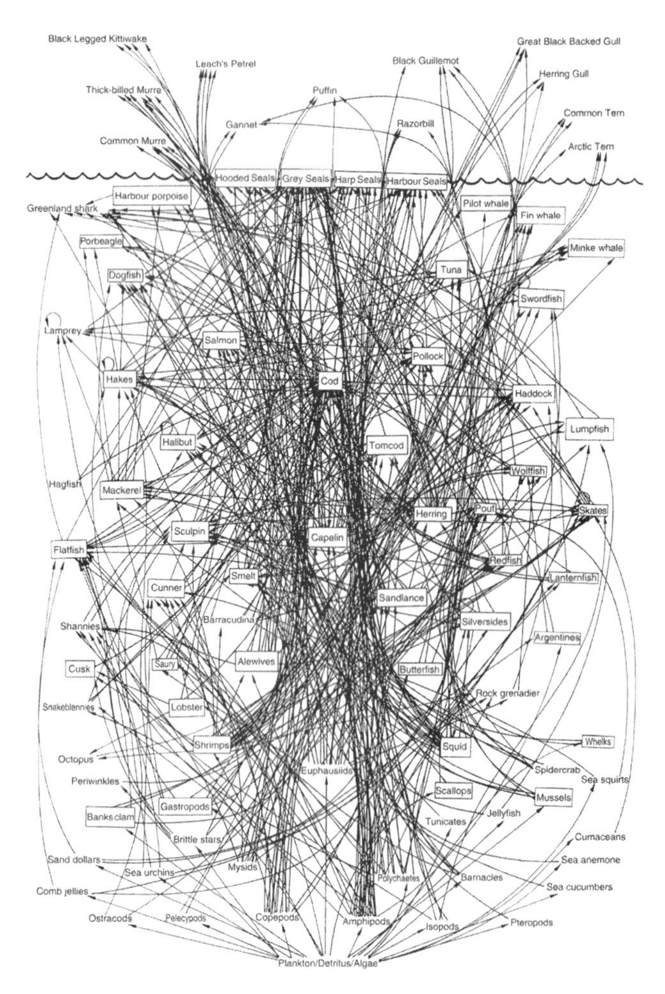


**FIGURE 9.3 An example of a food web, for the northwest Atlantic Ocean, illustrating the complex set of pathways that link species to one another.** Here and in Figure 9.4, arrows link food sources with the organisms that consume them, with the arrow pointed in the direction of the consumer. *(From the International Marine Mammal Association, www.imma.org/codvideo/foodwebpic.html)*

biological compositions of successive communities occupying a particular place. To some extent, the roster of species present at any given stage in a succession is thought to be a consequence of environmental changes caused by the organisms comprising earlier stages in the succession (Figure 9.4). However, the initiation of the first stage of a succession may depend either on physical events that remove much of the living biota from an area, such as a fire or hurricane, or on the geological development of new areas for colonization, such as a newly emergent oceanic island.

Clements viewed the final stage of a succession as the adult stage in the development of a kind of superorganism, which he termed a **sere.** Clements therefore considered a sere to be an entity that exhibits an ontogeny consisting of several highly predictable steps, following the physical emptying of a space and the initiation of recolonization of that space. Biotic interactions among individuals were seen by Clements as playing significant roles in this ontogeny, and he even suggested that different seres following similar developmental courses are bound together by something akin to phylogenetic relationships.

Against this backdrop, plant ecologists have gathered copious data on the spatial distributions of species in order to test different viewpoints on the intimacy of relationships between species within communities. Figure 9.5 illustrates the ways in which species distributions can speak to this issue. The abundance ("importance") distributions of several hypothetical species are graphed along a hypothetical **environmental gradient**—a

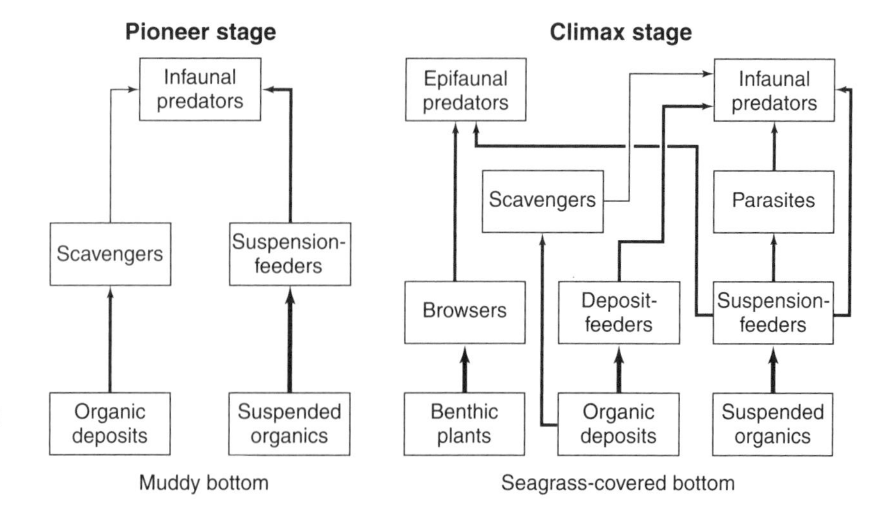


**FIGURE 9.4 Two stages in a succession recognized among ancient communities preserved in the Middle Miocene of Poland.** It is thought that a sea floor covered initially by mud (pioneer stage) became colonized by seagrass (climax state). With this colonization came an increase in faunal diversity, in the complexity of the food web, and in the diversity of energy pathways. *(Based on Hoffman 1977, 1979)*

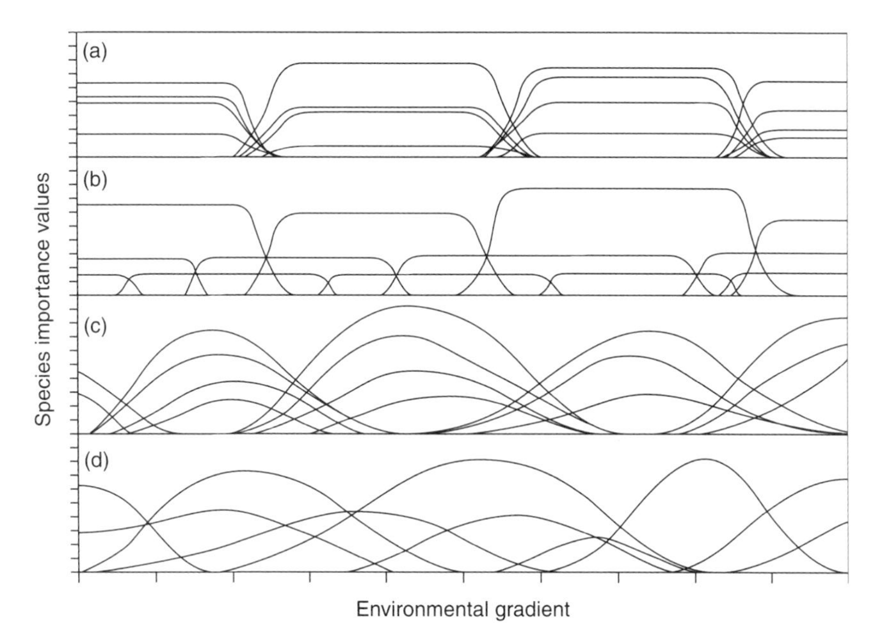


**FIGURE 9.5 Schematic depictions of species abundance distributions along a hypothetical environmental gradient.** Models range from a scenario in which the distributions of species were highly dependent on one another (top) to one in which they were independent of one another (bottom). See text for descriptions of parts (a) through (d). *(From Whittaker, 1975)*

transect along which environmental transitions are gradual and continuous rather than abrupt. In Figure 9.5a, despite the gradual environmental transitions, there is distinct zonation of species along the gradient, with well-defined, sharply bounded communities of species that develop close associations with one another, as evidenced by their parallel abundance patterns. Boundaries between communities are sharpened by **competitive exclusion,** which relates to the hypothesized tendency of closely related, potentially interactive species to avoid competing with one another for resources by cordoning off from one another different portions of ecological space.

In Figure 9.5b, competitive exclusion causes the distributions of individual species to be sharply bounded, but there is no tight linkage of individual species within communities. In Figure 9.5c, there is little competitive exclusion among species, so their abundance distributions are not sharply bounded, and abundances fall gradually away from presumed zeniths somewhere along the transect, where environmental conditions are most ideal for the species to thrive. However, biotic interactions do produce parallel abundance relationships among some species. Finally, in Figure 9.5d, biotic interactions do not govern the development of parallel relationships of species distributions to one another, nor do they serve to sharpen boundaries between species through competitive exclusion.

Thus, the four panels in Figure 9.5 illustrate a spectrum of possible scenarios, ranging from a system governed strongly by biotic interactions [part (a), a "Clementsian" perspective] to one in which species respond individually to physical attributes of the gradient [part (d), a "Gleasonian" perspective]. Empirical data from plant communities appear to suggest that Figure 9.5d is by far the most common of the four scenarios observed in nature (Figure 9.6). Based on these observations, the biologist R. H. Whittaker (1975) concluded:

> It is useful to recognize life-zones. . . . But the zones are continuous with one another. . . . The zones are kinds of communities [that humans] recognize . . . mainly by their dominant plants, within the continuous change of plant populations and communities along the elevation gradient. The zones can be compared to the colors [humans] recognize . . . and accept . . . as useful concepts, within the spectrum of wavelengths of light which are known to be continuous. [pp. 116–117]

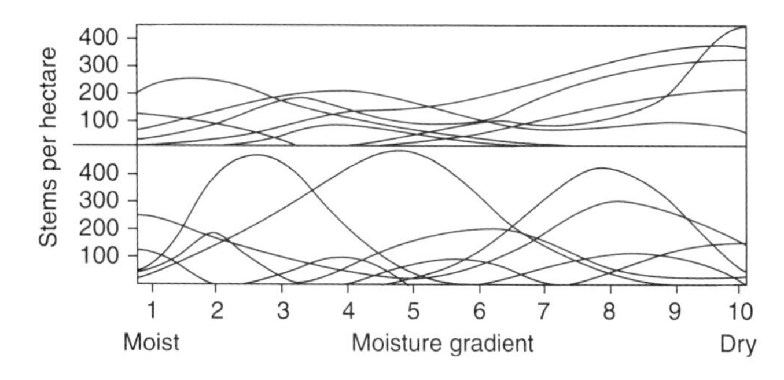


**FIGURE 9.6 Real distributions of plant species abundances along two moisture gradients in Oregon (top) and Arizona (bottom).** These patterns appear to mimic closely the hypothetical scenario illustrated in Figure 9.5d. *(From Whittaker, 1975)*

The metaphor of a continuous color spectrum does not apply in all instances, but it can be said with some confidence that there is little evidence in the present day for the kinds of biotic units implied by Figure 9.5a.

## 9.3 PALEOCOMMUNITIES

### Distributions of Fossil Taxa within and among Paleocommunities

As we have already suggested, the central prerequisite for most paleoecological investigations is identification and quantification of the organisms present at particular places in space and time. For this reason, paleontologists have long been interested in knowing whether it is possible to meaningfully recognize ancient communities in the fossil record. Of course, in most cases, direct information about the soft-bodied taxa present in an ancient community will not be available, thereby precluding an understanding of the entire roster of taxa present in most preserved paleoenvironments. Nevertheless, even when limited to studying the skeletonized portion of the biota, paleontologists have discovered that it is possible—and worthwhile—to identify and characterize the taxa that lived in ancient environments. And while we might not be able to determine definitively the soft-bodied taxa present in a paleocommunity, we can often infer the broad presence of certain groups that are not preserved, such as phytoplankton in marine paleocommunities throughout the Phanerozoic.

During the 1960s, Alfred Ziegler sought to recognize paleocommunities preserved in Silurian strata of Wales and elsewhere. He began by distinguishing five brachiopod-rich paleocommunities in the Welsh borderland that were arrayed from nearshore to deep water (Table 9.1; Figure 9.7), for which he recognized *characteristic* (i.e., abundant) and *associated* (i.e., usually present but less abundant) species. The similarity to the approach advocated earlier by Petersen in the study of present-day organisms is unmistakable. In fact, Ziegler noted that "it is clear that the animal community technique of the ecologists is applicable to fossil assemblages" (p. 272).

TABLE 9.1

Lists of the Principal Taxa Contained within Each of Five Communities Recognized in Silurian Strata of the Welsh Borderland

| Community Name | Characteristic Species | Associated Species |
|---|---|---|
| *Lingula* | *Lingula pseudoparallela*<br>*"Camarotoechia" decemplicata*<br>*"Nucula" eastnori* | *"Hormotoma"* sp.<br>*"Pterinia"* sp.<br>*Cornulites* sp. |
| *Eocoelia* | *Eocoelia* spp.<br>*"Leptostrophia" compressa*<br>*Dalmanites weaveri* | *Howellella crispa*<br>*Salopina* sp.<br>*"Pterinia"* sp. |
| *Pentamerus* | *Pentamerus* spp.<br>*Atrypa reticularis*<br>*Dalejina* sp. | *Eocoelia* spp.<br>*Howellella crispa* |
| *Stricklandia* | *Stricklandia* spp.<br>*Eospirifer radiatus*<br>*Atrypa reticularis* | *Resserella* sp. |
| *Clorinda* | *Clorinda* spp.<br>*Diocoelosia biloba*<br>*Cyrtia exporrecta*<br>*Skenidioides lewisi* | *Plectodonta millinensis*<br>*Coolinia applanata*<br>*Plectatrypa marginalis* |

*SOURCE:* Ziegler (1965), Adapted from Table 1

(a)

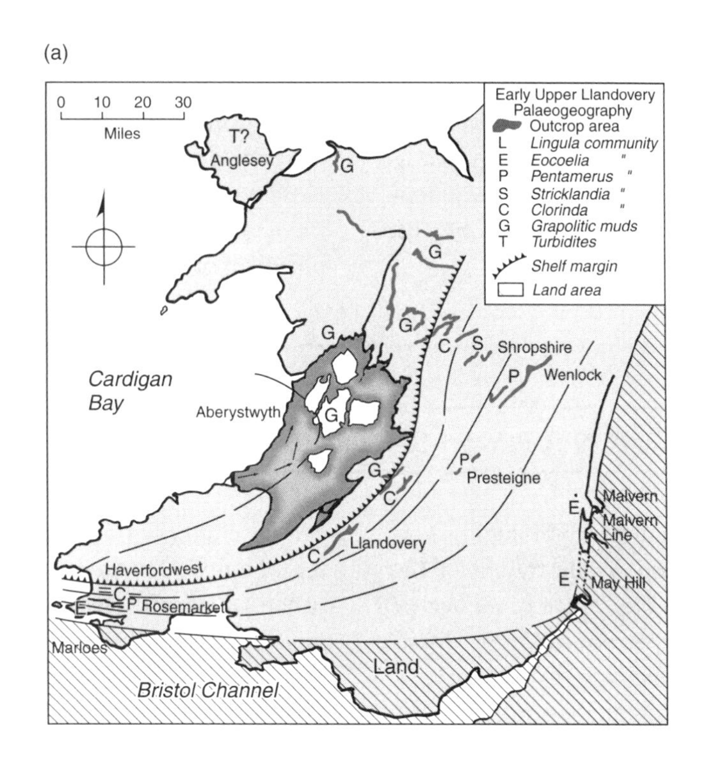


(b) (c)

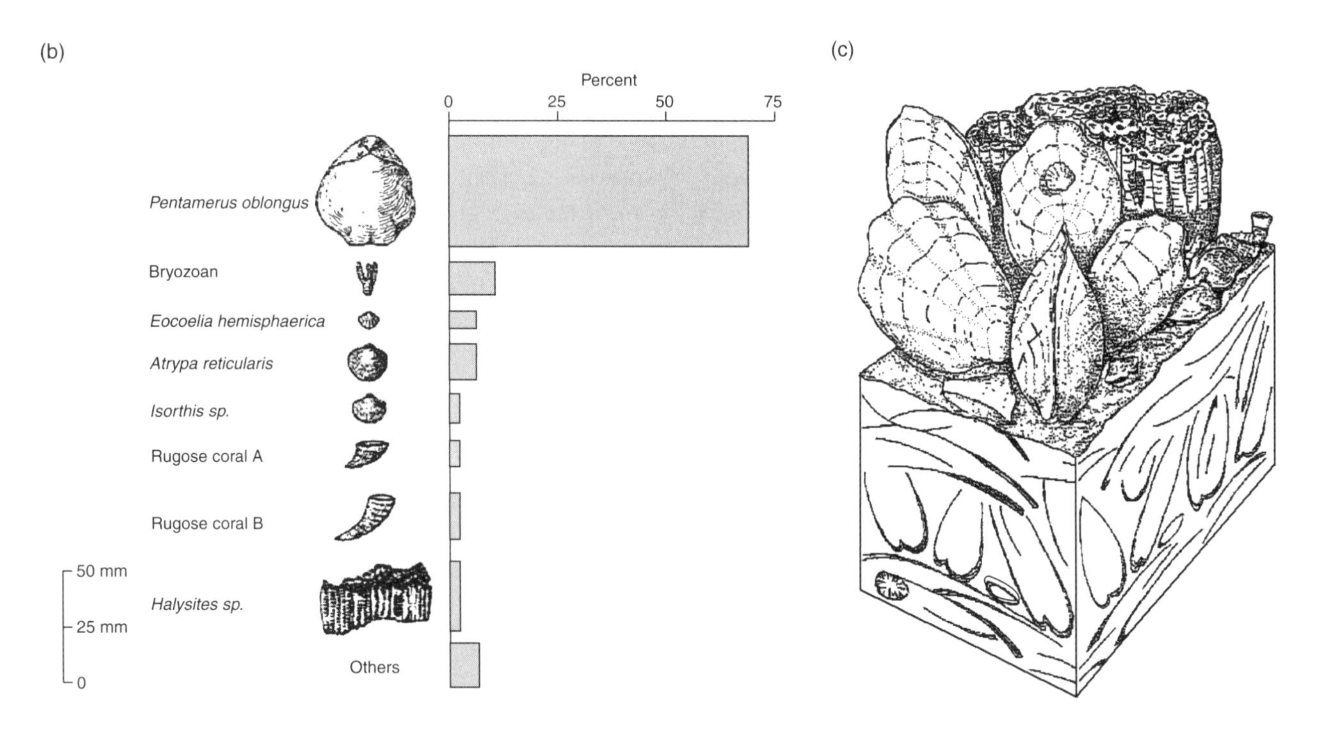


**FIGURE 9.7 Ziegler's delineation of Silurian communities, in the Welsh borderland and elsewhere.** (a) A paleogeographic map depicting the communities from the shoreline (lower right) into deep water (middle). (b) A bar graph illustrating the relative abundances of major taxa in a fossil assemblage that preserves the *Pentamerus* community, at a locality in Wales. (c) A sketch of the *Pentamerus* community, based on data from assemblages preserved in North American strata. *(a: From Ziegler, 1965; b, c: From Ziegler et al., 1968)*

Ziegler followed this by demonstrating that suites of paleocommunities that were rather similar to those documented in Wales could be recognized in Scandinavia and North America. This provided further credence to the view not only that paleocommunities could be recognized dependably, but also that they were coherent, repeatable units of organization.

Paleontologists now routinely analyze quantitatively the spatial distributions of ancient organisms within and among paleocommunities. As a result, they have observed patterns in the fossil record that conform closely to those recognized along present-day environmental gradients. In Chapter 6, we described the application of multivariate techniques to paleoecological

*Box 9.2*

## GRADIENT ANALYSIS IN THE FOSSIL RECORD

Figure 9.8 highlights the diagnosis of biological gradients in the fossil record and an illustration of the sometimes subtle effects that the choice of numerical techniques can have on our perception of the distributional patterns exhibited by taxa. In Figure 9.8a, cluster analysis (see Box 3.2) is applied to paleoecological data collected through bulk sampling of a stratigraphic interval of the Ordovician Martinsburg Formation at a locality in southwestern Virginia. Cluster analysis has been used widely in paleoecology for assessing variations among samples in their biological compositions (a so-called Q-mode analysis), as well as variations among taxa in their occurrences within samples (an R-mode analysis).

Computationally, the set of methodological steps is very similar to that applied to morphological data in the *Stegoceras* example presented in Chapter 3, except that in this case, the starting point is a set of samples, each containing taxa for which abundances were reported, rather than a set of specimens for which measurements of multiple morphological features are reported (see Box 3.2). The main output of cluster analysis is a dendrogram, a kind of graphical tree on which samples or taxa that are in close proximity have distributions that are more similar to one another than those that are comparatively remote from one another. The two dendrograms placed at right angles to one another in Figure 9.8a illustrate the tendency for certain clusters of taxa to recur within certain clusters of samples. In fact, as illustrated with the lettering of the sample clusters and the lines between them, it might be reasonable to view each lettered cluster as representing a separate paleocommunity, given the high level of consistency in taxonomic composition exhibited among the samples in each cluster.

However, paleontologists have come to appreciate that, because of its inherent tendency to produce discrete groupings of samples on a dendrogram, as well as the reduction of what may be complex relationships among samples down to the single dimension of a dendrogram, cluster analysis tends to mask what are actually more gradational transitions in the compositions of samples and the distributions of taxa comprising each community. This is illustrated in Figures 9.8b and 9.8c. In Figure 9.8b, the samples illustrated in the dendrogram were compared using *polar ordination* (PO), an ordination technique that is similar in intent to *detrended correspondence analysis* (DCA) (see Box 6.5).

In this example, the samples are arrayed as points in a two-dimensional space, determined by simultaneous consideration of the abundances of all the taxa that are present in the samples. Like DCA, PO seeks to identify the main axes of variation among samples, based on a numerical combination of taxa rather than a single taxon. Thus, the way in which samples are

**FIGURE 9.8 Gradient analysis of fossil assemblages in the Upper Ordovician Martinsburg Formation at a locality in southwestern Virginia.** (a) A two-way cluster analysis, depicting the tendency of certain taxa to be present in particular clusters of samples collected at one locality. The dendrogram on the left compares samples based on their taxonomic compositions. The dendogram at top compares taxa based on their occurrences within samples. In the middle grid, dots denote the presence of a taxon in a sample. Letters along the side denote major cluster of samples (see Figure 9.8b). *(From Springer & Bambach, 1985)*

data as a means of achieving high-resolution, regional correlations among localities. This approach was predicated on the recognition that boundaries between adjacent paleocommunities tend to be gradational rather than discrete, and, as we noted, the evaluation of taxonomic distributional patterns among communities is known as **gradient analysis.** A discussion of the diagnosis and interpretation of gradients in the fossil record is presented in Box 9.2.

While spatial variation within individual horizons (i.e., during a confined time interval) was not assessed in the study illustrated in Box 9.2, it has been assessed elsewhere. Figure 9.9 illustrates an investigation of variations in the compositions of bulk samples of fossil

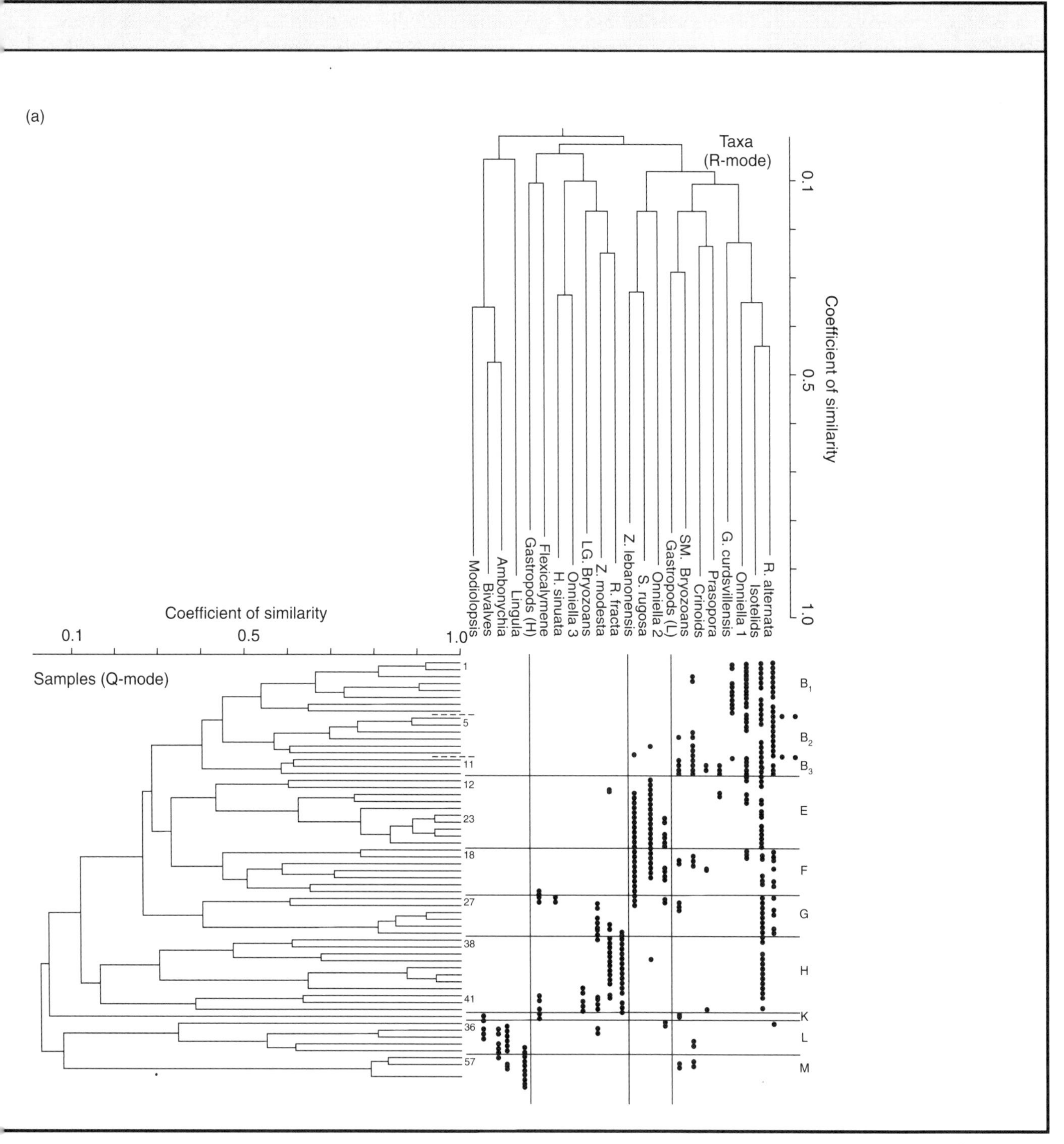


continued on next page

## *Box 9.2 (continued)*

arrayed along an axis may reflect variations in the abundances of not just a single taxon, but several taxa that vary similarly in their presence and abundance among samples. Moreover, different axes may highlight different aspects of variation of the data. Although only two axes are illustrated here, it is generally possible to derive additional axes for a data set that contains a large number of samples and taxa.

We can observe in Figure 9.8b that the samples identified by cluster analysis are not arrayed as discretely as suggested by the sample dendrogram in Figure 9.8a. Furthermore, when the abundances of several of the important taxa are graphed through the sampling interval (Figure 9.8c), a pattern is observed that is reminiscent of that depicted earlier for the spatial arrangement of plants along present-day environmental gradients (Figure 9.6). In both cases, there is broad overlap in the distributions and abundances of taxa rather than discrete boundaries. While the depiction in Figure 9.8c is stratigraphic instead of spatial, the broad distributions of these taxa suggest that this pattern would also be reflected in the spatial distributions among the communities that coexisted at any given time during the interval under investigation.

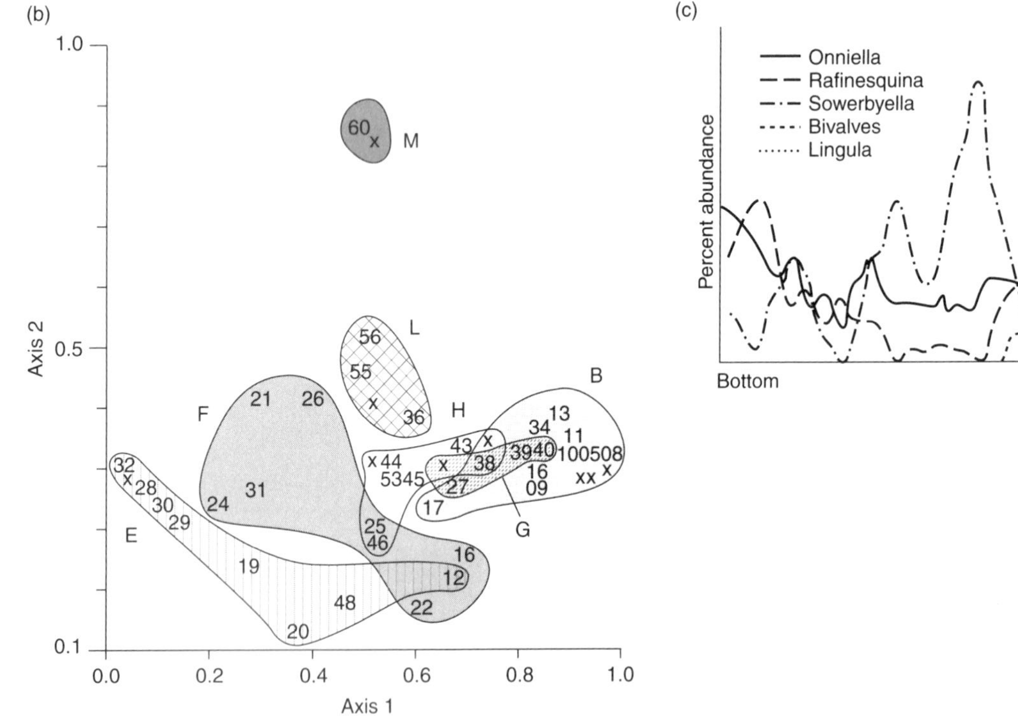


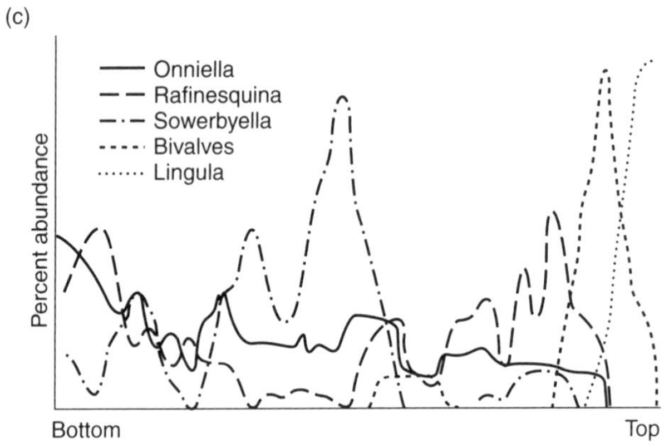


**FIGURE 9.8** ***(cont.)*** (b) An ordination analysis of the same samples as in part (a). Individual samples are denoted with numbers; x's mark the overlap of two or more samples. Lettered, demarcated groups of samples correspond to the lettered clusters in Figure 9.8a. Note the broad overlap among the groups of samples depicted in the dendrogram.
(c) Stratigraphic depictions of variations in the abundances of major taxa at the locality. Note the tendencies of the five taxa to vary individualistically. *(From Springer & Bambach, 1985)*

oysters collected at several spatial scales in a single horizon of the Upper Cretaceous Navesink Formation in New Jersey. The two localities were approximately 10 km apart, and, at each locality, outcrops were spaced several tens of meters apart. Four samples were collected from each outcrop, and adjacent samples were no more than a few meters apart (Figure 9.9a). The taxonomic data collected from each sample indicate that the abundances of taxa contained in adjacent samples were highly variable (Table 9.2). This suggests a pattern known as **patchiness,** which appears to be a general feature of spatial distributions within communities and paleocommunities at small scales. It may reflect the interplay of multiple factors, including the tendency of some species to aggregate in clumps, coupled with chance spatial variations in local environmental factors—such as the availability of substrates suitable for the settlement of larvae, seeds, spores, or other early developmental stages of organisms.

Ecologists have become interested in understanding the nature of interactions and genetic communication among members of the same species that are semi-isolated from one another in different patches or subpopulations. They now understand that the ecological persistence of a metapopulation [SEE SECTION 3.1]—the collective set of subpopulations—and the evolutionary persistence of the species itself may be contingent on the nature and extent of genetic interaction among the subpopulations.

While paleoecologists have begun to embrace metapopulation theory, particularly as it relates to the evolutionary persistence of species, the incorporation of metapopulation dynamics into paleoecology is still in its infancy. In part, this is because paleoecologists are still attempting to develop means of determining assuredly that the patchiness preserved in most fossil assemblages reflects actual biological patchiness in the once-living community of species rather than simply a combination of time averaging and post-mortem transport. Among marine biotas, there are indications from a few investigations that biological patchiness can persist through several generations of skeletal accumulation on a sea floor, even in the face of severe storms. But it remains to be determined whether this is a general feature of the fossil record.

Whatever its cause, the pervasiveness of patchiness points to the need to collect multiple samples at local scales because of the possibility that a single sample will provide a misleading sense of the overall biotic composition of the area. For example, of the 20 *Gryphaeostrea* individuals that were counted in the illustrated study (Table 9.2), 18 came from a single sample! Had this been the only sample collected at the locality, it would have obviously provided an inaccurate view of the relative importance of *Gryphaeostrea* at the locality. That said, when the samples at each locality are averaged together to determine locality-wide abundances (Figure 9.9b), the aggregate biotic compositions of the two localities are remarkably similar despite the 10-km distance between them.

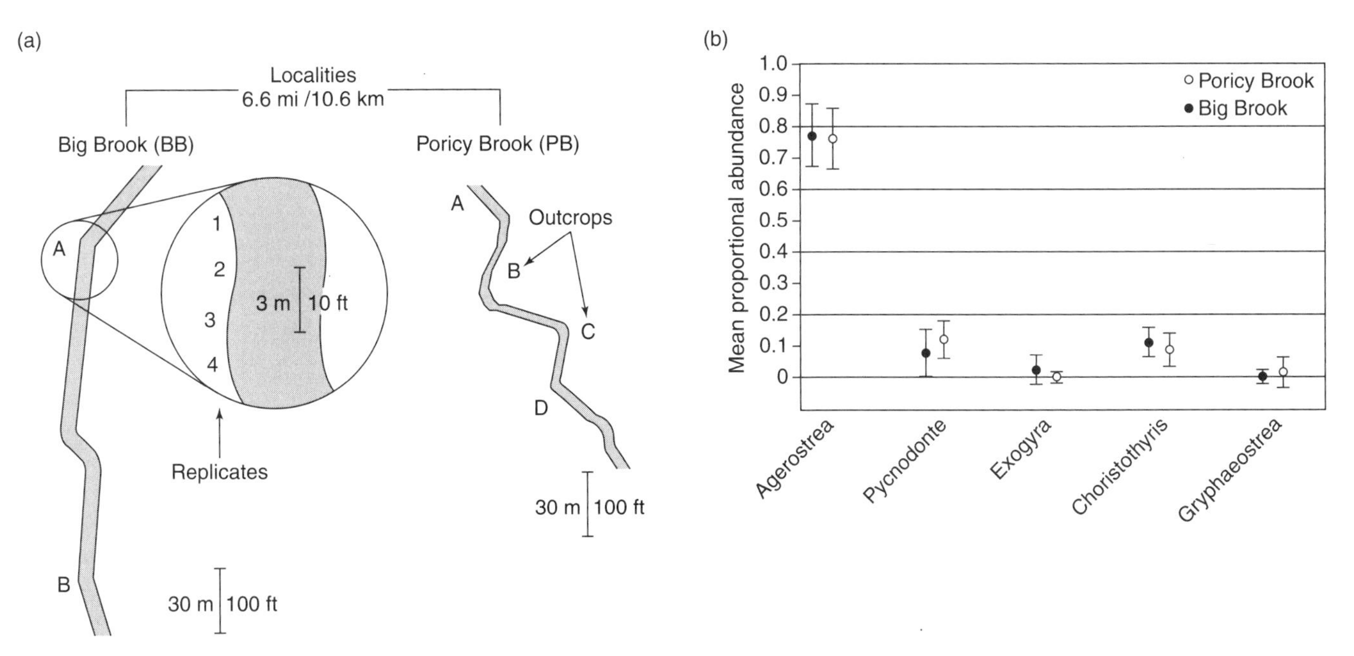


**FIGURE 9.9 An assessment of variation in the faunal contents of samples collected at several different spatial scales, in Upper Cretaceous strata of New Jersey.** (a) Samples were collected in replicates that were a few meters apart at outcrops spaced several tens of meters apart at two localities that were about 10 km apart. (b) Although the compositions of adjacent replicates were sometimes highly variable in composition (see Table 9.2), aggregate abundances for the two localities were comparable to one another for the five main species found in the study. *(From Bennington, 2003)*

TABLE 9.2

Faunal Data Matrix for a Set of Samples Collected from Upper Cretaceous Strata in New Jersey

| Sample ID | *Agerostrea* | *Pycnodonte* | *Exogyra* | *Choristothyris* | *Gryphaeostrea* | Total |
|---|---|---|---|---|---|---|
| BBA1 | 35 | 7 | 4 | 9 | 1 | 56 |
| BBA2 | 5 | 6 | 0 | 1 | 0 | 12 |
| BBA3 | 3 | 7 | 0 | 0 | 0 | 10 |
| BBA4 | 27 | 5 | 5 | 2 | 0 | 39 |
| BBB1 | 46 | 4 | 1 | 8 | 0 | 59 |
| BBB2 | 77 | 3 | 1 | 6 | 0 | 87 |
| BBB3 | 65 | 1 | 0 | 11 | 0 | 77 |
| BBB4 | 75 | 2 | 0 | 12 | 0 | 89 |
| PBA1 | 46 | 5 | 0 | 7 | 1 | 59 |
| PBA2 | 22 | 13 | 0 | 0 | 0 | 35 |
| PBA3 | 21 | 5 | 0 | 1 | 0 | 27 |
| PBA4 | 59 | 14 | 0 | 10 | 0 | 83 |
| PBB1 | 42 | 10 | 1 | 0 | 0 | 53 |
| PBB2 | 69 | 10 | 1 | 11 | 0 | 91 |
| PBB3 | 124 | 6 | 1 | 10 | 0 | 141 |
| PBB4 | 25 | 6 | 1 | 13 | 0 | 45 |
| PBC1 | 57 | 20 | 1 | 24 | 0 | 102 |
| PBC2 | 45 | 6 | 0 | 9 | 0 | 60 |
| PBC3 | 121 | 4 | 0 | 6 | 0 | 131 |
| PBC4 | 106 | 5 | 0 | 6 | 0 | 117 |
| PBD1 | 15 | 9 | 0 | 0 | 0 | 24 |
| PBD2 | 63 | 6 | 0 | 0 | 0 | 69 |
| PBD3 | 21 | 9 | 0 | 0 | 0 | 30 |
| PBD4 | 22 | 12 | 0 | 2 | 18 | 54 |
| **Total** | 1191 | 175 | 16 | 148 | 20 | 1550 |

*SOURCE:* Bennington (2003), Table 3

*NOTE:* These data were used to assess the degree of compositional variability evident among samples collected at different spatial scales. Key to sample ID: Localities: BB (Big Brook), PB (Poricy Brook); Outcrops: A, B, C, D; Replicate Samples: 1, 2, 3, 4.

## Regional Stratigraphic Distributions of Fossil Taxa

The study of paleocommunities has provided a biologically based foundation for the further development of paleoecology as a subdiscipline. Methods of data collection and analysis pioneered in this research have become standard practice for paleoecologists, and their importance has become particularly evident in the investigation of ecological dynamics associated with regional patterns of diversification and extinction.

In numerous investigations of stratigraphic transitions among regional biotas, a common theme seems to have emerged: A large percentage of the appearances and disappearances of taxa tend to be concentrated in narrow intervals that punctuate broader intervals of relatively stable composition. For example,

Carlton Brett and Gordon Baird (1995) investigated stratigraphic patterns of regional biotic change in Silurian and Devonian strata located across a large, interior portion of New York State. Brett and Baird compiled species-level faunal lists and found that biotas within several stratigraphically consecutive, region-wide intervals exhibited a high degree of compositional stability at the species level, with some 60 to 80 percent of species persisting through a typical interval (Table 9.3a).

These intervals each lasted some 3 to 8 million years and appear to have been punctuated by much shorter intervals during which there were dramatic transitions in species composition throughout the basin, with compositional turnovers on the order of 80 percent from one interval to the next (Table 9.3b). Lithological changes

TABLE 9.3

Calibration of Taxonomic Turnover Metrics for Silurian and Devonian Ecological Evolutionary (e-e) Subunits in the Appalachian Basin of New York

**(a)** *Persistence and Extinction Values of Four Well-Characterized Faunas of the Silurian to Middle Devonian Interval*

| Fauna | e-e Subunit | Duration (m.y.) | Persistence (%) | Extinction (%) |
|---|---|---|---|---|
| Hamilton | 10 | 5–6 | 80 | 5 |
| Onondaga | 9 | 6–7 | 78 | < 10 |
| Helderberg | 6 | 7–8 | 70 | < 10 |
| Up. Clinton–Lockport | 3 | 7–8 | 66 | 32 |

*SOURCE:* Brett & Baird (1995), Table 9.2

*NOTE:* The degree of persistence and extinction within four well-characterized e-e subunits. Persistence is measured as the percentage of species that range from the lowest to the highest parts of the e-e subunit. Extinction is measured as the percentage of species that disappear prior to the end of the subunit. Note that the majority of species persist through each of the subunits.

**(b)** *Appalachian Basin Silurian and Devonian Faunas, Showing Holdover and Carryover Indices*

| e-e Subunit | Fauna | Age | Duration (m.y.) | Carryover | Holdover |
|---|---|---|---|---|---|
| 10 | Hamilton–Tully | Givetian | 6–7 | 30/335 (9%) | 32/335 (10%) |
| 9 | Onondaga | Eifelian | 5 | 32/200 (16%) | 37/200 (19%) |
| 8 | Schoharie | Emsian | 5 | 37/125 (30%) | 10/125 (8%) |
| 7 | Oriskany | Pragian | 2 | 10/94 (11%) | 25/94 (27%) |
| 6 | Helderberg | Gedinnian | 6 | 25/130 (19%) | 7/130 (5%) |
| 5 | Keyser–Bertie | Pridolian | 2 | 7/54 (13%) | 14/54 (26%) |
| 4 | Salina | Late Ludlovian | 3–4 | 14/48 (29%) | 7/48 (15%) |
| 3 | Up. Clinton–Lockport | Late Llandovery–Wenlock | 7–8 | 7/149 (5%) | 30/149 (20%) |
| 2 | Lo. Clinton | Mid-Llandovery | 4 | 30/87 (34%) | 48/87 (55%) |
| 1 | Medina | E. Llandovery | 5 | 48/139 (35%) | ? |

*SOURCE:* Revised from Brett & Baird (1995), Table 9.5

*NOTE:* The degree of holdover and carryover from one e-e subunit to the next in the Appalachian Basin. Carryover is measured as the proportion of species within a subunit that carries over to the subsequent subunit. Holdover is measured as the proportion of species within a subunit that appeared in the subunit that preceded it. Note that, for the most part, these values tended not to exceed 25 percent, suggesting that there were substantial biotic transitions from one e-e subunit to the next.

associated with the turnover events suggest that the broad pattern of turnover was controlled by major, region-wide environmental transitions. The details of these transitions, including a determination of whether some or all of them are correlated with global-scale changes, are currently being investigated.

To underscore the potential significance of these intervals, Brett and Baird called them **ecological evolutionary (e-e) subunits,** a name derived from a set of broader designations known as **ecological evolutionary units** that some researchers had used previously to subdivide Phanerozoic diversity on a global scale. While there has been only limited evaluation of the morphological attributes of species within e-e subunits, it has been suggested that species morphologies remained fairly static from the bottom to the top of each e-e subunit, producing a pattern that was dubbed **coordinated stasis**—in recognition of the suggestion, if not the documentation, that biotic compositions within communities and the morphologies of constituent species remained fairly static throughout an e-e subunit.

The diagnosis of coordinated stasis is not a claim that biotic composition at any given location remained the same through the entire duration of an e-e subunit. Brett and Baird were careful to point out, for example, that several episodes of sea-level increase and decrease took place *during* the intervals represented by most e-e subunits, and that a range of paleocommunities inhabited environmental gradients that varied spatially as well as temporally. The main point, however, was that throughout a given e-e subunit interval, the compositions of biotas associated with paleocommunity gradients remained fairly stable. Therefore, when environmental conditions at a given locality reverted to an earlier state, a paleocommunity returned to the locality that was fairly indistinguishable from the earlier occurrence, provided that both occurrences were contained in the same e-e subunit.

The coordinated stasis hypothesis has been scrutinized carefully since its introduction, and several questions have arisen about the biological significance of the pattern. From a sampling standpoint, the rate of compositional change across e-e subunit boundaries may be artificially inflated because of the overprint of sequence stratigraphic architecture on the stratigraphic distributions of fossils [SEE SECTION 6.4]. It is probable that many transitions between e-e subunits coincided with boundaries between depositional sequences, which would tend to exaggerate the number of first and last appearances associated with these boundaries. In addition, while species composition may remain fairly stable, it has been shown that the relative abundances of species sometimes shift dramatically within the confines of an e-e subunit.

Finally, the extent to which appearances and disappearances within the Appalachian Basin constitute true originations and extinctions of **endemic** (localized) species, as opposed to the regional appearances and disappearances of widespread species that persist elsewhere, remains to be determined. Needless to say, this information is crucial for understanding the evolutionary significance of coordinated stasis. However, such data are not easy to acquire because they must be underpinned by definitive species-level identification.

Paleontologists have also sought to determine whether the pattern recognized by Brett and Baird transcended brachiopod-dominated faunas of the Silurian and Devonian of New York. Results have been mixed, even when considering other Paleozoic marine settings that were dominated taxonomically by a similar suite of organisms. In many cases, a broad pattern of relative compositional stability *has* been recognized for intervals that were somewhat protracted, in comparison to bounding intervals during which changes apparently took place more rapidly. However, the actual extent of stability within e-e subunits, the durations of e-e subunits, and the taxonomic levels at which stability is exhibited have all been shown to vary widely.

Because regional patterns of biotic variation are currently being investigated for a range of settings and taxa, there is reason to expect that paleontologists might soon be in a position to determine whether particular environments, biotas, or time intervals are characterized by their own unique levels of stability and turnover. This kind of cataloguing, in turn, will be useful for determining whether evolutionary rates vary in a predictable fashion as a function of environment. For example, it has been suggested that marine e-e subunits exhibit greater stability in shallow settings than in deeper settings. If this proves generally to be the case, then it would indicate that rates of taxonomic origination and extinction are greater in deep-water settings.

## 9.4 EVOLUTIONARY PALEOECOLOGY

The question of coordinated stasis, and the broader analysis of regional biotic patterns, reflect a growing focus in paleontology on the investigation of paleoecological changes through geologic time. Paleontologists have embraced this perspective in recent years, melding evolutionary and paleoecological approaches together in the study of evolutionary paleoecology. It is probably no coincidence that evolutionary paleoecology has blossomed in conjunction with the growing focus on Phanerozoic diversification, which was highlighted in Chapter 8. In fact, a prime objective of this research has been to explain the biotic transitions recognized in global diversity compilations.

### Ecospace Utilization through Time

As we noted earlier, it is not clear whether it is possible in most cases to reconstruct the fine-scale, spatial distributions of taxa in paleocommunities. However, if we can establish the life habits of these taxa based on principles discussed in Chapter 5, then it is possible to determine the ecospace within the local habitat that was likely utilized by the taxa in a paleocommunity.

In this context, an ecospace is a setting in which ecological conditions are suitable for the subsistence of a particular taxon or group of taxa. For example, in considering the fossils contained in strata that preserve a sand-covered sea floor, in many cases it is possible to determine, based primarily on their morphological characteristics, whether taxa occupied positions above, at, or below the surface of the sediment. Because they provide direct indications of the kinetic activity of individuals, trace fossils are also particularly helpful in making these assessments, especially with respect to activity at or below the sediment surface.

Among the striking features of Phanerozoic marine diversity are the biotic transitions among the three evolutionary faunas [SEE SECTION 8.4], each of which apparently achieved successively greater levels of total diversity. Given that the three faunas were characterized by rather different sets of animals, we might expect that, in a broad sense, they were living their lives in different ways. With this in mind, we can ask whether successive evolutionary faunas achieved higher levels of diversity than their predecessors because, collectively, their constituents occupied greater portions of ecospace. This issue can be addressed by attempting to categorize the different modes of life represented by the taxa extant at any given point in time and then determining whether there were changes through time in the number and variety of such modes. For this purpose, it has been convenient to use a unit of categorization known as a **guild,** which can be thought of here as a group of taxa that exploit the same class of environmental resources in a similar way without regard to taxonomic position.

Richard Bambach (1983, 1985) sought to calibrate the number of guilds represented within Phanerozoic communities, which he did by identifying and tabulating the number of guilds in some 193 fossil collections from the Ordovician, Silurian, Devonian, Carboniferous, and Neogene. His determination of guild membership for each species in a collection was based on (1) its basic physiological and morphological characteristic (i.e., its body plan); (2) its food sources; and (3) its space utilization (Figure 9.10).

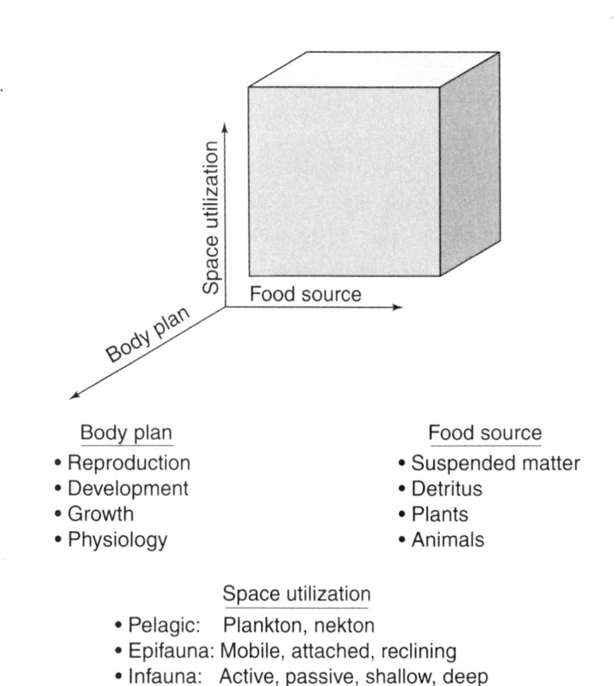


**FIGURE 9.10 Bambach's schematic representation of the three dimensions and related criteria that he used to categorize the ecological characteristics of taxa throughout the Phanerozoic.** He delineated guilds of organisms as taxa that were concentrated in a confined portion of this space because they possessed similar features based on the criteria listed on the diagram. *(From Bambach, 1983)*

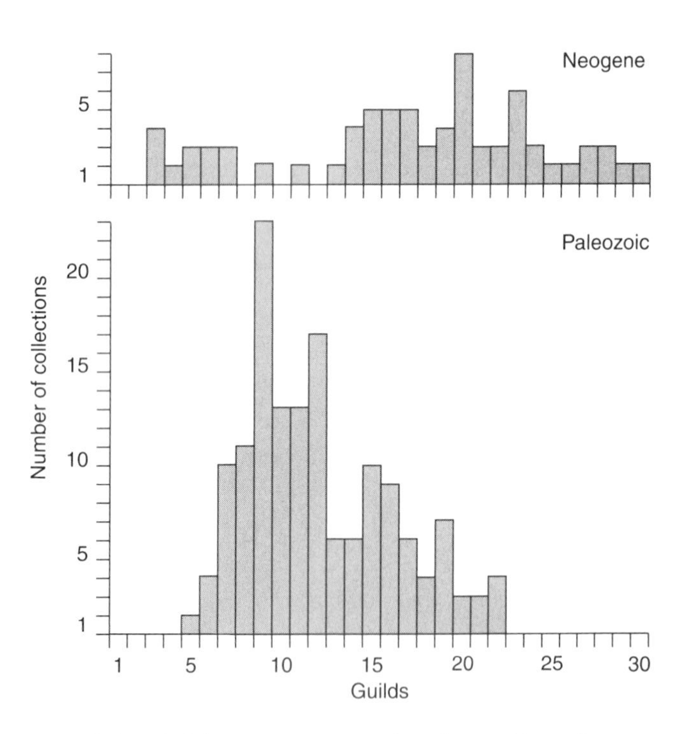


**FIGURE 9.11 Histograms comparing the number of guilds represented in Paleozoic collections, versus those in Neogene collections.** *(From Bambach, 1983)*

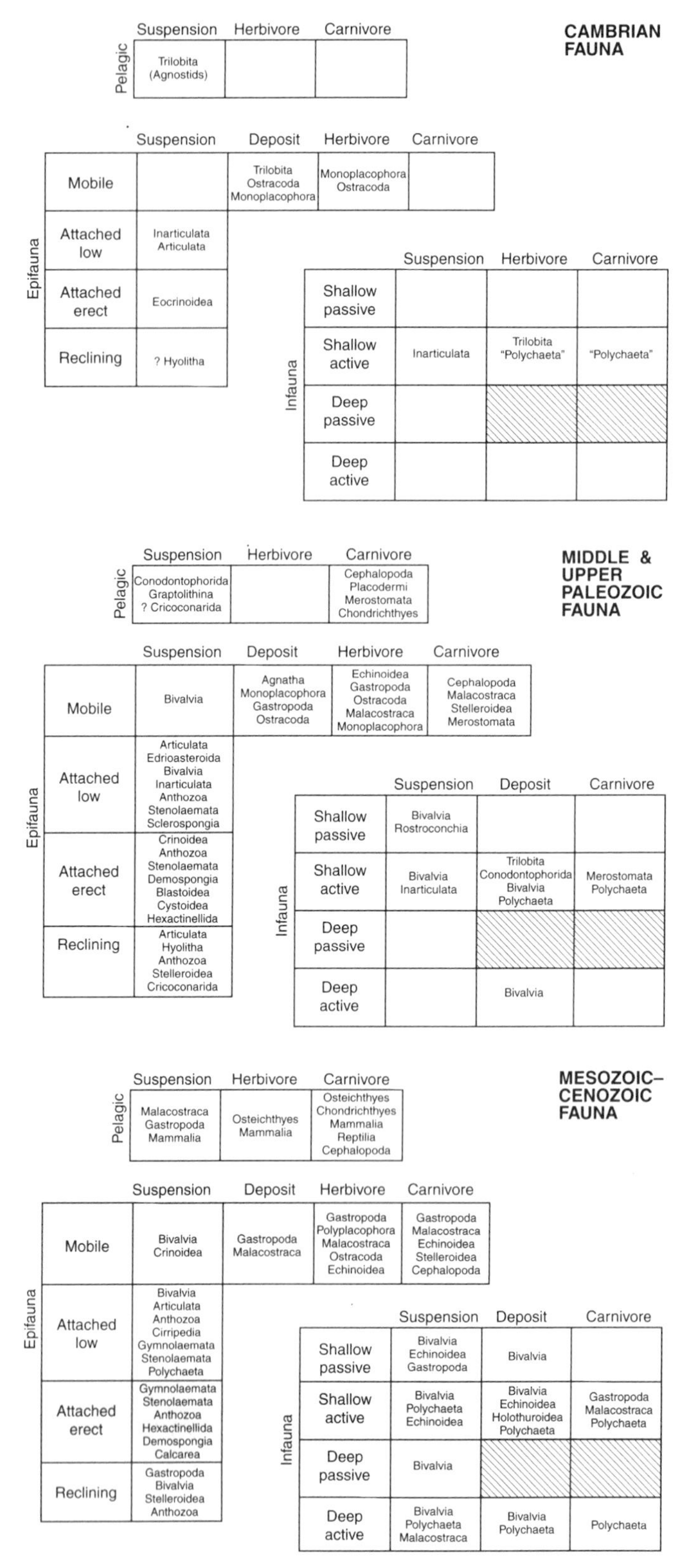


**FIGURE 9.12 Charts depicting the adaptive strategies of taxa that comprised biotas during the Cambrian, the Middle and Upper Paleozoic, and the post-Paleozoic.** Note the increasing diversity of strategies exhibited successively from one interval to the next. *(From Bambach, 1983)*

Bambach's results are summarized in Figures 9.11 and 9.12. Figure 9.11 indicates that the degree of ecospace utilization increased through the Phanerozoic: The median number of guilds grew from 11 in his Paleozoic assemblages to 18 in his Neogene assemblages. As Figure 9.12 indicates, this was associated with an increase in the variety of adaptive strategies exhibited by successive evolutionary faunas. Thus, it is plausible that the increasing level of diversity was indeed based ecologically on the growing amount of available ecospace.

A related pattern is recognized in the nature of tiering through the Phanerozoic among marine organisms that live on soft substrates, such as muddy, silty, or sandy bottoms. A **tier** is thought to be a fairly discrete level above, at, or below the sea floor where there is a concentration of organisms. In the case of infaunal organisms that live beneath the sea floor, it is possible to distinguish between deep and shallow sediment burrowers, which would therefore occupy deeper or shallower infaunal tiers. These assessments can be made based on the style and extent of active burrowing as indicated by trace fossils (Figure 9.13a) and the morphological features of preserved organisms (Figure 9.13b). For epifaunal organisms, which occupy the surface of the sea floor, skeletal features, such as the length

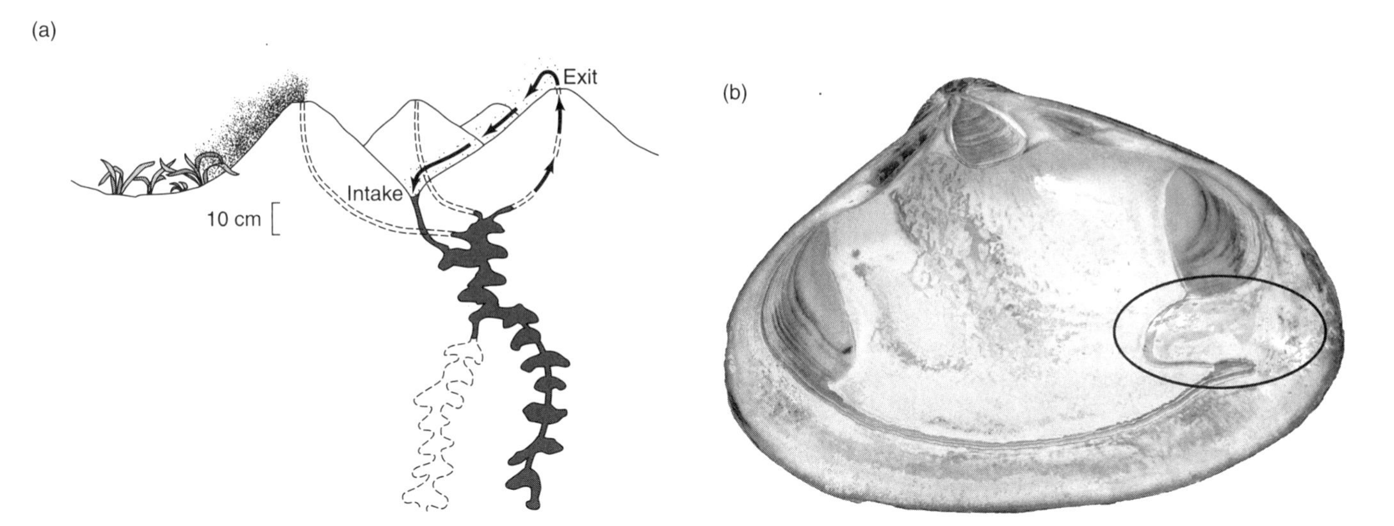


**FIGURE 9.13 Indications of an active, burrowing lifestyle.** (a) The burrow structure of *Callianassa rathbunae*, an active bioturbating shrimp that lives in Caribbean sea floors. Sediment enters the living chamber of the organism intake tubes located between mounds. *C. rathbunae* strips food off of these sediment grains and expels fine-grained sediment back to the sediment surface. These expelled sediments form the distinctive, volcano-shaped mounds that are illustrated here. (b) The interior of the shell of the bivalve *Spisula*, illustrating the position of the pallial sinus (circled), marking the position to which the siphons of this species are retracted. The possession of siphons, as well as other features of the shell, indicate that this species lives beneath the sediment–water interface. *(a: From Suchanek, 1983; b: The Academy of Natural Sciences)*

of the stem of an attached crinoid, can indicate occupancy of epifaunal tiers. There were several transitions in the extent of tiering among infaunal and epifaunal organisms throughout the Phanerozoic that, in some cases, appear to have paralleled changes in marine taxonomic diversity (Figure 9.14), [SEE SECTIONS 8.3 AND 8.4].

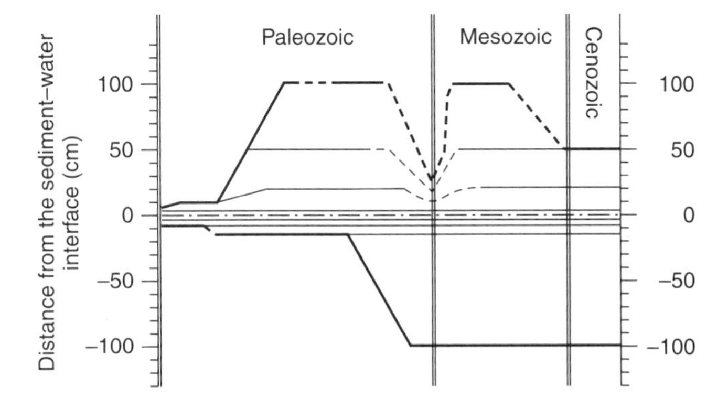


**FIGURE 9.14 A depiction of the history of epifaunal and infaunal tiering on soft substrates throughout the Phanerozoic.** Note that the pattern appears to parallel, in some respects, the Phanerozoic history of global marine diversity (Figure 8.5). Solid lines are based on direct observations and measurements of fossil organisms; dotted lines are inferred patterns, based on interpolation between solid lines. *(From Bottjer & Ausich, 1986)*

Some paleontologists have questioned whether marine organisms really were, or are, distributed into discrete tiers. It is clear, however, that the maximum distance above the sea floor occupied by epifaunal organisms and the maximum depth of infaunal penetration increased from the Cambrian to the mid-Paleozoic. It is also clear that the maximum depth of burrowing increased further during the Mesozoic. Interestingly, the maximum reach of attached, sedentary organisms above the sea floor appears to have decreased during the Cenozoic, which may relate to an increase in predation intensity. These organisms would have been particularly vulnerable to attack by a diversifying array of mobile predators, the importance of which we consider more fully below.

## Evolutionary Transitions Associated with Ecological Interactions

Ecologists have come to recognize that interactions among present-day organisms play significant roles in determining their distributions and abundances. Prominent among these are predator–prey interactions, which are governed by complex cycles of booms and busts in the abundances of predators and the organisms on which they prey. These relationships, and the interwoven

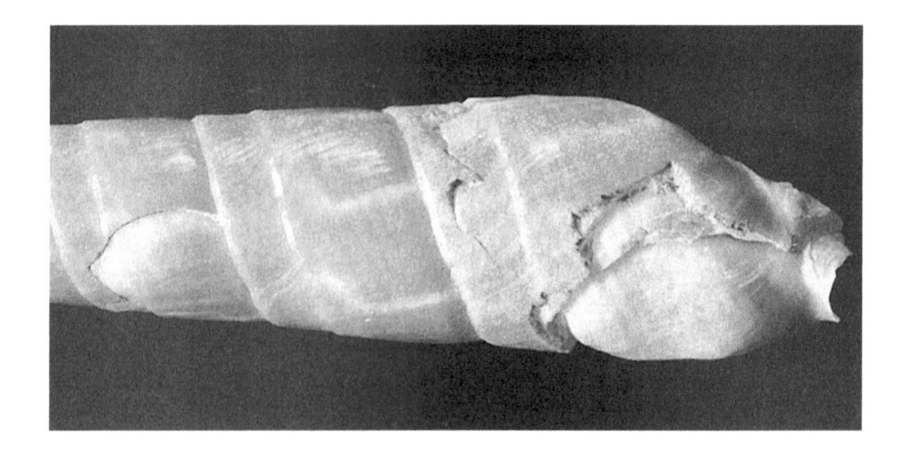

**FIGURE 9.15 A specimen of the gastropod *Terebra dimidiata*, from Guam.** This specimen preserves in its whorls the scars of several instances of nonlethal predation by shell-peeling crabs, followed by repairs to the shell. *(From Vermeij ,1987)*

strategies of predators and prey that govern them, are assuredly the products of extended intervals of evolution, and paleontologists have therefore assumed that similar kinds of interactions have been a pervasive part of ecosystems for as long as there have been predatory organisms. This, in turn, has motivated a general question for paleoecologists: Do the ecological interactions that can be shown to mediate the distributions and abundances of organisms in the present day also play significant roles in the mediation of evolutionary trends that we can observe through geologic time?

In the case of predator–prey interactions, one straightforward expectation is that if there is a diversification of predatory taxa that prey on a particular set of organisms, then in the course of evolution, morphological characteristics might evolve among prey taxa that increase their chances of surviving predatory attacks. Geerat Vermeij (1977, 1987) investigated this possibility extensively and has compiled evidence suggesting that these "arms races" between predator and prey lineages may be pervasive features of evolution. Among the most compelling cases are those involving gastropods and the predators that crushed their shells, which Vermeij tracked through the Mesozoic and Cenozoic Eras. He referred to the collective transitions associated with these arms races as the **Mesozoic Marine Revolution.**

There is ample evidence, in the form of damage and/or repair features on gastropod shells (Figure 9.15), that many gastropods are subjected to repeated episodes of predation during their lifetimes. Has this always been the case? As far back as the Devonian, there was a notable increase in the diversity of marine taxa capable of breaking shells, but, from the early Cretaceous through the Neogene, the diversification of these shell crushers appears to have been particularly pronounced, increasing at the family level by a factor of about 2.5 (Figure 9.16). This rate of increase was greater than that for marine animal families as a whole (see Figure 8.4a), which, for well-preserved organisms, increased by a factor of about 1.3 during the same interval. Therefore, even if the apparent rate of increase was amplified by the Pull of the Recent [SEE SECTION 8.3], the *percentage* of the marine biota that was made up of shell crushers appears to have increased substantially during the late Mesozoic and Cenozoic eras.

While gastropods were also apparently undergoing a significant diversification at that time, certain morphological features exhibited by gastropod taxa were in decline. For example, Vermeij noted that gastropods

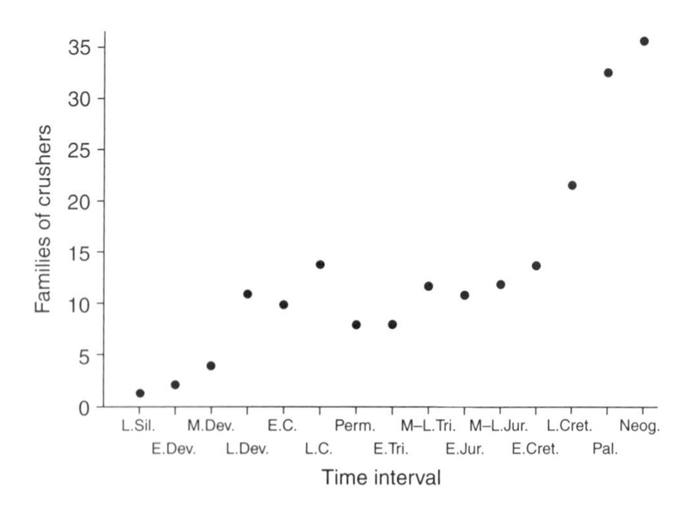


**FIGURE 9.16 A depiction of the Phanerozoic diversity of marine families with members that were capable of crushing shells.** Note the moderate increase in the mid-Paleozoic, followed by a sharp increase in the Late Mesozoic. This increase in diversity appears to exceed that of the biota as a whole during the same interval. *(From Vermeij, 1987)*

possessing an umbilicus—a hollow cavity associated with the axis of coiling (Figure 9.17a)—declined markedly as a percentage of total gastropod diversity in the Mesozoic and Cenozoic (Figure 9.17b). Vermeij also investigated the changing morphologies of gastropod apertures, the openings at the margins of gastropod shells. He documented a Mesozoic and Cenozoic increase in the percentage of gastropods with apertures that were narrow, or in which the shell region surrounding the aperture was notably thickened (Figures 9.17c and 9.17d).

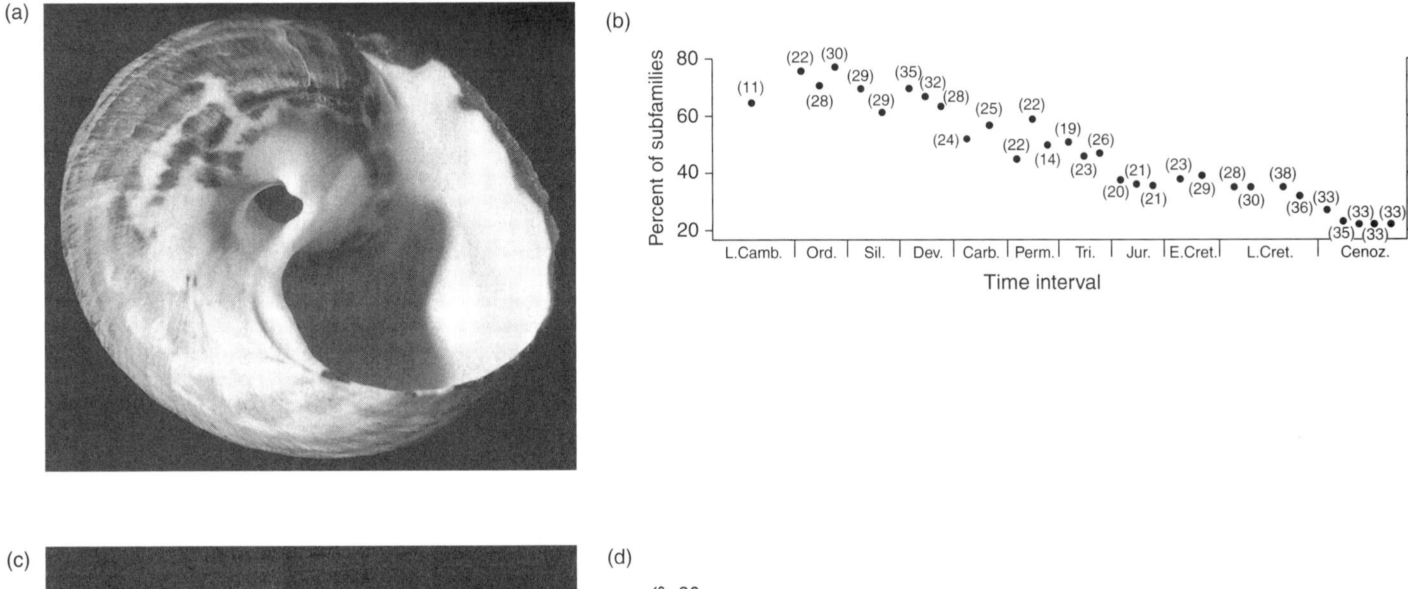


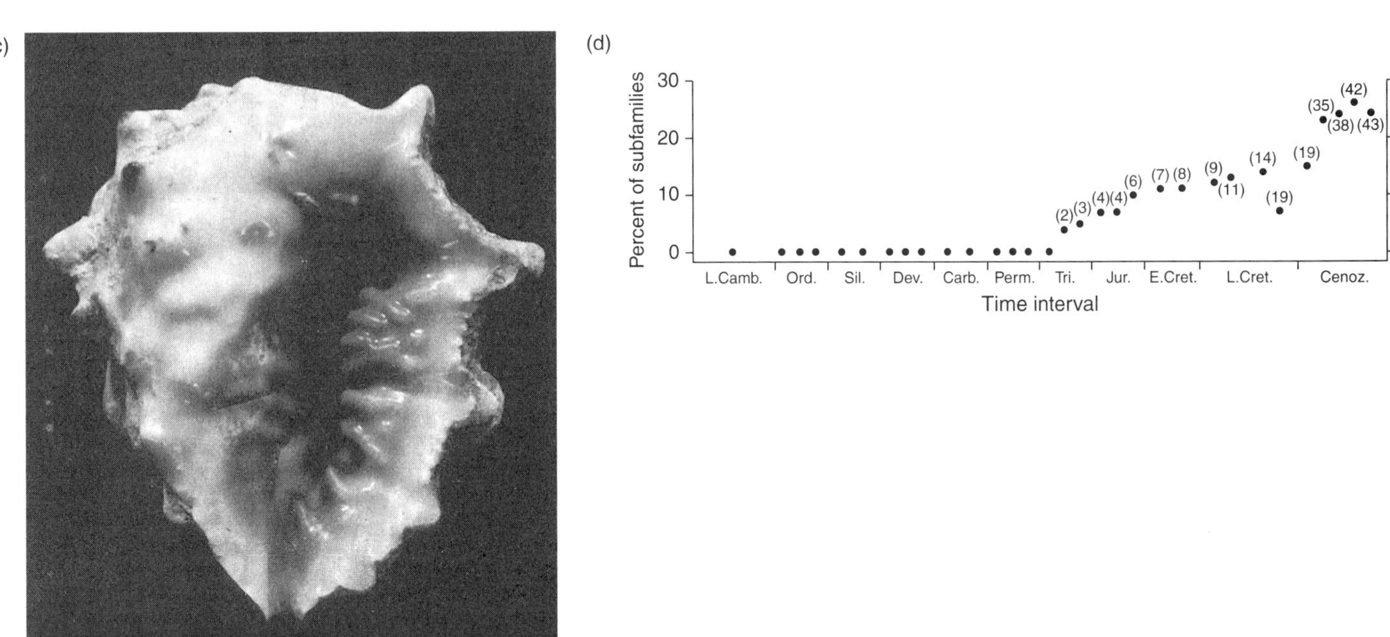


**FIGURE 9.17 Phanerozoic transitions in the morphology of gastropods.** (a) *Cittarium pica*, from Panama, a species with an umbilicus, a broad aperture, and a thin wall surrounding the aperture. These features are thought to render this species vulnerable to predation by shell crushers and peelers. (b) The percentage of gastropod species that were umbilicate, as a function of geologic time, in assemblages representing warm-water settings. (c) *Drupa morum*, from the northern Mariana Islands, a species with a narrow aperture and thickened walls surrounding the aperture. These and other features are thought to render this species less vulnerable to predation by shell crushers and peelers. (d) The percentage of species with thickened or narrowed apertures, as a function of geologic time. In (b) and (d), numbers in parentheses are the number of assemblages sampled. *(From Vermeij, 1987)*

Vermeij attributed all of these transitions to the diversification of predatory organisms, particularly crustaceans that possessed hammerlike appendages capable of crushing shells or strong claws capable of peeling them open. The possession of an umbilicus renders a gastropod more vulnerable to crushing [SEE SECTION 5.3], whereas the possession of a broad aperture associated with a thin shell renders it more vulnerable to being peeled open. Therefore, the patterns among gastropods illustrated in Figure 9.17 are precisely what one would expect in the face of increasing predation.

Morphological transitions do not constitute the only potential macroevolutionary responses to increased predation. Another possible response is for organisms to become restricted to physical environments that are beyond the reach of their predators. Just such a response has been suggested in the restriction during the Mesozoic and Cenozoic Eras of stemmed crinoids to deep water in association with the diversification of teleost fish and other organisms that prey on them (Meyer & Macurda, 1977). In the Paleozoic, stemmed crinoids were among the most prolific organisms in shallow-water marine communities. Today, however, mobile crinoids, which can take refuge in cracks and crevices on reefs and other hard substrates, tend to predominate in shallow water, particularly in reef settings. Moreover, the decline in the maximum height above the sediment surface of epifaunal tiering (see Figure 9.14) appears to be related to the loss of the longest-stemmed crinoids in shallow, marine settings, which may have been particularly vulnerable to predation.

## Onshore–Offshore Patterns of Diversification

It stands to reason that changes through time in the diversity of any group might be related closely to environmental transitions exhibited by the group. For example, we might expect the diversification of a higher taxon to be accompanied by an expansion in the range of environments occupied by its constituents, whereas we might expect environmental contraction to accompany a diversity decline. However, we should not take these kinds of associations for granted. It is certainly plausible that a taxon undergoing a major radiation could simply accommodate more individuals within the settings that it already occupies, particularly if other taxa have become extinct in those settings.

Jack Sepkoski and Peter Sheehan (1983) investigated the environmental signatures of the Cambrian Explosion and Ordovician Radiation, the global aspects of which were discussed in Chapter 8, by compiling a database of 102 species- and genus-level faunal lists from published data sources. These lists were all representative of faunal assemblages collected from Cambrian and Ordovician rocks in North America, and they were accompanied by sedimentological data that permitted broad determinations of the paleoenvironmental settings in which each assemblage was thought to have lived. In addition, it was possible to determine the stratigraphic position of each assemblage within the Cambrian or Ordovician. Based on this information, Sepkoski and Sheehan placed the assemblages on a two-dimensional grid, known as a **time–environment diagram,** on which the vertical dimension depicts time and the horizontal dimension depicts environmental position on a very broad, simplified gradient that ranges from intertidal to deep-water settings (Figure 9.18).

From the faunal data, Sepkoski and Sheehan assembled a matrix at the taxonomic level of order by tabulating the species richness of each order within each assemblage. Assemblage compositions were then compared using cluster analysis, the results of which showed that most assemblages could be classified in the resultant dendrogram into two broad clusters, with a much smaller number of assemblages contained in each of two other clusters (Figure 9.19). The compositional differences among the assemblages in each cluster were clear, with the two main clusters dominated by elements of the Cambrian (lower large cluster) and Paleozoic (upper large cluster) evolutionary faunas. The small cluster at the base of the dendrogram was dominated by a unique Early Cambrian biota, whereas the four samples in the cluster at the top of the diagram were rich in bivalves (i.e., elements of the Modern evolutionary fauna).

When the assemblages were demarcated on the time–environment diagram based on their membership in the four clusters (Figure 9.18), an onshore–offshore pattern emerged: The Cambrian Fauna appeared to have become restricted offshore through the Ordovician, while the Paleozoic Fauna diversified initially in nearshore settings and subsequently expanded offshore. This was thought to suggest a kind of paleoecological cohesiveness to the Cambrian and Paleozoic evolutionary faunas, given their apparent segregation from one another on Paleozoic sea floors. These findings were further amplified by extending the analysis of North American fossil assemblages to the rest of the Paleozoic (Sepkoski & Miller, 1985). In these later analyses, there were

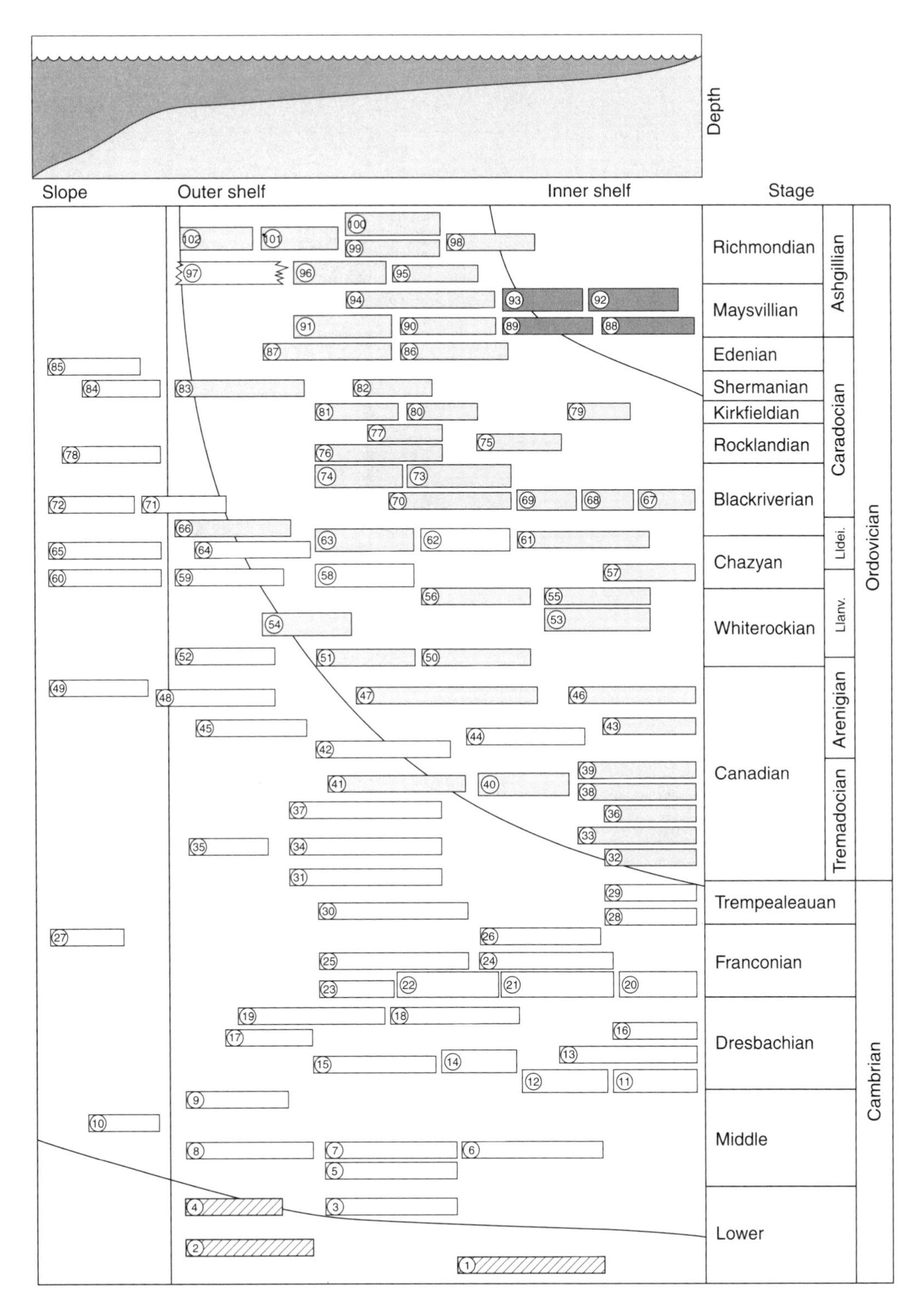


**FIGURE 9.18 Time–environment diagram depicting the positions in space and time of the 102 community-level samples from North America compiled by Sepkoski and Sheehan.** Note that the environmental dimension is a simple onshore–offshore scheme. Rectangles vary in width, according to the environmental breadths of the samples depicted and, in some cases, uncertainties thereof. Information on cluster membership from Figure 9.19 was superimposed using different stippling patterns on the rectangles. Note specifically that a cluster of samples in which trilobites were particularly diverse became limited mainly to deep-water settings during the Ordovician, as samples rich in articulated brachiopods and other elements of the Paleozoic evolutionary fauna dominated settings closer to shore. *(From Sepkoski & Sheehan, 1983)*

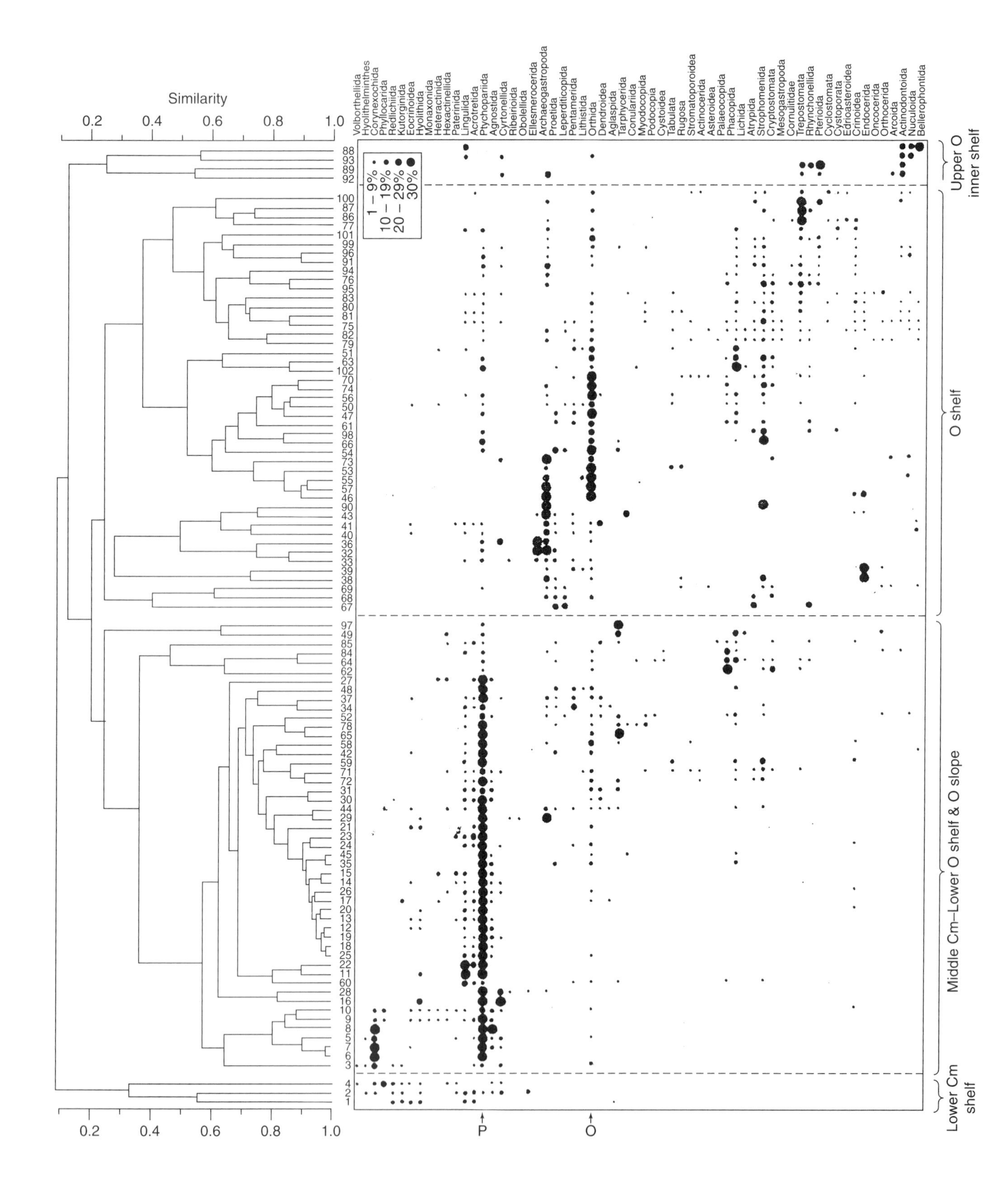

**FIGURE 9.19 Cluster analysis and sorted data matrix that compares the biotic compositions of the 102 community-level samples compiled by Sepkoski and Sheehan.** Sizes of dots reflect the relative contributions of taxa to each sample in which the taxa occurred. Most of the samples were grouped on the dendrogram into two broad clusters, with a few samples apportioned to two smaller clusters at the top and bottom of the dendrogram, based on their highly unique biotic compositions. Note that the lower large cluster is dominated taxonomically by ptychopariid trilobites, important members of the Cambrian Fauna. The upper large cluster is dominated most consistently by orthid brachiopods, important constituents of the Paleozoic Fauna, as well as several other taxa. The positions of ptychopariids (P) and orthids (O) are highlighted at the bottom. *(From Sepkoski & Sheehan, 1983)*

indications that the Modern Fauna expanded slowly and irregularly offshore throughout the mid- to late Paleozoic, as the Paleozoic Fauna became restricted primarily to settings away from the shoreline.

The general suggestion of an onshore–offshore vector to the diversification of higher taxa was extended to the Mesozoic and Cenozoic eras in a series of time–environment studies conducted by David Jablonski and David Bottjer (1983). Jablonski and Bottjer found that "Paleozoic-type" organisms living on the surface of soft sediment substrates—including inoceramid bivalves, oysters, and other nonbivalve elements—became restricted offshore, while a more "modern" biota, dominated by bivalves that burrowed beneath the sediment surface, diversified and expanded closer to shore. Subsequently, Jablonski and Bottjer (1991) conducted case studies for several other taxa, demonstrating some form of onshore–offshore diversification in various groups of bivalves, bryozoans, crinoids, and trace fossils. In addition, and perhaps most interestingly, Jablonski and Bottjer conducted an aggregate analysis of post-Paleozoic orders, recently updated by Jablonski (2005). They found that the first appearance of orders, considered to be a kind of taxonomic proxy for the origination of evolutionary novelties, took place overwhelmingly in nearshore through middle-shelf settings and were rare in outer-shelf and deep-water settings (Figure 9.20). Interestingly, there was no indication of environmental bias in origination at lower taxonomic levels, suggesting that the origination of orders is a distinct phenomenon from the origination of lower taxa.

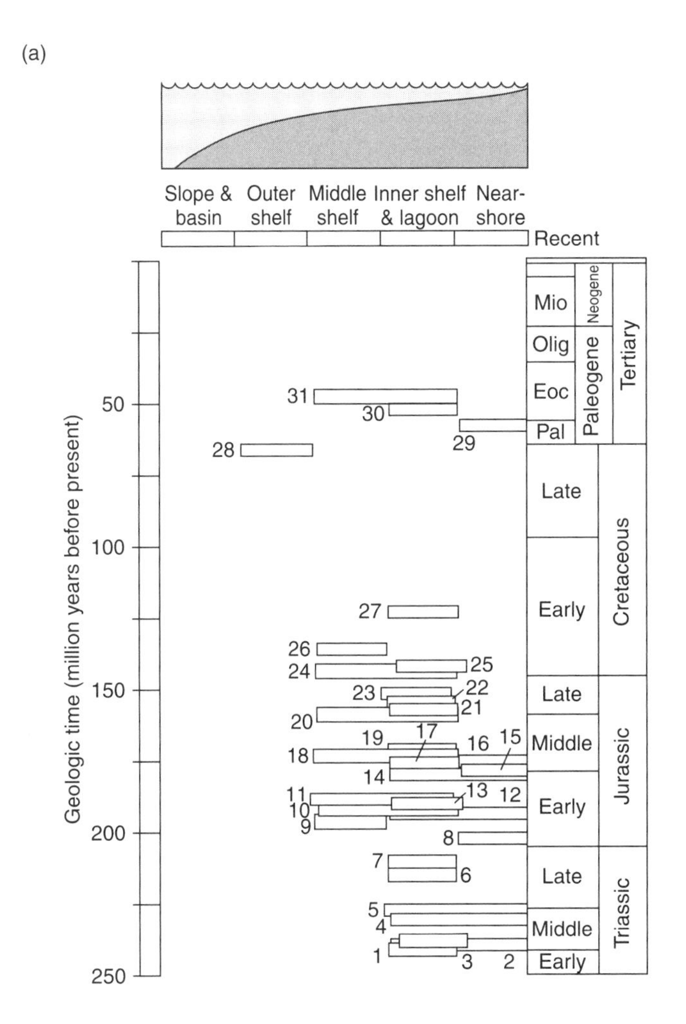


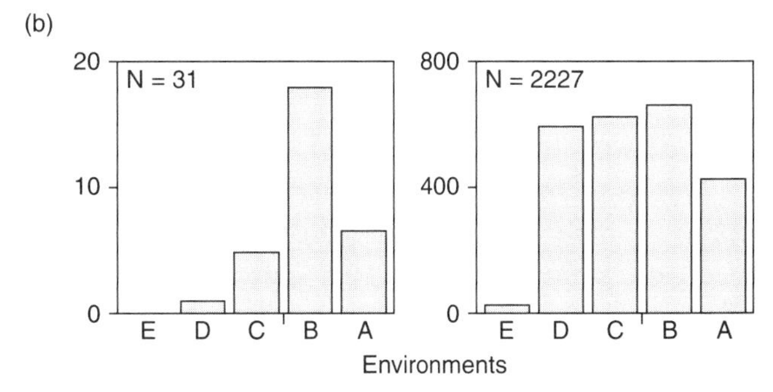


**FIGURE 9.20 Environmental patterns in the origin of post-Paleozoic marine invertebrate orders.** (a) Time–environment diagrams depicting the environmental locations of first appearances of well-preserved orders. [Key: 1, Encrinida; 2, Millericrinida; 3, Scleractinia; 4, Isocrinida; 5, Thecidida; 6, Pedinoida; 7, Tetralithistida; 8, Phymosomatoida; 9, Pygasteroida; 10, Cyrtocrinida; 11, Orthopsida; 12, Cephalaspidea; 13, Holectypoida; 14, Cassiduloida *sensu lato*; 15, Calycina (Salenioida); 16, Lithonida; 17, Disasteroida; 18, Arbacioida; 19, Lychniscosida; 20, Echinoneina; 21, Sphaerocoelida; 22, Cheilostomata; 23, Milleporina; 24, Spatangoida; 25, Holasteroida; 26, Temnopleuroida; 27, Coenothecalia (Helioporacea); 28, Stylasterina; 29, Clypeasteroida; 30, Echinoida; 31, Oligopygoida.] The vast majority of originations took place in nearshore through middle shelf settings. (b) Histograms contrasting the environments of origin of the 31 well-preserved orders versus the records for 2227 species of crinoids, echinoids, and cheilostome bryozoans. Whereas well-preserved orders originated preferentially in nearshore/inner shelf settings, no such pattern is apparent at the species level. (Key: A, nearshore; B, inner shelf and lagoon; C, middle shelf; D, outer shelf; E, slope and basin) *(From Jablonski, 2005)*

Collectively these studies argue for the existence of a paleoecologically mediated mechanism that transcends individual taxa and promotes both the origination of novelty and the initial diversification of higher taxa in nearshore environmental settings. That said, the onshore–offshore approach has motivated many studies intended to assess in greater detail the patterns just summarized, and to explore further the question of whether unique mechanisms are actually needed to explain the observations. With respect to the Paleozoic patterns that we described earlier, it has been demonstrated that neither the environmental segregation between evolutionary faunas nor the cohesiveness of individual evolutionary faunas is as straightforward as suggested in earlier research. For the most part, trilobites, which were major components of the Cambrian evolutionary fauna, did not actually become restricted to deep water during the Ordovician; rather, their numbers were "diluted" nearshore by elements of the diversifying Paleozoic Fauna (Westrop & Adrain, 1998). Similarly, the Ordovician distribution of bivalves was not limited to shallow water but was controlled largely by the distribution of terrigenous sediments [SEE SECTION 8.8], which sometimes extended into offshore settings.

Despite these caveats, it is clear that the investigation of diversification in an environmental context has begun to provide a significant conceptual advance to paleontology by treating the space dimension as an emergent characteristic of taxa that, like morphology, is subject to evolution. More generally, this research has motivated paleontologists to incorporate information about paleoenvironments directly and routinely into evolutionary analyses, and data relevant to these assessments are now compiled in large-scale paleontological databases [SEE SECTION 8.7].

## Ecological Interactions and Mass Extinctions

In Chapter 8, we considered the roles of mass extinctions in mediating major changes in taxonomic composition through the removal and replacement of incumbent taxa [SEE SECTION 8.6]. Given these changes, we might also expect major transitions in the nature of ecological interactions. To assess this possibility, Conrad Labandeira and colleagues (2002) investigated transitions in insect–plant interactions across the Late Cretaceous–Paleocene (K/T) mass-extinction boundary. Although the record of insect body fossils is rather limited throughout the study interval, there is abundant evidence of the presence and behavior of insects preserved on the fossil leaves of woody plants found in paleofloral assemblages. Labandeira and colleagues focused on assemblages preserved in southwestern North Dakota, in which they examined more than 13,000 fossil leaf specimens for signs of insect damage.

In these and other studies, Labandeira and his colleagues have been able to recognize more than 50 unique damage types caused by insects that can be grouped into eight categories, several of which are illustrated in Figure 9.21. In addition, by determining the degree of plant–host specificity observed for the same damage types in the present day, Labandeira and colleagues classified these damage types into three categories—generalized, intermediate, and specialized—reflecting the degree to which insects were selective in their choice of food.

When they partitioned their data to analyze stratigraphic transitions in leaf damage (Figure 9.22), Labandeira and colleagues found a significant decrease across the K/T boundary in the percentage of leaves damaged and in the number of damage types. Moreover, the decline in leaf damage was significantly greater among damage types classified as intermediate or specialized than among generalized damage types.

Importantly, Labandeira and colleagues observed a substantial Paleocene recovery near the top of their study interval in the number of damage types and in the percentage of leaves in a given assemblage that showed some form of damage. However, the recovery was far more pronounced among generalized damage types than among intermediate or specialized types, suggesting that: (1) Insects specialized for feeding on particular plant hosts were affected more profoundly than generalists during the K/T mass extinction; or (2) the plant hosts associated with specialized insects were strongly affected by the extinction; or (3) both. These findings are of obvious relevance to discussions about the biological selectivity of mass extinctions, discussed in Chapter 8 [SEE SECTION 8.6].

In addition to demonstrating a significant transition in the nature of ecological interactions in association with the K/T mass extinction, the study by Labandeira and colleagues also illustrates the general importance of fossil leaves as treasure troves of information about the evolution and behavior of herbivores throughout the Phanerozoic. We consider another example of the use of leaf-damage data in Chapter 10 [SEE SECTION 10.4].

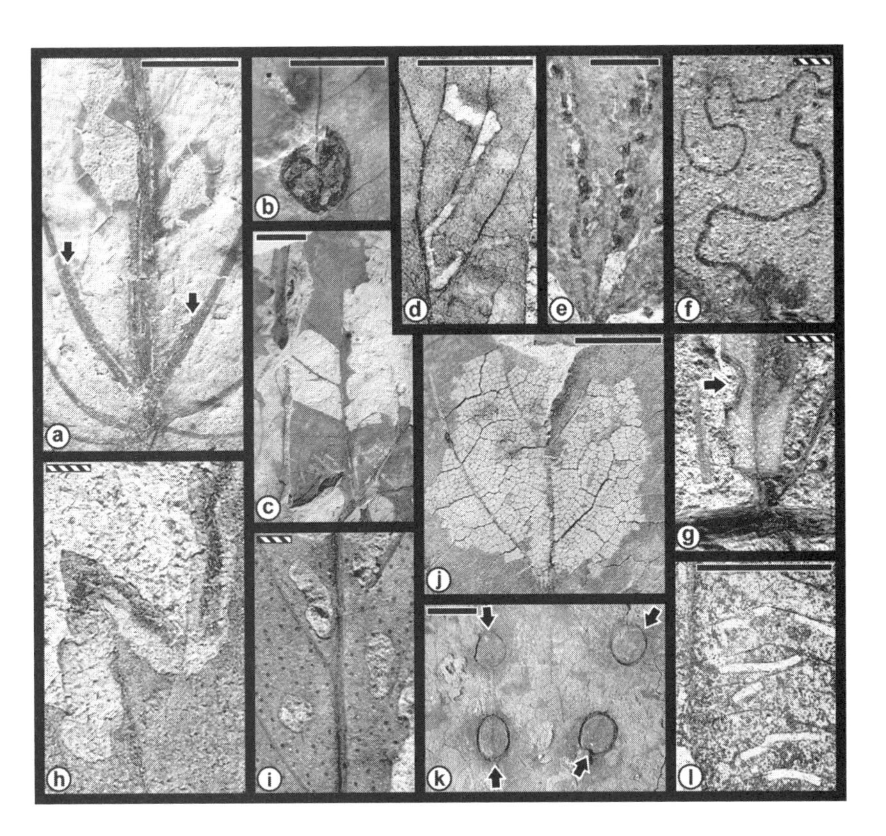


**FIGURE 9.21 Examples of damage to leaves caused by insects, from the K/T boundary interval in southwestern North Dakota.** Black scale bar, 1 cm; striped scale bar, 0.1 cm. (a) Two linear leaf mines, with oviposition (egg laying) at the sites of the arrows. (b) A gall (a bump caused by irregular tissue growth after damage by an insect). (c) Free feeding. (d) Skeletonization (chewing damage). (e) Multiple galls. (f) Serpentine leaf mine. (g) Cuspate margin feeding. (h) Serpentine leaf mine. (i) Hole feeding. (j) Skeletonization. (k) Insect impressions at sites of arrows. (l) Slot hole feeding. *(From Labandeira et al., 2002)*

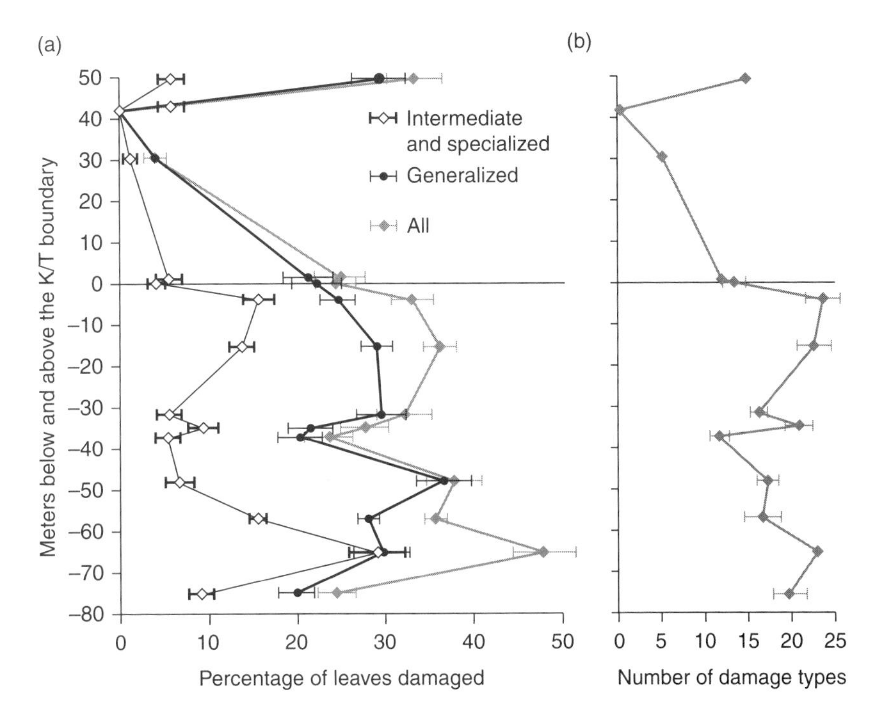


**FIGURE 9.22 Declines in the extent and diversity of insect-induced leaf damage across the K/T boundary in southwestern North Dakota, for horizons containing at least 200 leaf specimens.** (a) Decline in the percentage of leaves showing evidence of damage. Note that the decline is greater for damage caused by intermediate and specialized feeders. (b) Decline in the number of different damage types, in aggregate, exhibited across the K/T boundary. To make the comparison equitable from horizon to horizon, all samples have been rarefied (see Box 1.2) to 200 specimens. *(From Labandeira et al., 2002)*

## 9.5 NEW APPROACHES TO PALEOENVIRONMENTAL AND PALEOCLIMATIC RECONSTRUCTION

An underlying theme throughout this chapter has been the question of paleoenvironmental interpretation. While the characteristics of enclosing sediments can help us determine the environmental conditions that prevailed at the time that the sediments were deposited, there are many ways in which fossils can be used directly as data for reconstructing paleoenvironments.

At a basic level, determining the range of guilds represented by the constituents of a fossil assemblage may help us constrain the properties of the paleoenvironment represented by the strata in which fossils are found. For example, if it is determined that the majority of taxa preserved in an assemblage of fossil organisms were sedentary organisms with structures or skeletal features indicative of attachment to the substrate, this would imply that the substrate in question, be it on a sea floor or a forest floor, was firm enough to support this lifestyle.

The differing properties of fossil assemblages in close stratigraphic proximity may be diagnostic of important paleoenvironmental transitions, such as the degree of oxygenation on ancient sea floors. This approach was pioneered by Charles Savrda (1995) in the study of trace fossils, as illustrated in Figure 9.23. Savrda established a

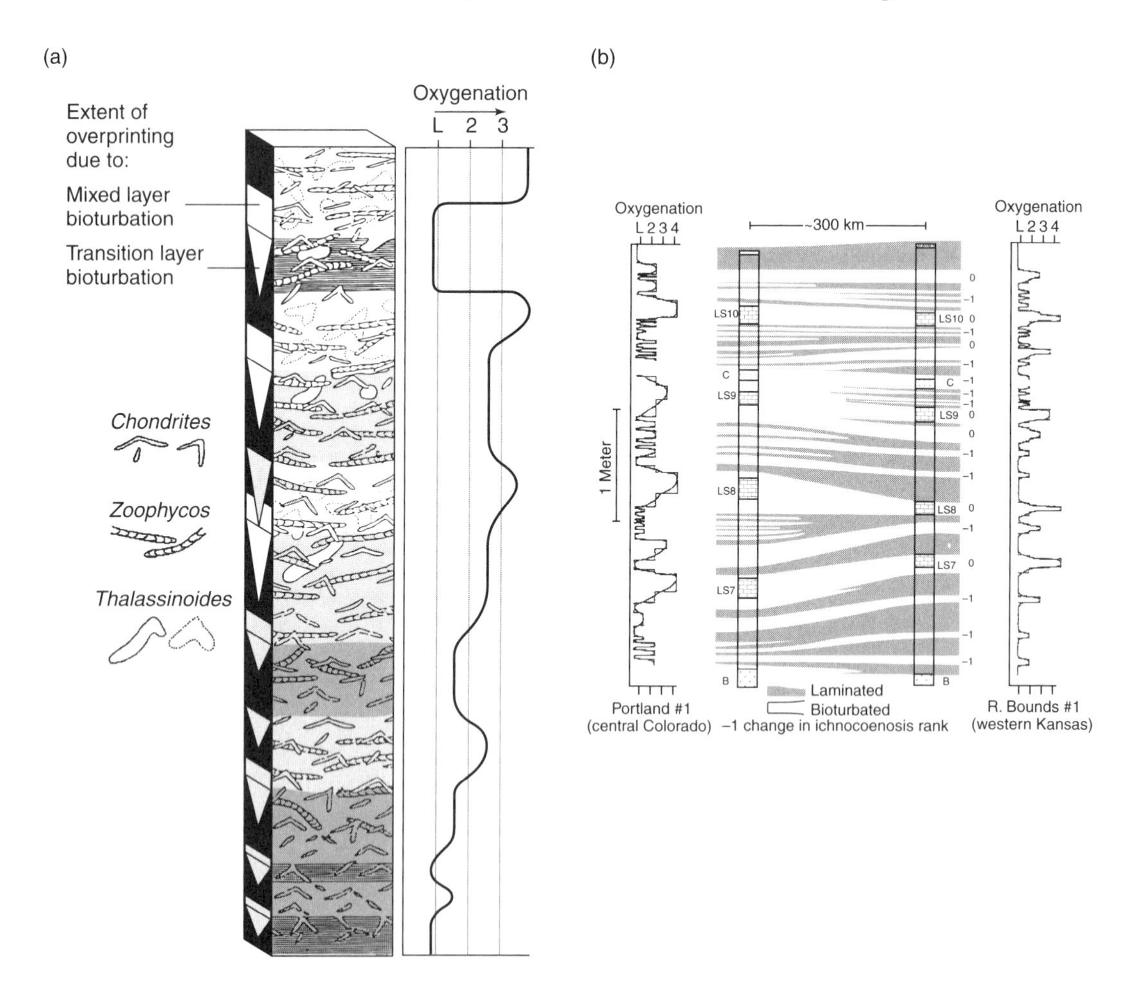


**FIGURE 9.23 The use of trace fossils to diagnose oxygenation levels on a sea floor.** (a) Hypothetical stratigraphic column illustrating variations in the concentration and nature of trace fossil assemblages in association with the degree of oxygenation of the sea floor. At low oxygen levels (L on the schematic graph), lamination is preserved, and only the producers of the trace fossil *Chondrites* survive. As oxygenation increases, the number and complexity of trace fossils also increase. (b) Oxygenation curves for two localities of the Lower Bridge Creek Limestone (Cretaceous) in western Kansas. The two localities have been correlated on the basis of two bentonites (preserved volcanic ash fall), marker beds (B and C), and a series of limestone beds. In general, the degree of bioturbation and associated oxygen levels decrease eastward (i.e., to the right). An eastward increase in the thickness and frequency of laminated intervals can also be observed. *(From Savrda, 1995)*

framework that relates the presence of particular trace fossils, and the general extent of bioturbation, to the amount of oxygen available in sea-floor sediments (Figure 9.23a).

Sediments in which there was little or no oxygen are characterized by the preservation of fine-scale horizontal features of primary bedding, known as laminations. With increasing levels of oxygenation, a more complex bottom fauna burrows through the sea floor (see Figure 1.20), obliterating laminations and generating a diverse set of trace fossils. This framework has been shown to be applicable broadly to the geological record (e.g., Figure 9.23b) and has been useful for recognizing fine-scale transitions in oxygenation levels, both stratigraphically and geographically.

## $\delta^{18}$O and the Mg/Ca Ratio

Beyond the characterization of the ecological properties of the taxa preserved within a fossil assemblage, paleontologists are now making use of a sophisticated set of geochemical techniques that evaluate directly the isotopic properties of growth increments, or bands, preserved in the hard parts of many organisms. While these analyses are, of course, limited to the subset of preserved hard parts that have not been altered chemically after the death of the organism, the temporal resolution of the paleoenvironmental transitions diagnosed with these procedures is unmatched by any other paleontological method.

In many instances, it is possible to analyze samples collected with a microscopic coring device, within and across the annual growth increments of an organism. There can be environmental and geographic variation among the individuals of a species in the yearly emplacement of the "lighter" and "darker" bands that make up an annual increment in many organisms (Figure 9.24a). However, the geochemical transitions measured across these bands are often unmistakable.

We have already seen an illustration of the utility of oxygen isotopic techniques in Chapter 2, in the analysis of heterochrony in the Jurassic oyster *Gryphaea* [SEE SECTION 2.3]. This analysis depended on the ability of

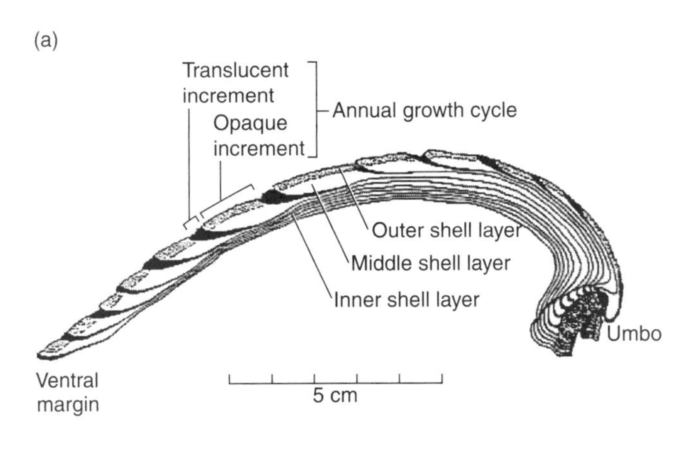


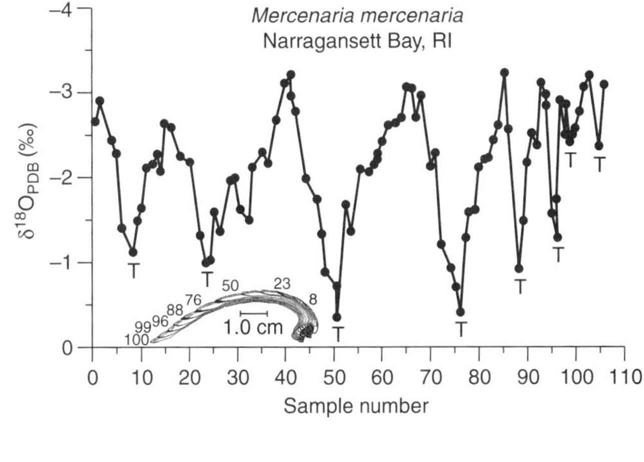


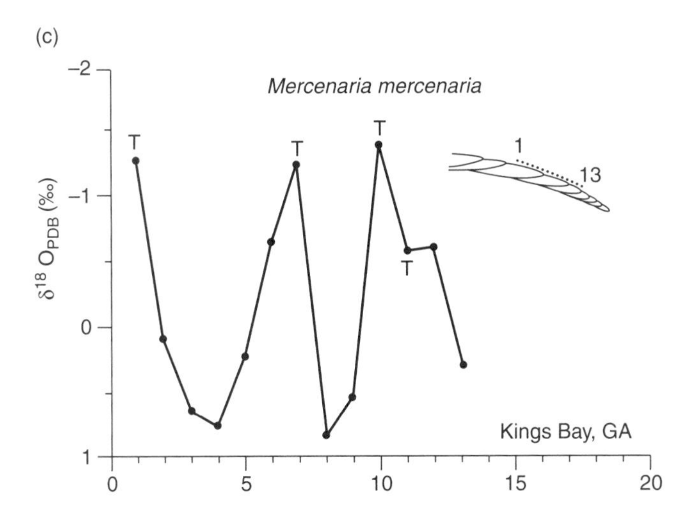


**FIGURE 9.24 Geochemical analysis of seasonality preserved in bivalve growth bands.** (a) Cross-sectional sketch through a valve of the bivalve *Mercenaria mercenaria*. Each annual growth band consists of a translucent and opaque increment. (b) A $\delta^{18}$O profile of a *M. mercenaria* valve from Narragansett Bay, Rhode Island, in which translucent intervals are shown to correspond with $\delta^{18}$O maxima (values that are the least negative, in this case), indicating that translucent layers correspond to the coldest times of year at the locality where the valve was collected. Sample numbers correspond to positions along the shell, as illustrated by the inset. (c) A more limited $\delta^{18}$O profile for a valve from Kings Bay, Georgia, in which translucent intervals correspond with $\delta^{18}$O minima, indicating in this case that translucent layers represent the *warmest* times of year. *(From Jones & Quitmyer, 1996)*

paleontologists to recognize seasonal and annual growth increments based on variations in the relative concentrations of two naturally occurring isotopes of oxygen, $^{16}O$ and $^{18}O$, which can be measured with an instrument known as a mass spectrometer. Although the concentrations of these two isotopes in skeletal material may be affected by several factors, there is a measurable association of the ratio of these two isotopes with the temperature in surrounding waters at the time of skeletal accretion.

Isotopic ratios are generally reported in terms of their deviations from standard substances, in parts per thousand (‰), as defined in the following equation:

$$\delta X = [(R_{\text{sample}}/R_{\text{standard}}) - 1] \times 10^3$$

where $\delta X$ is the enrichment (if positive) or depletion (if negative) of any high-mass isotope $X$ (e.g., $^{18}O$, in comparisons of $^{18}O$ and $^{16}O$), and $R$ is the high-mass to low-mass isotopic concentration ratio. Standard substances with which samples are compared in oxygen isotopic assessments include a belemnite cephalopod collected from the Upper Cretaceous Peedee Formation of the eastern United States (PDB) and present-day standard mean ocean water (SMOW).

Equations have been developed empirically for calcitic and aragonitic material that enable a researcher to estimate changes in the ambient temperature of a marine setting based on the $\delta^{18}O$ value determined for the sample. In a relative sense, a higher value for $\delta^{18}O$ (i.e., enrichment in $^{18}O$ relative to $^{16}O$) corresponds to a lower temperature at the time of skeletal accretion. As an example, the annual record of oxygen isotopic variation can be observed in a specimen of a Recent bivalve, *Mercenaria mercenaria*, collected from Narragansett Bay, Rhode Island, as illustrated in Figure 9.24b. Here, translucent (dark) increments correspond dependably to colder portions of the year, as evidenced by the enrichment in $^{18}O$ associated with these increments. Elsewhere, however, translucent increments have been shown to correspond to the warmest portions of the year (see Figure 9.24c), suggesting that paleontologists should be cautious in assessing the significance of light and dark banding. Nevertheless, the oxygen isotopic pattern appears to record faithfully the temperature profile on an annual basis, regardless of timing during the year of light and dark increments.

The utility of oxygen isotope profiles for assessing seasonality can be seen further in the example illustrated in Figure 9.25, which records $\delta^{18}O$ and temperature profiles based on the microsampling of successive growth bands of fish otoliths (ear stones) collected at localities spanning the Eocene–Oligocene boundary in the Gulf Coast region of the United States. A comparison by Linda Ivany and colleagues (2000) of otolith profiles for Eocene versus Oligocene specimens suggests that, while there was little change in summer temperatures across the boundary, there appears to have been a significant decrease in winter temperatures, indicating an overall increase in seasonality. It remains to be determined whether this apparent paleoclimatic change was associated with a major extinction that took place at that time, but the coincidence is intriguing.

Geologists are now making routine use of isotopic analyses in the study of ancient environments. Among the other isotopic suites that are of particular importance to paleontologists are the $^{13}C/^{12}C$ suite, which is useful for diagnosing the presence of photosynthetic activity [SEE SECTION 10.5], and the $^{34}S/^{32}S$ suite, which provides an indication of the degree of oxygenation.

The use of $\delta^{18}O$ as a paleo-thermometer is not limited to assessment of temperature variations recorded in the growth bands of individual skeletons or skeletal elements. It can also be used to diagnose an extended record of temperature change in a region based on intraspecific stratigraphic variations in the average isotopic compositions of samples of skeletal elements. Researchers typically assess $\delta^{18}O$ trends for more than one species because different species sometimes exhibit different stratigraphic $\delta^{18}O$ trends through the same interval. As we will see in Chapter 10 [SECTION 10.4], the demonstration of a similar stratigraphic pattern among several species provides compelling evidence that the pattern transcends taxonomic bias.

Complications in the interpretation of $\delta^{18}O$ trends can arise for other reasons, including variations through time in the volume of ice occurring on the earth. Because $^{16}O$ is lighter than $^{18}O$, water molecules containing $^{16}O$ evaporate more readily than those containing $^{18}O$, and it follows that $^{16}O$ will occur preferentially in the precipitation derived from evaporated water. Given that the growth of glaciers depends on the water supplied by this $^{16}O$-enriched precipitation, it stands to reason that $^{16}O$ will be sequestered preferentially in glacial ice, thereby causing a global enrichment in $^{18}O$ in sea water (i.e., increased $\delta^{18}O$) during intervals of increased glaciation. Of course, we would expect the earth to be cooler globally during an interval of glacial advance, and this cooling might be reflected in a positive excursion in $\delta^{18}O$ at a given locality. However, it is possi-

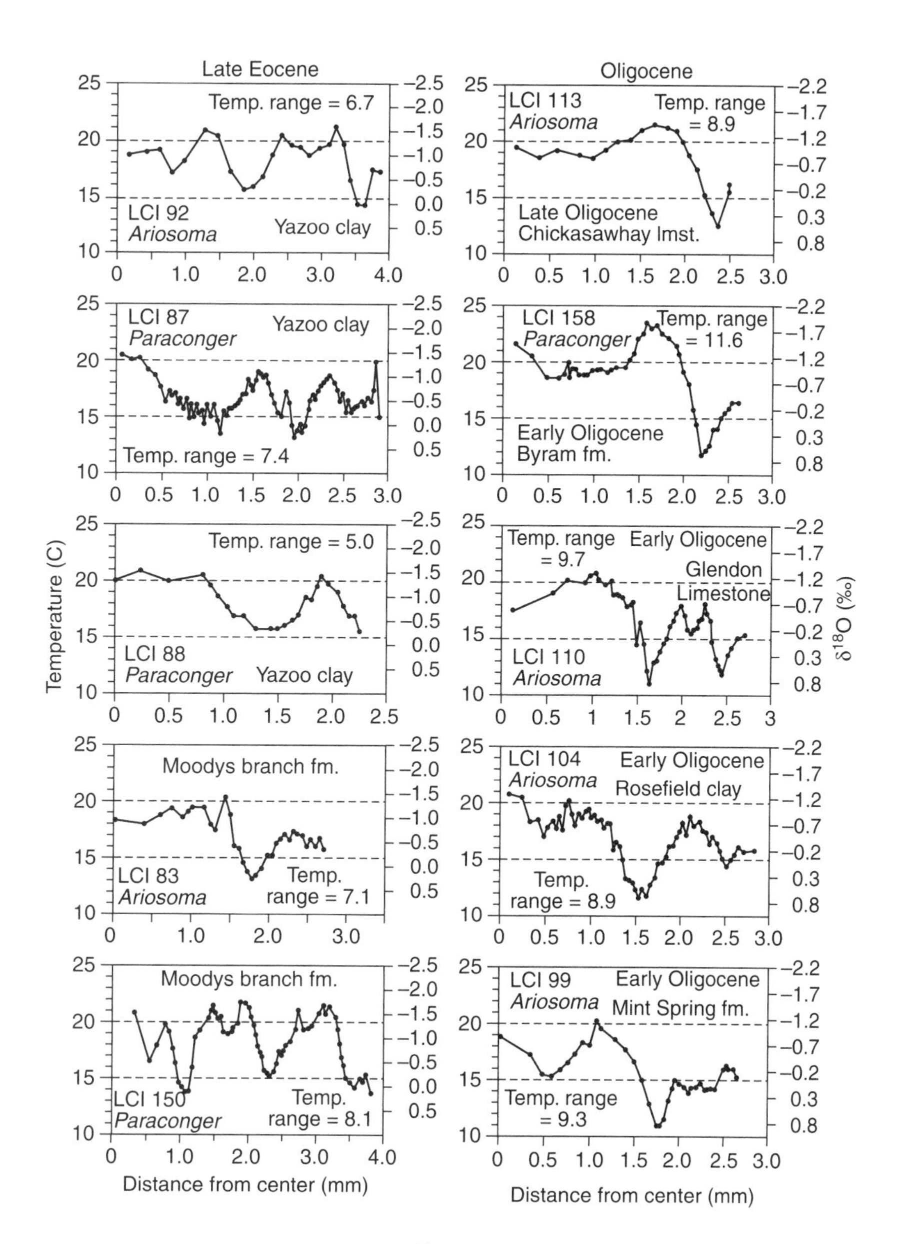


**FIGURE 9.25 Temperature records, based on $\delta^{18}O$ profiles from microsampling of several Late Eocene and Oligocene fish otoliths.** Note the tendency of Oligocene profiles to exhibit greater amplitudes, primarily because of colder temperatures during the winter. Summer temperatures remained fairly constant from the Late Eocene into the Oligocene. *(From Ivany et al., 2000)*

ble that excursions related to changes in ice volume will overwhelm patterns that would otherwise reflect local or regional temperature variations that did not parallel the advance or retreat of glaciers.

With this in mind, additional methods for estimating ancient temperatures have been developed that use other elements preserved in fossil skeletons. These should be viewed as supplements to assessments based on $\delta^{18}O$. One of the more promising of these approaches uses the ratio of magnesium to calcium (Mg/Ca) preserved in the skeletons of foraminifera. While not entirely free of its own complications, a strong empirical association with temperature has been recognized in the Mg/Ca ratios of several present-day benthic foraminiferal species. In general, the Mg/Ca ratio preserved in benthic foraminifera increases exponentially with increased temperature (Figure 9.26); temperature changes through time can therefore be diagnosed by this alternative means.

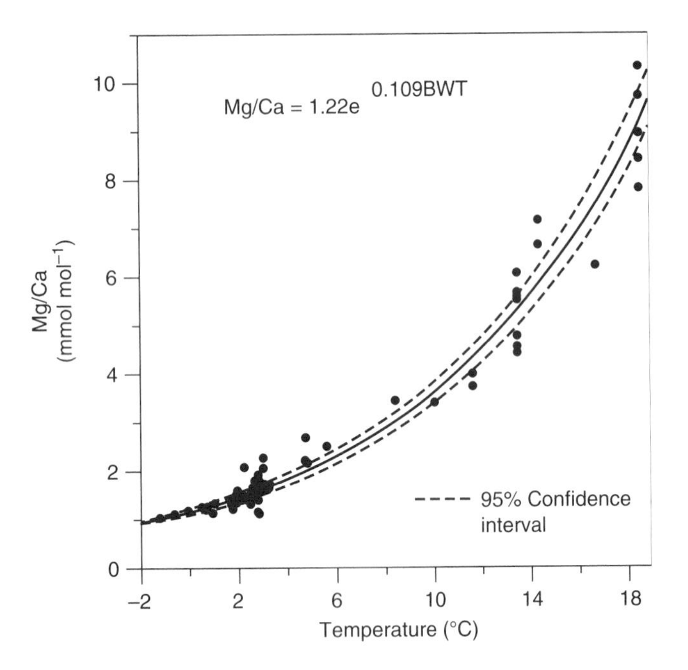


**FIGURE 9.26 An illustration of the strong empirical relationship between Mg/Ca ratios in the benthic foraminiferal genus *Cibicidoides* and present-day water temperature, based on an analysis of specimens collected from the tops of cores collected at several deep-sea locations.** An equation is fitted by statistical methods to the data, as illustrated here, which can then be used to estimate bottom-water temperature (BWT) based on Mg/Ca ratios observed for this taxon in older samples. *(From Martin et al., 2002)*

Pamela Martin and colleagues (2002) demonstrated a clear relationship between changes in bottom-water temperature estimated by this method and glacial–interglacial cycles during the past 350,000 years. Interestingly, for one deep-sea location in the tropical Atlantic, Martin and her colleagues were able to diagnose the interplay of two different water masses in association with glacial–interglacial cycles. Intervals of cooling coincided with regional decreases in the fraction of North Atlantic deep water (NADW) relative to the contribution of Antarctic bottom water (AABW; Figure 9.27).

## Paleoclimatic and Paleoatmospheric Estimates from Fossil Plants

The anatomical features of leaves are highly responsive to changes in ambient climatic and even atmospheric conditions. Taking advantage of several of these features, paleontologists have established a set of novel methods that provide unique information about ancient paleoenvironments. Collectively, these methods illustrate the amazing array of information available through the thoughtful analysis of fossil specimens.

Leaf margin analysis (LMA) and leaf area analysis (LAA) relate to the observation of significant, present-day

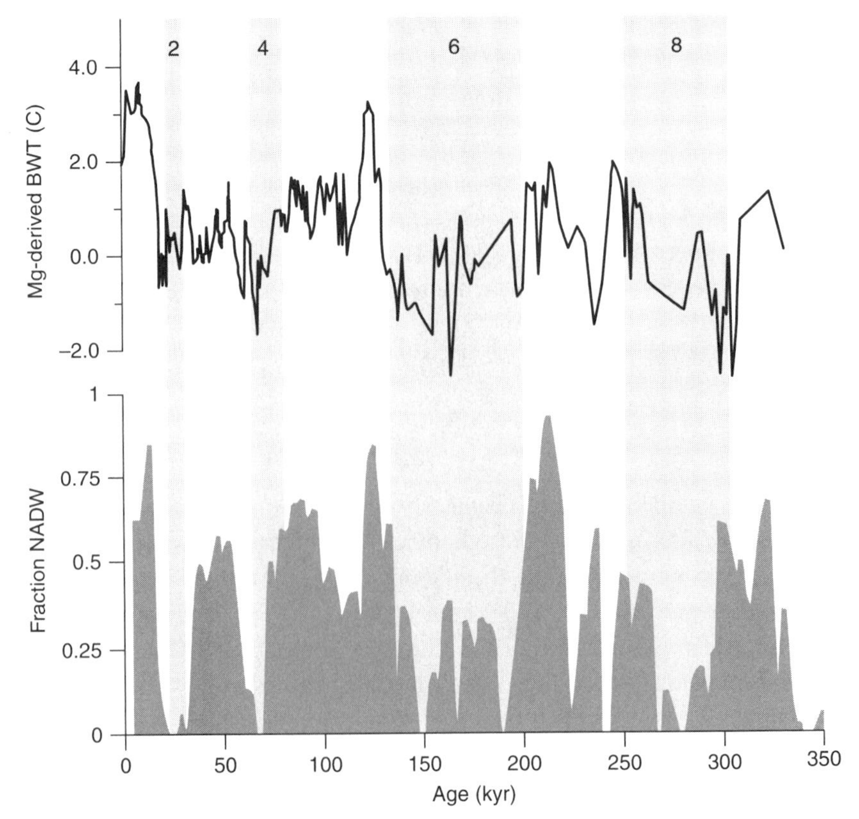


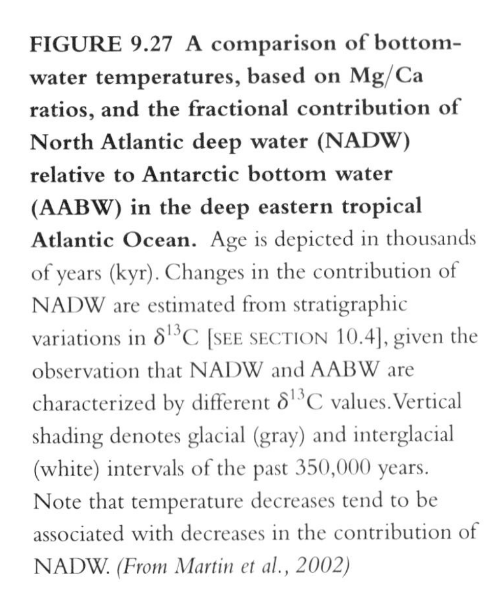

**FIGURE 9.27 A comparison of bottom-water temperatures, based on Mg/Ca ratios, and the fractional contribution of North Atlantic deep water (NADW) relative to Antarctic bottom water (AABW) in the deep eastern tropical Atlantic Ocean.** Age is depicted in thousands of years (kyr). Changes in the contribution of NADW are estimated from stratigraphic variations in $\delta^{13}C$ [SEE SECTION 10.4], given the observation that NADW and AABW are characterized by different $\delta^{13}C$ values. Vertical shading denotes glacial (gray) and interglacial (white) intervals of the past 350,000 years. Note that temperature decreases tend to be associated with decreases in the contribution of NADW. *(From Martin et al., 2002)*

(a)

(b)

(c)

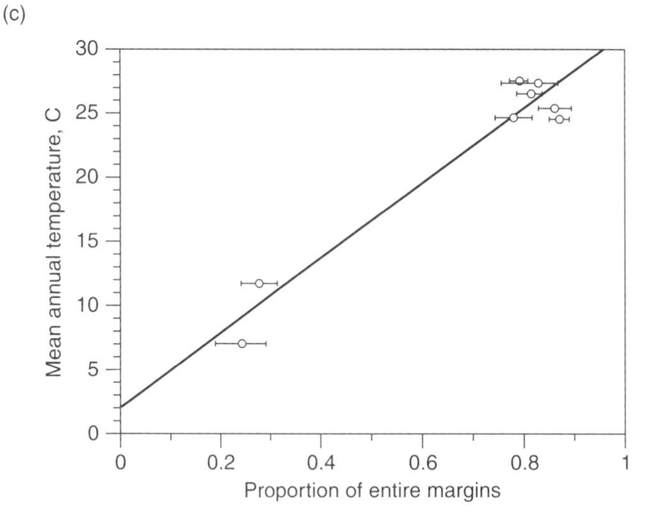


**FIGURE 9.28 The relationship between leaf margins and mean annual temperature (MAT).** (a) An example of a leaf with an entire (i.e., smooth) margin. (b) An example of a leaf with a toothed margin. Both specimens were collected in Wyoming from an Early Eocene interval characterized by very warm temperatures [SEE SECTION 10.4]. In (a) and (b), the scale bar is 5 cm. (c) Illustration of the relationship between the proportion of entire-margined leaves in samples collected from nine present-day regions and the MATs of these regions. *(a, b: Courtesy Scott Wing; c: From Wilf, 1997)*

relationships between variations in leaf morphology and climate. LMA is based on the recognition of a consistent, positive relationship between mean annual temperature (MAT) in a given region and the proportion of leaves in the flora of the region that have smooth, untoothed margins (so-called **entire-margined** leaves; Figure 9.28). Inversely, as MAT decreases, the proportional representation in the flora of leaves with toothed margins increases.

While more sophisticated methods have been developed for paleoclimatological reconstruction that make use of a much larger array of leaf characteristics, Peter Wilf (1997) has presented a compelling case that simple LMA is at least as effective as methods that rely on the broader array of characteristics. Similarly, Wilf and colleagues (1998) illustrated a strong positive relationship between mean (average) leaf area and mean annual precipitation (MAP) for a set of 50 samples collected from locations around the world (Figure 9.29), suggesting that LAA can be applied to fossil leaves to estimate paleoprecipitation. An application of LMA and LAA to the fossil record is discussed in Chapter 10 [SECTION 10.4].

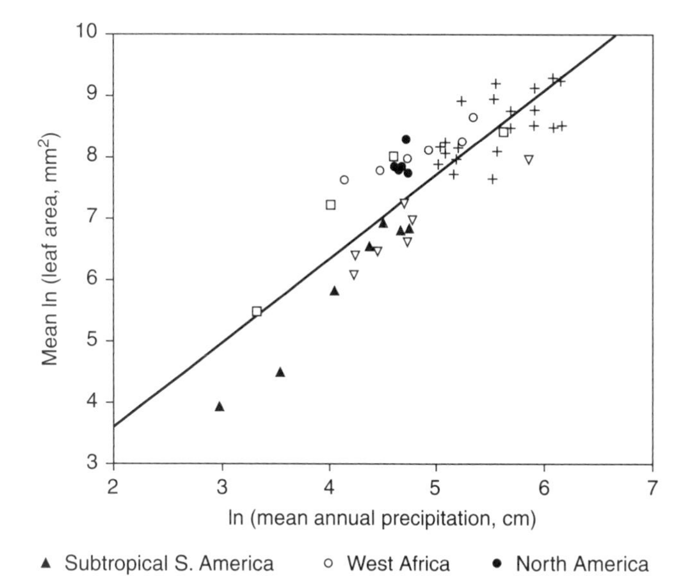


**FIGURE 9.29 Illustration of the relationship between mean annual precipitation (MAP) in several regions worldwide and the mean areas of the leaves contained in 50 samples from these regions.** *(From Wilf et al., 1998)*

Finally, botanists have discovered through empirical observations and experiments that the density of **stomata** (tiny pores involved in gas exchange) on the undersides of leaves is sensitive to $CO_2$ levels in the atmosphere: The number of stomata per unit area decreases as $CO_2$ levels increase. Recognizing concerns that global warming in the present day is linked to elevated atmospheric $CO_2$ levels, researchers have investigated historical changes in stomatal density (SD) and have discovered that, indeed, SD has decreased substantially during the past 200 years (Figure 9.30).

Given the relationship between SD and $CO_2$, paleobotanists have used SD as well as a related metric, the stomatal index (SI), to estimate changes to paleo-$CO_2$ levels during critical intervals in the Phanerozoic. The SI is measured as SD/(SD + epidermal cell density) $\times$ 100 and is less sensitive than simple SD to potential extraneous factors unrelated to $CO_2$. Generally, paleobotanists report both values, since the SI is derived directly from SD.

In an analysis of samples collected from southern Sweden and eastern Greenland that span the Triassic–Jurassic boundary, Jennifer McElwain and colleagues (1999) found significant decreases in SD and the SI through the Late Triassic and into the earliest Jurassic (Figure 9.31). Based on the extent of the decrease, McElwain and her colleagues estimated a $CO_2$ increase that, given the known effects of $CO_2$ as a **greenhouse** gas, implies a temperature rise of 3° to 4°C. This suggests a role for global warming in a major extinction and turnover of floral species recognized across the Triassic–Jurassic boundary.

The application of LMA, LAA, and stomatal analysis is not problem-free. For example, MAT may be underestimated when LMA is conducted on the leaves of plants that lived adjacent to rivers and lakes (Burnham et al., 2001). With the growing awareness of these biases, however, adjustments to analytical results can be made to account for them.

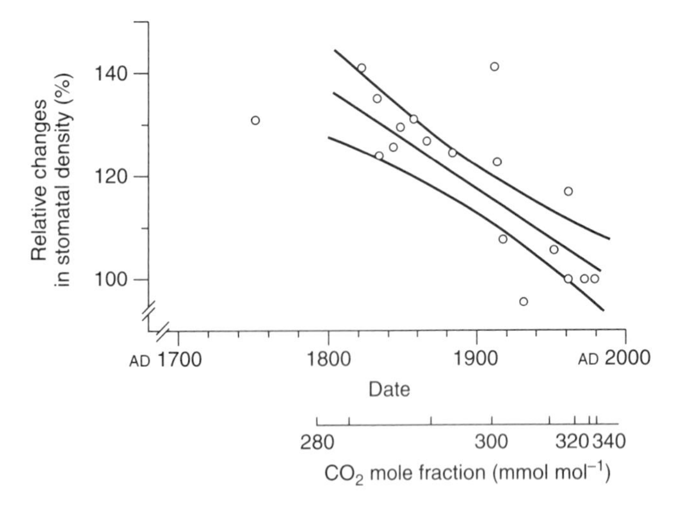


**FIGURE 9.30 Illustration of changes through historical time in the stomatal densities of individual specimens for leaves of five different plant species.** Measurements illustrated in the graph were collected from specimens that had been stored and dried in an herbarium over the past 200 years. The values are reported as percentages of present-day densities. The two outer lines are 95 percent confidence intervals on the central line, which were fitted to the data. Estimated changes in atmospheric $CO_2$, based on analyses of ice cores, are also presented in parallel with the $x$ (time) axis to illustrate the relationship with the decline in stomatal densities. *(From Woodward, 1987)*

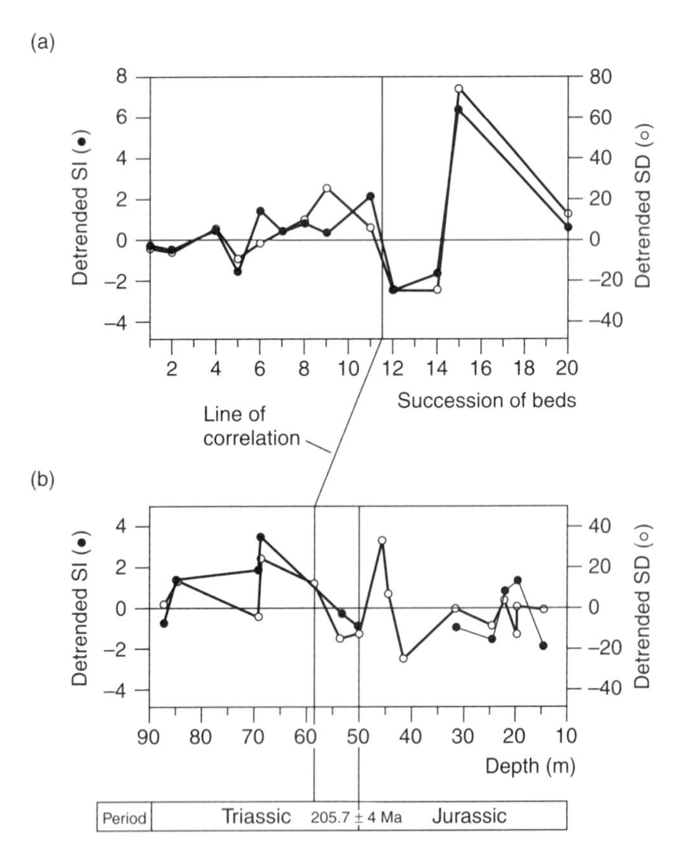


**FIGURE 9.31 Changes in stomatal density (SD) and the stomatal index (SI) in samples of fossil leaves collected across the Triassic–Jurassic boundary in (a) southern Sweden and (b) East Greenland.** A line of correlation is illustrated between the two regions and is based on the first occurrence of a new flora found in both regions. For both metrics, "detrended" values, which reflect deviations from the overall mean values of the genera analyzed in a given sample, are used in order to overcome bed-to-bed variations in the taxa analyzed. Ma signifies millions of years ago. *(From McElwain et al., 1999)*

## 9.6 PALEOBIOGEOGRAPHY

When we consider the environment in which an organism lives, it should be understood that we are speaking mainly of the physical and biological conditions of a particular space, rather than the geometric outlines or location of the space. It is common for an ecologist or paleoecologist to make reference, say, to a mud-covered sea floor or a rocky cliff face, and we expect these and other environments to be characterized by communities of organisms that are or were adapted for life in those settings. Of course, any environment is characterized not by a single physical or biological variable, but by a multitude of variables that interact synergistically. Thus, from an environmental standpoint, no two mud-covered areas of a sea floor are likely to be identical because, even if they are covered by mud of precisely the same grain size and mineralogy, other variables that define these settings, such as temperature or water depth or salinity, are likely to differ at least to some degree. Following the logic of H. A. Gleason that we discussed earlier, we should therefore not expect the taxonomic compositions of the communities present at any two locations to be identical.

In the same vein, we should not expect the species in any community to have identical geographic distributions. It stands to reason that the inherent uniqueness of every species would imply that the environmental requirements, and therefore the geographic distributions of each species, differ at least to some extent.

Assessments of the geographic, as opposed to environmental, distributions of taxa preserved in the fossil record bring us into the realm of paleobiogeography. Given the dynamism through time of the earth from a physical standpoint, we might expect the geographic distribution of any species to be equally dynamic throughout its evolutionary lifetime. In cases where the stratigraphic record provides opportunities to observe changes through time with very high stratigraphic–temporal resolution, this dynamism is, in fact, quite evident.

For example, paleobotanists have studied the stratigraphic records of plant pollen preserved in lake sediments deposited during the late Quaternary Period throughout the world. The abundance records for several species in a 160-m-thick sequence are depicted in Figure 9.32. While these patterns appear to resemble the stratigraphic relationships that we observed earlier for brachiopods in the Ordovician (Figure 9.8c), the temporal duration of the interval covered by the Quaternary data is far shorter, and the temporal resolution is much finer. In fact, it has been suggested that there is sufficient temporal resolution in these Quaternary sediments to recognize fine-scale ecological transitions. Note, for example, that times of dominance by trees, such as pines (*Pinus*) and oaks (*Quercus*), tend also to be characterized by reduced abundances of grasses (e.g., Gramineae),

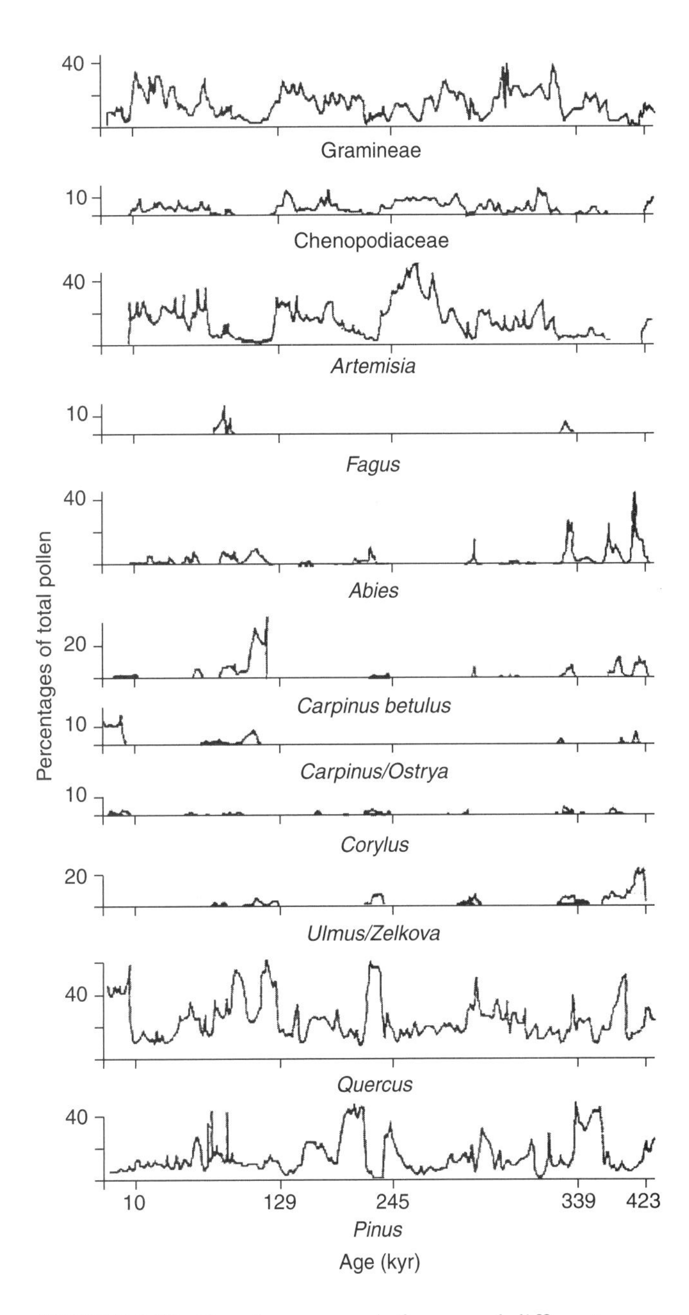


**FIGURE 9.32 Abundance records for several different pollen types in a 160-m sequence of Quaternary lake sediments, spanning some 400,000 years, collected in northwestern Greece.** Age is depicted in thousands of years (kyr). Note, in particular, the nearly opposite abundance tendencies of *Quercus* (oaks) and *Pinus* (pines). *(From Bennett, 1997)*

herbs, and shrubs (e.g., Chenopodiaceae and *Artemisia*). One such interval occurred just after 245,000 years ago. At other times, trees were subordinate to other groups, at least as indicated by the relative abundances of their pollen.

The dramatically shifting abundances through time apparent at the locality in Greece hint at significant geographic dynamism among the species preserved there. This is particularly evident in the mapping of plant distributions based on pollen data from Holocene sediments preserved throughout Europe, an example of which is shown for oak trees in Figure 9.33. In general, where the fossil record affords opportunities to investigate changes in the geographic ranges of species with resolutions finer than a few thousand years, these ranges have proven to be so dynamic that they have been likened by Keith Bennett, who has evaluated these patterns extensively, to the kneading of dough. By this, Bennett meant that the response of species to a major climatic shift was almost assuredly *not* to be reduced in abundance to the point of extinction, but to exhibit a kind of plasticity in their geographic ranges in rapid response to geographic shifts in climatic belts. Analyses of other biotas, including terrestrial vertebrates and marine molluscs, suggest not only that this is a common phenomenon among species, but also that the ranges of species tend to change independently of one another.

Given this dynamism and independence in the geographic ranges of species, we might be tempted to infer that the geographic ranges of different species are *never* similar to one another at *any* geographic scale. However, it can be demonstrated easily that this is not the case. Consider the Isthmus of Panama, the narrow land bridge that connects North and South America. Because it is a fully emergent feature (or at least was, until the construction of the Panama Canal), it serves as a very effective barrier to migration of marine species between the Atlantic and Pacific sides of the Isthmus. Thus, there is obviously a strong coincidence in the westernmost boundaries of the geographic ranges of many species that abut the Isthmus on the Atlantic side, just as there is a close relationship among the easternmost boundaries of species on the Pacific side.

Geologists who have studied the history of the Isthmus of the Panama know that it was not fully emergent until some 4 to 5 million years ago. Although the emergence was not like the rapid flipping of a switch—prior to which there was free interchange among marine biotas and after which there was a completely opaque barrier to movement between the two sides—it is certainly true that, once the barrier formed, populations and subpopulations of species were cut off from one another. This has resulted in suf-

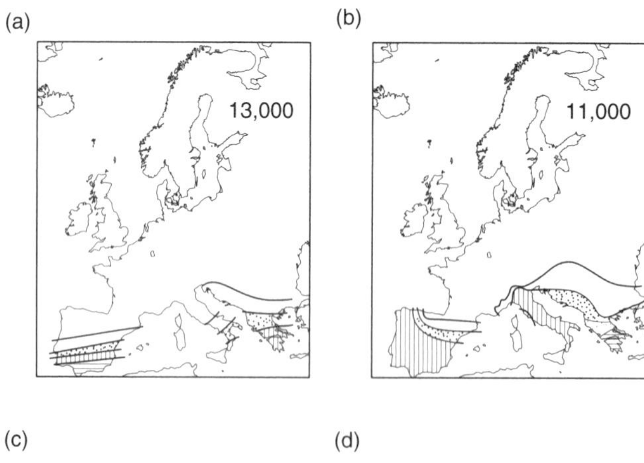


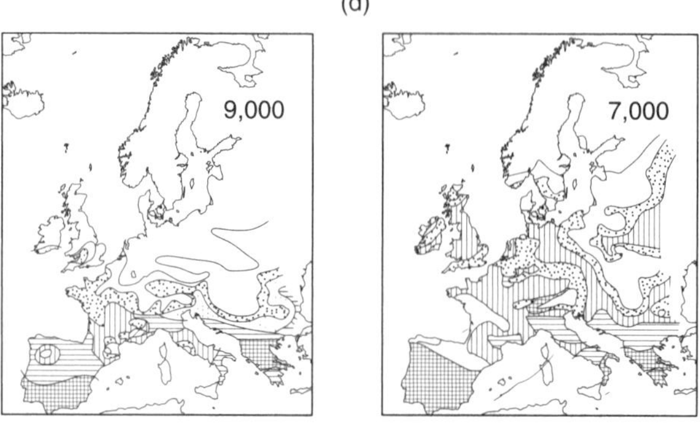


**FIGURE 9.33 Map of western Europe, showing the changing distribution and abundance of oak trees, as recorded in pollen records covering a span from 13,000 years ago to 7000 years ago.** *(From Bennett, 1997)*

ficient divergence on opposite sides of the Isthmus to cause the evolution of sister species.

The Isthmus of Panama is an example of a barrier to the migration and dispersal of organisms that simultaneously affects a large number of species. These kinds of barriers are quite common in nature and are not limited to land bridges that inhibit the movement of marine organisms. In fact, the *lack* of land bridge or some other connection among land masses effectively isolates terrestrial organisms from one another, particularly if they are incapable of airborne dispersal through flight or if they do not possess a developmental stage, such as a pollen or spore stage, during which dispersal by wind is possible.

Throughout the history of life, there are many well-documented examples of cases in which barriers to the movement of terrestrial organisms have come and gone. Some barriers involve the in-place development or loss of waterways through tectonic activity or changes in sea level. Well-known examples include the waterway between North and South America that existed prior to the formation of the Isthmus of Panama, or the waters that currently isolate several islands from one another off the coast of Southeast Asia, but which were previously part of a single landmass when sea level was lower during the Pleistocene.

Other barriers, however, have come and gone as a direct consequence of the movement of continents throughout geologic time. When land masses rift apart, as they have repeatedly in earth history, they carry with them terrestrial biotas that become isolated from one another. What may be less obvious, but is no less true, is that these same land masses are surrounded by shallow marine environments containing organisms that can also become isolated from one another because of the sheer distance that develops among regions that diverge geographically. This is particularly true among marine organisms that become separated by deep ocean basins, especially if they are bottom-dwellers that do not pass through planktonic larval stages during which they might be carried over large distances by ocean currents.

The development of barriers, whatever their scales or causes, results in the formation of **provinces**—groups of taxa with similar geographic ranges that, owing to the presence of the barriers, are geographically separated from other provinces that are similarly bounded. One of the main goals of paleobiogeography has been to understand the history of **provinciality.** Because of differences in the ecological characteristics of various taxa, including their abilities to disperse widely, it should be understood that at any given time, different taxa may be characterized by different degrees of provinciality; this must be taken into account when mapping out faunal provinces.

Nevertheless, from a practical standpoint, the delineation of ancient provinces has proven to be a valuable tool in determining the relative paleogeographic positions of ancient continents and smaller land masses known as **terranes.** As an example, there is a region in present-day Argentina known as the Precordillera, which contains a suite of Ordovician rocks and fossils that is unique in comparison to the strata contained in adjacent areas. Interestingly, this interval contains in its oldest portions a biota that is similar taxonomically to that preserved in rocks of the same age in parts of present-day North America. However, younger intervals in the Precordillera contain a biota that exhibits growing taxonomic similarities with strata preserved nearby, in present-day South America (Figure 9.34a). These data support the still-controversial view that the Precordillera was a terrane that became separated from the paleocontinent of Laurentia (an ancient portion of present-day North America) and subsequently moved southward, where it collided with the ancient Gondwannan supercontinent, which included South America (Figure 9.34b).

## 9.7 CONCLUDING REMARKS

Admittedly, this chapter has covered much ground, but there is a simple explanation for this: The subdiscipline of paleoecology has been expanding at a remarkable rate. In the 1960s, the study of paleocommunities was the major conceptual theme in paleoecology, but this was supplemented in the 1970s and 1980s by evolutionary paleoecological investigations that incorporated the geologic time dimension. During the 1980s and 1990s, paleoecology expanded again with the infusion of an ever-growing palette of techniques that provide sophisticated means for high-resolution paleoenvironmental reconstruction. Given the excitement generated by the continued development of these new methods, it is tempting to downplay the earlier themes. However, beyond the desire to describe the intellectual roots of the subdiscipline, there is another reason to consider anew paleocommunities and related topics: These remain as important conceptual themes in paleontology.

For example, the study of metapopulations, which we discussed briefly in this chapter, requires an understanding

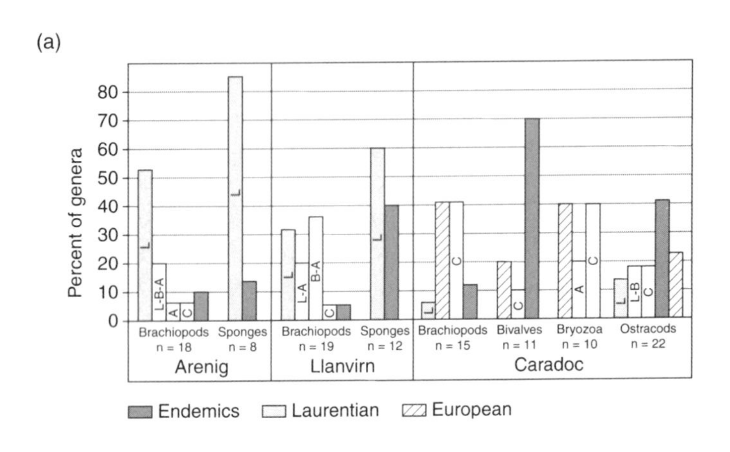


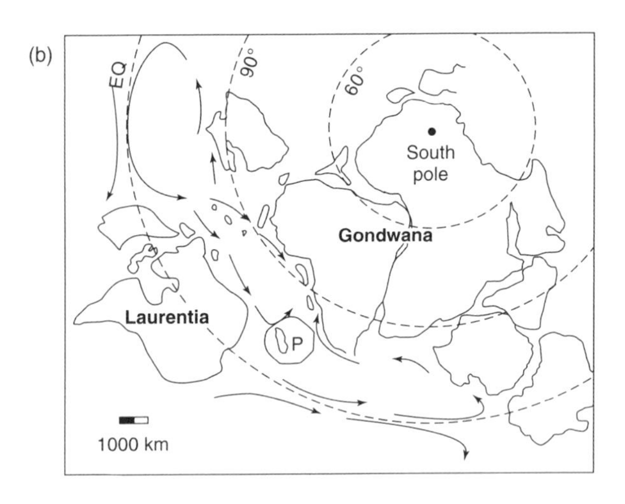


**FIGURE 9.34 An assessment of the paleobiogeography of the Precordillera region of South America.** (a) The changing biogeographic affinities of taxa preserved in the Precordillera region of South America from the Arenig (Early Ordovician) to the Caradoc (Middle and Late Ordovician). Note, in particular, the decreasing affinity with Laurentian forms. [Key to locations of genera (see Figure 8.25 for a complete depiction of these paleocontinents): L = Laurentia; L–A = Laurentia and Avalonia; L–B = Laurentia and Baltoscandia; L–B–A = Laurentia, Baltoscandia, and Avalonia; C = **cosmopolitan**] (b) Paleogeographic map for the Early Ordovician, depicting the position of the Precordillera (P, circled) roughly midway between Laurentia and the portion of Gondwana that includes present-day South America. *(From Benedetto et al., 1999)*

of the extent to which the fine-scale spatial distributions of organisms preserved in the fossil record can be shown to reflect a biological signal rather than post-mortem processes. As we have suggested in this chapter, there are reasons to believe that the patchiness evident in biological communities can be preserved in the fossil record, but we still have a long way to go before we can estimate the extent to which this is the case among taxa that lived in a variety of marine and terrestrial settings.

As we will see in Chapter 10, another unique new paleoecological theme is emerging in the study of recent, human-induced changes to present-day settings.

# SUPPLEMENTARY READING

Brown, J. H. (1995) *Macroecology*. Chicago, University of Chicago Press, 269 pp. [An overview of an emerging, large-scale approach to the investigation of present-day ecological distributions and properties, with significant ramifications for the study of paleoecology.]

DiMichele, W. A., Behrensmeyer, A. K., Olszewski, T. D., Labandeira, C. C., Pandolfi, J. M., Wing, S. L., and Bobe, R. (2004) Long-term stasis in ecological assemblages: Evidence from the fossil record. *Annual Review of Ecology, Evolution, and Systematics* **35**:285–322. [A comprehensive look at issues related to the biological significance of communities, including coordinated stasis.]

Fortey, R. A., and Cocks, L. R. M. (2003) Paleontological evidence bearing on global Ordovician–Silurian continental reconstruction. *Earth-Science Reviews* **61**:245–307. [An extensive example of the use of paleobiogeographic data in the assessment of paleocontinental positions.]

Jackson, J. B. C., Budd, A. F., and Coates, A. G. (eds.) (1996) *Evolution and Environment in Tropical America*. Chicago, University of Chicago Press, 425 pp. [A collection of papers that provide a comprehensive overview of the development of the Isthmus of Panama, from paleontological and geochemical perspectives, with an emphasis on paleobiogeography.]

Koch, P. L. (1998) Isotopic reconstruction of past continental environments. *Annual Review of Earth and Planetary Sciences* **26**:573–613. [An overview of principles involved in the use of carbon, oxygen, and nitrogen isotopes derived from terrestrial settings for environmental reconstruction.]

# *Chapter 10*

# MULTIDISCIPLINARY CASE STUDIES IN PALEONTOLOGY

## 10.1 PALEONTOLOGY AS AN INTEGRATIVE SCIENCE

While the procedures highlighted throughout this book will undoubtedly continue to evolve well into the future, many basic paleontological principles that have emerged in the past half-century will be with us for a long time. To cite but one example, consider the development of principles and methods for phylogenetic reconstruction. As noted earlier, paleontologists continue to investigate the relative merits of different procedures for phylogenetic analysis. At the same time, it is clear that the centrality of phylogenetic reconstruction has now been well established for a host of paleontological themes, including studies of classification [SEE SECTION 4.3], the tempo and mode of evolution [SECTION 7.3], and morphological diversification [SECTION 8.6].

Although these basic goals may be considered timeless, the problems to which they are applied are not. Paleontology has become an increasingly dynamic science, with a growing interdisciplinary focus on broadly based research questions, propelled by discoveries reported nearly every week in leading scientific journals. Although any attempt to review these themes will undoubtedly be somewhat outdated by the time this book is published, it is nevertheless useful to illustrate the application of paleontological principles and procedures to a diverse set of questions ranging from the investigation of the most primitive forms of life on earth—and perhaps elsewhere—to the assessment of human-induced pressures that living organisms have been forced to confront in the present day.

## 10.2 THE CAMBRIAN EXPLOSION OF MARINE LIFE

Many periods of evolutionary radiation and extinction (see Chapter 8) can be seen as simple increases in the rate of turnover of species, genera, and families. By contrast, there are certain unique episodes in the history of life that were of fundamental importance in permanently reshaping the biosphere. For animals, the most important of these is the so-called **Cambrian Explosion,** a time of great morphological and functional diversification during which most of the basic animal body plans—the phyla—appeared within just a few tens of millions of years (Figure 10.1). This statement applies, of course, to the skeletonized phyla; the later appearance of a number of soft-bodied phyla almost certainly reflects a failure of preservation.

Here we consider a few of the principal questions that surround the Cambrian Explosion:

1. What evidence does the fossil record provide of the basic sequence of events?
2. How is this evidence interpreted from an evolutionary standpoint?

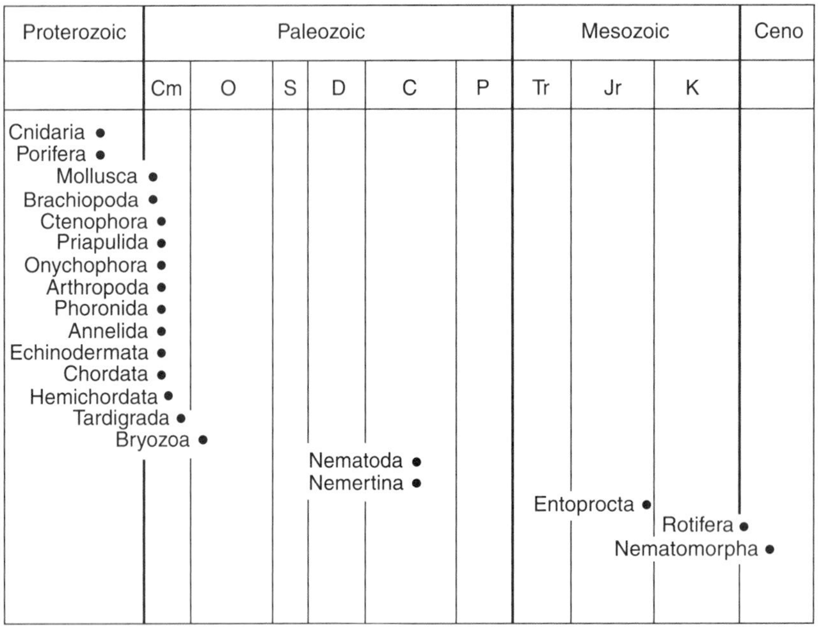

Proterozoic | Paleozoic | Mesozoic | Ceno
Cm | O | S | D | C | P | Tr | Jr | K

Cnidaria •
Porifera •
Mollusca •
Brachiopoda •
Ctenophora •
Priapulida •
Onychophora •
Arthropoda •
Phoronida •
Annelida •
Echinodermata •
Chordata •
Hemichordata •
Tardigrada •
Bryozoa •
Nematoda •
Nemertina •
Entoprocta •
Rotifera •
Nematomorpha •

**FIGURE 10.1 First appearances of animal phyla in the fossil record.** Most skeletonized phyla first appear in the Cambrian and Ordovician. Subsequent first appearances probably reflect sampling failure rather than later origins. *(From Valentine et al., 1999)*

3. Why did the radiation of animal phyla occur when it did, and why has such a profound proliferation of animal body plans not occurred since?

## The Fossil Record

Prokaryotes and algae were abundant and diverse through much of the Proterozoic Eon, but the oldest undoubted body fossils of animals are phosphatized embryos from the Neoproterozoic, around 570 million years old (Figure 1.15). Shortly after this, a diversity of biotic forms appears in the fossil record, many of which are of unknown phylogenetic affinities (Figure 10.2). These are informally referred to as the Ediacaran Fauna, after one of the classic collecting localities in Ediacara Hills, Australia. Many other problematic fossils have been described from other Proterozoic localities (for example, see Box 4.3). In addition, the oldest unambiguous trace fossils [SEE SECTION 1.2] come from the Neoproterozoic and are about 560 million years old. Possible traces and animal body fossils may be as old as 600 million years, but there is some uncertainty about these.

Figure 10.1 shows the approximate times at which the animal phyla first appear in the fossil record. Of the groups that first appear in the Cambrian, many—such as priapulids, onychophorans, and chordates—are soft-bodied, and others—such as arthropods—include mostly unmineralized forms. We are fortunate to have a Cambrian record of these phyla thanks to several deposits with exceptional fossil preservation. The best known of these is the Middle Cambrian Burgess Shale of British Columbia. As we noted in Chapter 1, the great majority of Burgess Shale species are soft-bodied, and the mineralized taxa seem to be fairly typical of shelly Cambrian deposits. A number of Burgess Shale forms have clear affinities with living phyla, and arthropods are especially common.

There are numerous problematic taxa, however. Every paleontologist has a favorite example: *Opabinia*, with its segmented body, five eyes, and hose-like snout terminating in what appears to be a grasping organ, is the most charming of beasts (Figure 10.3). Other unusual Burgess Shale animals include *Odontogriphus*, a flattened, annulated animal whose mouth was surrounded by a ring of toothlike structures that may have supported feeding tentacles (Figure 10.4); and *Hallucigenia*, an elongate animal with two prominent rows of dorsal protective spines (Figure 10.5). When the animal was first described, the spines of *Hallucigenia* were thought to be legs, but more complete material of this and related

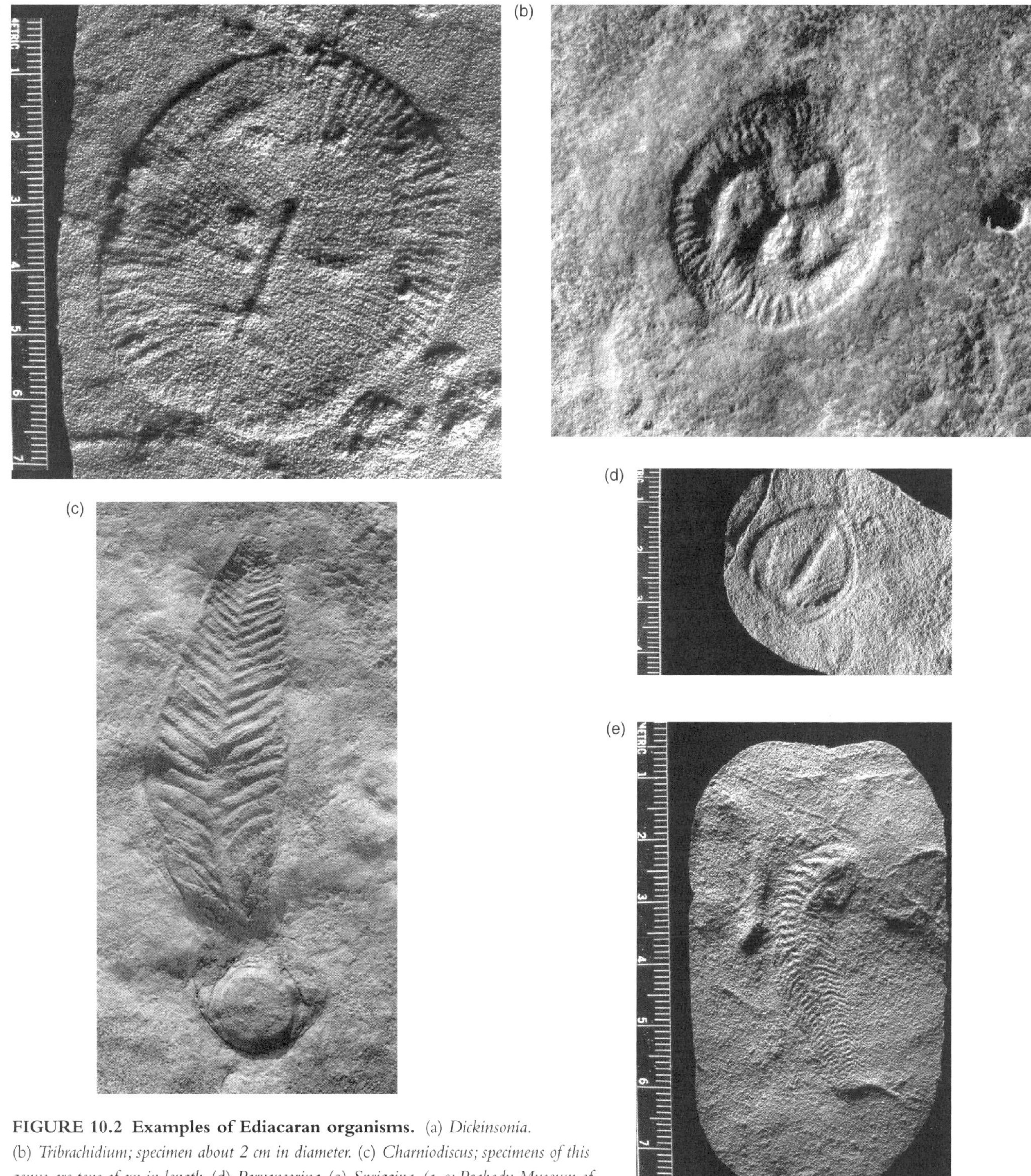

**FIGURE 10.2 Examples of Ediacaran organisms.** (a) *Dickinsonia.* (b) *Tribrachidium; specimen about 2 cm in diameter.* (c) *Charniodiscus; specimens of this genus are tens of cm in length.* (d) *Parvancorina.* (e) *Spriggina. (a, e: Peabody Museum of Natural History, Yale University; b: Palaeontological Museum, University of Oslo, Norway; c: South Australian Museum; d: W. K. Sacco/Peabody Museum of Natural History)*

forms from elsewhere in the world now suggests that the spines are dorsal and that *Hallucigenia* is a member of a group of animals known as Lobopodia after their characteristic legs.

The animal *Anomalocaris* illustrates well the difficulty that has attended the interpretation of Burgess Shale fossils. In 1892, C. D. Walcott described the genus *Anomalocaris* on the basis of a segmented fossil that he

(a)

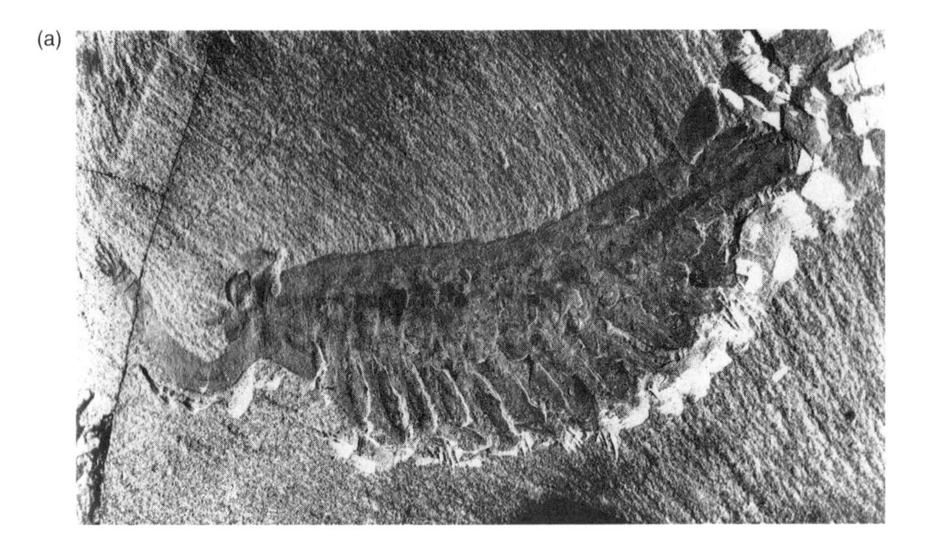

(b)

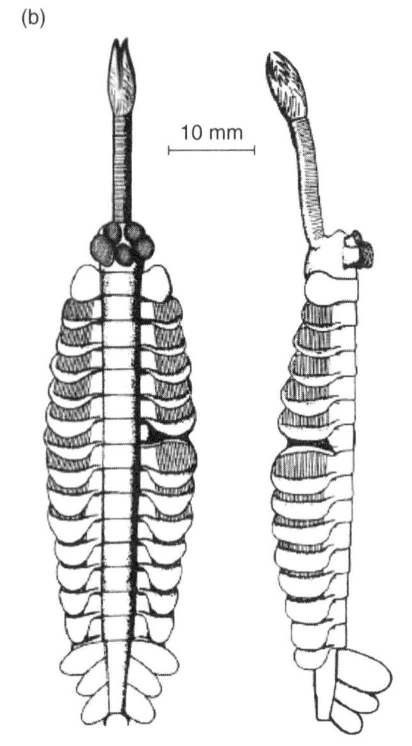


**FIGURE 10.3 The Burgess Shale animal *Opabinia*.** (a) Specimen in lateral view. (b) Reconstruction in dorsal and lateral views. *(From Whittington, 1985, reproduced with permission of the Minister of Public Works, Canada)*

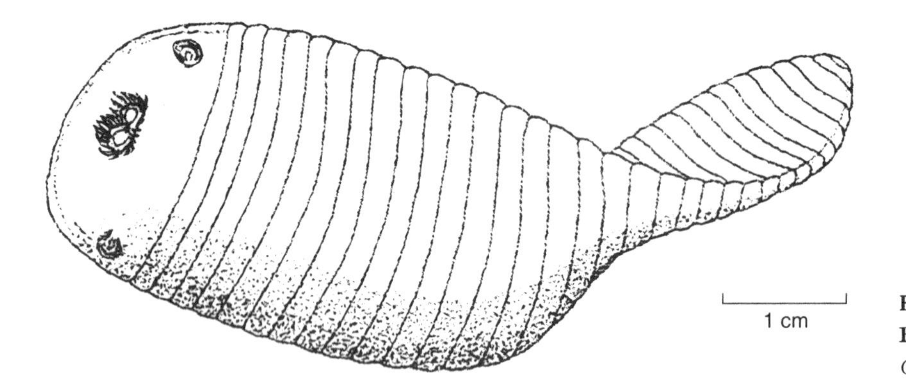


**FIGURE 10.4 Reconstruction of the Burgess Shale animal *Odontogriphus*.** *(From Conway Morris, 1976)*

(a)

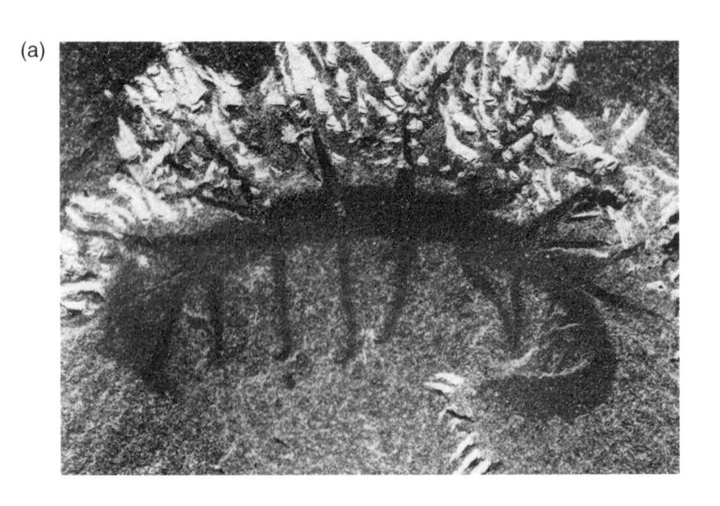

(b)

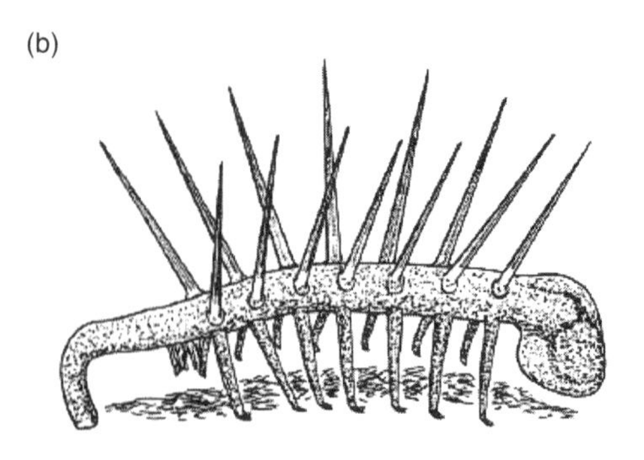

**FIGURE 10.5 The Burgess shale animal *Hallucigenia*.** (a) Specimen in lateral view. Field of view is about 1.4 cm wide. (b) Reconstruction. *(From Conway Morris, 1998)*

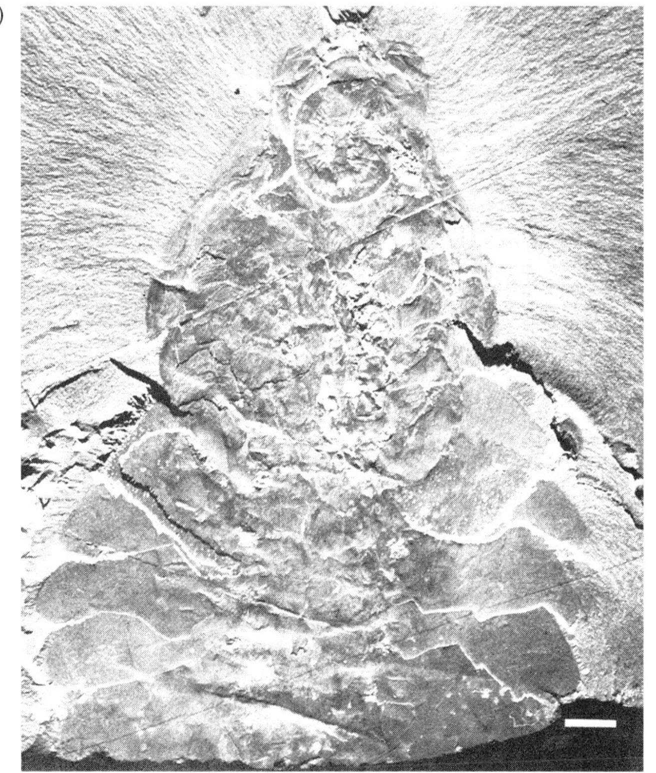

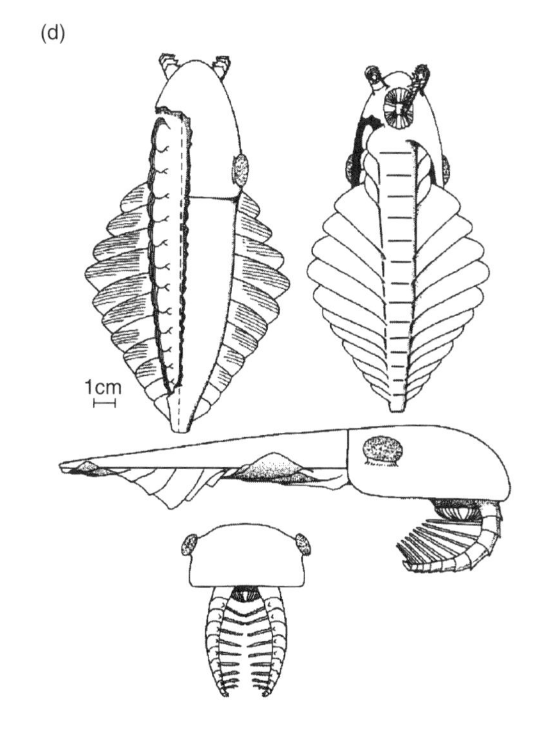


**FIGURE 10.6 The Cambrian animal *Anomalocaris*.** (a) Appendage, initially described as the body of an arthropod under the name *Anomalocaris*. (b) Mouth disc, initially described as a free-living animal under the name *Peytoia*. (c) Relatively complete specimen in ventral view, anterior toward the top. (d) Reconstruction. The scale bar is 1 cm. *(From Whittington, 1985, reproduced with permission of the Minister of Public Works, Canada)*

thought to be the body of an arthropod (Figure 10.6a). Later, in 1911, he reported a vaguely jellyfishlike fossil, a plated disk that he named *Peytoia* (Figure 10.6b). In the same year, he also described an incomplete, segmented body under the genus name *Laggania*. Only in the late 1970s was it revealed, through painstaking specimen preparation, that the *Anomalocaris* fossil is actually the limb of a much larger animal, of which *Laggania* is part of the body and *Peytoia* the mouth (Figure 10.6c). By the rule of priority [SEE SECTION 4.1], the entire animal is given the name *Anomalocaris*. Shown reconstructed in Figure 10.6d, this animal, with its grasping limb, plated mouth, and streamlined body, was evidently a predator.

Since the initial discovery of the Burgess Shale, other instances of exceptional preservation in the Cambrian have been found. Most notable are the Chengjiang Fauna in Yunnan, China, and the Sirius Passet Fauna of Greenland. Although these Early Cambrian deposits are older than the Burgess Shale, they reveal a similar array of organisms, and some genera have proven to have long stratigraphic ranges. *Anomalocaris*, for example, is found through much of the Lower and Middle Cambrian.

The fossils of the Burgess Shale are preserved as carbonized films and clay sheets and are thought to have been deposited under low-oxygen conditions. Internal organs are often found preserved by early diagenetic mineralization. Motivated in part by questions of Burgess Shale preservation, experimental taphonomic investigations [SEE SECTION 1.2] have been conducted on shrimps and other organisms with proteinaceous exoskeletons. These have helped demonstrate that reduced oxygen levels promote the fossilization of soft parts and lightly skeletonized hard parts, and have also pinpointed the reasons why this is the case. Under controlled laboratory conditions, reduced oxygen levels do not appreciably inhibit the decay of soft parts. Rather, they promote mineralization, including the early formation of pyrite within the interstices of soft parts [SEE SECTION 1.2]. Therefore, it is this process of early mineralization, rather than the preservation of soft parts per se, that is largely responsible for the exquisite fossils of the Burgess Shale.

## Evolutionary Interpretation of Cambrian Events

The fossil record suggests a brief early history of animals followed by a rapid radiation of the phyla. Although Proterozoic animals were unknown in Darwin's time, the sudden appearance of phyla had already been documented. Darwin and many after him have attributed this to a long gap in the stratigraphic record, during which the phyla as we know them evolved but were not preserved. This simple view can no longer be supported. For one thing, the trace fossils that are present in the Proterozoic are generally small and horizontal, indicating simple behavior, and they increase in size and complexity in the Cambrian. The kinds of animals that made complex traces in the Cambrian left no such traces in the Proterozoic. Moreover, Proterozoic rocks have been extensively studied, and they have yielded many body fossils (Figure 10.2). But there is no sign of brachiopods, trilobites, echinoderms, and so on. Because the conditions needed to preserve body fossils were present in the Proterozoic, these groups should have been preserved if they had existed in abundance. Thus, the unpreserved history of the animal phyla that appear in the Cambrian cannot extend very far into the Proterozoic.

This conclusion depends partly on what we mean by a *phylum*, however. Familiar characteristics that we generally use to distinguish the members of a phylum or other taxonomic group need not have been present at the initial time of lineage splitting that led to that group. Take the earliest bird, *Archaeopteryx* [SEE SECTION 4.2]. It has feathers but largely resembles nonavian theropod dinosaurs rather than modern birds in its skeletal features. It does not have a well-developed breastbone that would indicate strong wing muscles, its bones are much denser than those of modern birds, and it is fully toothed. Similarly, it is possible that many of the animal phyla—in the phylogenetic sense of the lineages that led up to brachiopods, echinoderms, molluscs, and so on—were present in the Proterozoic, but that they evolved their distinctive features only in the Cambrian. In other words, there may have been a significant delay between the evolution of stem and crown groups [SEE SECTION 4.3]. Which particular features the members of the stem lineages might have had is rather speculative. Arthropod ancestors without a cuticle or jointed limbs may have resembled segmented worms, and molluscan ancestors may have resembled flatworms in general aspect. At the same time, if one imagines a brachiopod without its lophophore, shell, and musculature that opens and closes the shell, there is very little left to imagine.

One line of evidence in support of the delay between phylum origin and the evolution of diagnostic anatomical features comes from molecular biology. When a lineage splits, the two branches evolve independently and therefore accumulate genetic differences. On average, the longer the time since their divergence, the greater will be the genetic difference between the lineages. If we knew the rate of divergence, then the amount of genetic difference between two living lineages would reveal the amount of time the lineages have been evolving separately. Because divergence rates vary among genes, among lineages, and over time, there is great uncertainty in calibrating these so-called **molecular clocks.** Thus, estimated divergence dates among the living phyla vary widely, from just under 600 million to well over 1 billion years ago. Nonetheless, even the youngest estimated divergence dates are in the Proterozoic and suggest an unpreserved history of some tens of millions of years.

It therefore seems plausible that the first appearance of animal phyla in the Cambrian was not mainly a divergence of lineages but rather a series of evolutionary events in which the skeletal and other anatomical traits that characterize the phyla were acquired over a period of just 10 to 20 million years. This suggests rapid rates of morphological evolution, which can be assessed by comparing the range of anatomical diversity or disparity [SEE SECTION 8.10] that evolved during the Cambrian with that which accumulated during the rest of the Phanerozoic.

The arthropods are one of several animal groups that rapidly increased in disparity during the Cambrian. Figure 10.7 portrays Cambrian and Recent arthropods ordinated along three axes that summarize multivariate morphological data in much the same way as principal components do [SEE SECTION 3.2]. The plot depicts 24 Cambrian taxa that are sufficiently well preserved for the morphological features to be observed, and 24 Recent taxa that were chosen to represent the living classes and subclasses. The Cambrian and Recent taxa are indicated in oblique and plain type, respectively. The most striking

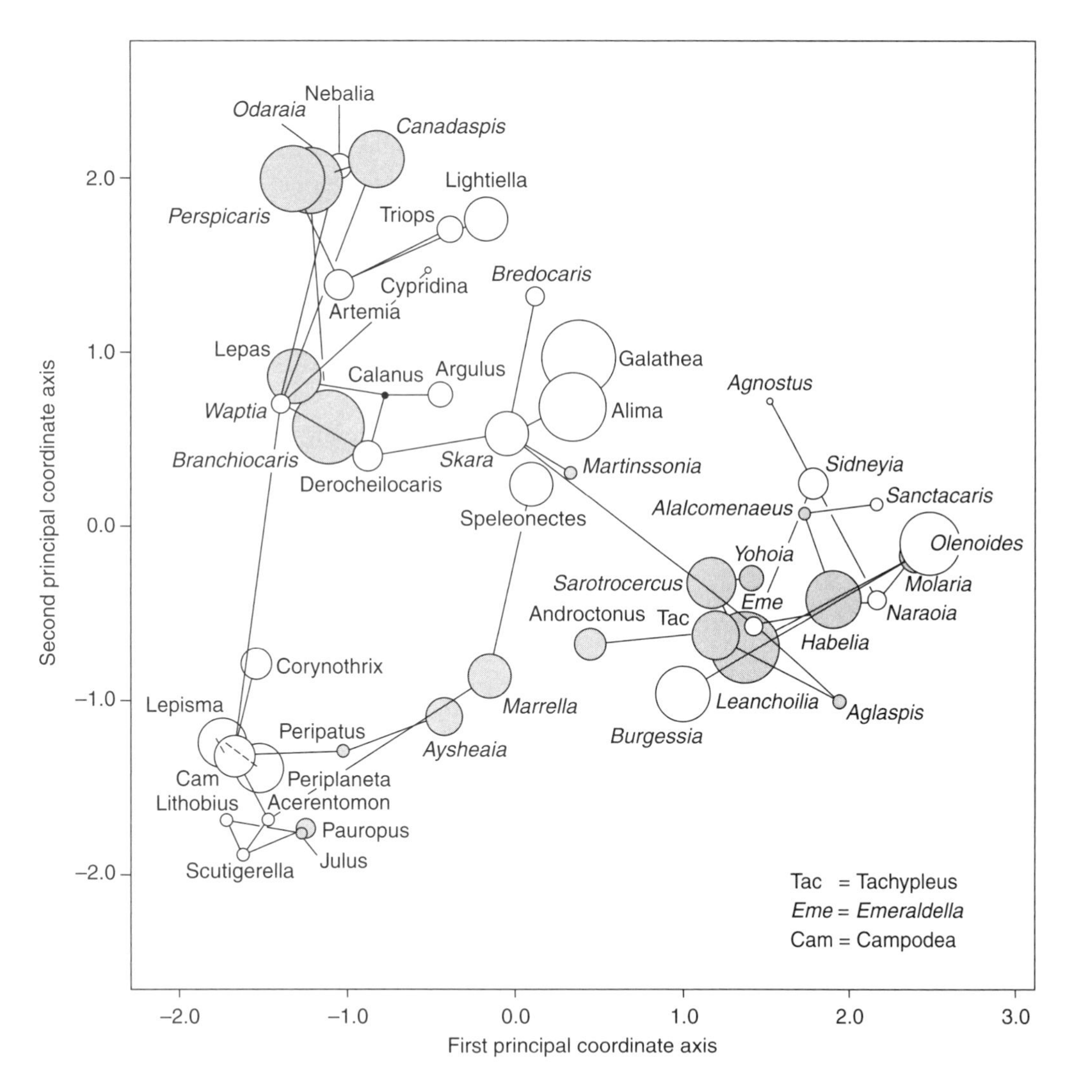


**FIGURE 10.7 Ordination of Cambrian and Recent arthropods based on multivariate morphological data and using a method similar to principal-component analysis [SEE SECTION 3.2].** Cambrian taxa are indicated in oblique type, Recent taxa in plain type. The original data from which these axes were formed consist of about 120 morphological characters having two or more character states. The third dimension in this plot is represented by the size and shading of the circles; dark circles are positive values while open circles are negative, and larger circles represent larger positive or negative values. In addition, lines are drawn between taxa that are most similar to each other on the basis of the entire set of characters that were studied. This plot shows that Cambrian marine arthropods were as diverse in morphology as are living forms. *(From Wills et al., 1994)*

feature of Figure 10.7 is that Cambrian and Recent taxa are about equally dispersed. When the morphological distances between taxa are measured [SEE SECTION 8.10], the disparity of Cambrian arthropods is indistinguishable from that of Recent forms. Cambrian arthropods evolved a range of anatomical form that was not exceeded subsequently in over 500 million years. This same conclusion has been drawn from similar studies of priapulid worms and a few other groups.

The work on arthropod morphology suggests that rates of morphological evolution were much greater in the Cambrian than afterwards, and that morphological diversity has not increased substantially since the Cambrian Explosion. In fact, this question is not quite settled. There are at least three reasons for continued debate. First, Cambrian and post-Cambrian disparity have been compared for relatively few animal phyla. Although selected classes have been studied, we still do not have phylum-level analogs of Figure 10.7 for molluscs, echinoderms, brachiopods, sponges, cnidarians, and so on.

Second, some studies of morphological diversity have simply compared the Cambrian and the Recent, with little attention paid to the intervening time. It is conceivable that arthropod disparity increased substantially over the past half billion years and then declined to a level comparable to that of the Cambrian. This would imply that post-Cambrian evolution was not as limited as suggested by Figure 10.7.

Third, the traits studied represent a limited part of the overall anatomy of the organisms in question, and they have comparatively little discriminatory power. For example, a shrimp and a lobster are essentially identical with respect to the kinds of traits that have been used to compare disparity in Cambrian and Recent arthropods. This coarse resolution, which is nearly inevitable when studying such a wide range of organisms [SEE SECTIONS 2.2 AND 5.3], has left a number of workers skeptical as to whether the true history of disparity has been adequately documented. In what follows, we accept for the sake of discussion that evolutionary innovation was unusually pronounced in the Cambrian, but it is important to bear in mind that there is still debate about this point.

## Reasons for the Cambrian Explosion

In explaining the Cambrian Explosion, just as in documenting the basic sequence of events, it is important to distinguish between the splitting of lineages and the evolution of novel body forms. With respect to lineages, why is it that phylum and class originations are concentrated early in animal history, as shown in Figures 8.29 and 10.1?

A simple possibility, suggested by paleontologist David Raup, comes from mathematical modeling of the shape of evolutionary trees. Consider the partial evolutionary tree of Figure 10.8a, simulated with a computer program that holds the per-capita rate of origination and extinction constant over time [SEE SECTION 7.2]. Of the many species produced by this simulation, 28 are still extant at the "present day." Each pair of species can be traced backward in time to the point at which the lineages leading up to the two species first diverged. For example, the two species farthest to the left within group A diverged just a single time increment before present. By similar reasoning, any pair consisting of one species from group A and one from group B can be traced back to a divergence at 21 time units before present. Likewise, any species from group A or B has a divergence time of 47 relative to any species from C or D, and so on.

With 28 living species, there are $28 \times 27 \div 2$, or 378 possible species pairs. [In general, if a group has $N$ species, then there are $N \times (N - 1) \div 2$ species pairs within the group. For two groups with $N_1$ and $N_2$ species, there are $N_1 \times N_2$ between-group pairs.] If we tally the frequency distribution of divergence times, as in Figure 10.8b, we find that the majority of them—greater than two-thirds—are at either 41 or 47 time units before present.

The expected distribution of divergence times can be calculated exactly for any given rate of origination and extinction. As it turns out, the tree in Figure 10.8a represents a case of unrealistically high rates [SEE SECTION 7.2]. Trees produced with empirically reasonable rates—with species durations on the order of 1 to 10 million years (m.y.)—yield living species whose divergences can be extended much farther back in time than those of Figure 10.8a. In fact, for a wide range of origination and extinction rates, more than 90 percent of all possible pairs of living species are expected to share divergence times greater than 500 m.y. ago. The distribution of divergence times is dictated by the geometry of evolutionary trees, with many extinct species and with just a few living ones that can trace their common ancestry to divergence events deep in the past.

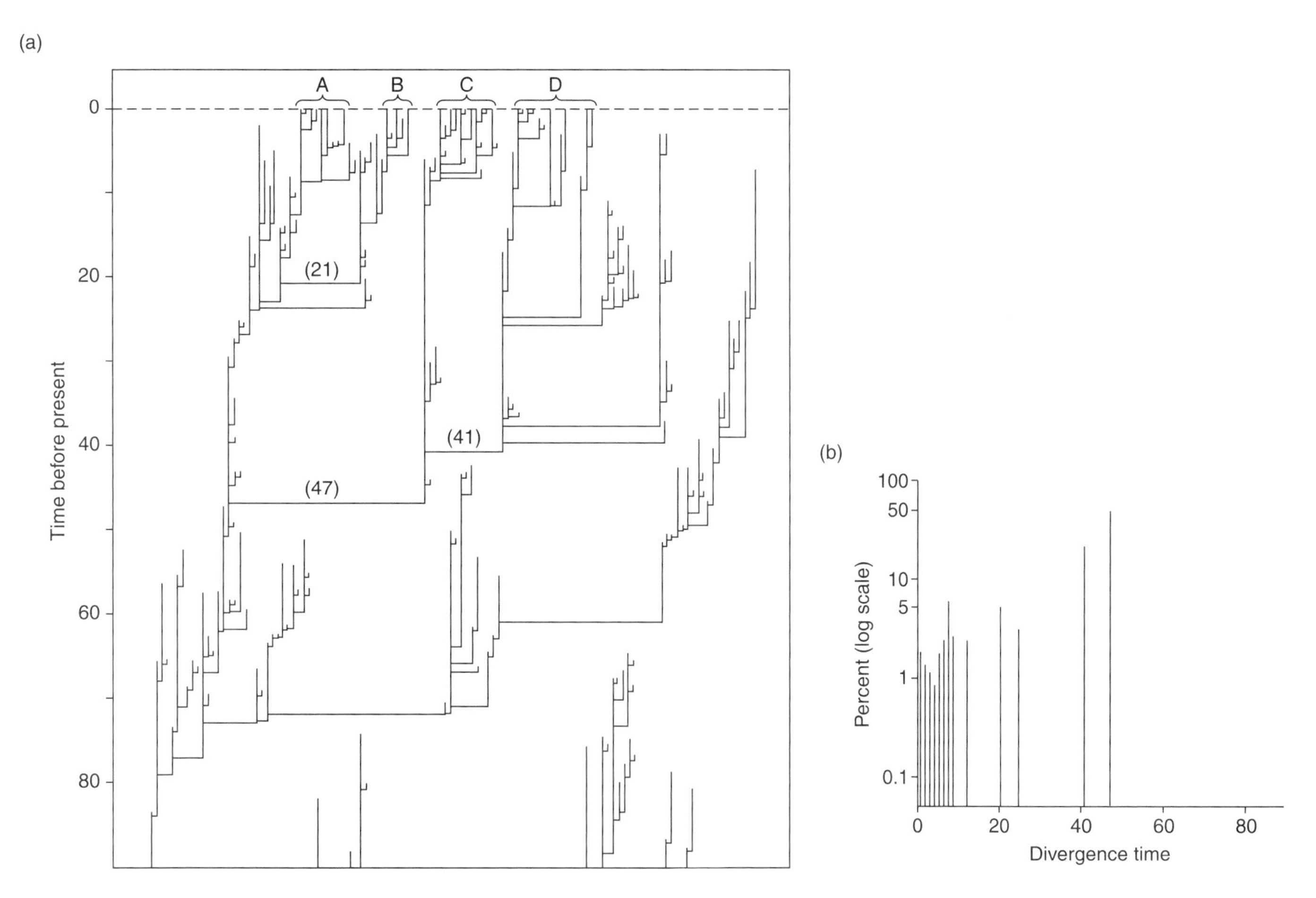


**FIGURE 10.8 Mathematical modeling of the expected divergence times of living species, assuming that origination and extinction rates are constant over time.** (a) A hypothetical evolutionary tree, with living species divided into four groups. Any species in group A and any species in group B share a latest common ancestor at 21 time units before present; this is the divergence time of the pair of species. Similar reasoning holds for other species pairs. (b) The frequency distribution of divergence times, showing that about two-thirds of all pairs of living species have divergence times of either 41 or 47 time units before present. *(From Raup, 1983)*

This result of the mathematical model is best appreciated by comparison with actual divergence times of living species. Although not known in detail, they can readily be sorted into those that involve pairs of species within the same phylum versus pairs of species in different phyla. Between-phylum pairs must have diverged at or before the time of divergence of their respective phyla, while within-phylum pairs must have diverged afterward. If we focus on the paleontologically important phyla that are still alive today, all except the Bryozoa had originated by the end of the Cambrian, about 500 m.y. ago, with the Bryozoa appearing in the Early Ordovician, by 480 m.y. ago. If most species pairs today are between-phylum pairs, then most species divergences are older than 480 m.y. It would then stand to reason that the early origin of the phyla could in principle be explained by the geometry of evolutionary trees.

Species diversity and numbers of species pairs are given in Table 10.1 for the living representatives of the main fossil groups. There are more than 180,000 living species in these groups, and nearly 17 billion species pairs. Of these pairs, nearly 80 percent are between-phylum pairs and therefore represent divergences older

**TABLE 10.1**

Number of Living Species in Paleontologically Important Groups and Number of Species Pairs that Represent Divergences within the Same Phylum and between Different Phyla

| Phylum | Number of Species | Within-Phylum Pairs |
|---|---|---|
| Annelida | 11,600 | $6.72 \times 10^8$ |
| Arthropoda | 50,000 | $1.25 \times 10^9$ |
| Brachiopoda | 325 | $5.26 \times 10^4$ |
| Bryozoa | 5000 | $1.25 \times 10^7$ |
| Chordata | 45,000 | $1.01 \times 10^9$ |
| Cnidaria | 9000 | $4.05 \times 10^7$ |
| Echinodermata | 6000 | $1.80 \times 10^7$ |
| Mollusca | 50,000 | $1.25 \times 10^9$ |
| Porifera | 5000 | $1.25 \times 10^7$ |
| Total species: | 181,925 | |
| Total species pairs: | $1.65 \times 10^{10}$ | |
| Total within-phylum pairs: | $3.66 \times 10^9$ | |
| Total between-phylum pairs: | $1.29 \times 10^{10}$ | |
| Percent of pairs within phyla: | 22% | |
| Percent of pairs between phyla: | 78% | |

*SOURCE:* Valentine (2004); Barnes, Calow, & Olive (1993). Species count for arthropods excludes terrestrial arachnids and insects.

than 480 m.y. We therefore see that the actual distribution of divergence times is roughly in accord with the expectations of the simple mathematical model. This suggests that branching geometry may largely explain the early origins of phylum-level groups. Similar reasoning can be applied to the early origins of classes as depicted in Figure 8.29.

Turning now to the morphological aspect of evolution, why was the Cambrian a time of profound anatomical innovation? Many animal phyla share fundamental genetic pathways for body patterning and other important aspects of anatomy. One class of genes involved are the *Hox* genes, which were discussed in Chapter 4 [SECTION 4.2]. Figure 10.9 shows the phylogenetic relationships among some of the animal phyla, based on genetic sequences that do not include these *Hox* genes. Superimposed on this phylogeny is the number of *Hox* genes present in the living representatives of each phylum. Cladistic parsimony [SEE SECTION 4.2] allows one to estimate the character states present in the common ancestor to two sister taxa. Among **BILATERIAN** phyla—namely, all animals except sponges, cnidarians, ctenophores, and some obscure groups—the smallest number of *Hox* genes is four. When evidence for secondary reduction in the nematode lineage is taken into

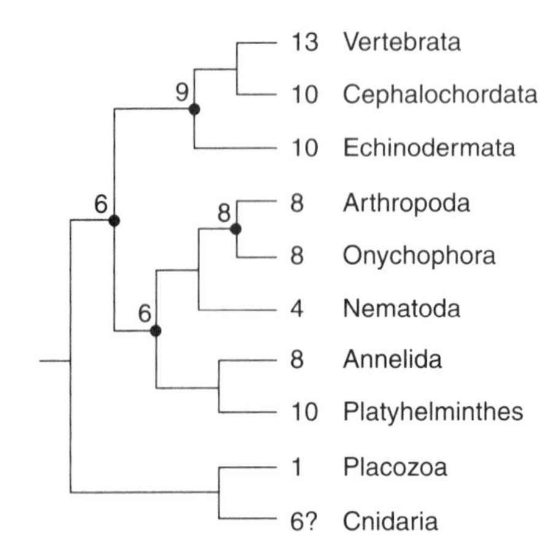


**FIGURE 10.9 One estimate of cladistic relationships among selected animal phyla and subphyla, with the number of *Hox* genes to the left of the group name.** The numbers at nodes indicate the minimal number of *Hox* genes in the common ancestor to the sister taxa connected by the node. *(From Valentine et al., 1999)*

account, it seems probable that the common ancestor to the bilaterians had at least six *Hox* genes. If the phyla that share the corresponding genetic pathways today diverged from one another in the Proterozoic, this would suggest that *Hox* genes and other major genetic pathways were already in place well before the Cambrian Explosion, implying that it is something other than the acquisition of these pathways that led to the Cambrian Explosion.

Two main proposals have been offered to explain the burst of evolutionary innovation in the Cambrian. The first is that there was some trigger, or the release of a constraint, in the physical environment. The second is that innovations in some lineages changed the ecological world in a way that promoted evolutionary responses in other lineages. These explanations are not mutually exclusive, and in both cases there is still considerable uncertainty about the exact details.

With respect to external environment, lower oxygen levels prior to the Neoproterozoic may have limited gas exchange, thereby placing an upper bound on the size that animals could attain [SEE SECTION 2.3]. There is geochemical evidence for an increase in oxygen through much of the Proterozoic Eon. A Neoproterozoic surge, which may be implicated in animal evolution, is suggested by the record of carbon isotopes in marine sediments. Carbon has two stable isotopes, $^{12}C$ and $^{13}C$. Lighter $^{12}C$ is preferentially incorporated into organic tissue during photosynthesis. Isotopically light organic carbon produced in the shallow oceans is transferred to deeper waters and to the sediment when it sinks (often as part of fecal pellets). If this organic carbon is buried, the oceans by default are enriched in $^{13}C$ because of the removal of $^{12}C$. Organic carbon and the carbon in precipitated carbonate minerals will then show the signal of heavier carbon. Observed increases in $^{13}C$ in the geologic record may therefore indicate increased burial of organic carbon. This in turn means that less oxygen would be consumed in the oxidation of carbon. Thus, an increase in $^{13}C$ could indicate an increase in available oxygen. Exactly such changes in the record of carbon isotopes are observed toward the end of the Proterozoic.

With respect to ecology, there is no firm evidence of predation in the Proterozoic, while bite marks, gut contents, and morphological features that make sense only as predatory adaptations show that predation was well established by the Middle Cambrian (Figure 10.6). The armored skeletons that characterize a number of Cambrian forms may have evolved in response to predation. Complex ecological relationships were clearly in place during the Cambrian, and such interactions between organisms would have provided a new suite of selective pressures that may have led to anatomical innovations.

As to why nothing as profound as the Cambrian Explosion has happened since, two principal explanations have been proposed. First is the idea that ecological opportunities as great as those present in the "empty" Cambrian world were never seen again (see Figure 8.29). Second is the concept that genetic and developmental systems of animals eventually became specialized or channeled in such ways that major innovations—of the kind that would lead to new phyla—were not possible after the Cambrian.

Because these two explanations make many similar predictions—for example, that evolutionary transitions in morphology should become smaller over time—the relative importance of the two has not yet been fully assessed. Testing these two possibilities against each other remains one of the most challenging and intriguing tasks as paleontologists continue to make sense of the Cambrian Explosion.

## 10.3 THE LATE PERMIAN EXTINCTION

Just as the Cambrian Explosion forever changed the course of animal evolution, one extinction event stands out because of the extent to which it profoundly altered life on earth. In the later part of the Permian Period, over 40 percent of the marine animal families and over 60 percent of the genera became extinct [SEE SECTION 8.6]. Numerous classes and orders also vanished, including trilobites, rugose and tabulate corals, blastoid echinoderms, rostroconch molluscs, and cystoporate bryozoans. Many of the clades that did survive were reduced to just a few lineages. Terrestrial vertebrates, insects, and some plant groups were also affected. These significant changes in biotic composition mark the end of the Paleozoic Era. Study of the Late Permian extinction has focused on three central questions:

1. Over how much time were the extinctions spread, and were they synchronous globally?
2. Can anything be learned from investigating the geographic locations of taxa that became extinct?
3. What caused the extinctions?

## Timing of Extinctions

Because species have limited geographic and environmental ranges, biostratigraphic correlation is always difficult on a global scale. It is even harder than usual in the Late Permian and Early Triassic because of a sparse distribution of outcrops. There are nonetheless a number of regions around the globe where the Permian–Triassic transition is preserved. These include the Armenian–Iranian border region, Kashmir, and southern China. In these and other areas, striking faunal changes are seen in the Late Permian, but does this imply that the extinctions were synchronous everywhere? Biostratigraphic correlation always requires the assumption of synchroneity to some extent [SEE SECTION 6.1], but it would obviously be circular to start with the assumption that a particular set of extinctions occurred simultaneously around the world, then use the assumption to correlate local stages with each other, and then to use the correlation to infer synchronous extinction.

Ideally, one would assess synchroneity with high-resolution numerical dating, but the requisite conditions are not met everywhere. The standard approach to this problem is therefore to develop a correlation scheme with one set of fossils, preferably those that have the desirable properties of index fossils [SEE SECTION 6.1], and to use the resulting correlation to study evolution and extinction in other groups of organisms. In the case of the Late Permian and Early Triassic, a global zonation has been constructed using the pelagic conodonts. This allows us to determine that, for data collected at the global scale, extinctions of marine genera were elevated in the Guadalupian Epoch and in the later part of the final stage of the Permian, the Changhsingian. Available data suggest that extinctions of terrestrial vertebrates may also have been concentrated in the Changhsingian.

While global analysis of marine genera indicates elevated extinction during the Changhsingian, the global data are not sufficiently well resolved to determine over precisely how much of the stage the extinctions were spread. To study the timing at a finer level of resolution, it is necessary to turn to local and regional data. Among the most thoroughly documented Late Permian through Early Triassic sections are those in southern China. Figure 10.10 shows the stratigraphic ranges of over 300 marine species in a series of sections at Meishan. A composite section was constructed by tracing distinctive clay beds among the individual

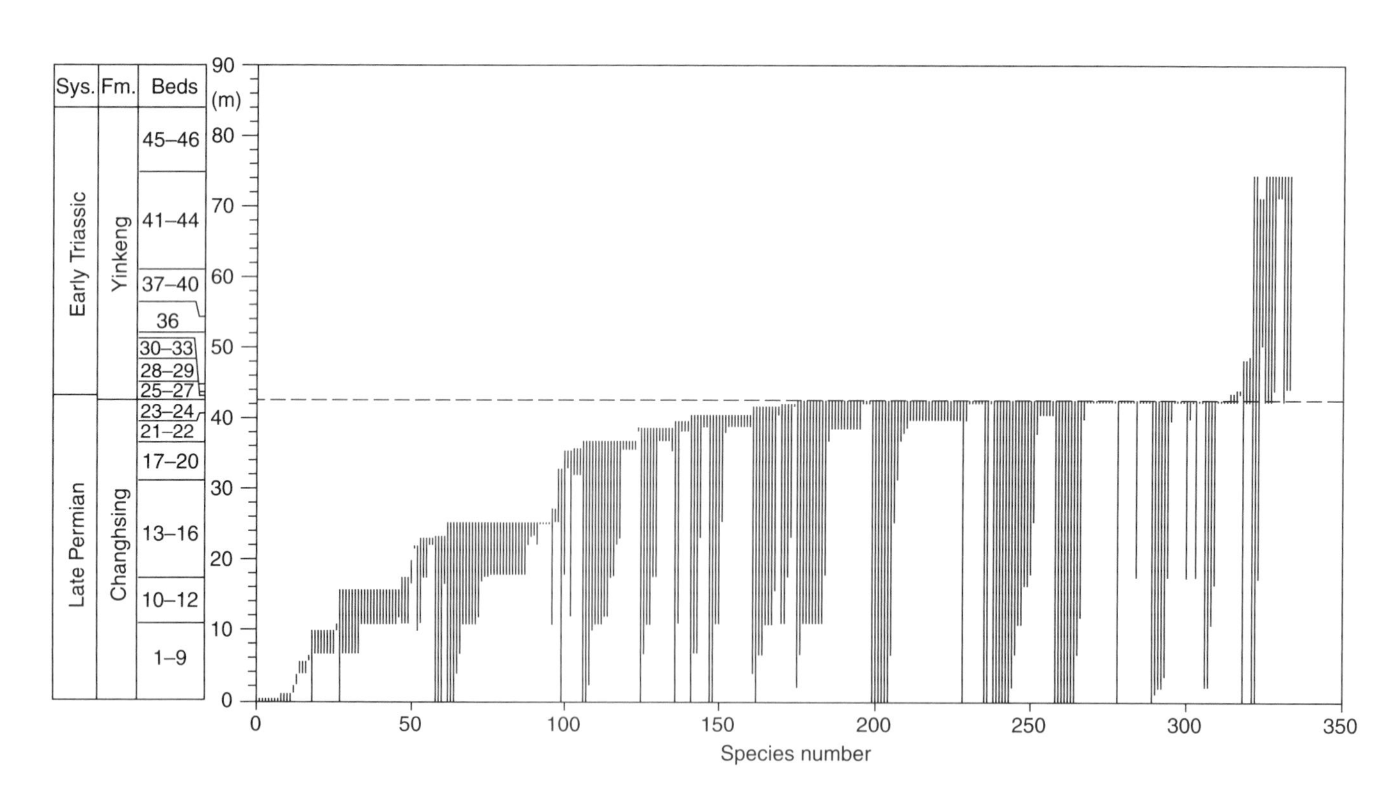


**FIGURE 10.10 Stratigraphic ranges of marine species at Meishan, southern China, shown relative to a composite of several sections.** The greatest concentration of last appearances is at the base of bed 25. *(From Jin et al., 2000)*

sections, and the ranges are given relative to this composite. This reveals that the last appearances of species are not evenly spread throughout the last stage of the Permian. Rather, they are most strongly concentrated near the bottom of bed 25.

Over how much time are the last appearances spread in the sections at Meishan? Fortunately, the Meishan beds are interbedded with volcanic ashes that can be radiometrically dated [SEE SECTION 6.1]. An ash below bed 21 has a radiometric date of 252.3 million years old, and one within bed 25 has a date of 251.4 million years old (Bowring et al., 1998). Therefore, most of the last appearances occurred within about a million years or less. If the rate of accumulation of beds 21–24 was even approximately uniform, the extinctions at the base of bed 25 must represent substantially less than 1 million years.

The outstanding question is whether extinctions were concentrated in an equally short time, and at the same stratigraphic level, throughout the globe. For the most part, biostratigraphic resolution is not fine enough to provide an answer. Hopefully, future radiometric dating at other boundary sections, similar to the work at Meishan, will enable this question to be addressed.

## Environmental Change, Biogeography, and Extinction Mechanisms

As noted in Chapter 8, to understand the causes of major changes in global biodiversity, it is helpful to consider them in a geographic context. During the Paleozoic Era, many of the continents familiar to us today were located in rather different places than they are at present [SEE SECTION 9.6]. In fact, several major present-day regions, such as Europe and China, are actually agglomerations of several smaller, Paleozoic paleocontinents that subsequently collided with one another. In addition, epicontinental seas—very broad, shallow seas that covered large expanses of continental interiors—were quite common in the Paleozoic. One of the most striking aspects of the transition from the Paleozoic to the post-Paleozoic is the decline and near-loss of these settings [SEE SECTION 1.4].

During the late Paleozoic and culminating in the Permian, many of the world's paleocontinents coalesced to form the supercontinent known as Pangea (Figure 10.11). In association with the formation of Pangea, many shallow-water areas around the world became fully emergent. It was once the prevailing view of

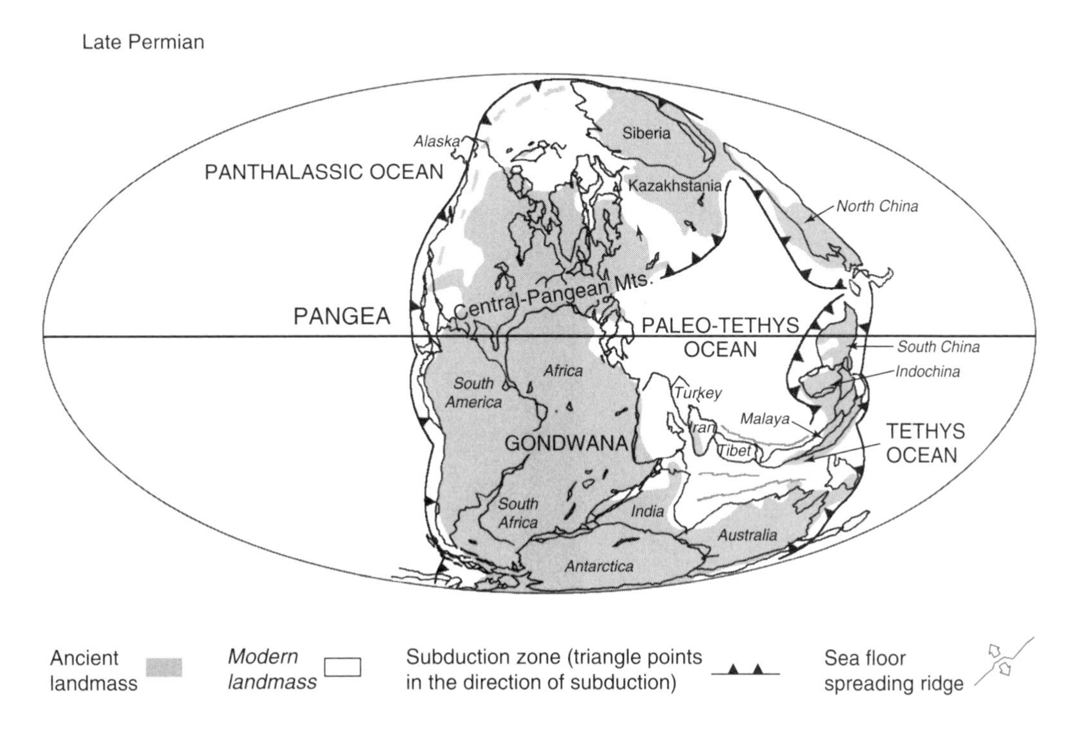


**FIGURE 10.11 Global paleogeographic map, showing the positions of paleocontinents during the Permian period.** There was a decline in the extent of epicontinental seas in the Permian, relative to much of the rest of the Paleozoic, in association with the formation of the supercontinent Pangea. *(From the Web site of* The Paleomap Project, *www.scotese.com/earth.htm)*

paleontologists that this loss of shallow-water area caused the Late Permian extinction, at least among marine organisms. Indeed, it appears that entire marine faunal provinces dried up during the Late Permian (Figure 10.12), so it is difficult to dismiss the possibility that the loss of these marine settings played a significant role in the Guadalupian extinction interval.

Given the likelihood that the end-Changhsingian extinction occurred very rapidly (Figure 10.10), however, the emerging view is that it was caused by some mechanism(s) more catastrophic than that associated with the comparatively slow draining of shallow-water areas over a period of perhaps several million years. A clear consensus has yet to emerge about the primary cause of this end-Permian event, but there is growing evidence that the boundary between the Permian and Triassic periods was marked by events that were catastrophic in nature, including a massive volcanic eruption in Siberia, widespread anoxia in marine settings, and, possibly, the impact of one or more large comets or asteroids. There was also a series of dramatic shifts in carbon isotope ratios that appear to rival those exhibited during the Neoproterozoic, but the nature and causes of these shifts are still not entirely understood and remain under active investigation.

If continued scrutiny supports the idea that there were two phases of extinction (Guadalupian and Changhsingian), this will lend credence to a growing perception that several, if not all, of the major Phanerozoic mass extinctions were more complex temporally than once thought. In the end, we may discover that mass extinctions occurred at times when several things went wrong at around the same time, resulting in an overlapping set of biotic responses to two or more mechanisms.

## 10.4 THE PALEOCENE–EOCENE THERMAL MAXIMUM

Grounded in the collection and analysis of extensive sets of deep-sea cores (e.g., Figure 9.1b), a detailed picture is beginning to emerge of the intimate relationship between the earth's climate and major biotic transitions throughout the Cenozoic Era. Geochemical analyses of the skeletal compositions of foraminifera and other organisms preserved in these cores have permitted the detection of high-resolution, stratigraphic transitions in $\delta^{18}O$.

As noted in Chapter 9 [SECTION 9.5], assessments of $\delta^{18}O$ transitions in deep-sea cores are based on strati-

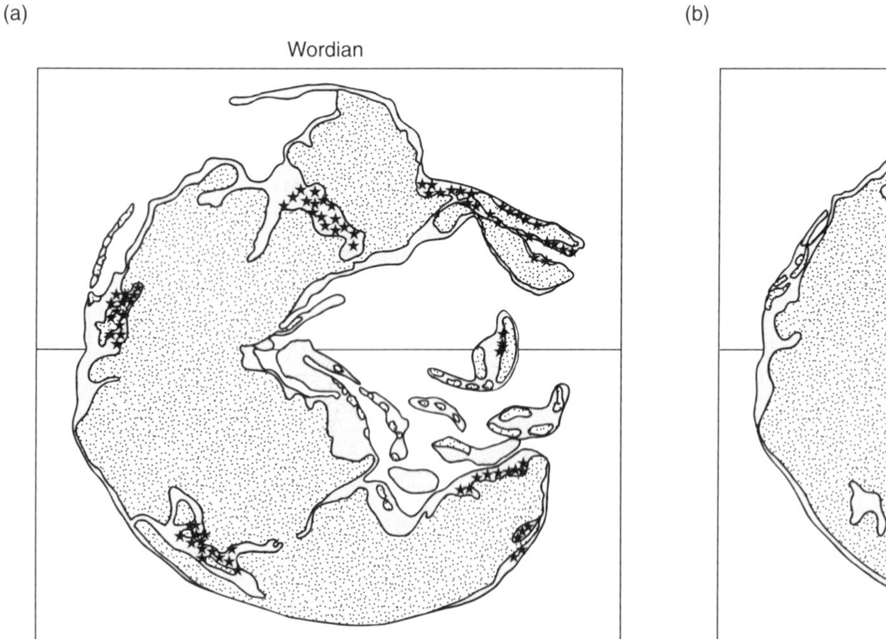


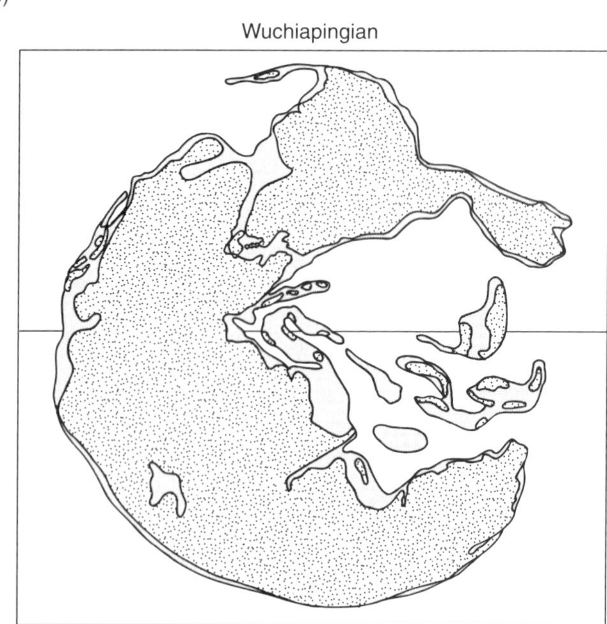


**FIGURE 10.12 Paleogeographic maps illustrating the loss of major brachiopod faunas [indicated with stars in part (a)] through the Late Permian.** Land areas are stippled, shallow-water areas are represented in grey, deeper water in white. In most cases, the loss of brachiopod faunas can be attributed to the loss of shallow-water areas that harbored them. Note that most of the starred regions in part (a) coincide with shallow-water areas that become emergent in part (b). *(From Shen & Shi, 2002)*

graphic variations in the average compositions of the skeletons of microorganisms. Because taxa differ in their propensities to fractionate oxygen, it is standard practice in these analyses to assess patterns separately for several species. If all or most of the species exhibit similar stratigraphic variations in $\delta^{18}O$, this demonstrates that the changes transcend the peculiarities of particular taxa.

In many cases, long- and short-term transitions in $\delta^{18}O$ can be recognized in correlated horizons around the world, indicating that they are diagnostic of global shifts, which, in turn, have been linked to major transitions in marine and terrestrial biotas recognized throughout the Cenozoic. James Zachos and colleagues (2001) have compiled a global record of deep-sea $\delta^{18}O$ for the Late Cretaceous through the Cenozoic. This provides evidence of major episodes of warming and cooling during the first half of the Cenozoic, as indicated by positive (cooling) and negative (warming) excursions in $\delta^{18}O$ [SEE SECTION 9.5]. Particularly pronounced among these episodes is a warming interval from the Paleocene into the Early Eocene, followed by a protracted interval of cooling through the rest of the Eocene.

Superimposed on the long-term $\delta^{18}O$ trend are a series of short-term spikes, the most prominent of which is a sharp, negative excursion near the Paleocene–Eocene Boundary, some 55 million years ago (Figure 10.13a). This excursion has now been recognized globally, in the tropics as well as at high latitudes, and has been documented for many species of foraminifera. Based on the size of the $\delta^{18}O$ excursion, Zachos and colleagues have estimated that the size of the sea-surface temperature increase was 8° to 10°C at high latitudes and 5° to 8°C in the tropics, although other studies have suggested that the difference in warming between low and high latitudes may not have been so great. Furthermore, the shape of the $\delta^{18}O$ excursion indicates that this remarkable increase, which is known as the *Paleocene–Eocene Thermal Maximum* (PETM), occurred in fewer than 10,000 years. The subsequent "recovery" from the PETM was somewhat more protracted, taking place over about 200,000 years, although the entire Early Eocene was apparently characterized by the warmest temperatures of the Cenozoic.

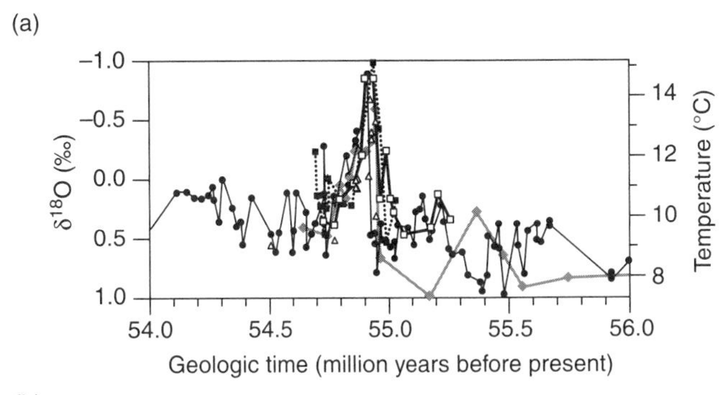


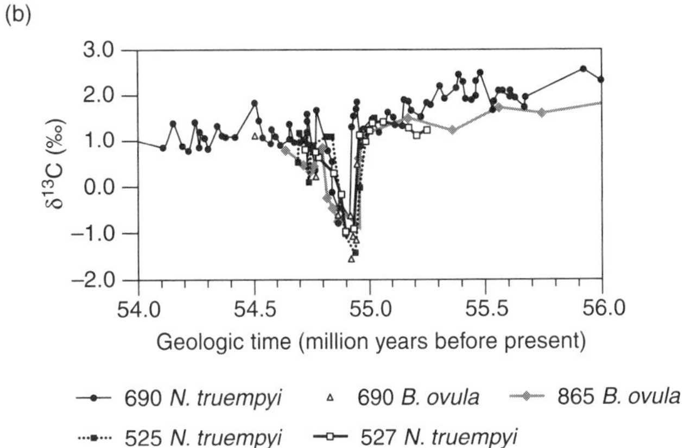


**FIGURE 10.13 Significant excursions in $\delta^{18}O$ and $\delta^{13}C$ near the Paleocene–Eocene boundary, as recorded by multiple foraminiferal species preserved in a series of deep-sea cores collected from sites in the southern Atlantic (525, 527, and 690) and western Pacific (865) oceans.** (a) $\delta^{18}O$. The temperature scale on the right is based on the establishment of an empirical relationship between the size of a $\delta^{18}O$ excursion and the temperature excursion associated with it, under the assumption (appropriate in this case) that the oceans were ice-free at the time. (b) $\delta^{13}C$, showing substantial decline. *(From Zachos et al., 2001)*

## Causes of the PETM

There is evidence suggesting that the PETM was associated with a catastrophic release of methane ($CH_4$) that had been sequestered previously in $CH_4$ hydrates, crystalline solids that effectively trap $CH_4$ in ice. In the present day, $CH_4$ hydrates are found in Arctic permafrost regions as well as in cold, marine settings, and the amount of $CH_4$ sequestered globally in this form is sufficiently extensive that it has even attracted the attention of researchers who view it as a major, new societal source of energy. Importantly, the $CH_4$ sequestered in $CH_4$ hydrates is known to be enriched in $^{12}C$, which, as we noted earlier in this chapter, is indicative of a biogenic origin. This is relevant to our consideration of the PETM because, in addition to the negative $\delta^{18}O$ excursion, the PETM is also characterized by a major negative excursion in $\delta^{13}C$—that is, an enrichment in $^{12}C$ (Figure 10.13b).

Of course, this does not *prove* that the release of $CH_4$ hydrates caused the $\delta^{13}C$ excursion or the PETM.

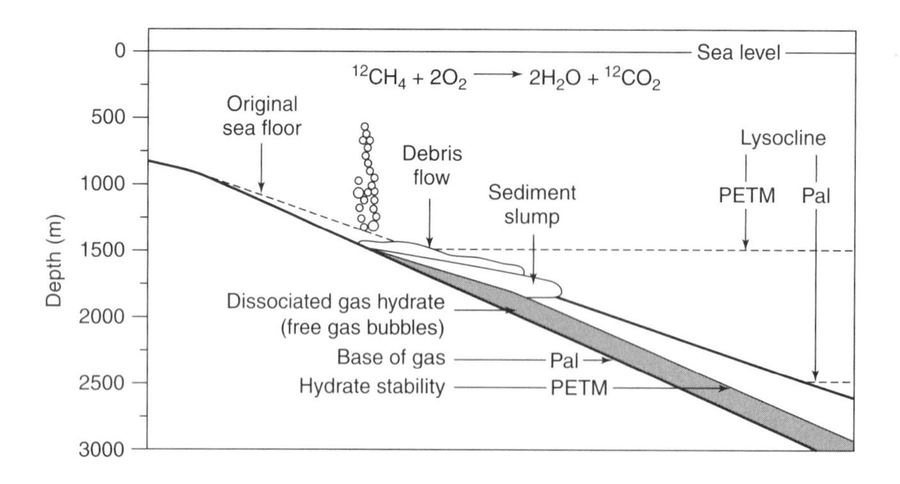


**FIGURE 10.14 Schematic representation of a proposed model for the catastrophic release of methane ($CH_4$) as the cause of the Paleocene–Eocene Thermal Maximum (PETM).** In the scenario illustrated, ongoing warming during the Late Paleocene (Pal) caused the release of methane gas bubbles that had previously been frozen. This induced major slumping that, in turn, caused a far more substantial release of methane. This is hypothesized to have caused a significant elevation of carbon dioxide ($CO_2$) levels in the atmosphere, inducing significant global warming and a shallowing of the lysocline. *(From Katz et al., 1999)*

However, to date, $CH_4$ hydrates are among the only plausible sources known for the large amount of $^{12}C$ required to account for an excursion of the size observed. Moreover, a viable model has been proposed by Gerald Dickens and colleagues (1995) to account for this scenario and its relationship to the PETM (Figure 10.14).

The following sequence of events is envisioned: (1) Some $CH_4$ that was sequestered in hydrates on the continental slope was released as gas bubbles when subjected to the moderately increased temperatures associated with the long-term warming trend that preceded the PETM; (2) this release, in turn, caused a collapse in the pore space between sedimentary grains on the sea floor, inducing catastrophic sediment slumping and the associated release of a much larger quantity of $CH_4$ at water depths ranging from about 900 to 2000 m, the zone in which the estimated temperature increase would have promoted $CH_4$ release (i.e., the dissociation of the hydrates); (3) mediated by bacteria, the released $CH_4$ reacted with oxygen, which produced significant quantities of carbon dioxide ($CO_2$; see the equation for this reaction in Figure 10.14).

The major infusion of $CO_2$ into the global system should have affected marine and terrestrial settings alike. In marine settings, the addition of $CO_2$ likely promoted the dissolution of calcium carbonate, manifested in part as a shallowing of the lysocline (the depth in the oceans beneath which carbonate dissolution greatly increases). This should have had a deleterious effect on calcium-carbonate–secreting organisms in these settings. Moreover, a major influx of $CO_2$ into the atmosphere would have induced an episode of global warming that significantly amplified the ongoing slow warming trend already underway. Given that $CH_4$ is, if anything, a more potent greenhouse gas than $CO_2$, global warming would have ensued even if the $CH_4$ had not been converted to $CO_2$. But under the hypothesized conditions, it seems likely that conversion to $CO_2$ indeed occurred.

Ideally, any test of this scenario should include physical and biological evidence of the postulated sequence of events. Such evidence was provided by Miriam Katz and colleagues (1999) based on data from a deep-sea site near the Bahamas, where a core was collected that includes the critical interval. Analysis of this core diagnosed a major decline in the number of benthic foraminiferal taxa from about 30 to fewer than 10, which coincides stratigraphically with negative excursions in $\delta^{18}O$ and $\delta^{13}C$. This lends credence to the suggestion that a combination of warming and enhanced dissolution of calcium carbonate caused the loss of foraminiferal taxa observed at this site, as well as elsewhere around the world. Furthermore, a layer of mud clasts was observed immediately below the $\delta^{13}C$ excursion, indicating a major disturbance just prior to the onset of the excursion. This offers tantalizing evidence for a perturbation like that predicted by the catastrophic release of $CH_4$.

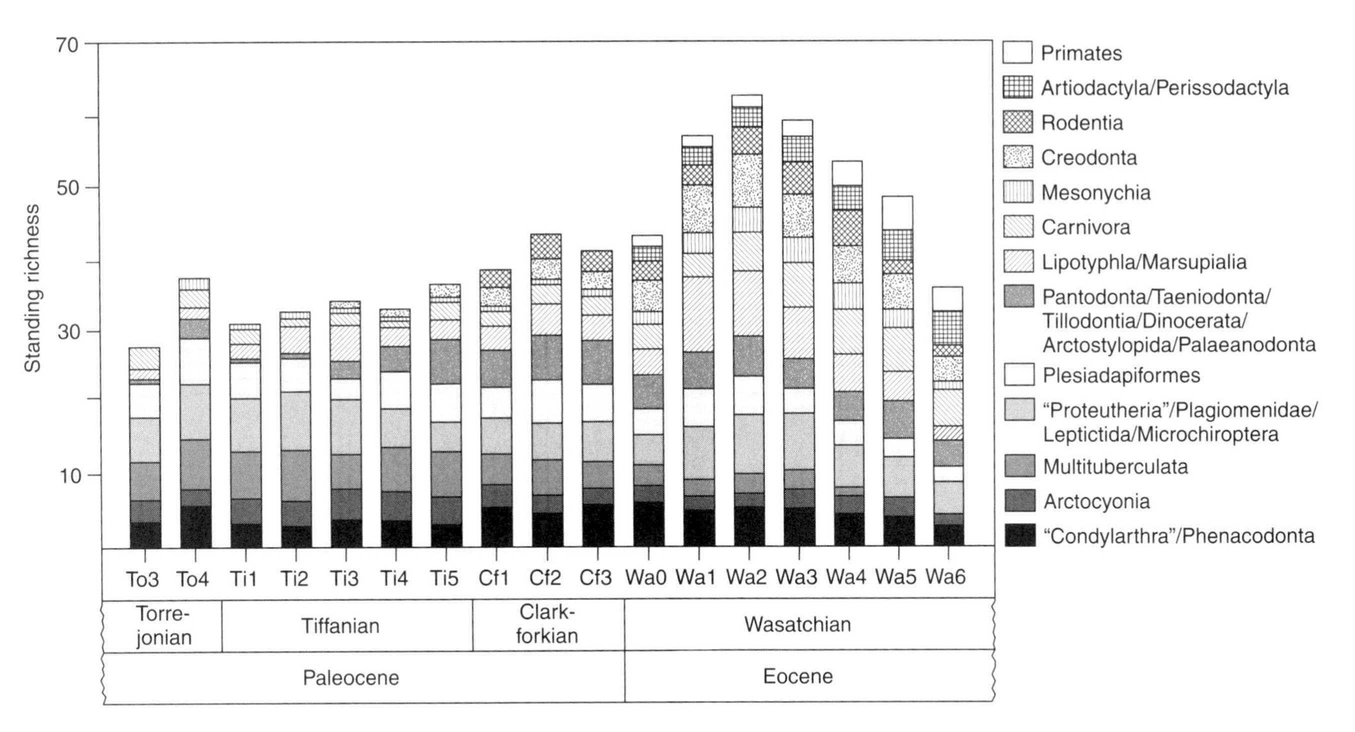


**FIGURE 10.15 Transitions in the genus richness of major mammalian orders across the Paleocene–Eocene boundary in Wyoming and Montana, based on assessments in 17 separate faunal zones.** Note that, as the boundary is crossed, primates, artiodactyls, and perissodactyls appear for the first time. *(From Maas et al., 1995)*

**Seismic profiles** collected through the region also show chaotic reflections at the horizon in question, indicating a set of disturbed sedimentary layers. Finally, analyses of $\delta^{13}C$ in Late Paleocene–Early Eocene terrestrial samples, such as the tooth enamel of herbivorous mammals and carbonates contained in ancient soil-horizons (Koch et al., 1992), demonstrate that geochemical changes associated with the PETM occurred in terrestrial and marine settings alike.

## Biological Effects

As suggested by the foraminiferal decline in the deep-sea core from the Bahamas, the PETM was marked by a major extinction of foraminifera. The biotic effects of the PETM were not limited to marine settings, however. A growing number of studies of terrestrial vertebrate and plant biotas have yielded evidence of profound biogeographic and paleoecological transitions that mark the interval, some of which reverberate today.

Among terrestrial vertebrates, analyses of the fossil mammalian record in Wyoming and Montana by Mary Maas and colleagues (1995) illustrate well that the PETM marks the first appearances of several modern mammalian orders (Figure 10.15) that would come to dominate terrestrial ecosystems thereafter, including primates and hoofed mammals (artiodactyls and perissodactyls). While the Wyoming and Montana occurrences likely reflect the first appearances of these taxa in North America, recent studies suggest that they originated elsewhere—quite possibly in Asia. Gabriel Bowen and colleagues (2002) analyzed carbon isotopic records from localities in Hunan Province, China, that also contain fossil primates, artiodactyls, and perissodactyls. They were able to recognize the negative $\delta^{13}C$ excursion that marks the PETM.

Strikingly, the first appearances of perissodactyls can be tied to strata that demonstrably precede the PETM, indicating that they occurred in Asia earlier than in North America. The first appearances in China of artiodactyls and primates appear to coincide with the PETM. This suggests that these taxa occurred at least as early in Asia as in North America, leaving open the possibility that, like perissodactyls, they originated in Asia and subsequently migrated to North America. In any case, it seems clear that the PETM had a lasting effect on the diversification and distribution of terrestrial mammals.

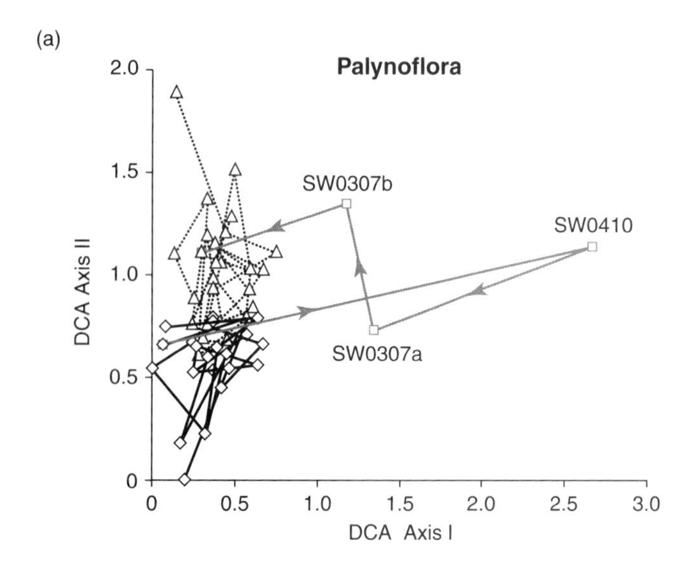


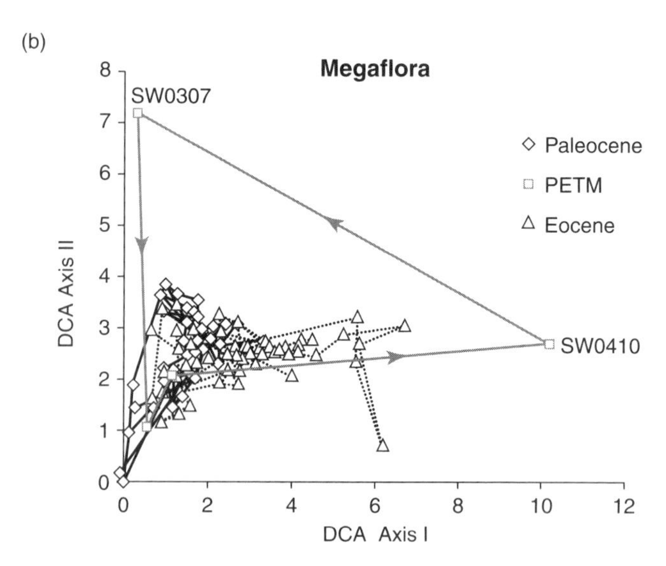


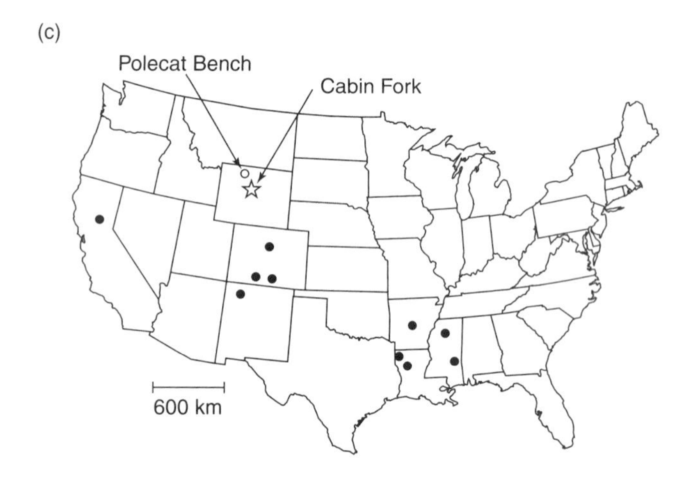


**FIGURE 10.16 Paleobotanical transitions associated with the PETM in northern Wyoming.** (a, b) Detrended correspondence analyses (DCA) of the palynoflora (e.g., pollen) and the megaflora (e.g., leaves and stems) contained in samples collected from below, within, and above the interval containing the PETM. Positions of samples with respect to DCA axes reflect differences in paleofloral composition. Labeled samples are for intervals contained within the PETM. Lines are drawn to connect samples in stratigraphic sequence, with arrows indicating the upward direction. Note that the compositions of samples within the PETM interval are decidedly different from those in intervals below and above the interval. (c) A map of the United States illustrating sites to the south and west of Wyoming containing Paleocene and Eocene floral elements that are restricted to the PETM interval at the two Wyoming localities labeled on the map. *(From Wing et al., 2005)*

That the PETM had a significant, but perhaps more ephemeral, impact on terrestrial plant biotas is suggested by an analysis of paleofloras that bracket the PETM in north-central Wyoming. The stratigraphic positions of samples collected throughout this interval were tied by Scott Wing and colleagues (2005) to strata located beneath, within, and above the PETM interval. Stratigraphic correlations were based on analyses of $\delta^{13}C$ (including the sharp negative excursion at the base of the PETM) and diagnostic elements of the mammalian fauna.

Wing and colleagues conducted separate analyses of the megaflora (macroscopic material, mainly leaves and stems) and palynoflora (microscopic material, mainly pollen). Detrended correspondence analyses [SEE SECTION 6.3] of samples collected throughout the interval revealed similar patterns for both sets of data (Figures 10.16a and 10.16b): Collections from within the PETM had compositions that were quite unusual when compared with those from below and above the PETM. The compositional shift was caused by the inclusion in these samples of non-native taxa that were known to have occurred earlier in more southerly regions (Figure 10.16c) and therefore represent a migration that was almost certainly related to the warming event [SEE SECTION 9.6]. As the PETM came to a close, floras reverted back to compositions that were similar, but generally not identical, to their pre-PETM counterparts.

Wing and colleagues also conducted leaf margin analysis (LMA) and leaf area analysis (LAA) [SEE SECTION 9.5] of fossil leaf specimens collected throughout the interval, through which they diagnosed a temperature increase of about 5°C, as well as a 40 percent decline in mean annual precipitation with the onset of the PETM. Near the end of the PETM, precipitation levels rebounded to normal values. The temperature change diagnosed with LMA in this terrestrial setting is comparable to that recognized for tropical sea-surface temperatures based on $\delta^{18}O$ measurements derived from foraminifera in deep-sea cores.

Furthermore, the PETM appears to have had a significant effect on ecological interactions among taxa, as demonstrated by an analysis of insect damage on fossil leaves [SEE SECTION 9.4] conducted by Peter Wilf and Conrad Labandeira (1999). In the present day, there is a marked increase in the rate of leaf predation toward the tropics, and Wilf and Labandeira hypothesized that, if this pattern relates to the warmer temperatures of the tropics (which is not a certainty), then we might expect to observe an increase in predation-related leaf damage during the PETM. To test this hypothesis, they compared the incidence of insect-produced leaf damage with fossil specimens from the Upper Eocene and Lower Paleocene collected from southwestern Wyoming.

The results demonstrate a significant increase in leaf predation by insects during the PETM (Figure 10.17). While the rate of predation increased substantially on leaves of Betulaceae (birch trees, a group whose leaves are known to be particularly palatable to insects in the present day), other plant groups were affected as well. In addition, Wilf and Labandeira observed an increase in the diversity of damage types exhibited by Eocene leaves.

Therefore, it is clear that the PETM profoundly affected terrestrial and marine life in ways that may be particularly instructive with respect to what we should expect in the face of present-day global warming. To be sure, there is still a long way to go in understanding the fabric of evolutionary transitions associated with the PETM (e.g., the development of a phylogenetic framework for the mammalian transition described earlier). Nonetheless, the range of ongoing investigations has already shown that a coupling of paleontological and geochemical data holds great promise for understanding the relationship between global climatic fluctuations and biotic transitions, and also for developing high-resolution correlations between marine transitions and their counterparts in the terrestrial realm.

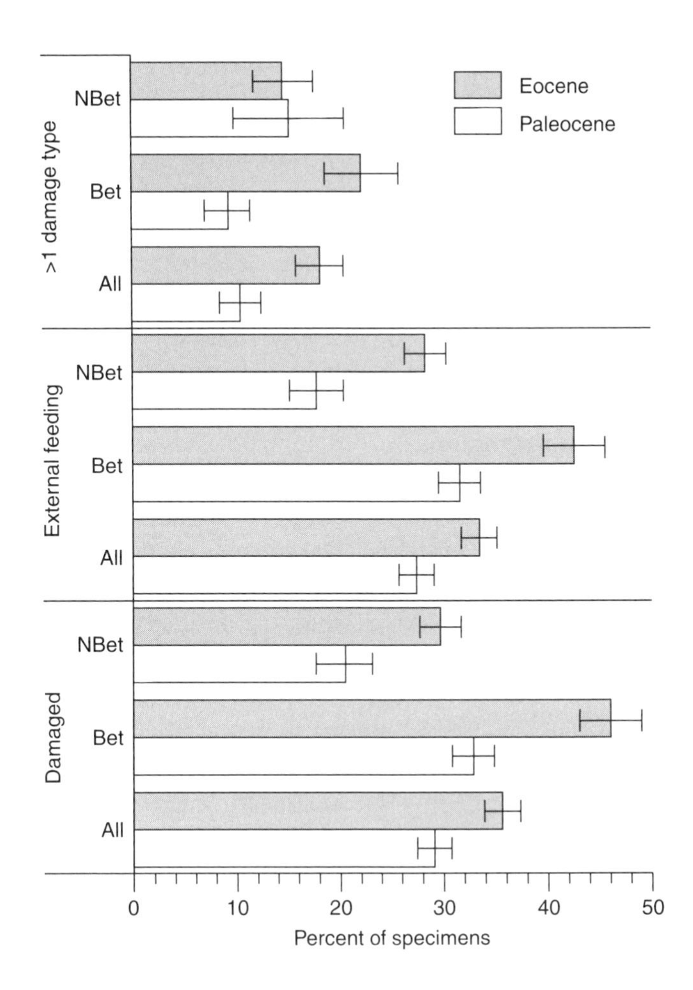


**FIGURE 10.17 Changes across the Paleocene–Eocene boundary in the percentages of fossil leaf specimens containing damage by insects, as indicated by an analysis conducted in southwestern Wyoming.** Bet and NBet are abbreviations for Betulaceae and non-Betulaceae. Predation is more pronounced in the Eocene, regardless of whether the analysis is restricted to the particularly ubiquitous and palatable Betulaceae (birches). The middle set of bars illustrates that much of the increase can be attributed to an increase in external feeding, as opposed to internal mining and other means of leaf damage. The upper set of bars illustrates that there is an increase in the percentage of leaves exhibiting more than one kind of damage. *(From Wilf & Labandeira, 1999)*

## 10.5 PLEISTOCENE MEGAFAUNAL EXTINCTIONS

The case of the Late Permian extinctions illustrates the importance of temporal resolution in documenting and understanding major evolutionary events. In that example, so remote in time, resolving a last appearance to the nearest 100,000 years and determining that the extinctions were concentrated in an interval of less than 1 million years are significant accomplishments. With respect to the most recent extinction event in the fossil record, that of large terrestrial vertebrates in the Late Pleistocene, it has often been possible to resolve the geologic ages of individual specimens and the disappearances of species to within 1000 years. This extreme precision has been essential in inferring the causes of extinction.

## The Nature of Late Pleistocene Extinctions

About 2.5 million years ago, the earth's climate entered its most recent glacial phase. We are now in an interglacial cycle but still overall in a glacial age. Toward the end of the last fully glacial interval—the end of the Pleistocene Epoch, about 10,000 years ago—a number of terrestrial mammals and other vertebrates became extinct in North America. What is most striking about these extinctions is that they are extremely selective. The so-called **megafauna**—conventionally said to be those animals with estimated body masses greater than about 40 kg—suffered significant extinction. In North America, for example, only about 14 of 50 megafaunal genera that were extant in the area north of Mexico during the Late Pleistocene survived into the Holocene (Figure 10.18). This translates to greater than 70 percent genus extinction—a level that would rival the most severe mass extinctions in the Phanerozoic, except that the Pleistocene event did not affect most groups of organisms.

Well-known victims of this extinction included mammoths, mastodons, ground sloths, giant beavers, and sabertooth cats. By contrast, small terrestrial mammals (Figure 10.18) and marine organisms were little affected at this time. Moreover, a comparable extinction has been documented elsewhere, including South America, and there is a substantial record of extinction of mammals and birds on oceanic islands. In Australia, there was a megafaunal extinction some 46,000 years ago.

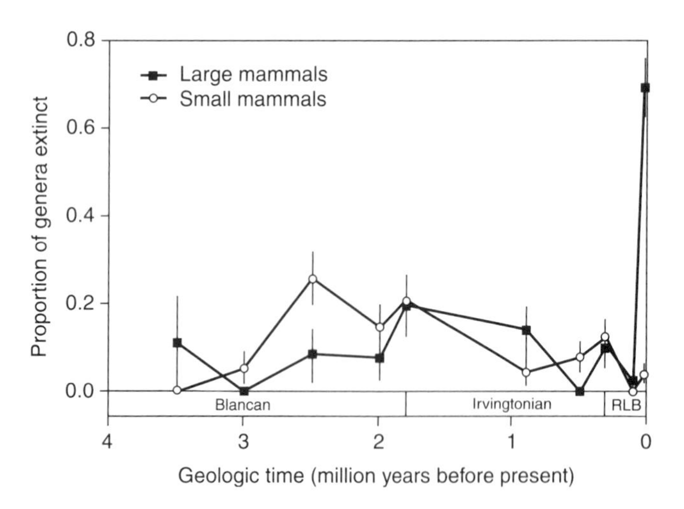


**FIGURE 10.18 Proportional extinction of North American mammal genera.** Stratigraphic intervals are North American land mammal ages; RLB denotes the Rancholabrean Age. The 4-million-year interval shown here is roughly equivalent to the post-Miocene. Error bars show 1 standard error on either side of extinction proportion (see Table 7.4). Curve with solid symbols, large mammals (44 kg or more); other curve, small mammals. Large mammals experienced much more extinction than small mammals at the end of the Rancholabrean, around 10,000 years ago. *(Data from Martin & Steadman, 1999)*

## The Role of Humans in Megafaunal Extinction

The two leading candidates to explain these extinctions are climate change and human influence, especially through overhunting. Changes in climate accompanying deglaciation would have altered vegetation on which herbivorous mammals feed, and could therefore have had broad effects on food chains. Although there is abundant evidence for climate change—for example, an increase in seasonality—it still leaves two questions: Why were large terrestrial vertebrates the principal victims, and why did numerous other climatic changes during the glacial epoch not result in extinctions that were similar in severity and selectivity?

That humans played a role in the Late Pleistocene megafaunal extinctions now seems undeniable to many researchers. Here we summarize some of the major evidence for this conclusion. One of the challenges in weighing the climatic and hunting hypotheses has been to generate testable predictions that are consistent with one but not the other. Later we will discuss one such prediction and the evidence that bears on it, but it is important to keep in mind that climatic and human influences are not mutually exclusive.

There have been four principal lines of evidence for the so-called **overkill hypothesis.** First, variation in the timing of extinction among continents is such that the extinctions generally follow the presence of humans in significant numbers. This is consistent with the hunting scenario but does not rule out climate change as a mechanism. Humans would presumably migrate in response to climate change, and the timing could, in principle, be coincidental.

Second, it is clear that prehistoric humans in North America and elsewhere hunted game and processed the meat. Some of the most compelling evidence comes from a series of detailed taphonomic, morphologic, and geochemical studies of proboscideans, especially mastodons and mammoths, carried out by Daniel Fisher and his co-workers. Figure 10.19 illustrates a typical example of an in-place assemblage of mastodon bones, which was unearthed during the digging of a farm pond in southeastern Michigan. One of the most striking

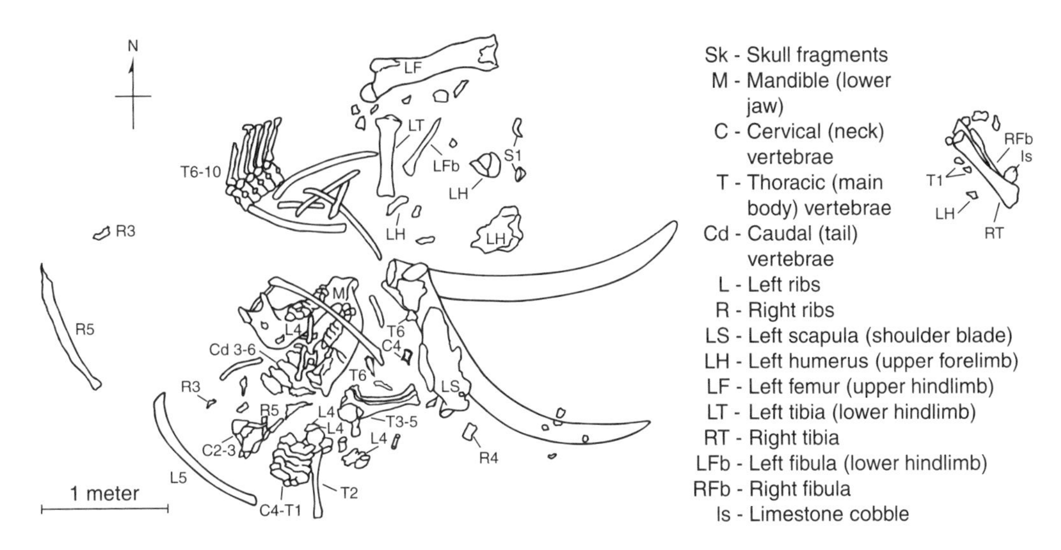


**FIGURE 10.19 Assemblage of mastodon bones from southeastern Michigan.** Groups of articulated bones are found together but are separate from other such groups, indicating butchery of the carcass. Letters denote different kinds of bones, and numbers indicate different elements in a series (such as ribs and vertebrae). The major elements represented here are shown in the key. The "ls" at far right denotes a limestone cobble. *(From Fisher, 1984)*

features of this assemblage is that groups of articulated bones that constitute whole sections of the mastodon are found together (e.g., the thoracic vertebrae T6–T10 in Figure 10.19) and that these large sections of the animal are separated by scattered bone fragments in no discernible order.

Based on observations of modern elephant carcasses, disarticulation of the joints tends to occur before tissues such as skin have broken down. For the bones of the mastodon to be displaced naturally would require the skin and other soft tissues to be degraded, but by the time this happened the bones would already have disarticulated from one another. In other words, under normal taphonomic conditions, it is not likely that groups of bones will remain articulated with one another and be displaced relative to other such groups. Thus, some other process is required. In this example, the presence of tool marks on facing surfaces of a number of joints indicates that certain joints were pried or wedged apart. This combination of observations makes sense only if the mastodon was butchered. Moreover, some bones show signs of burning—but only at limited points on the surface, indicating that most of the bone was still covered in flesh when it was exposed to fire. Thus, parts of the mastodon were also barbecued.

Many similar mastodon sites have been excavated, indicating that game processing was common. There is also taphonomic evidence that large sections of meat were deliberately submerged in ponds, sometimes with associated "clastic anchors" (sections of mastodon intestine filled with sediment), in order to preserve them. Actualistic experiments show that meat from a large mammal can remain edible—as one of the authors of this book can attest first hand—even after being submerged for several months.

Clearly, mastodons were butchered, but were the animals actively hunted, or perhaps only scavenged? The argument for hunting in the case of mastodons relies partly on reconstructing the season of death of individual animals. Figure 10.20 schematically shows growth

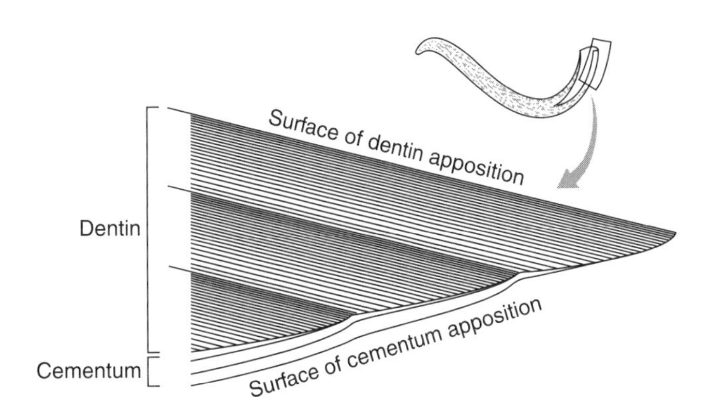


**FIGURE 10.20 Pattern of tusk growth typical of elephants, mastodons, and mammoths.** The upper drawing shows a longitudinally sectioned tusk. Material is accreted along the conical pulp cavity; thus, the older part of the tusk is at the tip (to the left). The lower drawing is an enlargement of a small part of the sectioned tusk. The numerous growth increments of dentin can be divided into three sets, interpreted to be annual. The finer-scale laminae are periodic, forming at roughly two-week intervals. *(From Fisher, 1996)*

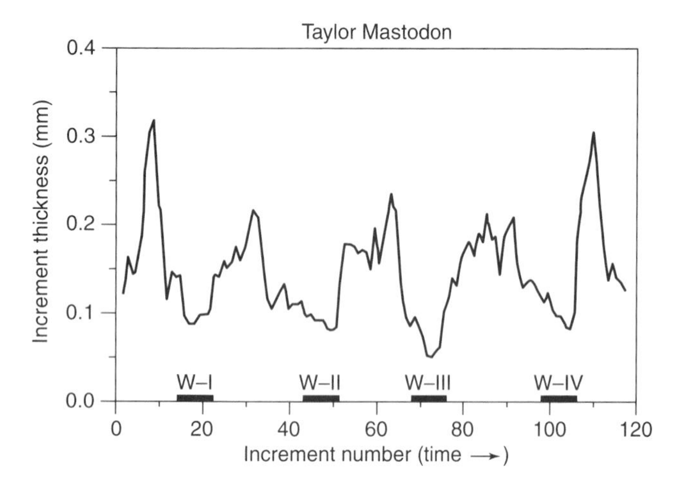


**FIGURE 10.21 Increment thicknesses (see Figure 10.20) in the tusk of a mastodon.** Thin layers marked W-I through W-IV correspond to winter growth. *(From Koch et al., 1989)*

bands in a mastodon tusk. By analogy with many living mammals, including elephants, thin, dark bands indicate the slow growth associated with winter. Therefore, the profile of increment thickness in Figure 10.21 is interpreted to show roughly four years' growth, with winters corresponding to thin layers. This inference can be tested with isotopic analysis of the growth increments [SEE SECTIONS 2.3 AND 9.5].

We have already seen that the ratio of $^{18}$O to $^{16}$O in skeletons of marine organisms is strongly affected by the ambient water temperature, with higher values (i.e., positive excursions in $^{18}$O) in colder water. By contrast, the ratio of $^{18}$O to $^{16}$O deposited in the apatite of tusks as they grow depends largely on the oxygen isotope ratio of the water the animal ingests; according to the general trend of precipitation in continental regions, this ratio is lower, rather than higher, during colder seasons. Figure 10.22 shows the oxygen isotopic profile of the tusk from Figure 10.21. Negative excursions in $\delta^{18}$O correspond with the thinnest growth increments, supporting the inference that thin increments were laid down in the winter.

Growth in the specimen detailed in Figures 10.21 and 10.22 ends between the summer peak and the winter trough in growth rate and in $\delta^{18}$O; it therefore appears to have died in autumn. This in itself says nothing about whether it was hunted or died a natural death, but season of death of a broader sample of animals may indicate whether the animals were hunted. The season of death of butchered animals, such as that in Figure 10.19, can be compared with that of animals that show no sign of butchery (and therefore probably died of natural causes). If the distribution of season of death of butchered animals does not match that of animals that died natural deaths, then the butchered animals are unlikely to represent natural deaths. In other words, they are likely to have been hunted rather than scavenged. In one study, Daniel Fisher and Paul Koch (1989) determined the season of death of six butchered and seven nonbutchered mastodons. The nonbutchered individuals died in the late winter and early spring, whereas the butchered individuals all died in mid- to late autumn. Thus, the butchered mastodons were unlikely to have died by natural causes and so were most likely hunted.

Analysis of the season of death of North American mammoths, estimated with oxygen isotopes as for the mastodon in Figure 10.22, also shows a pattern consistent with hunting. Prior to significant human presence, around 12,500 years ago, most of the deaths were in late winter and early spring. After that time, once evidence for human activity is clear, about half the deaths were in mid- to late autumn.

Evidence for hunting of other species is also found in the form of stone tools, including Clovis projectile points, associated with fossil skeletons. At least one report

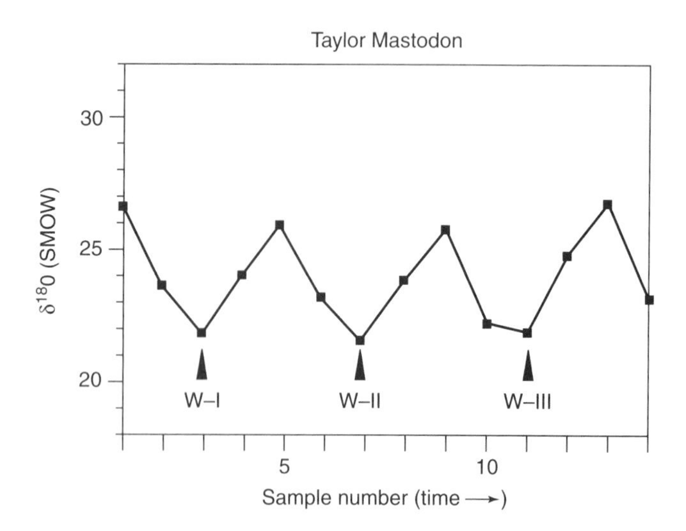


**FIGURE 10.22 Oxygen isotope ratios in selected growth increments of the tusk depicted in Figure 10.21.** Lighter oxygen corresponds to winter growth (W-I through W-III). Oxygen isotopes are compared with standard mean ocean water (SMOW). *(From Koch et al., 1989)*

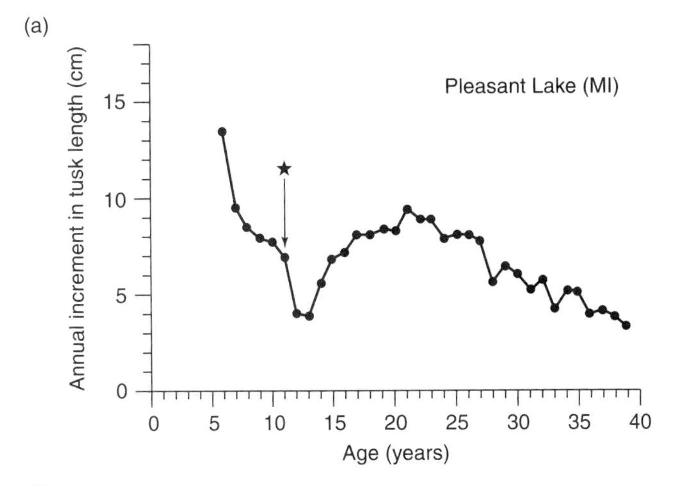


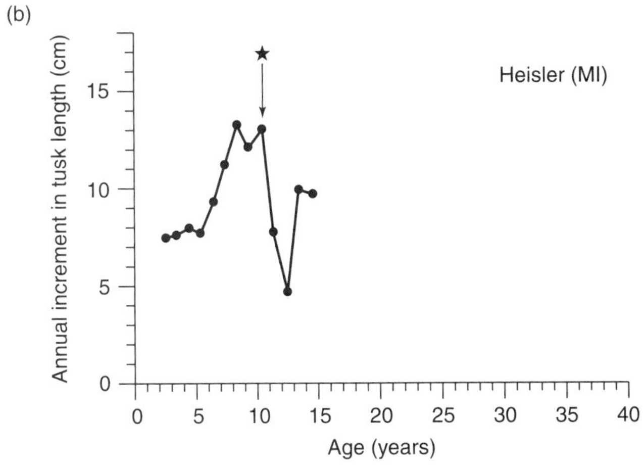


**FIGURE 10.23 Thickness of annual growth increments in the tusks of two male mastodons from two localities in Michigan, (a) Pleasant Lake and (b) Heisler.** By analogy with living elephants, sharp decline in growth rate, indicated by the asterisk, marks the onset of sexual maturity. *(From Fisher, 1996)*

indicates that a mammoth had points embedded in it but was not butchered and was not associated with other individuals that had been butchered. Evidently, then, this mammoth was attacked but subsequently escaped.

Thus, it is established that prehistoric humans hunted—but does this mean that hunting caused any of the megafaunal extinctions? The third, and most compelling, line of evidence that hunting was implicated in the extinctions comes in the form of a physiological response of prey species after the arrival of humans and before the time of the extinctions. Figure 10.23 shows the annual increments in tusk length for a couple of male mastodons from North America. At a certain point, each individual shows a drop in the increment size—in other words, a decline in growth rate—after which it takes a few years to return to the earlier growth rate. By analogy with living elephants, this is thought to mark the onset of sexual maturity in males, a time of great stress when they are evicted from the family group and must fend for themselves. Thus, the sharp drop in growth rate is used to determine the age of maturity of fossil males. Maturity in females is determined by the beginning of a 3-to-4-year cycle in growth rates that corresponds to the calving interval. In the two specimens of Figure 10.23, the decrease in growth rate occurs at an age of about 10 years.

The ability to determine growth rates and the age of sexual maturity enables a test of the hunting hypothesis against the climatic hypothesis, because the two hypotheses make different predictions about the probable effects on a stressed species and its response over time. The deleterious effects of climate change are supposed to be mediated through a change in vegetation consumed by many large mammals. If preferred food sources were less abundant, growth rates should have declined, and it is also likely that sexual maturation would have been delayed in response to poorer nutrition. If food sources had not deteriorated, but populations were instead stressed by hunting, then growth rates should not have declined. In fact, if populations were thinned by hunting, reduced competition within species may even have allowed faster growth.

It is also known from modern elephants that young males, whose maturation is inhibited by older males in musth ("rutting"), mature earlier if older males are removed from the population. Moreover, from an evolutionary standpoint, it would have been advantageous to mature at a younger age, to reproduce before being hunted. Figure 10.24 shows that mammoth growth rates just prior to the extinction were equal to or even a bit greater than the prehunting values. Figure 10.25 shows that the age of sexual maturation in mastodons declined after the onset of hunting. These observations fit the predictions of the overkill hypothesis rather than the climate hypothesis.

Note that the geological ages of specimens in Figures 10.24 and 10.25 are given in "years before present." Strictly speaking, these figures are referred to as *radiocarbon years* rather than *calendar years*. The relationship between radiocarbon years and calendar years is complicated, depending on such factors as the amount of radioactive $^{14}C$ in the atmosphere, which varies over time, and the ratio of $^{13}C$ to $^{12}C$ in the organism. The

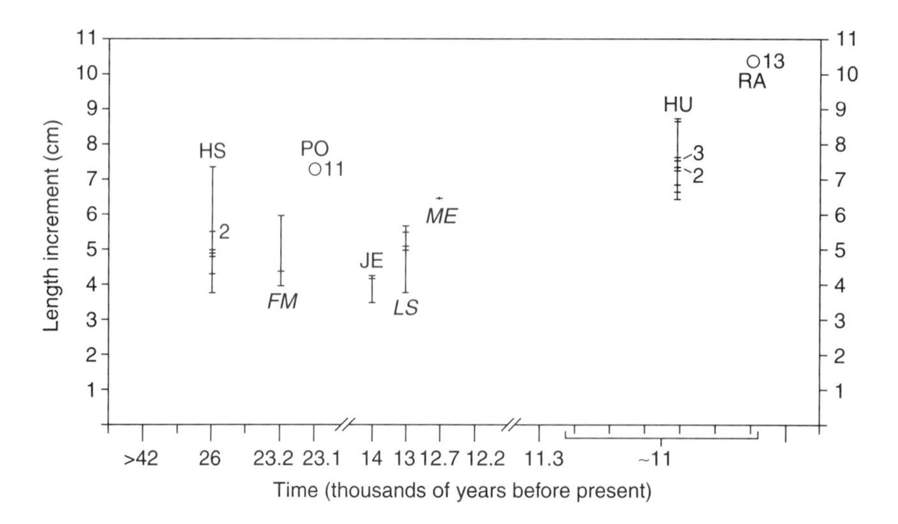


**FIGURE 10.24 Growth rates (measured as annual increments in tusk length) in North American mammoth specimens compared with geologic age in thousands of years before present.** Each tick mark shows the growth rate for a single year; the numbers next to certain tick marks indicate that multiple growth rates have the same value. Two-letter abbreviations are specimen codes; males are in plain font, females in oblique font. Circles show growth rates averaged over several years, with the number of years given to the right. Growth rates are somewhat higher just before extinction than they are prior to the onset of human hunting. *(From Fisher, 2002)*

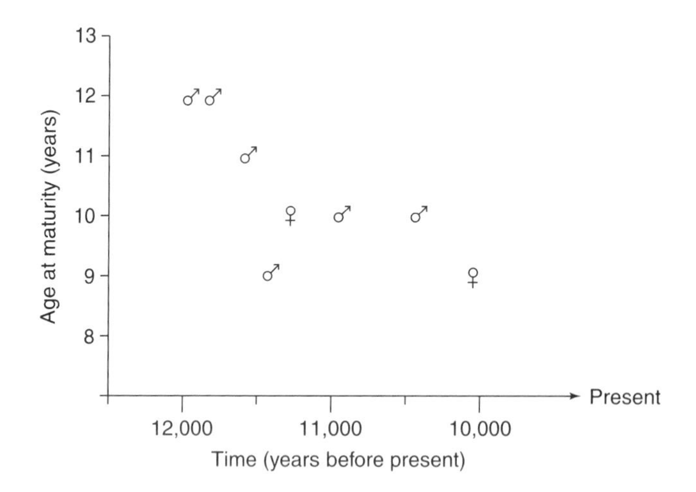


**FIGURE 10.25 Inferred age of sexual maturation in mastodons compared with geological age in thousands of years before present.** Maturity in males is inferred from a sharp decline in tusk growth rates (Figure 10.23). Maturity in females is inferred from the onset of a 3-to-4-year cycle in tusk growth rates that corresponds to the interval between successive offspring. Age of maturation declines after the onset of human hunting. *(Courtesy of Daniel C. Fisher)*

relative ordering of the geological ages in Figures 10.24 and 10.25 is approximately correct, however.

There is one last line of evidence in favor of the overkill hypothesis, which relates to the probable ecological response of various mammal species to human hunting. What is known of the ecology of different kinds of mammals allows a reasonable prediction of which species would have been most susceptible to population decline in response to hunting. With this in mind, John Alroy (2001) constructed a computer model of human hunting. This model has a few key elements: North America is seeded with a small number of humans of modest hunting ability; these humans harvest prey at random to satisfy their nutritional needs; and they gradually spread across the continent. Mammal species that serve as potential prey are characterized by factors such as geographic range and body mass, both of which can be estimated from fossil data. Body mass is especially important because it largely determines other factors that are incorporated into the model, such as birth and death rates and population density, and because the number of animals that must be killed to satisfy human nutritional needs depends on the amount of meat that can be obtained from an animal.

Simulations based on the ecological model break North America into 1°-by-1° cells. Within each cell in each year, the simulations determine the change in population size of each species (including humans) as the result of migration, birth, and death (including hunting). The computer model runs for 2500 virtual years, corresponding to the approximate time between the human colonization of North America in significant numbers and the time of the megafaunal extinctions. Over a wide range of model assumptions, the simulations predict with a fair degree of accuracy which species survive and which become extinct. In particular, they predict that large-bodied species will preferentially suffer extinction. *This happens even if humans do not hunt them selectively.* It is the small population size and low birth rate of large mammals that makes them most susceptible to extinction in response to hunting.

## 10.6 CONSERVATION PALEOBIOLOGY

Our discussion of the Pleistocene overkill hypothesis illustrates just one of the ways that humans may be affecting life on the earth. It has proven problematical to determine whether rates of human-induced extinction in the present day rival those observed in the fossil record for, say, the big five mass extinctions of the Phanerozoic [SEE SECTION 8.6]. In part, this is because taxa confined to limited areas in terrestrial settings, such as small islands, provide our most extensive data on the markedly elevated extinction rates of the present day. Terrestrial, endemic (localized) species would be more susceptible to extinction than, say, most marine species because, all else being equal, a marine species from an open oceanic setting is likely to be more widespread geographically—a property that would be expected to reduce its susceptibility to extinction [SEE SECTION 8.6]. Given the inherent bias of the fossil record toward marine species [SEE SECTION 1.4] and toward widespread species, it therefore stands to reason that ancient Phanerozoic extinction rates derived from the marine record, such as those reported in Chapter 8, may not be indicative of ancient extinction rates exhibited by endemic terrestrial species, the very group whose extinction rates are best understood today.

Nevertheless, it is clear that *Homo sapiens* has caused the demise of untold numbers of species over the past several centuries. Just as significantly, human agriculture, industry, and commerce have dramatically impacted ecosystems around the world. In helping to analyze the long-term variability of ecosystems in particular regions before and during intervals that appear to exhibit substantial influence by *H. sapiens*, paleontologists are providing a unique perspective on the significance of present-day environmental and ecosystem deterioration and, in particular, on the prospect that ecosystems and the species that comprise them can survive the current crisis. Because the themes addressed by this research are of broad societal concern and, in some cases, of practical concern to the people living in regions where the work is conducted, it has been suggested that this constitutes a new branch of paleontological investigation, aptly termed **conservation paleobiology.**

Ecosystems that we might otherwise view as having been relatively unsullied prior to some obvious, recent episode of deterioration can be shown, with the help of a longer-term perspective, to have already experienced much longer periods of human-induced decline. Consider, for example, the case of tropical coral reefs in the western Atlantic, including the Caribbean Sea. During the early 1980s, because of the spread of a waterborne pathogen, the nature of which is still not entirely understood, there was a sudden, region-wide mass mortality of the long-spined sea urchin *Diadema antillarum*, an extremely prolific grazer that played a central role in preventing macroalgae from overgrowing the surfaces of living corals. At about that time, the major framework-building corals on western Atlantic reefs, *Acropora palmata* (elk horn coral) and *A. cervicornis* (stag horn coral) also went into a period of sharp decline from which they have yet to emerge, and most reefs throughout the western Atlantic are currently nearly bereft of living acroporids.

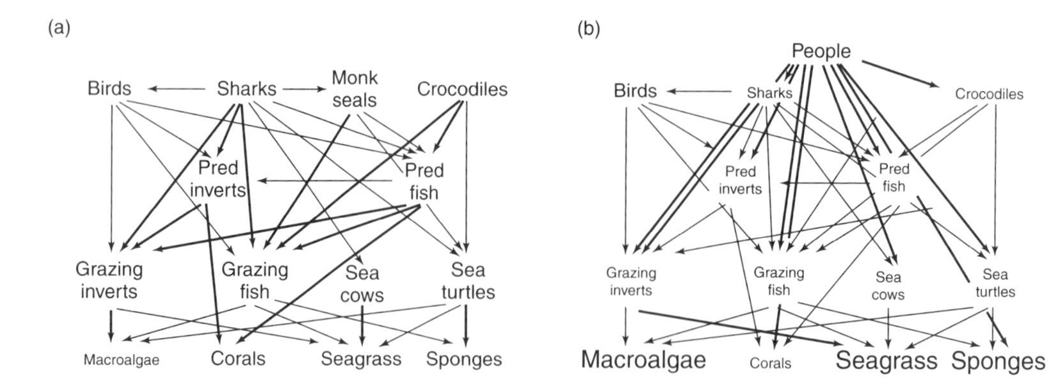


**FIGURE 10.26 Illustrations of the effect of overhunting by humans on the food web for Caribbean coral reefs.** (a) The food web prior to the onset of overhunting. (b) The food web after the onset of overhunting. Changes in abundances are indicated by change in font (larger fonts imply greater abundance). Note that monk seals were entirely eradicated. *(From Jackson et al., 2001)*

It is not clear that increased overgrowth by macroalgae and, in turn, the demise of *Diadema*, can be implicated as the primary cause of the decline of these corals; some researchers have argued for the importance of disease, perhaps associated with global warming, as an agent of coral decline (see below). But there can be no question that the loss of this major algal grazer has had a profound impact on reef ecosystems. What is not readily appreciated, however, is that, prior to the 1980s, these reefs had already suffered a significant loss that enhanced their vulnerability to algal overgrowth once *Diadema* declined. Historical records suggest that a host of large fish and other marine vertebrates, some of which were important grazers of macroalgae (Figure 10.26), had already been hunted to the point of near-disappearance long before the decline of *Diadema*. Thus, these reefs were already far removed from their natural states prior to the demise of *Diadema*.

In this case, the evidence for the demise of large marine vertebrates comes primarily from historical records rather than from paleontological data. Nevertheless, paleontological research has been pivotal in offering a deeper temporal perspective on the nature of Late Cenozoic reefs throughout the western Atlantic and Caribbean. An important finding that has emerged from this work is that acroporids have dominated Caribbean reefs since the Late Pleistocene, and that their rapid decline over such a broad region is without precedent throughout at least the past several thousand years. Richard Aronson and colleagues (2002) have evaluated a transition that took place on the reefs of Belize in the late 1980s and early 1990s from *Acropora cervicornis* to *Agaricia tenuifolia*, a coral with a leafy appearance reflected by its common name, the "lettuce coral."

The reign of *Agaricia tenuifolia* was itself temporary because this species experienced a major decline in Belize during 1998. Nevertheless, by collecting a series of cores from the sea floor (Figure 10.27) throughout a nearly 400-km$^2$ area off the coast of Belize, some of which penetrated records that were determined with radiometric techniques to be as old as 3000 years, Aronson and colleagues found that an extensive, regional transi-

**FIGURE 10.27 The contents of a core sample collected in Belize, illustrating the transition (shown right to left) from *Acropora cervicornis* to *Agaricia tenuifolia* in the late 1980s and early 1990s.** *Agaricia tenuifolia* is distinctively platelike in appearance and contrasts strongly with the sticklike appearance of *Acropora cervicornis*. *(From Aronson et al., 2002)*

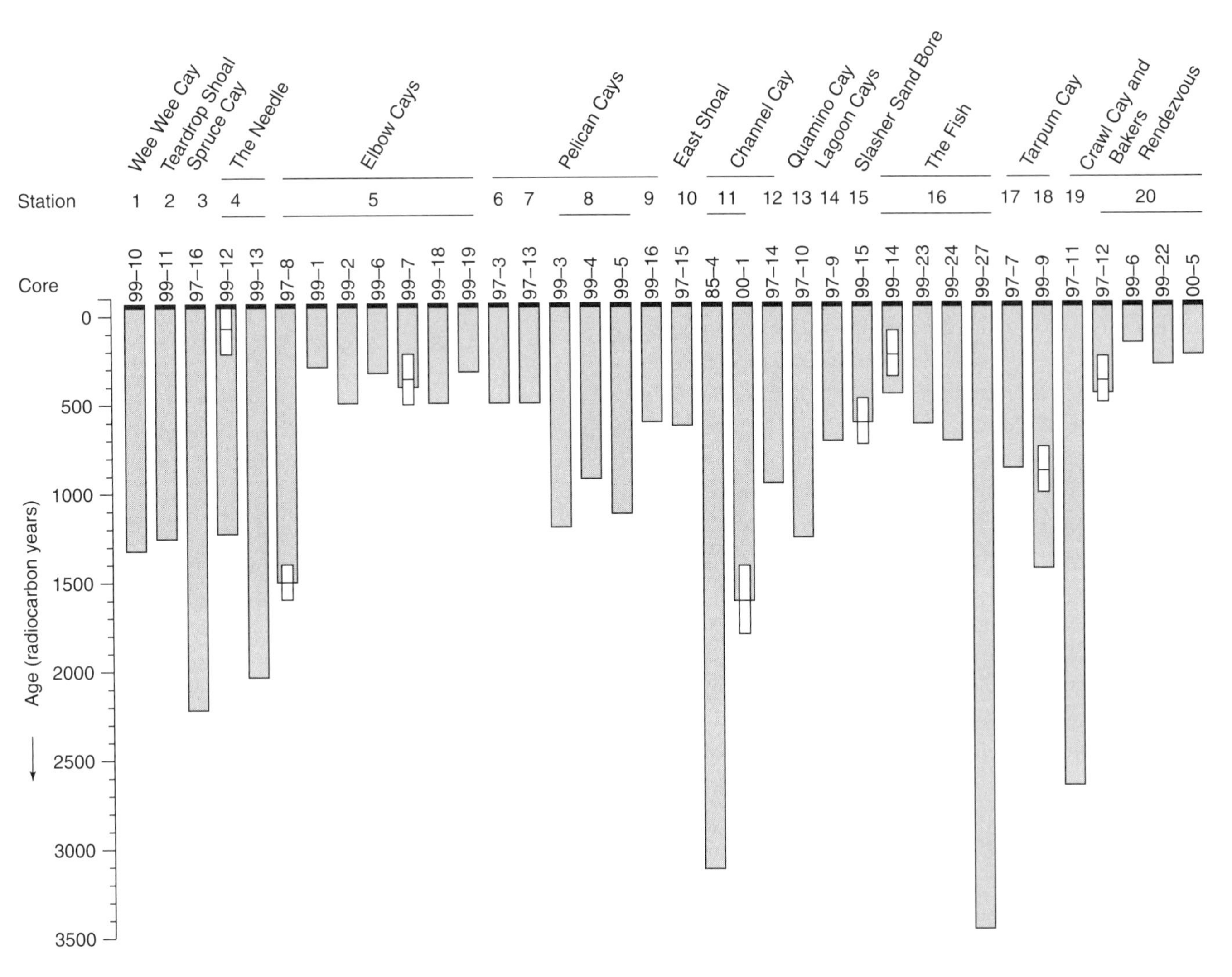


**FIGURE 10.28 Representations of cores collected from reef localities in Belize.** Gray fill represents dominance by *Acropora cervicornis*; black bars at the core tops and within the cores represent dominance by *Agaricia tenuifolia* (white rectangles surrounding the bars within cores are 95 percent confidence intervals on the dates of *Agaricia tenuifolia* "events" based on radiocarbon dating). Note the scarcity of *Agaricia tenuifolia* events prior to the area-wide transition at the top. *(From Aronson et al., 2002)*

tion from *Acropora cervicornis* to *Agaricia tenuifolia* was without precedent in the area (Figure 10.28). There were a few previous instances of temporary dominance by *Agaricia tenuifolia* indicated in some of the cores, but these were found to be far more patchy in space and time than the recent episode.

Processes implicated by some biologists in the decline of corals on present-day reefs, such as spontaneous events of bleaching (the expulsion by corals of their resident photosynthetic symbionts, known as *zooxanthellae*), may be associated with the current episode of global warming; this, in turn, has been linked to an atmospheric increase in the levels of greenhouse gases, notably $CO_2$ emitted from the burning of fossil fuels. For this reason, there is growing concern that corals and other species will be adversely affected over the long term by global warming. After all, there are indications that, in other Cenozoic intervals, coral biotas have responded noticeably to episodes of warming and cooling [SEE SECTION 8.8]. Interestingly, however, recent analyses suggest that at least some species are capable of responding to global climatic changes through geographic range shifts into cooler waters. William Precht and Richard Aronson (2004) reviewed the present-day and historical distributions of acroporid corals along the coast of the state of Florida. They noted that in just the past few decades, there has been a northward, 50-km expansion in the geographic ranges of *Acropora* species. Fossil data available for the same area demonstrate that some 6000 years ago, *Acropora* underwent a similar range expansion in association with a short-term increase in regional sea-surface temperatures. Taken together, these and similar analyses of the geographic distributions of corals suggest a degree of dynamism in geographic ranges reminiscent of that exhibited in response to climatic fluctuations throughout the Quaternary by terrestrial plants, insects, and other organisms [SEE SECTIONS 7.3 AND 9.6].

In a somewhat different vein, data garnered from the accumulated skeletal remains of bivalves have been instrumental in calibrating the loss of biological productivity on the Colorado River Delta in the Gulf of California, related to the establishment of a series of dams, beginning around 1930. There, the subfossil record of mollusc shells is being investigated to address an important question: How has the establishment of dams affected the population densities of living organisms and, therefore, the level of biological productivity? The damming of the river, and the associated cutoff of new sediment input to the delta, resulted in the exhumation by currents and waves of shells that were previously buried in intertidal muds. Subsequently, these shells became concentrated in a series of beach ridges, intertidal bars, and islands.

By assessing the total volume of these concentrations, coupled with an estimation of the number of shells contained within an average cubic meter of sediment in the study area, Michal Kowalewski and colleagues (2000) determined that at least $2 \times 10^{12}$ dead individuals were present throughout the study area! Remarkably, when Kowalewski and colleagues determined the absolute ages of 125 shells collected for analysis, they found that none were produced by individuals that lived after the year 1950. When they sorted the data into 50-year time increments (Figure 10.29), they observed a steady falloff in the number of shells starting with those dating back to earlier than 1800 AD. This is what we might reasonably expect, given that older shells would have been exposed to taphonomic processes for longer periods of time. Thus, it is highly likely that the actual number of shells produced in the study area over the past millennium (the approximate interval represented by the shells contained in the sample; Figure 10.26) was substantially greater than the number found there today.

The vast majority (greater than 85 percent) of the shells in the vicinity of the Colorado Delta belong to a single species, *Mulinia coloradoensis*, the mean length of which is about 30 mm. Oxygen isotope profiles of shells

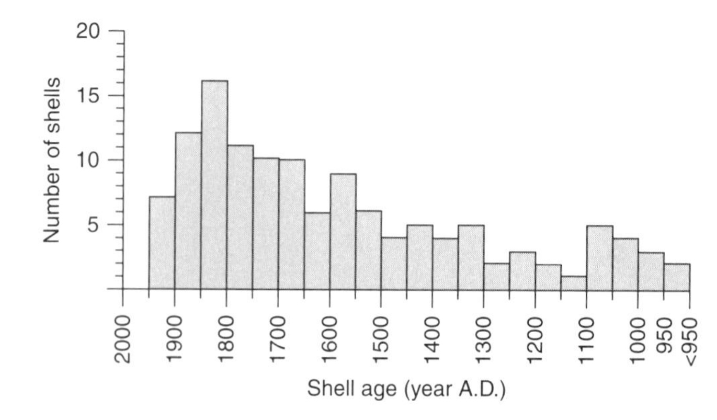


**FIGURE 10.29 Age-frequency distribution of 125 shells of the species *Chione fluctifraga* collected from the vicinity of the Colorado River delta.** *(From Kowalewski et al., 2000)*

of this species [SEE SECTION 9.5] permit the recognition of seasonal cycles of growth and demonstrate that a typical individual required at least three years to reach average size. Thus, over the past millennium, it can be estimated that about 333 generations of this species lived in the study area.

Armed with this information, Kowalewski and colleagues estimated that, at a minimum, some $6 \times 10^9$ members of this species $(2 \times 10^{12}/333)$ were alive at any given time during the past millennium. Finally, given that *M. coloradoensis* is a brackish water species limited to environments that were affected by the influx of water from the river, and considering the area covered by the intertidal zone in the vicinity of the delta, the area in which these shells were produced measured some $1.2 \times 10^8$ m$^2$ at most. Given the previous estimate of the standing population size, it can therefore be estimated that the average standing density of living individuals at any point in time over the past millennium prior to the damming of the Colorado River was about 50 individuals per square meter $(6 \times 10^9/1.2 \times 10^8)$.

While these calculations are admittedly somewhat coarse, it should be understood that they err on the side of underestimating past abundance. In all likelihood, the pre-1950 density of living individuals, including smaller standing populations of other bivalves, was much greater than the number that Kowalewski and colleagues reported. More important, these estimates exceed current densities of living bivalves in the study area by at least an order of magnitude! Restoration of intermittent river flow over the past 20 years has had virtually no effect on these living densities, and the research by Kowalewski and colleagues has helped demonstrate the magnitude of the reduction to biological productivity that resulted from the damming of the river.

Collectively, these examples illustrate an important new way in which paleontologists are influencing present-day discussions in the public policy arena. In the future, it is likely that this role will continue to expand, particularly in marine coastal regions worldwide, where human influence is particularly profound and, at the same time, the subfossil record is easily accessible. By mining the subfossil record, paleontologists are able to significantly extend the temporal reach of historical studies in threatened regions.

## 10.7 ASTROBIOLOGY

It seems fitting to conclude this book with a topic that is quite literally expanding the frontiers of paleontology, in this case off of the planet! In the search for evidence of life outside of the earth, which falls under the heading of **astrobiology,** it might seem surprising that scientists would look to the fossil record for guidance, but there are good reasons why this has been the case. While it is reasonable, and certainly a lot of fun, to speculate on the nature of alien life forms, for the moment the only actual data that we have about the existence of life comes from the DNA- and **RNA**-based forms that we have on earth. And, in the direct search for evidence of life outside of the earth, we are limited for the present to our solar system.

At the moment that this chapter is being written, two rovers are still rolling around on the surface of Mars, far exceeding their expected life spans, sampling rocks and conducting geochemical analyses that have already provided compelling evidence of the previous existence of a considerable volume of standing water and the presence of evaporites (Figure 10.30). This lends strong support to the view that had been developing for some time that Mars was once significantly warmer and wetter than it is at present. Furthermore, satellite images of the

(a)

(b)

(c)

**FIGURE 10.30 Photographs taken in 2004 by the NASA rover *Opportunity*, highlighting evidence that there was once standing water at the landing site.** (a) Microscopic concretions (arrows), usually attributed to chemical precipitation in a watery medium (photo is approximately 1.5 cm in width). (b) Panoramic view of a rock outcrop, showing evidence of sedimentary stratification. (c) Photomicrograph of vugs, which are cavities left by the dissolution of erosion of minerals, such as sulfates, that are thought to have previously grown as crystals at these sites. *(NASA Opportunity website: www.nasa.gov/vision/universe/solarsystem/opportunity_water.html)*

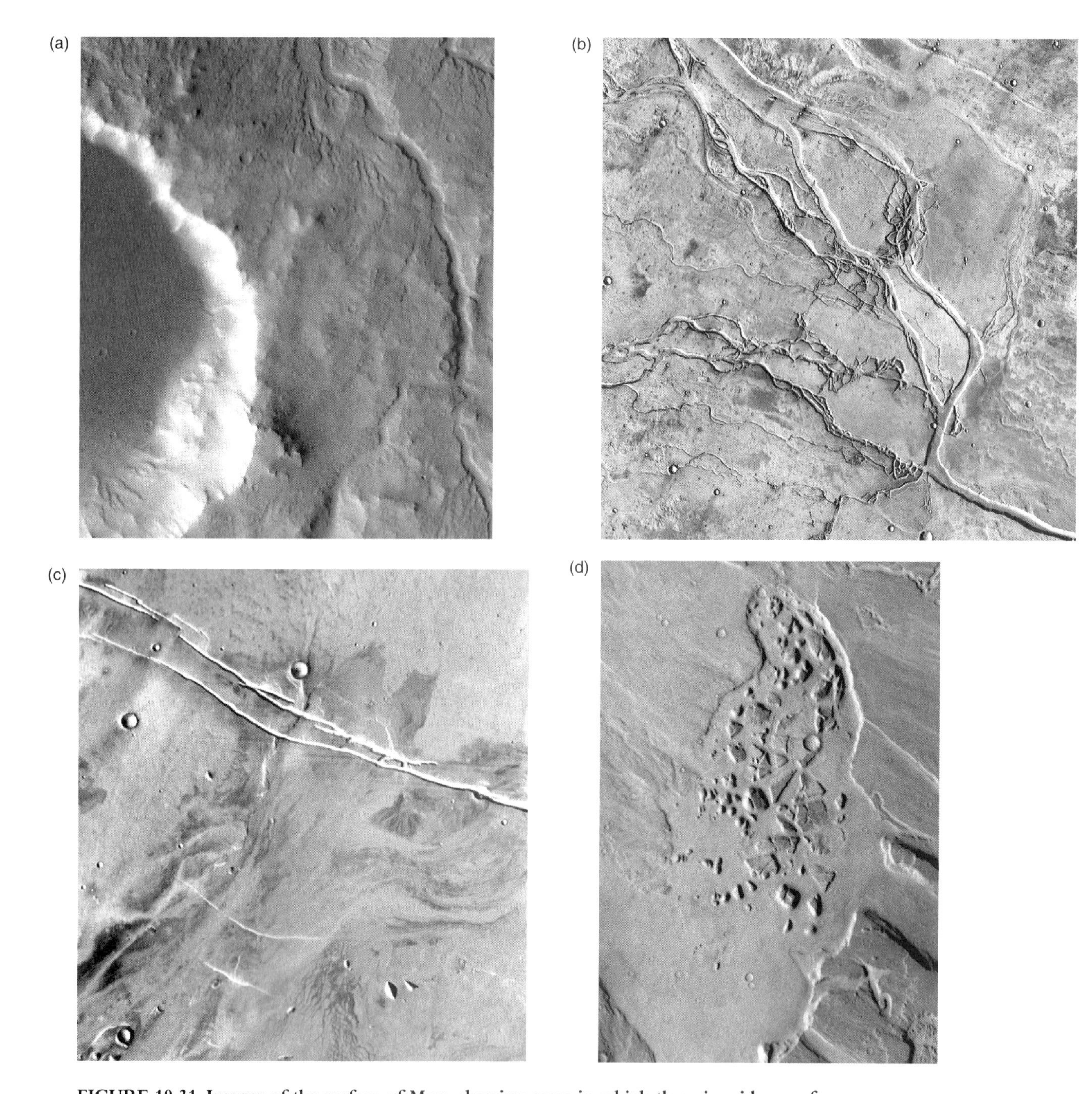


**FIGURE 10.31 Images of the surface of Mars, showing cases in which there is evidence of hydrothermal activity, water movement associated with heating.** (a) Water flow (right) linked to an impact (left). (b) A series of channels likely cut by water on the margin of Elysium Mons, the second-largest volcanic terrane on Mars. (c) Water flow in association with the opening of a fissure at Athabasca Valles. (d) Chaotic features and associated water flow at Mangala Valles; the chaotic features are thought to be related to a collapse produced by the melting of ice or removal of groundwater. [Images were captured by the Thermal Emission Imaging System (THEMIS) on NASA's Mars Odyssey Mission; *From NASA/JPL/Arizona State University*].

Martian surface reveal significant evidence not only of the past movement of water, but also of the existence of past hydrothermal activity (Figure 10.31).

Interestingly, sites of present-day hydrothermal activity on earth, such as areas associated with the activity of geysers and hot springs, are hotbeds for some of the most primitive, microbial life forms known on the planet, including prokaryotic organisms classified as **EUBACTERIA,** as well as those belonging to another, less well-known, group called the **ARCHAEA,** which seem

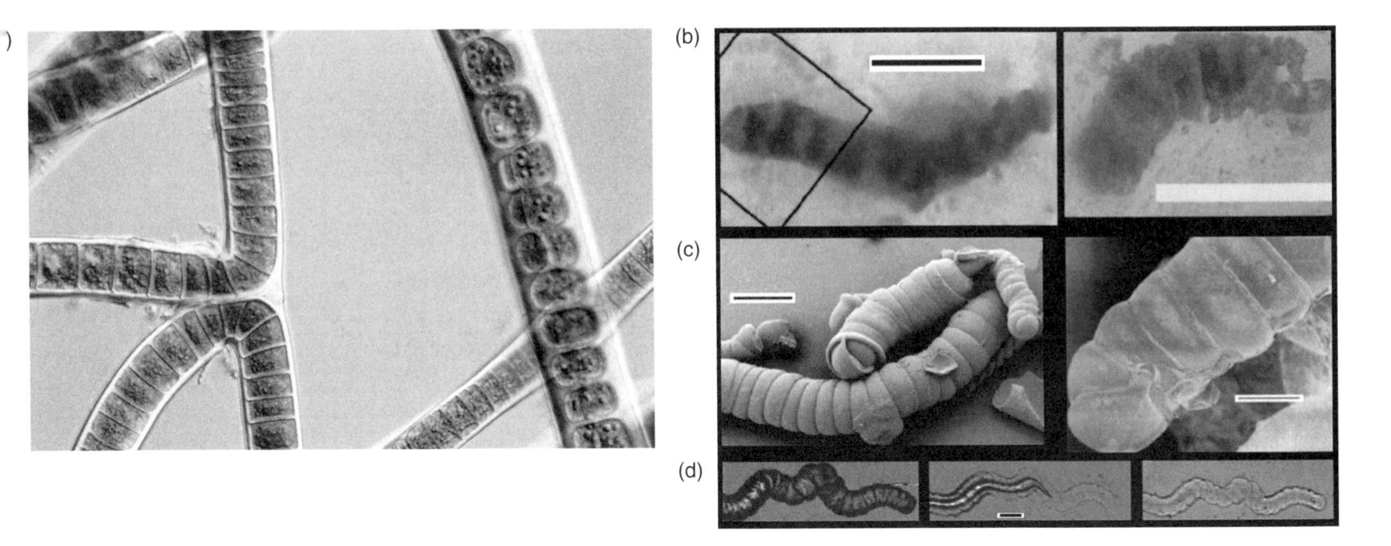


**FIGURE 10.32 A montage of illustrations, highlighting the difficulty of identifying cyanobacteria in ancient rocks.** (a) Modern cyanobacteria; such filaments are typically tens of $\mu$m in diameter. (b) Images of what were thought to be ancient cyanobacteria from the Archean Warrawoona Chert of western Australia (~3.5 billion years old), recognized previously as the oldest, direct fossil evidence of life on earth. (c) Microscopic filaments, produced in a laboratory, of silica-coated carbonate crystals, combining materials and conditions that may have been available during the Archean. (d) Progressive dissolution and hollowing, with acid, of synthetic filaments. Compare the image on the left, in particular with the images in part (b). Scale bars in parts (b) through (d) are 40 $\mu$m. *(a: Biophoto Associates/Photo Researchers, Inc.; b–d: From Garcia-Ruiz et al., 2003)*

particularly adept at living in extreme, seemingly harsh, environments. Some Archaea, for example, thrive off of carbon dioxide and hydrogen, producing methane in the process.

For these reasons, one wonders whether there was once a window of opportunity for the origin and evolution of primitive life forms on Mars—it is this possibility that has motivated much of the current interest in exploring and sampling the Martian surface. This is where paleontology comes in. While nobody can reasonably expect to see a brachiopod or clam in any of the close-up photographs snapped by the Mars rovers, there is the possibility that one day, more primitive, almost certainly microbial, evidence of past life will be discovered. In the study of the Precambrian fossil record in particular, paleontologists have shown an ever-improving ability to recognize even tiny microfossils representing some of the most primitive life forms on the planet [SEE SECTION 1.2].

Not surprisingly, the search for the earliest, most archaic forms of life on earth has not been without its pitfalls. Evidence from the biogeochemical record, which provides information about the metabolic by-products of life, now suggests that primitive life forms existed on earth as far back as 3.7 billion years ago. And additional, indirect evidence of life is found in rocks that are nearly that old, in the form of **stromatolites.** Stromatolites are layered structures produced by the trapping of sediment and the precipitation of calcium carbonate within filamentous mats formed mainly by prokaryotic cyanobacteria, other bacteria, and later, eukaryotic algae in aqueous settings during the Precambrian. Today, stromatolites are restricted mainly to environments that are inhospitable to grazing organisms that feed on the mats and thereby prevent stromatolitic development. However, during the Precambrian, before the evolution and diversification of multicellular life, stromatolites were far more common in marine settings and are therefore ubiquitous features of the Precambrian marine record.

Of course, paleontologists would like to find direct fossil evidence of the early inhabitants of earth, and there is evidence of the existence of possible prokaryotic cyanobacteria in rocks dated at 3.5 billion years old, deposited in western Australia (Figures 10.32a and 10.32b). However, in a series of experiments seeking to replicate geochemical conditions at this site of ancient hydrothermal activity, Juan M. Garcia-Ruiz and colleagues (2003) produced an assemblage of microscopic objects that look remarkably similar to the putative prokaryotes collected from the site (Figures 10.32c and 10.32d). While it is possible that Australian specimens are, indeed, the fossil remains of primitive forms of life, the Garcia-Ruiz experiments remind us that it is also possible for nature to produce inorganic structures that look decidedly lifelike [SEE SECTION 1.2].

| Physiographic setting | Vent wall and floor | Vent margin, proximal spring and channels | Median outflow channels and terrace ponds | Spring mound interfluves | Median to distal terracettes and pools | Distal ponds, marshes and pond margins |
|---|---|---|---|---|---|---|
| Idealized profile (not to scale: scale is variable according to date of discharge and local topography) | 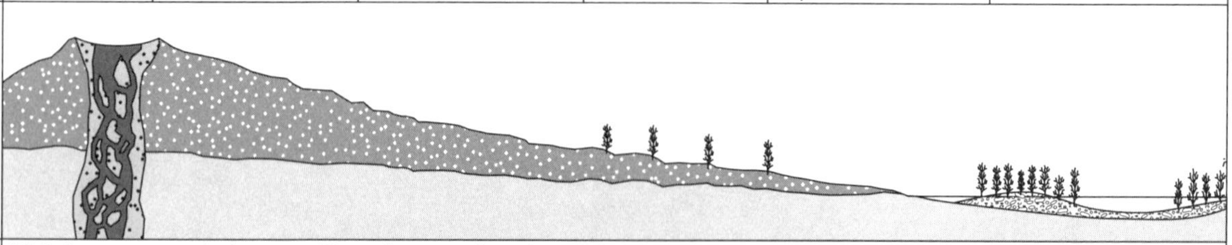 | | | | | |
| Relative hydrodynamic conditions | High flow rates, explosive discharge, brecciation | Sheet flow; ponding of water behind microterracettes; pools of a few centimeters depth | | | | Low-energy pools and waterlogged soils |
| Fabric | Massive to finely laminated, pisolitic, and brecciated cherts. Alternating layers of microgranular quartz and megaquartz | Thinly bedded cherts and chert breccias. Irregular laminae of fine-grained microcrystalline quartz alternating with fine- to medium-grained megaquartz | | Thinly bedded cherts with palisade fabrics. Framework includes patches of megaquartz with "bush-like" filamentous microfossils | | Reddish-brown cherts with irregular to mottled bedding and scattered plant fossils. Dominated by medium- to coarse-grained, inclusion-rich megaquartz and isolated patches of microcrystalline quartz |
| Biofacies | Geyserites and vent wall and floor deposits | High temperature stratiform stromatolites with streamers and fenestrae | Medium temperature stratiform stromatolites with palisade fabric | Spring mound interfluves with molds / silicified stems of plants | Medium temperature pools with "pseudocolumnar stromatolites" | Low temperature ponds with molds and silicified stems of plants, and microbial overgrowths |
| Biota (fossilized or inferred) | Thermophilic and hyperthermophilic organisms | Thermophilic, finely filamentous, sometimes tufted or erect, cyanobacteria (cf. *Phormidium*) and other bacteria | Mats of vertically oriented cyanobacteria (cf. *Calothrix*) and local lenses of randomly oriented cyanobacteria (cf. *Phormidium*) | Semi-aquatic, herbaceous, lycophytes and sphenophytes | Pseudocolumnar stromatolites and "shrub-like" features formed by coarsely filamentous cyanobacteria | Herbaceous lycophyte and sphenophyte plants with radial microbial crusts |
| Palaeo-temperature | >59C | 59–30C | <30C | About ambient | <30C | About ambient |

Tuff, agglomerate — Massive to weakly bedded chert, and chert veins — Thin-bedded chert — Pool deposits and soil — Pools

**FIGURE 10.33 A summary of physical, lithological, and paleobiological features associated with an environmental gradient at the site of a Devonian thermal spring in Australia.** These determinations draw on comparisons with present-day thermal springs. *(From Walter et al., 1998)*

Nevertheless, there are now numerous instances in which preserved prokaryotic and simple eukaryotic fossil organisms have been recovered from Precambrian and younger rocks, sometimes in conjunction with the collection of biogeochemical data that bolsters, at least indirectly, the claim that the microfossils in question actually are what they appear to be. For the moment, some of the most compelling direct evidence for fossil microbial life in association with ancient hot springs comes not from the Precambrian, but from Devonian strata in eastern Australia. There, microfacies are preserved over lateral distances of 100 m in association with a temperature gradient related to hot spring activity (Figure 10.33). Within these microfacies, a variety of microscopic filaments thought to represent various cyanobacteria and other microbes are preserved, although the precise identities of these organisms remain uncertain (Figure 10.34). Based on evidence of the occurrences of Archaea and Eubacteria in present-day hot springs, it appears that some of the organisms found in the Devonian deposits lived at temperatures well in excess of 50°C (122°F)!

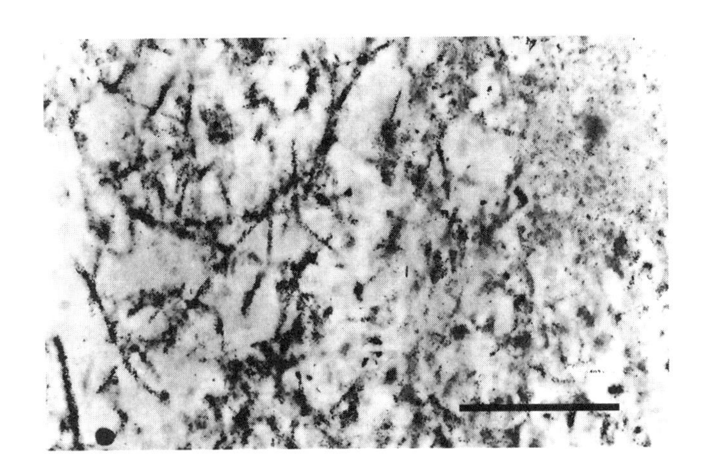

**FIGURE 10.34 A photomicrograph of a thin section illustrating tubular molds of what is thought to be a filamentous microorganism, from the site of a Devonian thermal spring in Australia.** The scale bar at the lower right is 100 micrometers. *(From Walter et al., 1996)*

The preservation of these fossils provides a basis for the emerging view of some astrobiologists (or, perhaps more appropriately, *astropaleontologists*) that, in future attempts to sample the Martian surface, we should pay special attention to sites where there is clear evidence not only of the presence of standing water, but also for the existence of hydrothermal activity (Farmer and Des Marais, 1999). Hydrothermal environments are likely to figure prominently in future scientific missions to Mars.

## 10.8 CONCLUDING REMARKS

As the foregoing examples illustrate, paleontology sits at the interface of a number of scientific fields, including geology, geochemistry, ecology, evolutionary biology, and even astronomy. It draws from these fields and influences them in turn. In that sense, paleontology is like nearly all areas of natural science in having become more interdisciplinary in recent years, and it follows that paleontologists need to be more broadly trained than ever before. While one can safely predict that this trend will continue, the particular research questions that will motivate paleontology in the coming years are hard to predict, just as they would have been hard to predict 10 or 20 years ago. We certainly have some ideas as to where the science will be headed next, but, realistically, these are little more than hunches. We can only look to the future with great anticipation!

# SUPPLEMENTARY READING

Aubry, M.-P., Lucas, S., and Berggren, W. A. (eds.) (1998) *Late Paleocene–Early Eocene Climatic and Biotic Events in the Marine and Terrestrial Records*. New York, Columbia University Press, 513 pp. [Overview of the Paleocene–Eocene Thermal Maximum; provides perspective for the many later papers published on this subject.]

Briggs, D. E. G., Erwin, D. H., and Collier, F. J. (1994) *The Fossils of the Burgess Shale*. Washington D.C., Smithsonian Institution Press, 238 pp. [Richly illustrated overview of Burgess Shale fossils.]

Conway Morris, S. (1998) *The Crucible of Creation*. Oxford, U.K., Oxford University Press, 242 pp. [An evolutionary interpretation of early animals, with particular focus on the Burgess Shale.]

Erwin, D. H. (2006) *Extinction: How Life on Earth Nearly Ended 250 Million Years Ago*. Princeton, N.J., Princeton University Press, 296 pp. [Overview of biological and geological data surrounding the Permian–Triassic extinction event, as well as possible causes.]

Gould, S. J. (1989) *Wonderful Life*. New York, Norton, 347 pp. [Discussion of Burgess Shale organisms and some possible evolutionary interpretations.]

Knoll, A. H. (2003) *Life on a Young Planet: The First Three Billion Years of Evolution on Earth*. Princeton, N.J., Princeton University Press, 277 pp. [Comprehensive overview of Precambrian research and implications for astrobiology.]

MacPhee, R. D. E. (ed.) (1999) *Extinctions in Near Time*. New York, Kluwer Academic, 394 pp. [Series of papers on extinction in the past 100,000 years.]

Squyres, S. W., and Knoll, A. H. (eds.) (2005) *Sedimentary Geology at Meridiani Planum, Mars. Earth and Planetary Science Letters*, vol. 240, no. 1, pp. 1–190. [Series of articles on the characteristics and origins of some martian sedimentary rocks.]

Valentine, J. W. (2004) *On the Origin of Phyla*. Chicago, University of Chicago Press, 614 pp. [Thorough treatment of the nature of animal phyla and their evolution.]

Whittington, H. B. (1985) *The Burgess Shale*. New Haven, Conn., Yale University Press. [Emphasis on anatomical interpretation of Burgess Shale organisms.]

# GLOSSARY

**absolute rate** A measure of the morphological rate of evolution that expresses the amount of change in a trait per unit time, not scaled to the magnitude of the trait (see **relative rate**).
**actuopaleontology** Study of present-day post-mortem processes to gain insight into taphonomy and other aspects of paleontology.
**adaptation** The fit between an organism's phenotype on the one hand and its environment and way of life on the other; the evolutionary mechanisms and pathways that produce adaptive traits in a lineage.
**alleles** Alternative forms of the same gene.
**allometry** A kind of anisometric growth in which one component ($X$) increases as a constant power ($a$) of another component ($Y$), so that $Y$ is proportional to $X^a$; anisometric growth that results when two components have different relative growth rates, each of which remains constant during growth.
**allopatric** A term used to describe populations with disjunct geographic ranges (see **sympatric**).
**anagenetic change** Change reflecting evolution within single, species-level lineages (see **cladogenetic change**); phyletic change.
**analogous trait** A similar trait shared by two or more organisms that is thought to result from convergence (see **convergent trait**).
**analytical time averaging** Aggregation of fossil material of different ages by combining samples from different stratigraphic levels (see **time averaging**).
**anisometric growth** Pattern of development in which shape changes as size increases (see **isometric growth**).
**apatite** A phosphate mineral, $Ca_5(PO_4,CO_3)_3(F,OH,Cl)$, important in several groups, including vertebrates and brachiopods.
**apomorphy** In phylogenetic analysis, a trait that represents evolutionary change relative to the primitive condition in some group (see **plesiomorphy**).
**appearance event ordination (AEO)** A method of constructing a composite standard of first and last appearances of taxa using data from two or more outcrops, based on F/L statements (see **composite standard reference section, F/L statements**).
**aragonite** A thermodynamically less stable form of calcium carbonate, $CaCO_3$ (see **calcite**).
**arithmetic mean** A statistical measure of central tendency in a sample of values, equal to $\Sigma x_i/n$, where $n$ is the sample size and $x_i(i = 1, \ldots, n)$ are the observed values.
**astogeny** Growth and development of a colonial organism.
**astrobiology** The study of, and search for, life beyond the earth.
**autapomorphy** A derived character uniquely possessed by one taxon, not shared with others.
**available** Said of a taxonomic name that is erected according to the rules of nomenclature and is validly published.
**average** Arithmetic mean.

**background extinction** Extinction of taxa not associated with mass extinction (see **mass extinction**).
**baseline** The reference axis for calculating shape coordinates.
**bathymetric** Pertaining to water depth.
**begging the question** See **circular reasoning.**
**bentonite** A clay layer that is the remnant of a volcanic ash fall.
**biodiversity** General concept referring to the variety of kinds of organisms (see **disparity, diversity, morphological diversity, taxonomic richness**).
**biological species concept** Definition of a species as a group of interbreeding populations that are reproductively isolated from other groups.
**biomechanical analysis** Study of biological function with reference to principles of mechanics and engineering and to the physical properties of biologic materials (see **experimental approach, paradigm approach**).
**biostratigraphy** The study of the fossil contents of rock, usually for the purpose of correlation [see **correlation (biostratigraphy)**].
**biostratinomy** Processes of fossilization that affect a dead organism prior to burial, usually mechanical in nature.
**bioturbation** Reworking of sediments resulting from the activities of organisms.
**biozone** A body of rock defined by its fossil content; usually viewed as representing a particular interval of time based on that content (see **guide fossil, index fossil, overlapping range, zone, zone fossil**).

**body fossils** The remains of actual parts of organisms.

**boundary-crossing taxa (boundary crossers)** Used as a measure of taxonomic richness, a count of the taxa that are inferred to be present in two successive time intervals, and thus extant at the boundary between them.

**bulk sample** A volume of fossil-laden rock or sediment collected for an analysis of its fossil content.

**calcite** A thermodynamically stable mineral form of calcium carbonate, $CaCO_3$ (see **aragonite**).

**Cambrian Explosion** An episode of great morphological and functional diversification approximately 540 million years ago, during which most of the basic animal body plans appeared within just a few tens of millions of years.

**Cambrian Fauna** The marine evolutionary fauna that dominated the oceans during the Cambrian period. (see **evolutionary faunas, Modern Fauna, Paleozoic Fauna**).

**camera lucida** An optical device that facilitates drawing by allowing one to see the image of one's hand superimposed on the image of a specimen.

**carbonates** A class of minerals containing the carbonate ion $CO_3^{2-}$.

**carbonization** Preservation of organic tissue through extreme distillation, leaving a highly carbon-rich residue.

**cast** Positive impression formed when a mold or trace is filled with sediment or other material.

**categorical variable** Nominal variable.

**cellulose** A polysaccharide that is the main component of cell walls in plants and algae.

**centroid size** A measure of size in morphometric analysis based on the distances of each landmark from the centroid or algebraic mean position of all landmarks.

**character distribution** In phylogenetic analysis, the pattern of character states, typically represented as numerical codes, within a set of species.

**chitin** A polysaccharide that is a major constituent of arthropod cuticles.

**chronospecies** Segments of a single evolving lineage that are operationally designated as distinct species on the basis of accumulated morphological differences (see **phyletic speciation**).

**circular reasoning** See **begging the question.**

**clade** Group consisting of two or more species that share a common ancestor, together with all the other descendants of that common ancestor; a monophyletic group.

**cladogenetic change** Evolutionary change involving lineage splitting (see **anagenetic change, speciation**).

**cladogram** A branching diagram that portrays proximity of evolutionary relationship without an explicit temporal dimension (see **evolutionary tree**).

**classification (multivariate analysis)** The delineation of groups using clustering techniques.

**classification (systematics)** The organization of species and higher taxa into a hierarchical system of named categories.

**clonal** Said of an organism that consists of genetically identical subunits that generally resemble individual organisms or function as if they were individuals.

**clustering techniques** Multivariate methods aimed at determining the number of groups represented by a sample of specimens and the composition of each group (see **multivariate analysis**).

**cohort survivorship analysis** A style of taxonomic survivorship analysis in which groups of taxa with a common time of first appearance are followed forward in absolute time to track the number remaining.

**collagen** A protein that is a major constituent of connective tissue in animals.

**community** In ecology, a regularly recurring combination of numerically common species (many other definitions have been proposed in the biological literature).

**community paleoecology** The study of the ecology of ancient communities preserved in the fossil record (see **ecology, paleoecology**).

**competitive exclusion** The principle that there cannot be a stable co-existence of two species that utilize the same resources.

**composite standard reference section (composite standard, CS)** The sequence of events (generally FADs and LADs) preserved at two or more localities combined together using any of a family of numerical methods [see **appearance event ordination (AEO), constrained optimization (CONOP), first appearance datum (FAD), graphic correlation (GRAPHCOR), last appearance datum (LAD), ranking and scaling (RASC)**].

**composite standard units (CSU)** Intervals of time on a composite standard reference section delineated by correlating the events contained on a composite standard reference section directly to stratigraphic intervals depicted on a geologic timescale (see **composite standard reference section**).

**concurrent range** In biostratigraphy, an overlap recognized in the stratigraphic ranges of two or more taxa.

**congeneric** Said of two or more species belonging to the same genus.

**conservation paleobiology** Application of paleontological data and analyses to present-day problems related to conservation biology and human-induced environmental change.

**constrained optimization (CONOP)** A method of constructing an optimal composite standard of first and last appearances of taxa with a minimum number of contradictions based on the sequences preserved at two or more localities [see **composite standard reference section, ranking and scaling (RASC)**].

**constructional morphology** An approach to morphological analysis that simultaneously considers how aspects of form reflect phylogenetic inheritance, functional demands, and structural limitations (see **functional factor, historical factor, structural factor**).

**convergent trait** A trait that is shared in two or more lineages, each of which evolved it independently.

**coordinate transformation** Study of shape variation by considering how one form would have to be geometrically transformed to yield a different form (see **thin-plate spline**).

**coordinated stasis** Pattern of regional stratigraphic distribution of fossils characterized by extended intervals of relative stability in taxonomic composition and punctuated by brief intervals of significant change (see **ecological evolutionary subunit**).

**correlation (biostratigraphy)** Determination of the age-equivalency of a stratigraphic interval at two or more localities.

**correlation (statistics)** The strength and direction of association between two measured variables.

**correlation coefficient** A statistical measure of correlation.

**cosmopolitan** Said of a taxon that is present in many geographic regions (see **endemic**).

**cross-sectional analysis** Study of ontogeny by comparing a series of specimens, each at a different size or growth stage (see **longitudinal analysis**).

**crown group** A monophyletic group containing a set of living species, the latest common ancestor of these species, and all descendants of that common ancestor (see **plesion, stem group**).

**cryptic species** Sibling species.

**death assemblage** The set of species in a subfossil or fossil accumulation (see **life assemblage**).

**deep homology** A phenomenon in which structures appear to have evolved independently in different lineages but nonetheless arise in ontogeny through the action of homologous genes.

**dendrogram** A branching diagram that links similar entities together into groups and separates them from other groups (see **clustering techniques**).

**depth tolerance** In modeling the distribution of a taxon along a marine depth gradient, the degree of spread in its distribution; a greater spread implies a greater depth tolerance.

**derived character** In phylogenetic analysis, an apomorphy.

**determinate growth** Pattern of development in which a mature stage is reached and structural growth stops even though the organism continues to live (see **indeterminate growth**).

**developmental constraint** Broadly speaking, any interaction between available genetic variation and developmental processes that results in nonrandom variation in phenotype.

**diagenesis** In taphonomy, processes of fossilization that affect a dead organism after burial, usually chemical in nature.

**digitization** Electronic storage of images or data.

**directed speciation** Postulated mechanism of phylogenetic trends, in which change is concentrated at speciation events and descendant species tend to shift in a preferred morphological direction relative to their ancestors.

**discrimination** Practical recognition of distinct populations and species; study of the factors that make them distinct.

**disparity** An aspect of biodiversity reflecting morphological differences among species.

**distillation** Preferential removal of volatile elements, such as hydrogen and oxygen, from organic compounds subjected to heat and pressure, leaving a more carbon-rich residue (see **carbonization**).

**diversity** Synonymous with **biodiversity.**

**DNA** Deoxyribonucleic acid, the principal molecule that carries genetic information.

**ecological evolutionary (e-e) subunits** Regional intervals of prolonged stability in taxonomic composition (see **coordinated stasis**).

**ecological evolutionary units** Global intervals of prolonged stability in taxonomic composition.

**ecology** The study of interrelationships between organisms and the environments in which they live.

**ecophenotypic** Said of phenotypic variation that is attributable to environmental rather than genetic variation.

**ecospace** Conceptualization of a space in which ecological conditions are suitable for the subsistence of a given taxon.

**eigenvalue** A number expressing how much of the variation in a multivariate data set is contained within a principal component or other synthetic variable.

**endemic** Said of a taxon restricted to a particular geographic region (see **cosmopolitan**).

**entire-margined leaves** Leaves that have smooth, untoothed outer margins.

**epicontinental sea (epeiric sea)** Extension of an ocean that overlies a continent or continental shelf (e.g., present-day Hudson Bay).

**era** A broad division of the geologic timescale that is intermediate in scale between an eon and a period.

**evolution** Heritable change in the aggregate genotypic and phenotypic composition of populations and species over time.

**evolutionary faunas** Groups of higher taxa, generally delineated at the class level, that comprised the majority of the global biota during particular intervals of geologic time; generally applied to marine biotas (see **Cambrian Fauna, Modern Fauna, Paleozoic Fauna**).

**evolutionary morphology** General study of the diversity and nonrandom distribution of form in the history of life (see **constructional morphology, functional morphology, theoretical morphology**).

**evolutionary paleoecology** The study of paleoecological transitions in geologic time and of the relationship between ecology and major evolutionary transitions (see **paleoecology**).

**evolutionary tree** A branching diagram that portrays ancestral–descendant relationships over actual time and may also depict other aspects of evolution, such as morphological change (see **cladogram**).

**evolutionary trend** A pattern of evolution in which the direction of change is persistent over an appreciable span of time.

**experimental approach** A style of biomechanical analysis in which a postulated function is considered for an observed structure, and direct experimentation with the structure, or a mathematical or physical model of it, is used to assess the capacity of the structure to perform the function (see **paradigm approach**).

**exponential survivorship** Pattern of taxonomic survivorship that results when the per-capita rate of extinction is constant. The natural logarithm of the number of taxa that endure at least to a given age declines linearly with age, and the slope of this decline is equal to the extinction rate.

**extinction** Evolutionary termination of a lineage; the last appearance of a taxon.

**extinction-rate bias** A postulated mechanism for phylogenetic trends, in which species sorting drives the trend and the sorting is in turn caused by a difference in extinction rates.

**facies** The characteristics of sedimentary rocks or the environments they represent.

**first appearance datum (FAD)** The base of the stratigraphic range of a taxon at a locality [see **last appearance datum (LAD)**].

**F/L statements** The recognition that, if the stratigraphic ranges of two taxa overlap, the first appearance datum (FAD) of one taxon must predate the last appearance datum (LAD) of the other taxon [see **appearance event ordination (AEO), first appearance datum (FAD), last appearance datum (LAD)**].

**flooding surface** In sequence stratigraphy, a horizon marked by rapid deepening.

**form taxonomy** A taxonomic system that classifies organisms or isolated parts based on their morphology, without necessarily seeking to recognize biologically meaningful species.

**fossil** Any physical or chemical remains or traces of past life.

**functional factor** In constructional morphology, the aspects of form that reflect immediate adaptation to functional needs.

**functional morphology** The study of biologic form with respect to organismic function.

**gap (stratigraphic gap)** An interval within the stratigraphic range of a taxon from which no fossil remains of the taxon are known.

**gene flow** Exchange of genetic information between populations via migration and interbreeding.

**gene pool** The aggregate genetic composition of a population.

**genetic drift** Evolutionary change that results from chance fluctuations in the relative proportions of different genotypes and phenotypes within a population, rather than natural selection.

**genetic mutation** Spontaneous change in an organism's genetic material.

**genotype** The genetic composition of an organism, encoded in its DNA sequences (see **phenotype**).

**global completeness** Paleontological completeness measured on a global scale; the proportion of all taxa within some group that have left some known fossil remains (see **local completeness**).

**gradient analysis** The assessment of the distributions and abundances of taxa in relation to one another along environmental gradients.

**graphic correlation (GRAPHCOR)** The graphical construction of a composite standard of first and last appearances of taxa using data initially from two localities, typically followed by the sequential addition to the composite sequence of information from additional sections (see **composite standard reference section**).

**greenhouse gas** A gas that absorbs infrared radiation emanating from the earth's surface, thereby contributing to the warming of the atmosphere. Examples include carbon dioxide and methane.

**guide fossil** A synonym for index fossil and zone fossil.

**guild** A group of taxa that exploit a similar set of resources.

**habitat tracking** A postulated mechanism for stasis in the face of environmental change, in which the geographic distributions of populations and species shift to track local conditions to which they are adapted.

**half-life** The median duration of a group of species or other taxa.

**hard parts** Mineralized or otherwise hardened components of organisms (see **soft parts**).

**heritable** Said of phenotypic variation that is underlain by genetic variation and so is passed from parents to offspring.

**heterochrony** Evolutionary change in the timing of ontogeny.

**highstand systems tract (HST)** In sequence stratigraphy, the upper portion of an idealized depositional sequence, overlying the maximum flooding surface and characterized by renewed progradational stacking of parasequences or higher-order sequences [see **lowstand systems tract (LST), maximum flooding surface, parasequence, sequence, transgressive systems tract (TST)**].

**historical factor** In constructional morphology, the aspects of form possessed by a lineage because they evolved in an ancestor and were passed on, whether or not they are subject to immediate natural selection within the lineage in question.

**holophyletic group** Monophyletic group.

**holotype** A single specimen designated as the name bearer for a species.

**homologous trait (homology)** A trait that is shared, with possible modification, in two or more species because they inherited it from a common ancestor.

**homonyms** Identical names that denote different species or higher taxa.

**incongruence** Inconsistency that results when some characters imply a particular pattern of phylogenetic relationships while other characters imply a different pattern.

**indeterminate growth** Pattern of development in which growth continues throughout the life of the organism (see **determinate growth**).

**index fossil** A fossil taxon (usually a species) that is useful for defining a biozone; generally characterized by widespread geographic distribution and limited stratigraphic range (see **biozone, guide fossil, stratigraphic range, zone, zone fossil**).

**individualistic concept** The view that communities are loose aggregations of species that occur together in a particular place because of shared preferences for aspects of the local environment.

**ingroup** In phylogenetic analysis, the group of taxa among which one seeks to reconstruct evolutionary relationships (see **outgroup**).

**internal mold** A steinkern.

**interspecific allometry** The relationship between two components of form or body parts, as analyzed among many species.

**isometric growth** Pattern of development in which shape remains constant as size increases (see **anisometric growth**).

**isotopes** Alternative forms of a chemical element having the same number of protons in the nucleus but different numbers of neutrons and therefore different atomic masses.

**keratin** A proteinaceous component of horns, claws, bills, and feathers.

**Lagerstätte** (plural **Lagerstätten**) Sedimentary deposit with exceptionally good preservation of organic and skeletal remains.

**landmarks** In morphometric analysis, reference points on specimens used to measure their form.

**last appearance datum (LAD)** The top of the stratigraphic range of a taxon at a locality [see **first appearance datum (FAD)**].

**life assemblage** The set of species in a living community (see **death assemblage**).

**lignin** An organic compound that is the main constituent of conductive tissue in vascular plants.

**likelihood** A statistical measure of empirical support for a particular hypothesis. It is proportional to the probability, under an assumed model (e.g., an evolutionary model), that the data would be observed if the hypothesis were true.

**line of correlation (LOC)** In graphic correlation, a line that depicts the sequence of FADs and LADs shared by two outcrops, or between an outcrop and the composite standard [see **composite standard reference section, graphic correlation (GRAPHCOR)**].

**lineage** A temporally continuous series of populations or species related by common descent.

**lineage segment** In phylogenetic analysis, a sample of an evolving lineage, generally encompassing only a short span of time rather than the entire duration of the lineage.

**live–dead comparison** A comparison between life assemblages and death assemblages in a given area to assess the fidelity of the subfossil record.

**loading** A number expressing the strength of correlation between a principal component or other synthetic variable and an original measured variable; the contribution of the original variable to the principal component.

**local completeness** Paleontological completeness measured on a local scale; the proportion of all taxa within some group, and within a specified local area, that have left some known fossil remains (see **global completeness**).

**logistic equation** An equation that produces a sigmoidal (s-shaped) curve, used to test hypotheses that global diversity cannot increase indefinitely and, instead, is subject to equilibrial constraints.

**long-branch attraction** In phylogenetics, the artificial grouping of distantly related lineages resulting from their having mutually high rates of evolution and consequently an increased probability of attaining derived character states independently.

**longitudinal analysis** Study of ontogeny by following a single organism as it grows and develops (see **cross-sectional analysis**).

**lowstand systems tract (LST)** In sequence stratigraphy, the lower portion of an idealized depositional sequence, stratigraphically above the sequence boundary and characterized by progradational stacking of parasequences or higher-order sequences [see **highstand systems tract (HST), maximum flooding surface, parasequence, sequence, transgressive systems tract (TST)**].

**Lyellian proportion** Of an assemblage of species in the geologic past, the proportion that are still extant today.

**macroevolution** Evolutionary change at or above the species level, including phylogenetic trends and changes in the relative diversity of different clades over geologic time.

**mass extinction** The extinction of a large number of species in a short amount of time.

**maximum flooding surface** In sequence stratigraphy, the horizon recording the greatest water depth in a depositional sequence, marking the transition from retrogradational to progradational stacking.

**maximum-likelihood estimate** The particular hypothesis that corresponds to the highest possible likelihood for a given set of observations and an assumed model (e.g., a model of evolution).

**mean** Arithmetic mean.

**median** A statistical measure of central tendency in a sample of values, corresponding to the point that delineates the lower half from the upper half of the values.

**megafauna** Term applied collectively to relatively large terrestrial animals with body masses greater than about 40 kg.

**meristem** In plants, a zone of generation of new cells.

**meristic characters** Biological traits represented by a count of the number of similar elements, such as segments.

**Mesozoic Marine Revolution** The large-scale set of morphological transitions exhibited by marine organisms such as gastropods during the Mesozoic era in response to a major diversification of shell-crushing predators.

**metapopulation** A group of populations of the same species, between which migration and gene flow may occur only sporadically.

**microevolution** Evolutionary change within populations and species.

**Modern Fauna** The marine evolutionary fauna that dominated the oceans during the Mesozoic and Cenozoic eras (see **Cambrian Fauna, evolutionary faunas, Paleozoic Fauna**).

**mold** A kind of body fossil that represents a negative impression, generally of hard parts.

**molecular clock** Hypothetical constancy in the rate of evolution of DNA and other biomolecules over time, often used to estimate times of divergence of living groups from one another based on the magnitude of their accumulated genetic differences.

**monophyletic group** A group of species that consists of a common ancestor and all of its descendants; a clade; in earlier usage, the term included both paraphyletic groups and clades (see **paraphyletic group, polyphyletic group**).

**monotypic** Said of a genus with only one species, or a higher taxon with only one constituent lower-level taxon within it.

**morphological diversity** Aspect of biodiversity concerning the variety of form rather than the number of species (see **taxonomic richness**).

**morphological parsimony debt** The extent to which a postulated cladogram or evolutionary tree implies evolutionary steps beyond the minimal number theoretically required for the number of derived character states (see **stratocladistics**).

**morphological rate of evolution** The rate of change of anatomical traits, typically represented by quantitative measures.

**morphological species (morphospecies)** Two or more species discriminated on the basis of morphology rather than genetic or reproductive evidence.

**morphology** The study of biological form and structure; also the form and structure themselves.

**morphospecies** Species recognized on the basis of morphological attributes rather than genetic, behavioral, reproductive, or other criteria.

**multivariate analysis** A body of methods for simultaneously studying numerous measures of form or other variables, often with the goal of summarizing variation in fewer dimensions than the number of variables measured.

**natural selection** Preferential contribution of certain genotypes to future generations because these genotypes tend to produce phenotypic traits that lead to greater survival and fecundity.

**negative allometry** Pattern of allometry in which the *Y*-component increases with a lower relative growth rate than the *X*-component.

**node** A point on a cladogram where branches join, representing a common ancestor.

**nominal variable** A biological variable that takes on only particular values, such as male versus female, which have no natural ordering.

**novelty** In systematics, an apomorphy.

**objective synonyms** Different taxonomic names that are based on the same type specimen or specimens (see **subjective synonyms**).

**occupied** Said of a formal taxonomic name that is in use (see **available**).

**ontogeny** Growth and development of an organism.

**opal** Noncrystalline, hydrous silica, $SiO_2 \cdot H_2O$, important in certain kinds of fossils such as diatoms and petrified wood.

**open nomenclature** The practice followed when a newly described taxon is not placed within higher taxa because its evolutionary relationships are too uncertain.

**ordinal variable** A biological variable that takes on only particular values—such as absent, rare, common, and abundant—which have a natural ordering.

**ordination** Graphical or other representation of the positions of specimens or samples relative to one another, with respect to either measured variables or synthetic variables formed from them (see **multivariate analysis**).

**Ordovician Radiation** A major diversification of marine animals during the Ordovician Period, most notably among higher taxa of the Paleozoic Fauna.

**organic matrix** Unmineralized organic fabric in which mineral components of skeletons are embedded (see **hard parts, soft parts**).

**origination** Evolutionary establishment of a new lineage from an existing one; the first appearance of a taxon.

**outgroup** In phylogenetic analysis, a taxon taken to be outside the ingroup but closely related to it, and often used to establish the polarity of characters within the ingroup.

**overkill hypothesis** The hypothesis that the late Pleistocene extinction of many or most large-bodied terrestrial vertebrates was caused by human hunting.

**overlapping range** In biostratigraphy, a synonym of concurrent range.

**paedomorphosis** A form of heterochrony in which development evolves so that descendant adults resemble ancestral juveniles (see **peramorphosis**).

**paleobiogeography** The study of the ancient geographic distributions of fossil taxa.

**paleoecology** The study of the interrelationships between ancient organisms and the paleoenvironments in which they lived.

**paleontological completeness** The proportion of taxa extant within an interval of time that are sampled from that interval; the proportion of taxa sampled at least once over their entire durations.

**paleontology** The study of ancient life.

**Paleozoic Fauna** The marine evolutionary fauna that dominated the oceans during the Paleozoic era, after the Cambrian (see **Cambrian Fauna, evolutionary faunas, Modern Fauna**).

**paradigm approach** A style of biomechanical analysis in which a postulated function is considered for an observed structure, and the plausibility of the function is assessed by comparing the observed structure with the hypothetical structure that is ideally suited for that function (see **experimental approach**).

**paraphyletic group** A group of species that consists of a common ancestor and some but not all of its descendants (see **monophyletic group, polyphyletic group**).

**parasequence** In sequence stratigraphy, a shallowing-upward cycle, bounded at the top by a flooding surface.

**paratype** A specimen other than the holotype, which is formally designated by the author of a species as having been used in the description of the species.

**parsimonious** Said of scientific hypotheses that require comparatively few assumptions to explain observed data; said of a cladogram that implies relatively few evolutionary steps to explain the character data.

**patchiness** A common pattern observed in living and fossil assemblages, in which the spatial distributions of the individuals comprising a species are uneven.

**peak abundance** In modeling the distribution of a taxon along a marine depth gradient, the maximum abundance of a species at its preferred depth.

**peramorphosis** A form of heterochrony in which the development of the descendant proceeds farther than that of the ancestor, with the result that earlier stages of the descendant resemble ancestral adults (see **paedomorphosis**).

**per-capita extinction rate** Rate of extinction of existing taxa relative to the number of taxa at risk.

**per-capita origination rate** Rate of origination of new taxa relative to the number available to give rise to them.

**periodic extinction** The hypothesis that mass extinctions since the Permian have recurred at regular intervals of approximately 26 million years.

**permineralization** A fossilization process in which minerals precipitate from solution in the spaces within skeletal material (see **petrifaction**).

**petrifaction** A fossilization process in which organic material is converted to mineral material (see **permineralization**).

**pH** Measure of acidity or alkalinity of a solution, generally equal to the negative base-10 logarithm of the hydrogen ion concentration, with lower pH being more acidic.

**phenogram** A dendrogram based on phenotypic data.

**phenotype** The observable traits of an organism; its form, structure, physiology, biochemistry, and behavior (see **genotype**).

**phenotypic plasticity** The tendency for a single genotype to produce different phenotypes in different environments.

**phosphates** A class of minerals containing the phosphate ion $PO_4^{3-}$.

**phyletic** Pertaining to evolutionary change within a single species-level lineage.

**phyletic gradualism** An evolutionary model in which most change is anagenetic rather than associated with lineage splitting, and in which change occurs throughout the history of a lineage.

**phyletic speciation** Origin of nominal species by the accumulation of change within a single lineage rather than by lineage splitting.

**phylogenetic** Pertaining to evolutionary change within an entire clade.

**phylogenetic classification** A system in which only monophyletic higher taxa can be formally named.

**phylogenetic code** A proposed replacement for existing codes of nomenclature that would adhere to principles of phylogenetic classification.

**phylogenetic factor** See **historical factor.**

**phylogenetic species concept** Definition of a species as a group of populations having a shared evolutionary history independent of other species.

**phylogenetics** The branch of systematics that seeks to infer evolutionary relationships among species and higher taxa.

**plesiomorphy** In phylogenetic analysis, a trait that represents the primitive condition in some group (see **apomorphy**).

**plesion** A monophyletic group within a stem group (see **crown group, stem group**).

**point counting** Assessment of the abundance of a taxon by determining the frequency with which it occurs at a fixed number of predetermined points on a given surface.

**polarity** In phylogenetic analysis, whether a trait is primitive or derived (see **apomorphy, plesiomorphy**).

**polymorphs** Morphologically distinct forms that are nonetheless part of a single species.

**polyphyletic group** A group of species whose members do not derive from a single common ancestor within the group (see **monophyletic group, paraphyletic group**).

**polysaccharides** A class of polymeric organic compounds with the general chemical formula $C_n(H_2O)_{n-1}$.

**population** In sexually reproducing species, a group of individuals of the same species that live close enough together that they have ample opportunity for interbreeding (see **metapopulation**).

**pore water** Water that occurs between grains of sediment or between the grains of a sedimentary rock.

**positive allometry** Pattern of allometry in which the *Y*-component increases with a higher relative growth rate than the *X*-component.

**preferred depth** In modeling the distribution of a taxon along a marine depth gradient, the center of a distribution, where a taxon exhibits its peak abundance.

**principal-component analysis** A kind of multivariate analysis that forms synthetic axes from measured variables, ordinates specimens with respect to these synthetic axes, and gives insight into patterns of correlation among the variables.

**principle of similitude** Principle of biological scaling, according to which shape must change, generally in a physically predictable way, if function is to be maintained.

**priority** Taxonomic rule whereby the senior (first published) of objective synonyms must be retained and the junior synonyms rejected.

**proportional rate** A relative rate.

**province** In biogeography or paleobiogeography, a group of taxa that share a similar geographic distribution, or the region defined by such a group.

**provinciality** In biogeography or paleobiogeography, said of the distribution of taxa into provinces.

**pseudofossils** Inorganic structures that resemble biological remains and are sometimes mistaken for fossils.

**Pull of the Recent** The propensity of diversity curves to be artificially inflated as the present day is approached.

**punctuated anagensis** Evolutionary model in which most change is anagenetic rather than associated with lineage splitting, and in which change is concentrated in a few episodes of rapid evolution interspersed with longer periods of stasis.

**punctuated equilibrium** Evolutionary model in which most change occurs in association with lineage splitting and in which the intervals between splitting events are characterized by stasis.

**quadrat** A template that is usually square or rectangular in shape, used to assess the presence and abundance of taxa within a fixed space on a surface.

**quantitative variable** A biological variable, such as length or mass, that can take on any value on a continuum.

**random walk** Evolutionary model in which changes in a trait value are conceived of as being drawn at random from a constant statistical distribution of evolutionary steps, and in which each evolutionary change is independent of all past and future changes.

**range chart** A graphical depiction of the stratigraphic ranges of a set of fossil taxa preserved at a locality or within a region.

**range-through assumption** In the construction of diversity curves, the assumption that a fossil taxon ranges through the entire interval between its first and last appearances in the fossil record.

**rank** The level of a taxon within a classification (species, genus, family, and so on).

**ranked variable** An ordinal variable.

**ranking and scaling (RASC)** Similar to constrained optimization, a method of constructing, with a minimum number of contradictions, an optimum composite standard of first and last appearances of taxa based on the sequences preserved at two or more localities, but with preference given to sequences of events that occur most frequently among different localities [see **composite standard reference section, constrained optimization (CONOP)**].

**rarefaction** A statistical procedure to estimate the number of species or other taxa that would have been found if a smaller number of individuals had been sampled.

**recombination** Reassortment of existing genetic material into new genotypes, through the formation of sex cells and sexual reproduction.

**recrystallization** Fossilization process in which skeletal material that is subjected to elevated temperature and pressure converts spontaneously to a thermodynamically more stable form, characterized by an altered microstructure.

**relative growth rate** Rate of growth expressed in terms of proportional increase in size per unit time.

**relative rate** A measure of the morphological rate of evolution that expresses proportional change per unit time, or the amount of change in a trait per unit time relative to the magnitude of the trait (see **absolute rate**).

**replacement** Fossilization process in which new minerals, precipitating from solution, substitute for original skeletal material (see **permineralization**).

**RNA** Ribonucleic acid, used in translating the genetic information in DNA to synthesize proteins.

**rooted network** A branching diagram that portrays proximity of evolutionary relationship and indicates the direction of character polarity (see **unrooted network**).

**score** The position of a specimen or sample on a principal component or other ordination axis.

**seismic profile** A map of discrete horizons in the subsurface, such as beneath a veneer of sediment cover on a sea floor, developed by the artificial generation of shock waves in a given study area.

**sequence** In sequence stratigraphy, a cyclical unit bounded at its base and top by a sequence boundary.

**sequence boundary** In sequence stratigraphy, a depositional hiatus or a non-eroded interval that correlates to it, which marks the stratigraphic transition from one sequence to the next.

**sequence stratigraphy** The study of depositional sequences in the stratigraphic record and the processes that produced them.

**sere** In an ecological succession, the combined set of stages that are recognized from the initial, pioneer stage to the final, climax stage.

**series** In the stratigraphic subdivision of the geological record, a unit that is intermediate between a stage and a system.

**shape** Aspect of form reflecting the relative proportions of different features; a measure of form that is dimensionless.

**shape coordinate** A measure of shape based on forming a triangle from a set of three landmarks, two of which serve as a reference axis and the third of which contains the shape information (see **baseline**).

**sibling species** Closely related species that are genetically and/or behaviorally distinct but which are difficult to discriminate on the basis of morphology.

**Signor–Lipps effect** The tendency of taxonomic loss at extinction boundaries to appear, because of incomplete sampling, more gradual in the fossil record than it actually was during the extinction event.

**silica** An important component of many minerals and other inorganic compounds, $SiO_2$ (see **opal**).

**singletons** In the study of taxonomic diversity through a set of stratigraphic intervals, taxa that are restricted to a single interval.

**sister clades** A pair of clades that are mutually more closely related to each other than either is to any other clade.

**sister species** A pair of species that are mutually more closely related to each other than either is to any other species.

**size** Aspect of form reflecting the physical extent of an organism.

**size-specific growth rate** Relative growth rate.

**soft parts** Components of organisms that lack mineralization or other hardening (see **hard parts**).

**speciation** Origin of new species by the splitting of lineages into reproductively isolated branches.

**speciation-rate bias** A postulated mechanism for phylogenetic trends, in which species sorting drives the trend and the sorting is in turn caused by a difference in the rate of origination of new species.

**species selection** A form of species sorting in which species-level properties, such as geographic range, determine the tendency of species to produce daughter species or to become extinct.

**species sorting** Preferential accumulation of species with certain phenotypes or in certain branches of the evolutionary tree because of differences in speciation and/or extinction rates.

**stabilizing selection** A form of natural selection in which intermediate phenotypes are favored relative to those above or below the average.

**stage** In the stratigraphic subdivision of the geological record, a unit that is intermediate between a zone and a series.

**standard deviation** A statistical measure of dispersion in a sample of values, equal to the positive square root of the variance.

**standard error** A measure of the uncertainty in a sample statistic, equal to the standard deviation of the theoretical probability distribution of the sample statistic.

**stasis** The absence of appreciable evolutionary change within a lineage.

**steinkern** Body fossil in the form of hardened sediment that has filled the empty skeleton of an organism.

**stem group** A paraphyletic group consisting of a clade less the crown group of that clade (see **crown group, plesion**).

**stratigraphic parsimony debt** The extent to which a postulated evolutionary tree requires the existence of lineage segments that are not actually observed (see **stratocladistics**).

**stratigraphic range** Temporal or stratigraphic extent between the observed first and last appearances of a taxon.

**stratocladistics** A method of incorporating stratigraphic information into the construction of evolutionary trees by jointly minimizing the combination of morphological parsimony debt and stratigraphic parsimony debt.

**stratophenetics** A phylogenetic method in which species are linked between successive stratigraphic intervals on the basis of overall morphological similarity rather than by the possession of synapomorphies.

**strictly monophyletic group** A monophyletic group.

**stromatolites** Laminar structures consisting of layers of sediment trapped by organic (largely cyanobacterial) mats.

**structural factor** In constructional morphology, the aspects of form that reflect necessary consequences of physical principles and material properties.

**subfossil** A term referring to death assemblages that are relatively recent in origin and have undergone relatively little degradation or diagenesis.

**subjective synonyms** Names established for different type specimens and subsequently judged by a worker to belong to one species.

**succession** Predictable changes through time in the biological compositions of the communities occupying a particular place.

**survivorship table** A representation of taxonomic survivorship data in which, for each given age class, the number of taxa having the corresponding duration is tabulated.

**symmetric random walk** A special case of a random walk in which the trait value has an equal probability of increasing or decreasing at each time increment.

**sympatric** A term used to describe populations with overlapping geographic ranges (see **allopatric**).

**synapomorphy** A derived character shared by two or more species.

**synonyms** Two or more different names applied to the same taxon.

**syntypes** Several specimens that together serve as the name bearers for a species.

**system** In the stratigraphic subdivision of the geological record, a unit that is intermediate between a series and an erathem; corresponds to a period of geologic time.

**systematics** The study of the diversity of organisms and the relationships among them.

**taphofacies** Suites of sedimentary rock characterized by preservational features of the fossils they contain.

**taphonomic control** A search for the presence of a taphonomically similar group of organisms to determine whether an absence of a taxon is biologically meaningful, as opposed to reflecting a failure of sampling.

**taphonomy** The study of fossilization processes.

**taxonomic rate of evolution** The rate at which new lineages or taxa originate and existing lineages or taxa become extinct.

**taxonomic richness** A measure of biodiversity equal to the number of taxa.

**taxonomic standardization** The procedure of adopting a consistent approach to taxonomy and revising existing data accordingly.

**taxonomic survivorship analysis** The statistical study of the durations of taxa in geologic time.

**taxonomy** The theory and practice of describing and classifying organisms.

**terminal** Said of a species or lineage segment that does not give rise to any descendants.

**terrane** A small land mass or a small part of a larger land mass that is geologically distinctive from the main part of the large land mass; it is therefore thought to have originated elsewhere and then migrated to its current position.

**terrigenous** A term referring to sediment derived from the erosion of a land mass.

**theoretical morphology** Study of the distribution of form in the history of life in comparison to the spectrum of theoretically possible forms.

**thin-plate spline** A mathematical approach to coordinate transformation in which shape differences are measured at specified landmarks and interpolated between the landmarks.

**time averaging** Accumulation of fossil material over a span of time, yielding an assemblage of organisms that were not contemporaneous.

**time–environment diagram** A two-dimensional depiction of the environmental distributions of taxa through a temporal interval of interest.

**trace fossils** Remains, such as trails and boreholes, that represent activities of organisms.

**trade-off** A compromise resulting from two or more conflicting functional or other demands that cannot be simultaneously optimized.

**transect** A means of sampling taxa that involves the use of a stretched line or chain, or that is conducted along a linear trajectory.

**transgressive systems tract (TST)** In sequence stratigraphy, the central portion of an idealized depositional sequence, stratigraphically above the lowstand systems tract and marking an increase in the rate of sea-level rise that is characterized by retrogradational stacking of parasequences or higher-order sequences [see **highstand systems tract (HST), lowstand systems tract (LST), parasequence, sequence**].

**type genus** The genus that serves as the formal name bearer for a newly described family.

**type species** The species that serves as the formal name bearer for a newly described genus.

**type specimen** Any specimen that serves as a formal name bearer for a species (see **holotype, paratype, syntypes**).

**unrooted network** A branching diagram that portrays proximity of evolutionary relationship without indicating the direction of character polarity (see **rooted network**).

**variance** A statistical measure of dispersion in a sample of values, equal to $\Sigma(x_i - \overline{x})^2/(n - 1)$, where $n$ is the sample size, $x_i(i = 1, \ldots, n)$ are the observed values, and $\overline{x}$ is the sample mean.

**zone** A synonym for biozone.

**zone fossil** A synonym for index fossil or guide fossil.

# BIBLIOGRAPHY


Adrain, J. M., and Westrop, S. R. (2000) An empirical assessment of taxic paleobiology. *Science* **289:**110–112.

Aldridge, R. J. (ed.) (1987) *Palaeobiology of Conodonts.* Chichester, U.K., Ellis Horwood, 180 pp.

Allison, P. A. (1988) The role of anoxia in the decay and mineralization of proteinaceous macro-fossils. *Paleobiology* **14:**139–154.

Alroy, J. (1994a) Appearance event ordination: A new biochronologic method. *Paleobiology* **20:**191–207.

Alroy, J. (1994b) *Quantitative Mammalian Biochronology and Biogeography of North America* [unpublished Ph.D. dissertation]. Chicago, University of Chicago, 941 pp.

Alroy, J. (2001) A multispecies overkill simulation of the end-Pleistocene megafaunal mass extinction. *Science* **292:** 1893–1896.

Alroy, J. (2004) Are Sepkoski's evolutionary faunas dynamically coherent? *Evolutionary Ecology Research* **6:**1–32.

Alvarez, L. W., Alvarez, W., Asaro, F., and Michel, H. V. (1980) Extraterrestrial cause for the Cretaceous–Tertiary extinction—Experimental results and theoretical interpretation. *Science* **208:**1095–1108.

Aronson, R. B., MacIntyre, I. G., Precht, W. F., Murdoch, T. J.T., and Wapnick, C. M. (2002) The expanding scale of species turnover events on coral reefs of Belize. *Ecological Monographs* **72:**233–249.

Bambach, R. K. (1973) Tectonic deformation of composite-mold fossil Bivalvia (Mollusca). *American Journal of Science* **273-A:**409–430.

Bambach, R. K. (1983) Ecospace utilization and guilds in marine communities through the Phanerozoic. *In* M. J. S. Tevesz and P. L. McCall (eds.), *Biotic Interactions in Recent and Fossil Benthic Communities.* New York, Plenum Press, pp. 719–746.

Bambach, R. K. (1985) Classes and adaptive variety: The ecology of diversification in marine faunas through the Phanerozoic. *In* J. W. Valentine (ed.), *Phanerozoic Diversity Patterns: Profiles in Macroevolution.* Princeton, N.J., Princeton University Press, pp. 191–253.

Bambach, R. K. (1999) Energetics in the global marine fauna: A connection between terrestrial diversification and change in the marine biosphere. *Geobios* **32:**131–144.

Barnes, R. S. K., Calow, P., and Olive, P. J. W. (1993) *The Invertebrates: A New Synthesis*, 2nd ed. Oxford, U.K., Blackwell Scientific, 488 pp.

Baumiller, T. K. (1993) Survivorship analysis of Paleozoic Crinoidea: Effect of filter morphology on evolutionary rates. *Paleobiology* **19:**304–321.

Baumiller, T. K., and Plotnick, R. E. (1989) Rotational stability in stalked crinoids and the function of wing plates in *Pterotocrinus depressus. Lethaia* **22:**317–326.

Beck, C. B. (1962) Reconstructions of *Archaeopteris*, and further consideration of its phylogenetic position. *American Journal of Botany* **39:**373–382.

Benedetto, J. L., Sánchez, T. M., Carrera, M. G., Brussa, E. D., and Salas, M. J. (1999) Paleontological constraints on successive paleogeographic positions of Precordillera terrane during the Ordovician. Geological Society of America Special Paper 336, pp. 21–42.

Bennett, K. D. (1997) *Evolution and Ecology: The Pace of Life.* Cambridge, U.K., Cambridge University Press, 208 pp.

Bennington, J. B. (2003) Transcending patchiness in the comparative analysis of paleocommunities: A test case from the Upper Cretaceous of New Jersey. *Palaios* **18:**22–33.

Benton, M. J. (1995) Diversity and extinction in the history of life. *Science* **68:**252–258.

Benton, M. J. (1996) On the nonprevalance of competitive replacement in the evolution of tetrapods. *In* D. Jablonski, D. H. Erwin, and J. H. Lipps (eds.), *Evolutionary Paleobiology.* Chicago, University of Chicago Press, pp. 185–210.

Boag, P. T. (1983) The heritability of external morphology in Darwin's ground finches (*Geospiza*) on Isla Daphne Major, Galápagos. *Evolution* **37:**877–894.

Boardman, R. S., Cheetham, A. H., and Rowell, A. J. (eds.) (1987) *Fossil Invertebrates.* Palo Alto, Calif., Blackwell, 713 pp.

Bodenbender, B. E. (1996) Patterns of crystallographic axis orientation in blastoid skeletal elements. *Journal of Paleontology* **70:**466–484.

Bodenbender, B. E., and Ausich, W. I. (2000) Skeletal crystallography and crinoid calyx architecture. *Journal of Paleontology* **74:**52–66.

Bookstein, F. L. (1991) *Morphometric Tools for Landmark Data.* New York, Cambridge University Press, 435 pp.

Bookstein, F. L., Chernoff, B., Elder, R. L., Humphries, J. M., Jr., Smith, G. R., and Strauss, R. W. (1985) *Morphometrics in Evolutionary Biology* (Academy of Natural Sciences of Philadelphia Special Publication Number 12). Philadelphia, The Academy of Natural Sciences of Philadelphia, 277 pp.

Bottjer, D. J., and Ausich, W. I. (1986) Phanerozoic development of tiering in soft substrata suspension-feeding communities. *Paleobiology* **12:**400–420.

Bottjer, D. J., Etter, W., Hagadorn, J. W., and Tang, C. M. (2002) *Exceptional Fossil Preservation.* New York, Columbia University Press, 403 pp.

Bowen, G. J., Clyde, W. C., Koch, P. L., Ting, S., Alroy, J., Tsubamoto, T., Wang, Y., and Wang, Y. (2002) Mammalian dispersal at the Paleocene/Eocene boundary. *Science* **295:**2062–2065.

Bowring, S. A., Erwin, D. H., Jin, Y. G., Martin, M. W., Davidek, K., and Wang, W. (1998) U/Pb zircon geochronology and tempo of the end-Permian mass extinction. *Science* **280:**1039–1045.

Bowring, S. A., Grotzinger, J. P., Isachsen, C. E., Knoll, A. H., Pelechaty, S. M., and Kolosov, P. (1993) Calibrating rates of early Cambrian evolution. *Science* **261:**1293–1298.

Boyce, C. K., Cody, G. D., Fogel, M. L., Hazen, R. M., Alexander, C. M. O'D., and Knoll, A. H. (2003) Chemical evidence for cell wall lignification and the evolution of tracheids in Early Devonian plants. *International Journal of Plant Sciences* **164:**691–702.

Brett, C. E., and Baird, G. C. (1986) Comparative taphonomy: A key to paleoenvironmental interpretation based on fossil preservation. *Palaios* **1:**207–227.

Brett, C. E., and Baird, G. C. (1995) Coordinated stasis and evolutionary ecology of Silurian to Middle Devonian faunas in the Appalachian Basin. *In* D. H. Erwin and R. L. Anstey (eds.), *New Approaches to Speciation in the Fossil Record.* New York, Columbia University Press, pp. 285–315.

Brower, J. C. (1999) A new pleurocystitid rhombiferan echinoderm from the Middle Ordovician Galena Group of northern Iowa and southern Minnesota. *Journal of Paleontology* **73:**129–153.

Bruton, D. L., and Haas, W. (2003) The puzzling eye of *Phacops. Special Papers in Palaeontology* **70:**349–361.

Budd, A. F. (2000) Diversity and extinction in the Cenozoic history of Caribbean reefs. *Coral Reefs* **19:**25–35.

Burnham, R. J., Pitman, N. C. A., Johnson, K. R., and Wilf, P. (2001) Habitat-related error in estimating temperatures from leaf margins in a humid tropical forest. *American Journal of Botany* **88:**1096–1102.

Campbell, K. S. W. (1957) A Lower Carboniferous brachiopod-coral fauna from New South Wales. *Journal of Paleontology* **31:**34–94.

Carney, J. L., and Pierce, R. W. (1995) Graphic correlation and composite standard databases as tools for the exploration biostratigrapher. *In* K. O. Mann and H. R. Lane (eds.), *Graphic Correlation.* Tulsa, Okla., SEPM Special Publication No. 53, pp. 23–43.

Carrano, M. T. (1998a) Locomotion in non-avian dinosaurs: Integrating data from hindlimb kinematics, in vivo strains, and bone morphology. *Paleobiology* **24:**450–469.

Carrano, M. T. (1998b) *The Evolution of Dinosaur Locomotion: Functional Morphology, Biomechanics, and Modern Analogs* [unpublished Ph.D. dissertation]. Chicago, University of Chicago, 424 pp.

Carroll, R. L. (1988) *Vertebrate Paleontology and Evolution.* New York, W. H. Freeman and Company, 698 pp.

Cavicchioli, R. (2002) Extremophiles and the search for extraterrestrial life. *Astrobiology* **2:**281–292.

Chapman, R. E., Galton, P. M., Sepkoski, J. J., Jr., and Wall, W. P. (1981) A morphometric study of the cranium of the pachycephalosaurid *Stegoceras. Journal of Paleontology* **55:**608–618.

Cheetham, A. H. (1986a) Branching, biometrics, and bryozoan evolution. *Proceedings of the Royal Society of London* B **228:**151–171.

Cheetham, A. H. (1986b) Tempo of evolution in a Neogene bryozoan: Rates of morphologic change within and across species boundaries. *Paleobiology* **12:**190–202.

Clarkson, E. N. K. (1998) *Invertebrate Palaeontology and Evolution*, 4th ed. Oxford, U.K., Blackwell, 452 pp.

Clarkson, E. N. K., and Levi-Setti, R. (1975) Trilobite eyes and the optics of Descartes and Huygens. *Nature* **254:**663–667.

Clements, F. E. (1916) *Plant Succession: An Analysis of the Development of Vegetation.* Carnegie Institution of Washington Publication 242, 512 pp.

Cloud, P. E. (1973) Pseudofossils: A plea for caution. *Geology* **1:**123–127.

Clyde, W. C., and Gingerich, P. D. (1994) Rates of evolution in dentition of Early Eocene *Cantius*: Comparison of size and shape. *Paleobiology* **20:**506–522.

Conway Morris, S. (1976) A new Cambrian lophophorate from the Burgess Shale of British Columbia. *Palaeontology* **19:**199–222.

Conway Morris, S. (1998) *The Crucible of Creation.* Oxford, U.K., Oxford University Press, 242 pp.

Cooper, G. A., and Grant, R. E. (1976) Permian brachiopods of West Texas, IV. *Smithsonian Contributions to Paleobiology* **21:**1923–2607.

Cooper, R. A., Crampton, J. S., Raine, J. I., Gradstein, F. M., Morgans, H. E. G., Sadler, P. M., Strong, C. P., Waghorn, D., and Wilson, G. J. (2001) Quantitative biostratigraphy of the Taranaki Basin, New Zealand: A deterministic and probabilistic approach. *American Association of Petroleum Geologists Bulletin* **85:**1469–1498.

Courtillot, V., and Gaudemer, Y. (1996) Effects of mass extinctions on biodiversity. *Nature* **381:**146–148.

Darwin, C. R. (1859) *On the Origin of Species by Means of Natural Selection.* London, John Murray, 502 pp.

Davis, M., Hut, P., and Muller, R. A. (1984) Extinction of species by periodic comet showers. *Nature* **308:**715–717.

Dickens, G. R., O'Neil, J. R., Rea, D. K., and Owen, R. M. (1995) Dissociation of oceanic methane hydrate as a cause of the carbon isotope excursion at the end of the Paleocene. *Paleoceanography* **10:**965–971.

Dixon, O. A. (1989) Species definition in heliolitine corals of the lower Douro Formation (Upper Silurian), Canadian Arctic. *Journal of Paleontology* **63:**819–838.

Dominguez, P., Jacobson, A. G., and Jefferies, R. P. S. (2002) Paired gill slits in a fossil with a calcite skeleton. *Nature* **417:**841–844.

Droser, M. L., and Bottjer, D. J. (1993) Trends and patterns of Phanerozoic ichnofabrics. *Annual Review of Earth and Planetary Sciences* **21:**205–225.

Eble, G. J. (1999) Originations: Land and sea compared. *Geobios* **32:**223–234.

Eldredge, N., and Gould, S. J. (1972) Punctuated equilibria: An alternative to phyletic gradualism. *In* T. J. M. Schopf (ed.), *Models in Paleobiology.* San Francisco, Freeman, Cooper, and Company, pp. 82–115.

Erwin, D. H., Valentine, J. W., and Sepkoski, J. J., Jr. (1987) A comparative study of diversification events—The early Paleozoic versus the Mesozoic. *Evolution* **41:**1177–1186.

Fahn, A. (1982) *Plant Anatomy*, 3rd ed. New York, Pergamon Press, 544 pp.

Falkowski, P. G., Katz, M. E., Knoll, A. H., Quigg, A., Raven, J. A., Schofield, O., and Taylor, F. J. R. (2004) The evolution of modern eukaryotic phytoplankton. *Science* **305:**354–360.

Farmer, J. D., and Des Marais, D. J. (1999) Exploring for a record of ancient Martian life. *Journal of Geophysical Research* **104:**26977–26995.

Farrell, J. R. (1992) The Garra Formation (Early Devonian: Late Lochkovian) between Cumnock and Larras Lee, New South Wales, Australia: Stratigraphic and structural setting, faunas and community sequence. *Palaeontographica Abteilung A* **222:**1–41.

Ferson, S., Rohlf, F. J., and Koehn, R. K. (1985) Measuring shape variation of two-dimensional outlines. *Systematic Zoology* **34:**69–78.

Fisher, D. C. (1977) Functional morphology of spines in the Pennsylvanian horseshoe crab *Euproops danae. Paleobiology* **3:**175–195.

Fisher, D. C. (1984) Taphonomic analysis of late Pleistocene mastodon occurrences: Evidence of butchery by North American Paleo-Indians. *Paleobiology* **10:**338–357.

Fisher, D. C. (1991) Phylogenetic analysis and its application in evolutionary paleobiology. *In* N. L. Gilinsky and P. W. Signor (eds.), *Analytical Paleobiology.* Short Courses in Paleontology, Number 4. Knoxville, Tenn., The Paleontological Society, pp. 103–122.

Fisher, D. C. (1996) Extinction of proboscideans in North America. *In* J. Shoshani and P. Tassy (eds.), *The Proboscidea.* Oxford, U.K., Oxford University Press, pp. 296–315.

Fisher, D. C. (2002) Season of death, growth rates, and life history of North American mammoths. *In* D. West (ed.), *Proceedings of the International Conference on Mammoth Site Studies.* Publications in Anthropology, No. 22. Lawrence, Kans., University of Kansas, pp. 121–135.

Flessa, K. W., and Imbrie, J. (1973) Evolutionary pulsations: Evidence from Phanerozoic diversity patterns. *In* D. H. Tarling and S. K. Runcorn (eds.), *Implications of Continental Drift to the Earth Sciences*, volume 1. London, Academic Press, pp. 245–285.

Foote, M. (1988) Survivorship analysis of Cambrian and Ordovician trilobites. *Paleobiology* **14:**258–271.

Foote, M. (1991) Morphologic and taxonomic diversity in a clade's history: The blastoid record and stochastic simulations. *Contributions from the Museum of Paleontology, University of Michigan* **28:**101–140.

Foote, M. (1992) Paleozoic record of morphological diversity in blastozoan echinoderms. *Proceedings of the National Academy of Sciences USA* **89:**7325–7329.

Foote, M. (1993) Discordance and concordance between morphological and taxonomic diversity. *Paleobiology* **19:**184–205.

Foote, M., and Raup, D. M. (1996) Fossil preservation and the stratigraphic ranges of taxa. *Paleobiology* **22:**121–140.

Foote, M., and Sepkoski, J. J., Jr. (1999) Absolute measures of the completeness of the fossil record. *Nature* **398:**415–417.

Fortey, R. A. (1985a) Pelagic trilobites as an example of deducing the life-habits of extinct arthropods. *Transactions of the Royal Society of Edinburgh: Earth Sciences* **76:**219–230.

Fortey, R. A. (1985b) Gradualism and punctuated equilibria as competing and complementary theories. *Special Papers in Palaeontology* **33:**17–28.

Fortey, R., and Chatterton, B. (2003) A Devonian trilobite with an eyeshade. *Science* **301:**1689.

Futuyma, D. J. (1987) On the role of speciation in anagenesis. *American Naturalist* **130:**465–473.

Gál, J., Horváth, G., Clarkson, E. N. K., and Haiman, O. (2000) Image formation by bifocal lenses in a trilobite eye? *Vision Research* **40:**843–853.

Garcia-Ruiz, J. M., Hyde, S. T., Carnerup, A. M., Christy, A. G., Van Kranendonk, M. J., and Welham, N. J. (2003) Self-assembled silica-carbonate structures and detection of ancient microfossils. *Science* **302:**1195–1197.

Gilbert, J. J. (1966) Rotifer ecology and embryological induction. *Science* **151:**1234–1237.

Gilbert, J. J. (1967) *Asplanchna* and postero-lateral spine production in *Brachionus calyciflorus. Archiv für Hydrobiologie* **64:**1–62.

Gingerich, P. D. (1974) Stratigraphic record of Early Eocene *Hyopsodus* and the geometry of mammalian phylogeny. *Nature* **248:**107–109.

Gingerich, P. D. (1983) Rates of evolution: Effects of time and temporal scaling. *Science* **222:**159–161.

Gingerich, P. D. (1993) Quantification and comparison of evolutionary rates. *American Journal of Science* **293-A:**453–478.

Gingras, M. K., MacMillan, B., Balcom, B. J., Saunders, T., and Pemberton, S. G. (2002) Using magnetic resonance imaging and petrographic techniques to understand the textural attributes and porosity distribution in *Macaronichnus*-burrowed sandstone. *Journal of Sedimentary Research* **72:**552–558.

Gleason, H. A. (1926) The individualistic concept of the plant association. *Bulletin of the Torrey Botanical Club* **53:**7–26.

Goldman, D., Mitchell, C. E., and Joy, M. P. (1999) The stratigraphic distribution of graptolites in the classic upper Middle Ordovician Utica Shale of New York State: An evolutionary succession or a response to relative sea-level change? *Paleobiology* **25:**273–294.

Gould, S. J. (1967) Evolutionary patterns in pelycosaurian reptiles: A factor-analytic study. *Evolution* **21:**385–401.

Gould, S. J. (1982) The meaning of punctuated equilibrium and its role in validating a hierarchical approach to macroevolution. *In* R. Milkman (ed.), *Perspectives on Evolution.* Sunderland, Mass., Sinauer Associates, pp. 83–104.

Gould, S. J. (1985) The paradox of the first tier—An agenda for paleobiology. *Paleobiology* **11:**2–12.

Gould, S. J. (2002) *The Structure of Evolutionary Theory.* Cambridge, Mass., Harvard University Press, 1433 pp.

Gradstein, F. M., Ogg, J. G., and Smith, A. G. (eds.) (2004) *A Geologic Timescale 2004.* Cambridge, U.K., Cambridge University Press, 589 pp.

Grant, R. E. (1980) The human face of the brachiopod. *Journal of Paleontology* **54:**499–507.

Greenstein, B. J. (1991) An integrated study of echinoid taphonomy: Predictions for the fossil record of four echinoid families. *Palaios* **6:**519–540.

Grimes, S. T., Brock, F., Rickard, D., Davies, K. L., Edwards, D., Briggs, D. E. G., and Parkes, R. J. (2001) Understanding fossilization: Experimental pyritization of plants. *Geology* **29:**123–126.

Grotzinger, J. P., Watters, W. A., and Knoll, A. H. (2000) Calcified metazoans in thrombolite-stromatolite reefs of the terminal Proterozoic Nama Group, Namibia. *Paleobiology* **26:**334–359.

Haldane, J. B. S. (1949) Suggestions as to quantitative measurement of rates of evolution. *Evolution* **3:**51–56.

Hallam, A. (1986) The Pliensbachian and Tithonian extinction events. *Nature* **319:**765–768.

Hansen, T. A. (1980) Influence of larval dispersal and geographic distribution on species longevity in neogastropods. *Paleobiology* **6:**193–207.

Hansen, T. A. (1982) Modes of larval development in early Tertiary neogastropods. *Paleobiology* **8:**367–372.

Harland, W. B., Armstrong, R. L., Cox, A. V., Craig, L. E., Smith, A. G., and Smith, D. G. (1990) *A Geologic Time Scale 1989.* Cambridge, U.K., Cambridge University Press, 263 pp.

Harland, W. B., Cox, A. V., Llewellyn, P. G., Pickton, C. A. G., Smith, A. G., and Walters, R. (1982) *A Geologic Time Scale.* Cambridge, U.K., Cambridge University Press, 131 pp.

Hoffman, A. (1977) Synecology of macrobenthic assemblages of the Korytnica Clays (Middle Miocene; Holy Cross Mountains, Poland). *Acta Geologica Polonica* **27:**227–280.

Hoffman, A. (1979) A consideration upon macrobenthic assemblages of the Korytnica Clays (Middle Miocene; Holy Cross Mountains, central Poland). *Acta Geologica Polonica* **29:**345–352.

Holland, S. M. (1995) The stratigraphic distribution of fossils. *Paleobiology* **21:**92–109.

Holland, S. M. (1999) The new stratigraphy and its promise for paleobiology. *Paleobiology* **25:**409–416.

Holland, S. M. (2003) Confidence limits on fossil ranges that account for facies changes. *Paleobiology* **29:**468–479.

Hunt, A. S. (1967) Growth, variation, and instar development of an agnostid trilobite. *Journal of Paleontology* **41:**203–208.

Hunt, G. (2004a) Phenotypic variation in fossil samples: Modeling the consequences of time-averaging. *Paleobiology* **30:**426–443.

Hunt, G. (2004b) Phenotypic variance inflation in fossil samples: An empirical assessment. *Paleobiology* **30:**487–506.

Imbrie, J. (1956) Biometrical methods in the study of invertebrate fossils. *Bulletin of the American Museum of Natural History* **108:**211–252.

Ivany, L. C., Patterson, W. P., and Lohmann, K. C. (2000) Cooler winters as a possible cause of mass extinctions at the Eocene/Oligocene boundary. *Nature* **407:**887–890.

Jablonski, D. (1986) Background and mass extinctions—The alternation of macroevolutionary regimes. *Science* **231:**129–133.

Jablonski, D. (1998) Geographic variation in the molluscan recovery from the end-Cretaceous extinction. *Science* **279:**1327–1330.

Jablonski, D. (2005) Evolutionary innovations in the fossil record: The intersection of ecology, development, and macroevolution. *Journal of Experimental Zoology, Part B. Molecular and Developmental Evolution* **304B:**504–519.

Jablonski, D., and Bottjer, D. J. (1983) Soft-bottom epifaunal suspension-feeding assemblages in the Late Cretaceous: Implications for the evolution of benthic paleocommunities. *In* M. J. S. Tevesz and P. L. McCall (eds.), *Biotic Interactions in Recent and Fossil Benthic Communities.* New York, Plenum Press, pp. 747–812.

Jablonski, D., and Bottjer, D. J. (1991) Environmental patterns in the origins of higher taxa: The post-Paleozoic fossil record. *Science* **252:**1831–1833.

Jablonski, D., and Lutz, R. A. (1980) Molluscan larval shell morphology: Ecological and Paleontological applications. *In* D. C. Rhoads and R. A. Lutz (eds.), *Skeletal Growth of Aquatic Organisms: Biological Records of Environmental Change.* New York, Plenum, pp. 323–327.

Jablonski, D., Roy, K., Valentine, J. W., Price, R. M., and Anderson, P. S. (2003) The impact of the pull of the Recent on the history of marine diversity. *Science* **300:**1133–1135.

Jackson, J. B. C. (2001) What was natural in the coastal oceans? *Proceedings of the National Academy of Sciences USA* **98:**5411–5418.

Jackson, J. B. C., and Cheetham, A. H. (1990) Evolutionary significance of morphospecies: A test with cheilostome Bryozoa. *Science* **248:**579–583.

Jackson, J. B. C., and Cheetham, A. H. (1994) Phylogeny reconstruction and the tempo of speciation in cheilostome Bryozoa. *Paleobiology* **20:**407–423.

Jackson, J. B. C., and Cheetham, A. H. (1999) Tempo and mode of speciation in the sea. *Trends in Ecology and Evolution* **14:**72–77.

Jackson, J. B. C, Kirby, M. X., Berger, W. H., Bjorndal, K. A., Botsford, L. W., Bourque, B. J., Bradbury, R. H., Cooke, R., Erlandson, J., Estes, J. A., Hughes, T. P., Kidwell, S., Lange, C. B., Lenihan, H. S., Pandolfi, J. M., Peterson, C. H., Steneck, R. S., Tegner, M. J., and Warner, R. R. (2001) Historical overfishing and the recent collapse of coastal ecosystems. *Science* **293:**629–638.

Jackson, R. T. (1912) Phylogeny of the Echini with a revision of Paleozoic species. *Boston Society of Natural History Memoir* **7:**1–443.

Jerison, H. J. (1969) Brain evolution and dinosaur brains. *American Naturalist* **103:**575–588.

Jin, Y. G., Wang, Y., Wang, W., Shang, Q. H., Cao, C. Q., and Erwin, D. H. (2000) Pattern of marine mass extinction near the Permian–Triassic boundary in South China. *Science* **289:**432–436.

Jones, B., Renault, R. W., and Rosen, M. R. (2001) Taphonomy of silicified filamentous microbes in modern geothermal sinters—Implications for identification. *Palaios* **16:**580–592.

Jones, D. S., and Gould, S. J. (1999) Direct measurement of age in fossil *Gryphaea*: The solution to a classic problem in heterochrony. *Paleobiology* **25:**158–187.

Jones, D. S., and Quitmyer, I. R. (1996) Marking time with bivalve shells: Oxygen isotopes and season of annual increment formation. *Palaios* **11:**340–346.

Katz, M. E., Pak, D. K., Dickens, G. R., and Miller, K. G. (1999) The source and fate of massive carbon input during the Latest Paleocene Thermal Maximum. *Science* **286:**1531–1533.

Kemple, W. G., Sadler, P. M., and Strauss, D. J. (1989) A prototype constrained optimization solution to the time correlation problem. *In* F. P. Agterberg and G. F. Bonham-Carter (eds.), *Statistical Applications in the Earth Sciences.* Ottawa, Canada, *Geological Survey of Canada Papers* 89-9, pp. 417–425.

Kemple, W. G., Sadler, P. M., and Strauss, D. J. (1995) Extending graphic correlation to many dimensions: Stratigraphic correlation as constrained optimization. *In* K. O. Mann and H. R. Lane (eds.), *Graphic Correlation.* Tulsa, Okla., SEPM Special Publication No. 53, pp. 65–82.

Kenrick, P., and Crane, P. R. (1997) *The Origin and Early Diversification of Land Plants: A Cladistic Study.* Washington, D.C., Smithsonian Institution Press, 441 pp.

Kerr, R. A. (1987) Asteroid impact gets more support. *Science* **236:**666–668.

Kidwell, S. M. (2001) Preservation of species abundance in marine death assemblages. *Science* **294:**1091–1094.

Kidwell, S. M., and Baumiller, T. (1990) Experimental disintegration of regular echinoids: Roles of temperature, oxygen, and decay thresholds. *Paleobiology* **16:**247–271.

Kingsolver, J. G., and Koehl, M. A. R. (1985) Aerodynamics, thermoregulation, and the evolution of insect wings: Differential scaling and evolutionary change. *Evolution* **39:**488–504.

Knoll, A. H. (2003) *Life on a Young Planet.* Princeton, N.J., Princeton University Press, 277 pp.

Knoll, A. H., and Barghoorn, E. S. (1977) Archean microfossils showing cell division from the Swaziland System of South Africa. *Science* **198:**396–398.

Koch, P. L., Fisher, D. C., and Dettman, D. (1989) Oxygen isotope variation in the tusks of extinct proboscideans: A measure of season of death and seasonality. *Geology* **17:**515–519.

Koch, P. L., Zachos, J. C., and Gingerich, P. D. (1992) Correlation between isotope records in marine and continental reservoirs near the Paleocene/Eocene boundary. *Nature* **358:**319–322.

Kowalewski, M. (2001) Applied marine paleoecology: An oxymoron or reality? *Palaios* **16:**309–311.

Kowalewski, M., Avila Serrano, G. E., Flessa, K. W., and Goodfriend, G. A. (2000) Dead delta's former productivity: Two trillion shells at the mouth of the Colorado River. *Geology* **28:**1059–1062.

Kummel, B., and Steele, G. (1962) Ammonites from the *Meekoceras gracilitatus* Zone at Crittenden Springs, Elko County, Nevada. *Journal of Paleontology* **36:**638–703.

Labandeira, C. C., Johnson, K. R., and Wilf, P. (2002) Impact of the terminal Cretaceous event on plant–insect associations. *Proceedings of the National Academy of Sciences USA* **99:**2061–2066.

Lande, R. (1979) Quantitative genetic analysis of multivariate evolution, applied to brain:body size allometry. *Evolution* **33:**402–416.

Leaf Architecture Working Group. (1999) *Manual of Leaf Architecture.* Washington, D.C., Smithsonian Institution, 67 pp.

Levi-Setti, R. (1975) *Trilobites: A Photographic Atlas.* Chicago, University of Chicago Press, 213 pp.

Linck, O. (1954) Die Muschelkalk-Seelilie *Encrinus liliiformis. Aus der Heimat* **62:**225–235.

Linnaeus, C. (1758) *Systema Naturae per Regna tria Naturae, Secundum Classes, Ordines, Genera, Species cum Characteribus, Differentiis, Synonymis, Locis,* 10th ed. Stockholm, Laurentii Salvii.

Lyell, C. (1833) *Principles of Geology,* volume III. London, John Murray, 383 pp.

Lyon, A. G., and Edwards, D. (1991) The first zosterophyll from the Lower Devonian Rhynie Chert, Aberdeenshire. *Transactions of the Royal Society of Edinburgh: Earth Sciences* **82:**323–332.

Maas, M. C., Anthony, M. R. L., Gingerich, P. D., Gunnell, G. F., and Krause, D. W. (1995) Mammalian generic diversity and turnover in the Late Paleocene and Early Eocene of the Bighorn and Crazy Mountain basins, Wyoming and Montana (USA). *Palaeogeography, Palaeoclimatology, Palaeoecology* **115:**181–207.

MacArthur, R. H., and Wilson, E. O. (1967) *The Theory of Island Biogeography.* Princeton, N.J., Princeton University Press, 203 pp.

MacDonald, K. B. (1969) Quantitative studies of salt marsh mollusc faunas from the North American Pacific coast. *Ecological Monographs* **39:**33–60.

MacFadden, B. J. (1985) Patterns of phylogeny and rates of evolution in fossil horses: Hipparions from the Miocene and Pliocene of North America. *Paleobiology* **11:**245–257.

MacFadden, B. J. (1986) Fossil horses from "Eohippus" (*Hyracotherium*) to *Equus*: Scaling, Cope's Law, and the evolution of body size. *Paleobiology* **12:**355–369.

Martin, P. A., Lea, D. W., Rosenthal, Y., Shackleton, N. J., Sarnthein, M., and Papenfuss, T. (2002) Quaternary deep sea temperature histories derived from benthic foraminiferal Mg/Ca. *Earth and Planetary Science Letters* **198:**193–209.

Martin, P. S., and Steadman, D. W. (1999) Prehistoric extinctions on islands and continents. *In* R. D. E. MacPhee (ed.), *Extinctions in Near Time.* New York, Kluwer Academic, pp. 17–55.

Mayr, E. (1942) *Systematics and the Origin of Species.* New York, Columbia University Press, 334 pp.

McElwain, J. C., Beerling, D. J., and Woodward, F. I. (1999) Fossil plants and global warming at the Triassic–Jurassic boundary. *Science* **285:**1386–1390.

McGhee, G. R., and McKinney, F. K. (2000) A theoretical morphologic analysis of convergently evolved erect helical colony form in the Bryozoa. *Paleobiology* **26:**556–577.

McGhee, G. R., and McKinney, F. K. (2003) Evolution of erect helical colony form in the Bryozoa: Phylogenetic, functional, and ecological factors. *Biological Journal of the Linnean Society* **80:**235–260.

McKinney, F. K. (1992) Competitive interactions between related clades: Evolutionary implications of overgrowth interactions between encrusting cyclostome and cheilostome bryozoans. *Marine Biology* **114:**645–652.

Meyer, D. L., and Macurda, D. B., Jr. (1977) Adaptive radiation of the comatulid crinoids. *Paleobiology* **3:**74–82.

Miller, A. I. (2001) Ordovician radiation. *In* D. E. G. Briggs and P. R. Crowther (eds.), *Palaeobiology II.* Oxford, U.K., Blackwell Scientific, pp. 49–52.

Miller, A. I., and Sepkoski, J. J., Jr. (1988) Modeling bivalve diversification: The effect of interaction on a macroevolutionary system. *Paleobiology* **14:**364–369.

Miller, A. I., Holland, S. M., Meyer, D. L., and Dattilo, B. F. (2001) The use of faunal gradient analysis for intraregional correlation and assessment of changes in sea-floor topography in the type Cincinnatian. *Journal of Geology* **109:**603–613.

Moore, R. C., and Teichert, C. (eds.) (1978) *Treatise on Invertebrate Paleontology, Part T, Echinodermata 2.* Boulder, Colo. and Lawrence, Kans., The Geological Society of America and The University of Kansas, 1027 pp.

Morris, J. (1854) *A Catalogue of British Fossils.* The Author, London.

Niklas, K. J. (1994a) *Plant Allometry.* Chicago, University of Chicago Press, 395 pp.

Niklas, K. J. (1994b) Morphological evolution through complex domains of fitness. *Proceedings of the National Academy of Sciences USA* **91:**6772–6779.

Niklas, K. J. (1997) *The Evolutionary Biology of Plants.* Chicago, University of Chicago Press, 449 pp.

Niklas, K. J., and Kerchner, V. (1984) Mechanical and photosynthetic constraints on the evolution of plant shape. *Paleobiology* **10:**79–101.

Nixon, K. C., and Wheeler, Q. D. (1991) An amplification of the phylogenetic species concept. *Cladistics* **6:**211–223.

Norris, R. D. (1991) Biased extinction and evolutionary trends. *Paleobiology* **17:**388–399.

Norris, R. D., Kroon, D., Klaus, A., et al. (1998) Blake Nose. *Proceedings of the Ocean Drilling Program, Initial Reports* 171B, 83 pp.

Novack-Gottshall, P. M., and Miller, A. I. (2003) Comparative geographic and environmental diversity dynamics of gastropods and bivalves during the Ordovician Radiation. *Paleobiology* **29:**576–604.

Olson, E. C. (1972) *Diplocaulus parvus* n. sp. (Amphibia: Nectridea) from the Chickasha Formation (Permian: Guadalupian) of Oklahoma. *Journal of Paleontology* **46:**656–659.

Padian, K., and Clemens, W. A. (1985) Terrestrial vertebrate diversity: Episodes and insights. *In* J. W. Valentine (ed.), *Phanerozoic Diversity Patterns: Profiles in Macroevolution.* Princeton, N.J., Princeton University Press, pp. 41–96.

Pannella, G., and MacClintock, C. (1968) Biological and environmental rhythms reflected in molluscan shell growth. *In* D. B. Macurda (ed.), *Paleobiological Aspects of Growth and Development, a Symposium.* Menlo Park, Calif., Paleontological Society, *Paleontological Society Memoir* **2:**64–80.

Paul, C. R. C. (1982) The adequacy of the fossil record. *In* K. A. Joysey and A. E. Friday (eds.), *Problems of Phylogenetic Reconstruction* (Systematics Association Special Volume No. 21). London, U.K., Academic Press, pp. 75–117.

Paul, C. R. C., and Smith, A. B. (1984) The early radiation and phylogeny of echinoderms. *Biological Reviews* **59:**443–481.

Petersen, C. G. J. (1915) On the animal communities of the sea bottom in the Skagerak, the Christiania Fjord and the Danish waters. *Report of the Danish Biological Station to the Board of Agriculture* **23:**3–28.

Phillips, J. (1860) *Life on the Earth.* Cambridge, U.K., Macmillan, 224 pp.

Precht, W. F., and Aronson, R. B. (2004) Climate flickers and range shifts in reef corals. *Frontiers in Ecology and the Environment* **2:**307–314.

Purvis, A. (1995) A composite estimate of primate phylogeny. *Philosophical Transactions of the Royal Society of London* **348:**405–421.

Raup, D. M. (1966) Geometric analysis of shell coiling: General problems. *Journal of Paleontology* **40:**1178–1190.

Raup, D. M. (1967) Geometric analysis of shell coiling: Coiling in ammonoids. *Journal of Paleontology* **41:**43–65.

Raup, D. M. (1975) Taxonomic diversity estimation using rarefaction. *Paleobiology* **1:**333–342.

Raup, D. M. (1976a) Species diversity in the Phanerozoic: A tabulation. *Paleobiology* **2:**279–288.

Raup, D. M. (1976b) Species diversity in the Phanerozoic: An interpretation. *Paleobiology* **2:**289–297.

Raup, D. M. (1977) Stochastic models in evolutionary palaeontology. *In* A. Hallam (ed.), *Patterns of Evolution as Illustrated by the Fossil Record.* Amsterdam, Elsevier, pp. 59–78.

Raup, D. M. (1979) Biases in the fossil record of species and genera. *Bulletin of the Carnegie Museum of Natural History* **13:**85–91.

Raup, D. M. (1983) On the early origins of major biologic groups. *Paleobiology* **9:**107–115.

Raup, D. M. (1985) Mathematical models of cladogenesis. *Paleobiology* **11:**42–52.

Raup, D. M. (1991) A kill curve for Phanerozoic marine species. *Paleobiology* **17:**37–48.

Raup, D. M. (1992) Large-body impact and extinction in the Phanerozoic. *Paleobiology* **18:**80–88.

Raup, D. M., and Crick, R. E. (1981) Evolution of single characters in the Jurassic ammonite *Kosmoceras. Paleobiology* **7:**200–215.

Raup, D. M., and Jablonski, D. (1993) Geography of end-Cretaceous marine bivalve extinctions. *Science* **260:**971–973.

Raup, D. M., and Sepkoski, J. J., Jr. (1982) Mass extinctions in the marine fossil record. *Science* **215:**1501–1503.

Raup, D. M., and Sepkoski, J. J., Jr. (1984) Periodicity of extinctions in the geologic past. *Proceedings of the National Academy of Sciences USA* **81:**801–805.

Raup, D. M., and Sepkoski, J. J., Jr. (1986) Periodic extinction of families and genera. *Science* **231:**833–836.

Rice, S. H. (1998) The bio-geometry of mollusc shells. *Paleobiology* **24:**133–149.

Romer, A. S. (1970) *The Vertebrate Body*, 4th ed. Philadelphia, W. B. Saunders, 601 pp.

Rosenzweig, M. L., and McCord, R. D. (1991) Incumbent replacement: Evidence for long-term evolutionary progress. *Paleobiology* **17:**202–213.

Rudwick, M. J. S. (1961) The feeding mechanism of the Permian brachiopod *Prorichtofenia. Palaeontology* **3:**450–471.

Sadler, P. M., and Cooper, R. A. (2004) Calibration of the Ordovician timescale. *In* B. D. Webby, F. Paris, M. L. Droser, and I. G. Percival (eds.), *The Great Ordovician Biodiversification Event.* New York, Columbia University Press, pp. 48–51.

Savrda, C. E. (1995) Ichnologic applications in paleoceanographic, paleoclimatic, and sea-level studies. *Palaios* **10:**565–577.

Schäfer, W. (1972) *Ecology and Paleoecology of Marine Environments* (Translated by Inngard Oertel, edited by G. Y. Craig). Chicago, University of Chicago Press, 568 pp.

Schopf, J. W., and Walter, M. R. (1983) Archean microfossils: New evidence of ancient microbes. *In* J. W. Schopf (ed.), *Earth's Earliest Biosphere: Its Origin and Evolution.* Princeton, N.J., Princeton University Press, pp. 214–239.

Seilacher, A. 1970. Arbeitskonzept zur Konstruktions-Morphologie. *Lethaia* **3:**393–396.

Sepkoski, J. J., Jr. (1978) A kinetic model of Phanerozoic taxonomic diversity I. Analysis of marine orders. *Paleobiology* **4:**223–251.

Sepkoski, J. J., Jr. (1979) A kinetic model of Phanerozoic taxonomic diversity II. Early Phanerozoic families and multiple equilibria. *Paleobiology* **5:**222–251.

Sepkoski, J. J., Jr. (1981) A factor analytic description of the Phanerozoic marine fossil record. *Paleobiology* **7:**36–53.

Sepkoski, J. J., Jr. (1984) A kinetic model of Phanerozoic taxonomic diversity III. Post-Paleozoic families and multiple equilibria. *Paleobiology* **10:**246–267.

Sepkoski, J. J., Jr. (1997) Biodiversity: Past, present, and future. *Journal of Paleontology* **71:**533–539.

Sepkoski, J. J., Jr. (1998) Rates of speciation in the fossil record. *Philosophical Transactions of the Royal Society of London, Series B* **333:**315–326.

Sepkoski, J. J., Jr. (2002) A compendium of fossil marine animal genera. *Bulletins of American Paleontology* **363:**1–560.

Sepkoski, J. J., Jr., and Miller, A. I. (1985) Evolutionary faunas and the distribution of Paleozoic marine communities in space and time. *In* J. W. Valentine (ed.), *Phanerozoic Diversity Patterns: Profiles in Macroevolution.* Princeton, N.J., Princeton University Press, pp. 153–190.

Sepkoski, J. J., Jr., and Sheehan, P. M. (1983) Diversification, faunal change, and community replacement during the Ordovician Radiations. *In* M. J. S. Tevesz and P. L. McCall (eds.), *Biotic Interactions in Recent and Fossil Benthic Communities.* New York, Plenum Press, pp. 673–717.

Sepkoski, J. J., Jr., McKinney, F. K., and Lidgard, S. (2000) Competitive displacement among post-Paleozoic cyclostome and cheilostome bryozoans. *Paleobiology* **26:**7–18.

Shaw, A. B. (1964) *Time in Stratigraphy.* New York, McGraw-Hill, 365 pp.

Sheehan, P. M. (2001) The Late Ordovician mass extinction. *Annual Review of Earth and Planetary Sciences* **29:**331–364.

Sheehan, P. M., Coorough, P. J., and Fastovsky, D. E. (1996) Biotic selectivity during the K/T and Late Ordovician extinction events. *In* G. Ryder, D. Fastovsky, and S. Gartner (eds.), *The Cretaceous–Tertiary Event and Other Catastrophes in Earth History.* Geological Society of America Special Paper 307, pp. 477–489.

Shen, S.-z., and Shi, G. R. (2002) Paleobiogeographical extinction patterns of Permian brachiopods in the Asian-western Pacific region. *Paleobiology* **28:**449–463.

Shuto, T. (1974) Larval ecology of prosobranch gastropods and its bearing on biogeography and paleontology. *Lethaia* **7:**239–256.

Signor, P. W., III, and Lipps, J. H. (1982) Sampling bias, gradual extinction patterns, and catastrophes in the fossil record. *In* L. T. Silver and P. H. Schulz (eds.), *Geological Implications of Impacts of Large Asteroids and Comets on the Earth.* Geological Society of America Special Paper 190, pp. 291–296.

Simpson, G. G. (1953) *The Major Features of Evolution.* New York, Columbia University Press, 434 pp.

Smith, A. B. (1994) *Systematics and the Fossil Record: Documenting Evolutionary Patterns.* Oxford, U.K., Blackwell Scientific, 223 pp.

Smith, J. P. (1932) Lower Triassic ammonoids of North America. *U.S. Geological Survey Professional Paper* 167, pp. 1–199.

Smith, L. H. (1998) Species level phenotypic variation in lower Paleozoic trilobites. *Paleobiology* **24:**17–36.

Sober, E. (2004) The contest between parsimony and likelihood. *Systematic Biology* **53:**644–653.

Sohl, N. F. (1960) Archaeogastropoda, Mesogastropoda and stratigraphy of the Ripley, Owl Creek, and Prairie Bluff Formations. *U.S. Geological Survey Professional Paper* 331-A, 151 pp.

Sorgenfrei, T. (1958) Molluscan assemblages from the marine Middle Miocene of South Jutland and their environments. *Geological Survey of Denmark*, II Ser. No. 79, 503 pp.

Speyer, S. E., and Brett, C. E. (1986) Trilobite taphonomy and Middle Devonian taphofacies. *Palaios* **1:**312–327.

Springer, D. A., and Bambach, R. K. (1985) Gradient versus cluster analysis of fossil assemblages: A comparison from the Ordovician of southwestern Virginia. *Lethaia* **18:**181–198.

Squyres, S. W., and Knoll, A. H. (eds.) (2005) *Sedimentary Geology at Meridiani Planum, Mars. Earth and Planetary Science Letters*, vol. 240, no. 1, pp. 1–190.

Stanley, S. M. (1972) Functional morphology and evolution of bysally attached bivalve mollusks. *Journal of Paleontology* **46:**165–212.

Stanley, S. M. (1973) An explanation for Cope's Rule. *Evolution* **27:**1–26.

Stanley, S. M. (1975) A theory of evolution above the species level. *Proceedings of the National Academy of Sciences USA* **72:**646–650.

Stanley, S. M. (1999) The myth of the diversity plateau: High extinction rates, not ecological saturation, stifled the diversification of Paleozoic marine life. *Geological Society of America Abstracts with Programs* **31**(7):A-337.

Stanley, S. M., and Yang, X. (1994) A double mass extinction at the end of the Paleozoic Era. *Science* **266:**1340–1344.

Stanley, S. M., Addicott, W. O., and Chinzei, K. (1980) Lyellian curves in paleontology: Possibilities and limitations. *Geology* **8:**422–426.

Stewart, W. N. (1983) *Paleobotany and the Evolution of Plants.* Cambridge, U.K., Cambridge University Press, 405 pp.

Suchanek, T. H. (1983) Control of seagrass communities and sediment distribution by *Callianassa* (Crustacea, Thalassinidea) bioturbation. *Journal of Marine Research* **41:**281–298.

Taylor, T. N., and Taylor, E. L. (1993) *The Biology and Evolution of Fossil Plants.* Englewood Cliffs, N.J., Prentice Hall, 982 pp.

Thompson, D'A. W. (1942) *On Growth and Form.* Cambridge, U.K., Cambridge University Press, 1116 pp.

Towe, K. M. (1987) Fossil preservation. *In* R. S. Boardman, A. H. Cheetham, and A. J. Rowell (eds.), *Fossil Invertebrates.* Palo Alto, Calif., Blackwell Scientific, pp. 36–41.

Tree of Life Web Project [*http://www.tolweb.org*, accessed 10 May 2006].

Ubukata, T. (2000) Theoretical morphology of hinge and shell form in Bivalvia: Geometric constraints derived from space conflict between umbones. *Paleobiology* **26:**606–624.

Unrug, R., Unrug, S., Ausich, W. I., Bednarczyk, J., Cuffey, R. J., Mamet, B. L., and Palmes, S. L. (2000) Paleozoic age of the Walden Creek Group, Ocoee Supergroup, in the western Blue Ridge, southern Appalachians: Implications for evolution of the Appalachian margin of Laurentia. *Geological Society of America Bulletin*, Vol. 112, No. 7, pp. 982–996.

Valentine, J. W. (1970) How many marine invertebrate fossil species? A new approximation. *Journal of Paleontology* **44:**410–415.

Valentine, J. W. (1989) How good was the fossil record? Clues from the California Pleistocene. *Paleobiology* **15:**83–94.

Valentine. J. W. (2004) *On the Origin of Phyla.* Chicago, University of Chicago Press, 614 pp.

Valentine, J. W., Jablonski, D., and Erwin, D. H. (1999) Fossils, molecules, and embryos: New perspectives on the Cambrian explosion. *Development* **126:**851–859.

Valentine, J. W., Jablonski, D., Kidwell, S., and Roy, K. (2006) Assessing the fidelity of the fossil record by using marine bivalves. *Proceedings of the National Academy of Sciences USA* **103:**6599–6604.

Valentine, J. W., Tiffney, B. H., and Sepkoski, J. J., Jr. (1991) Evolutionary dynamics of plants and animals: A comparative approach. *Palaios* **6:**81–88.

Van Wagoner, J. C., Mitchum, R. M., Campion, K. M., and Rahmanian, V. D. (1990) *Siliciclastic Sequence Stratigraphy in Well Logs, Cores, and Outcrops: Concepts for High-Resolution Correlation of Time and Facies.* Tulsa, Okla., American Association of Petroleum Geologists, Models in Exploration Series, No. 7, 55 pp.

Vermeij, G. J. (1977) The Mesozoic marine revolution: Evidence from snails, predators, and grazers. *Paleobiology* **3:**245–258.

Vermeij, G. J. (1987) *Evolution and Escalation: An Ecological History of Life.* Princeton, N.J., Princeton University Press, 527 pp.

Wagner, P. J. (1995) Systematics and the fossil record. *Palaios* **10:**383–388.

Walossek, D., and Müller, K. J. (1994) Pentastomid parasites from the lower Palaeozoic of Sweden. *Transactions of the Royal Society of Edinburgh: Earth Sciences* **85:**1–37.

Walter, M. R., Des Marais, D., Farmer, J. D., and Hinman, N. W. (1996) Lithofacies and biofacies of mid-Paleozoic thermal spring deposits in the Drummond Basin, Queensland, Australia. *Palaios* **11:**497–518.

Walter, M. R., McLoughlin, S., Drinnan, A. N., and Farmer, J. D. (1998) Palaeontology of Devonian thermal spring deposits, Drummond Basin, Australia. *Alcheringa* **22:**285–314.

Webby, B. D., Cooper, R. A., Bergström, S. M., and Paris, F. (2004) Stratigraphic framework and time slices. *In* B. D. Webby, F. Paris, M. L. Droser, and I. G. Percival (eds.), *The Great Ordovician Biodiversification Event.* New York, Columbia University Press, pp. 41–47.

Westrop. S. R., and Adrain, J. M. (1998) Trilobite alpha diversity and the reorganization of Ordovician benthic marine communities. *Paleobiology* **24:**1–16.

Whittaker, R. H. (1975) *Communities and Ecosystems*, 2nd ed. New York, Macmillan, 385 pp.

Whittington, H. B. (1957) The ontogeny of trilobites. *Biological Reviews* **32:**421–469.

Whittington, H. B. (1985) *The Burgess Shale.* New Haven, Conn., Yale University Press, 151 pp.

Wilf, P. (1997) When are leaves good thermometers? A new case for Leaf Margin Analysis. *Paleobiology* **23:**373–390.

Wilf, P., and Labandeira, C. C. (1999) Response of plant–insect associations to Paleocene–Eocene warming. *Science* **284:**2153–2156.

Wilf, P., Wing, S. L., Greenwood, D. R., and Greenwood, C. L. (1998) Using fossil leaves as paleoprecipitation indicators: An Eocene example. *Geology* **26:**203–206.

Williams, A. (1957) Evolutionary rates of brachiopods. *Geological Magazine* **94:**201–211.

Williams, A., Carlson, S. J., Brunton, H. C., Holmer, L. E., and Popov, L. (1996) A supra-ordinal classification of the Brachiopoda. *Philosophical Transactions of the Royal Society of London* B **351:**1171–1193.

Wills, M. A., Briggs, D. E. G., and Fortey, R. A. (1994) Disparity as an evolutionary index: A comparison of Cambrian and Recent arthropods. *Paleobiology* **20:**93–130.

Wing, S. L., Harrington, G. J., Smith, F. C., Bloch, J. L., Boyer, D. M., and Freeman, K. H. (2005) Transient floral change and rapid global warming at the Paleocene–Eocene boundary. *Science* **310:**993–996.

Woodward, F. I. (1987) Stomatal numbers are sensitive to increases in $CO_2$ from pre-industrial levels. *Nature* **327:**617–618.

Xiao, S., and Knoll, A. H. (2000) Phosphatized animal embryos from the Neoproterozoic Doushantuo Formation at Weng'an, Guizhou, South China. *Journal of Paleontology* **74:**767–788.

Zachos, J, Pagani, M, Sloan, L., Thomas, E., and Billups, K. (2001) Trends, rhythms, and aberrations in global climate 65 Ma to present. *Science* **292:**686–693.

Ziegler, A. M. (1965) Silurian marine communities and their environmental significance. *Nature* **207:**270–272.

Ziegler, A. M., Cocks, L. R. M., and Bambach, R. K. (1968) The composition and structure of Lower Silurian marine communities. *Lethaia* **1:**1–27.

# INDEX

**Note:** Page numbers followed by f indicate figures; those followed by t indicate tables; and those followed by b indicate boxed material.

## C

# Paleontologically Important Groups of Organisms

**prokaryotes** [typically unicellular; cells small (1–10 microns), lacking nucleus and other membrane-bound organelles; DNA in single chromosome]
- **Domain Eubacteria**
  - **Cyanobacteria** [photosynthetic; probably ancestral to chloroplasts (photosynthetic organelles in eukaryotes)]
- **Domain Archaea** [some molecular similarities to eukaryotes]
  - **Kingdom Crenarchaeota** [includes many forms that live at extremely high temperatures and acidities, and that obtain energy from hydrogen–sulfur reactions]
  - **Kingdom Euryarchaeota** [includes many methanogenic (methane-producing) forms and those that live at very high salinities]

**Domain Eukarya** (eukaryotes) [cells larger than prokaryotes; DNA in nucleus; membrane-bound organelles such as mitochondria and chloroplasts; evolutionary origin via symbiotic association of prokaryotic cells]
- **Kingdom Fungi** [multicellular; relying on other organisms as food source (heterotrophic); external digestion; important as decomposers in ecosystems; closely related to animals]
  - **Division Ascomycota** (yeast and relatives)
  - **Division Basidiomycota** (mushrooms and relatives)
  - **Division Chytridiomycota** [large, spore-bearing bodies, using whiplike extensions (flagella) for motility]
  - **Division Zygomycota** (bread molds and relatives)
- **Kingdom Animalia** (animals) [multicellular; heterotrophic; cell mobility; development via ball of one cell layer (blastula), which, in most groups, invaginates at blastopore to produce stage with two cell layers]
  - **Bilateria** [bilateral symmetry; three embryonic cell layers]
    - **protostomes** [blastopore becomes mouth during development; generally with characteristic geometry of early-stage cell division (spiral cleavage)]
      - **Ecdysozoa** [animals that molt]
        - **Phylum Arthropoda** [segmented body; jointed limbs; chitinous exoskeleton, sometimes mineralized]
          - **Superclass Chelicerata** [uniramous (one-branched) limbs; two body regions (prosoma, opisthosoma); grasping limbs (chelicerae)]
            - **Class Arachnida** (spiders, scorpions)
            - **Class Merostomata** (†eurypterids, horseshoe crabs)
          - **Superclass Crustacea** [biramous (two-branched) limbs; three body regions (head, thorax, abdomen); two pair of antennae]
            - **Class Cirripedia** (barnacles)
            - **Class Malacostraca** (lobsters, crabs, shrimps)
            - **Class Ostracoda** [smaller; laterally compressed; bivalved, sometimes calcareous, carapace over most of body]
          - **Superclass Hexapoda** [uniramous limbs; three body regions (head, thorax, abdomen); thorax with three segments, six limbs, wings]
            - **Class Insecta** (insects)
          - **Superclass Myriapoda** (centipedes, millipedes) [long trunk with many segments and uniramous limbs]
          - **Class †Trilobita** [three body lobes (lengthwise); three body regions (head, thorax, pygidium); well calcified; biramous limbs]
      - **Lophotrochozoa** [group of phyla united largely on basis of molecular phylogenetic evidence]
        - **Eutrochozoa** [group of phyla united by spiral cleavage and other developmental similarities]
          - **Phylum Mollusca** [muscular foot and dorsal visceral mass; calcium carbonate shells secreted by mantle]
            - **Class Bivalvia** (clams, oysters, mussels, scallops) [two articulated shells; headless; gills used for filter feeding]
            - **Class Cephalopoda** (†ammonoids, nautiloids, squids, octopuses) [one shell, planispiral, chambered, buoyant; arms or tentacles]
            - **Class Gastropoda** (snails, slugs) [one shell, generally helical; torted body]
            - **monoplacophorans** [one shell, caplike; body untorted; paired, serial organs]
              - **Class †Helcionelloida**
              - **Class Tergomya**
            - **Class Polyplacophora** (chitons) [eight dorsal plates, articulating, covering body]
            - **Class †Rostroconchia** [single continuous shell resembling paired shells of bivalves]
            - **Class Scaphopoda** (tusk shells) [one long, tusk-shaped shell, open at both ends]
          - **Phylum Annelida** [segmented worms; fossils include hard, proteinaceous dental elements (scolecodonts), calcium carbonate tubes, and tubes made of agglutinated sediment grains]
            - **Class Polychaeta** [protrusions (parapodia) bearing chitinous bristles (setae)]
        - **Lophophorata** [group of phyla characterized by geometry of early-stage cell division (radial cleavage), tentacular feeding organ (lophophore), and other features]
          - **Phylum Brachiopoda** [bivalved, calcareous or organic-phosphatic shells, sometimes articulated (having a hinge structure)]
            - **Subphylum Linguliformea** [organic-phosphatic shells; inarticulated]
              - **Class Lingulata**
              - **Class †Paterinata**
            - **Subphylum Craniiformea** [calcitic shell; no attachment organ (pedicle); mostly inarticulated forms]
              - **Class Craniata**
            - **Subphylum Rhynchonelliformea** (calcitic shells, most articulated; many with calcitic lophophore supports; pedicle)
              - **Class †Chileata** [generally inarticulated]
              - **Class †Obolellata** [simple shell articulations]
              - **Class †Kutorginata** [simple shell articulations]
              - **Class †Strophomenata** [often more complex shell articulations; some forms with lophophore supports]
              - **Class Rhynchonellata** [often more complex shell articulations; many with well-developed lophophore supports]
          - **Phylum Bryozoa** [colonial; some forms calcified]
            - **Class Gymnolaemata** [some calcified; zooids box-shaped or cylindrical]
            - **Class Stenolaemata** [calcified; zooids typically cylindrical]
    - **Deuterostomia** [blastopore posterior during development, often becoming anus; generally with radial cleavage]
      - **Phylum Chordata** [dorsal rod (notochord); pharyngeal slits]
        - **Subphylum Vertebrata** [spinal column]
          - **jawless vertebrates**
            - **Superclass Agnatha** (lampreys and hagfishes)
            - **Class †Conodonta** [toothlike, phosphatic elements]
            - **Class †Pteraspidomorphi** [scaly tail; armored head]
          - **Gnathostomata** [jawed vertebrates]
            - **Class †Acandthodii** [spines supporting fins; body armor, spines]
            - **Class Chondrichthyes** (sharks, rays, skates) [cartilaginous endoskeleton]
            - **Class †Placodermi** [armored head and thorax]
            - **Osteichthyes** (bony fishes) [phosphatic endoskeleton]
              - **Class Actinopterygii** [ray-finned fishes; includes most of the familiar bony fishes]
              - **Class Sarcopterygii** (lungfishes, coelacanths) [lobe-finned fishes; includes ancestors of tetrapods]